Lecture Notes in Computer Science 6509

Commenced Publication in 1973
Founding and Former Series Editors:
Gerhard Goos, Juris Hartmanis, ɛ

Weili Wu
Ovidiu Daescu (Eds.)

Combinatorial Optimization and Applications

4th International Conference, COCOA 2010
Kailua-Kona, HI, USA, December 18-20, 2010
Proceedings, Part II

 Springer

Volume Editors

Weili Wu
University of Texas at Dallas
Department of Computer Science
Richardson, TX 75083, USA
E-mail: weiliwu@utdallas.edu

Ovidiu Daescu
University of Texas at Dallas
Department of Computer Science
Richardson, TX 75080, USA
E-mail: daescu@utdallas.edu

Library of Congress Control Number: 2010939794

CR Subject Classification (1998): F.2, G.2.1, G.2, C.2, E.1, I.3.5

LNCS Sublibrary: SL 1 – Theoretical Computer Science and General Issues

ISSN   0302-9743
ISBN-10   3-642-17460-4 Springer Berlin Heidelberg New York
ISBN-13   978-3-642-17460-5 Springer Berlin Heidelberg New York

springer.com

© Springer-Verlag Berlin Heidelberg 2010
Printed in Germany

Typesetting: Camera-ready by author, data conversion by Scientific Publishing Services, Chennai, India
Printed on acid-free paper   06/3180

Preface

The 4th Annual International Conference on Combinatorial Optimization and Applications (COCOA 2010) took place in Big Island, Hawaii, USA, December 18–20, 2010. Past COCOA conferences were held in Xi'an, China (2007), Newfoundland, Canada (2008) and Huangshan, China (2009).

COCOA 2010 provided a forum for researchers working in the areas of combinatorial optimization and its applications. In addition to theoretical results, the conference also included recent works on experimental and applied research of general algorithmic interest. The Program Committee received 108 submissions from more than 23 countries and regions, including Australia, Austria, Canada, China, Denmark, France, Germany, Hong Kong, India, Italy, Japan, Korea, Mexico, New Zealand, Poland, Slovak Republic, Spain, Sweden, Switzerland, Taiwan, UK, USA, Vietnam, etc.

Among the 108 submissions, 49 regular papers were selected for presentation at the conference and are included in this volume. Some of these papers will be selected for publication in a special issue of the *Journal of Combinatorial Optimization*, a special issue of *Theoretical Computer Science*, a special issue of *Optimization Letters*, and a special issue of *Discrete Mathematics, Algorithms and Applications* under the standard refereeing procedure.

We thank all authors for submitting their papers to the conference. We are grateful to all members of the Program Committee and all external referees for their work within demanding time constraints. We thank the Organizing Committee for their contribution to making the conference a success. We also thank Jiaofei Zhong and Donghyun Kim for helping us create and update the conference website and maintain the Springer Online Conference Service system and Shawon Rahman for helping in local arrangements.

Finally, we thank the conference sponsors and supporting organizations for their support and assistance. They are the University of Texas at Dallas, the University of Hawaii at Hilo, and the National Science Foundation of USA.

December 2010

Weili Wu
Ovidiu Daescu

Organization

COCOA 2010 was organized by the department of Computer Science, the University of Texas at Dallas, in cooperation with the University of Hawaii at Hilo.

Executive Committee

General Co-chairs   Ding-Zhu Du (University of Texas at Dallas, USA)
Panos M. Pardalos (University of Florida, USA)
Bhavani Thuraisingham (University of Texas at Dallas, USA)
PC Co-chairs   Ovidiu Daescu (University of Texas at Dallas, USA)
Weili Wu (University of Texas at Dallas, USA)
Local Chair   Shawon Rahman (University of Hawaii at Hilo, Hawaii)

Program Committee

Farid Alizadeh	Rutgers University, USA
Mikhail (Mike) J. Atallah	Purdue University, USA
Giorgio Ausiello	Università di Roma, Italy
Piotr Berman	Penn State University, USA
Vladimir Boginski	University of Florida, USA
Annalisa De Bonis	Università di Salerno, Italy
Sergiy Butenko	Texas A&M University, USA
Gruia Calinescu	Illinois Institute of Technology, USA
Gerard Jennhwa Chang	National Taiwan University, Taiwan
Zhi-Zhong Chen	Tokyo Denki University, Japan
Chuangyin Dang	City University of Hong Kong, Hong Kong
Vladimir Deineko	The University of Warwick, UK
Zhenhua Duan	Xidian University, China
Omer Egecioglu	University of California, Santa Barbara, USA
Dan Hirschberg	University of California, USA
Tsan-sheng Hsu	Academia Sinica, Taiwan
Hejiao Huang	Harbin Institute of Technology, China
Wonjun Lee	Korea University, South Korea
Asaf Levin	The Technion, Israel
Yingshu Li	Georgia State University, USA
Guohui Lin	University of Alberta, Canada
Liying Kang	Shanghai University, China
Naoki Katoh	Kyoto University, Japan
Ilias S. Kotsireas	Wilfrid Laurier University, Canada

Anastasia Kurdia	Smith College, USA
Mitsunori Ogihara	University of Miami, USA
Jack Snoeyink	The University of North Carolina at Chapel Hill, USA
Ileana Streinu	Smith College, USA
Vitaly Strusevich	University of Greenwich, UK
Zhiyi Tan	Zhejiang University, China
Doreen Anne Thomas	University of Melbourne, Australia
Alexey A. Tuzhilin	Moscow State University, Russia
Amy Wang	Tsinghua University, China
Caoan Wang	Memorial University of Newfoundland, Canada
Feng Wang	Arizona State University, USA
Lusheng Wang	City University of Hong Kong, Hong Kong
Wei Wang	Xi'an Jiaotong University, China
Weifan Wang	Zhejiang Normal University, China
Chih-Wei Yi	National Chiao Tong University, Taiwan
Alex Zelikovsky	George State University, USA
Cun-Quan Zhang	West Virginia University, USA
Huaming Zhang	University of Alabama in Huntsville, USA
Louxin Zhang	National University of Singapore, Singapore
Xiao Zhou	Tohoku University, Japan

Referees

Ferdinando Cicalese	Salvatore La Torre	Gaolin Milledge
Paolo D'Arco	Yuan-Shin Lee	Seth Pettie
Gianluca De Marco	Chung-Shou Liao	J.K.V. Willson
Natallia Katenka	Hongliang Lu	Wei Zhang
Donghyun Kim	Hsueh-I Lu	Zhao Zhang
Stefan Langerman	Martin Milanič	Jiaofei Zhong

Table of Contents – Part II

Table of Contents – Part I

Coverage with k-Transmitters
in the Presence of Obstacles

Brad Ballinger[1], Nadia Benbernou[2], Prosenjit Bose[3], Mirela Damian[4,*],
Erik D. Demaine[5], Vida Dujmović[6], Robin Flatland[7], Ferran Hurtado[8,**],
John Iacono[9], Anna Lubiw[10], Pat Morin[11], Vera Sacristán[12,**],
Diane Souvaine[13], and Ryuhei Uehara[14]

[1] Humboldt State University, Arcata, USA
brad.ballinger@humboldt.edu
[2] Massachusetts Institute of Technology, Cambridge, USA
nbenbern@mit.edu
[3] Carleton University, Ottawa, Canada
jit@scs.carleton.ca
[4] Villanova University, Villanova, USA
mirela.damian@villanova.edu
[5] Massachusetts Institute of Technology, Cambridge, USA
edemaine@mit.edu
[6] Carleton University, Ottawa, Canada
vida@cs.mcgill.ca
[7] Siena College, Loudonville, USA
flatland@siena.edu
[8] Universitat Politècnica de Catalunya, Barcelona, Spain
Ferran.Hurtado@upc.edu
[9] Polytechnic Institute of New York University, New York, USA
jiacono@poly.edu
[10] University of Waterloo, Waterloo, Canada
alubiw@uwaterloo.ca
[11] Carleton University, Ottawa, Canada
morin@scs.carleton.ca
[12] Universitat Politècnica de Catalunya, Barcelona, Spain
vera.sacristan@upc.edu
[13] Tufts University, Medford, USA
dls@cs.tufts.edu
[14] Japan Advanced Institute of Science and Technology, Ishikawa, Japan
uehara@jaist.ac.jp

Abstract. For a fixed integer $k \geq 0$, a k-transmitter is an omnidirectional wireless transmitter with an infinite broadcast range that is able to penetrate up to k "walls", represented as line segments in the plane. We develop lower and upper bounds for the number of k-transmitters that are necessary and sufficient to cover a given collection of line segments, polygonal chains and polygons.

* Supported by NSF grant CCF-0728909.
** Partially supported by projects MTM2009-07242 and Gen. Cat. DGR 2009SGR1040.

W. Wu and O. Daescu (Eds.): COCOA 2010, Part II, LNCS 6509, pp. 1–15, 2010.
© Springer-Verlag Berlin Heidelberg 2010

1   Introduction

Illumination and guarding problems generalize the well-known art gallery problem in computational geometry [15,16]. The task is to determine a minimum number of guards that are sufficient to guard, or "illuminate" a given region under specific constraints. The region under surveillance may be a polygon, or may be the entire plane with polygonal or line segment obstacles. The placement of guards may be restricted to vertices (*vertex* guards) or edges (*edge* guards) of the input polygon(s), or may be unrestricted (*point* guards). The guards may be *omnidirectional*, illuminating all directions equally, or may be represented as *floodlights*, illuminating a certain angle in a certain direction.

Inspired by advancements in wireless technologies and the need to offer wireless services to clients, Fabila-Monroy et al. [10] and Aichholzer et al. [2] introduce a new variant of the illumination problem, called *modem* illumination. In this problem, a guard is modeled as an omnidirectional wireless modem with an infinite broadcast range and the power to penetrate up to k "walls" to reach a client, for some fixed integer $k > 0$. Geometrically, walls are most often represented as line segments in the plane. In this paper, we refer to such a guard as a *k-transmitter*, and we speak of *covering* (rather than illuminating or guarding). We address the general problem introduced in [10,2], reformulated as follows:

> *k-Transmitter Problem: Given a set of obstacles in the plane, a target region, and a fixed integer $k > 0$, how many k-transmitters are necessary and sufficient to cover that region?*

We consider instances of the k-transmitter problem in which the obstacles are line segments or simple polygons, and the target region is a collection of line segments, or a polygonal region, or the entire plane. In the case of plane coverage, we assume that transmitters may be embedded in the wall, and therefore can reach both sides of the wall at no cost. In the case of polygonal region coverage, we favor the placements of transmitters *inside* the region itself; therefore, when we talk about a *vertex* transmitter, the implicit assumption is that the transmitter is placed just inside the polygonal region, and so must penetrate one wall to reach the exterior.

1.1   Previous Results

For a comprehensive survey on the art gallery problem and its variants, we refer the reader to [15,16]. Also see [9,7,4] for results on the *wireless localization* problem, which asks for a set of 0-transmitters that need not only cover a given region, but also enable mobile communication devices to prove that they are inside or outside the given region. In this section, we focus on summarizing existing results on the k-transmitter problem and a few related issues.

For $k = 0$, the k-transmitter problem for simple polygons is settled by the Art Gallery Theorem [5], which states that $\lfloor \frac{n}{3} \rfloor$ guards are sufficient and sometimes necessary to guard a polygonal region with n vertices. Finding the minimum number of 0-transmitters that can guard a given polygon is NP-hard [14,15]. For $k > 0$, Aichholzer et al. [10,2] study the k-transmitter problem in which the target region is represented as a monotone polygon or a monotone orthogonal polygon with n vertices. They show that $\frac{n}{2k}$ k-transmitters are sufficient, and $\lceil \frac{n}{2k+4} \rceil$ k-transmitters are sometimes necessary[1] to cover a monotone polygon. They also show that $\lceil \frac{n}{2k+4} \rceil$ k-transmitters are sufficient and necessary to cover any monotone orthogonal polygon. The authors also study simple polygons, orthogonal polygons and arrangements of lines in the context of very powerful transmitters, i.e, k-transmitters where k may grow as a function of n. For example, they show that any simple polygon with n vertices can always be covered with one transmitter of power $\lceil \frac{2n+1}{3} \rceil$, and this bound is tight up to an additive constant. In the case of orthogonal polygons, one $\lceil \frac{n}{3} \rceil$-transmitter is sufficient to cover the entire polygon. The problem of covering the plane with a single k-transmitter has been also considered in [12], where it is proved that there exist collections of n pairwise disjoint equal-length segments in the Euclidean plane such that, from any point, there is a ray that meets at least $2n/3$ of them (roughly). While the focus in [10,2,12] is on finding a small number of high power transmitters, our focus in this paper is primarily on lower power transmitters.

The concept of visibility through k segments has also appeared in other works. Dean et al. [8,13,11] study *vertical bar k-visibility*, where k-visibility goes through k segments. Aichholzer et al. [1] introduce and study the notion of *k-convexity*, where a diagonal may cross the boundary at most $2(k-1)$ times.

1.2 Our Results

We consider several instances of the k-transmitter problem. If obstacles are disjoint orthogonal segments and the target region is the entire plane, we show that $\lceil \frac{5n+6}{12} \rceil$ 1-transmitters are always sufficient and $\lceil \frac{n+1}{4} \rceil$ are sometimes necessary to cover the target region. If the target region is the plane and the obstacles are lines and line segments that form a guillotine subdivision (defined in §2.2), then $\frac{n+1}{2}$ 1-transmitters suffice to cover the target region. We next consider the case where the obstacles consist of a set of nested convex polygons. If the target region is the boundaries of these polygons, then $\lfloor \frac{n}{7} \rfloor + 3$ 2-transmitters are always sufficient to cover it. On the other hand, if the target region is the entire plane, then $\lfloor \frac{n}{6} \rfloor + 3$ 2-transmitters suffice to cover it, and $\lfloor \frac{n}{8} \rfloor + 1$ 2-transmitters are sometimes necessary. All these results (detailed in §2) use point transmitters, with the implicit assumption that transmitters on a boundary segment are embedded in the segment and can reach either side of the segment at no cost.

In Section 3 we move on to the case where the target region is the interior of a simple polygon. In this case, we restrict the placement of vertex and edge transmitters to the interior of the polygon. We show that $\frac{n}{6}$ 2-transmitters are sometimes necessary to cover the interior of a simple polygon. In Section 3.2 we introduce a class of spiral polygons, which we refer to as *spirangles*, and show that $\lfloor \frac{n}{8} \rfloor$ 2-transmitters are sufficient, and sometimes necessary, to cover the interior

---

[1] The bound $\lceil n/(2k+2) \rceil$ stated in Theorem 7 from [2] is a typo.

of a spirangle polygon. In the case of arbitrary spiral polygons, we derive an upper bound of $\lfloor \frac{n}{4} \rfloor$ 2-transmitters, matching the upper bound for monotone polygons from [2].

2   Coverage of Plane with Obstacles

We begin with the problem of covering the entire plane with transmitters, in the presence of obstacles that are orthogonal segments (§2.1), a guillotine subdivision (§2.2), or a set of nested convex polygons (§2.3). There is no restriction on the placement of transmitters (on or off a segment). In the case of a transmitter located on a segment itself, the assumption is that the segment does not act as on obstacle for that transmitter, in other words, that the transmitter has the power of a k-transmitter on both sides of the segment.

2.1   Orthogonal Line Segments

In this section the set of obstacles is a set of n disjoint orthogonal line segments and the target region is the whole plane. Czyzowicz et al. [6] proved that $\lceil (n+1)/2 \rceil$ 0-transmitters always suffice and are sometimes necessary to cover the plane in the presence of n disjoint orthogonal line segments. We generalize this to k-transmitters. Our main ideas are captured by the case of 1-transmitters, so we begin there:

Theorem 1. *In order to cover the plane in the presence of n disjoint orthogonal line segments, $\lceil (5n+6)/12 \rceil$ 1-transmitters are always sufficient and $\lceil (n+1)/4 \rceil$ are sometimes necessary.*

Proof. The lower bound is established by n parallel lines—a single 1-transmitter can cover only 4 of the $n + 1$ regions.

For the upper bound, the main idea is to remove from the set of segments, S, a set of segments that are *independent* in the sense that no covering ray goes through two of them consecutively. We then take a set of conventional transmitters for the remaining segments. By upgrading these transmitters to 1-transmitters we cover the whole plane with respect to the original segments S.

We now fill in this idea. We will assume without loss of generality that the segments have been extended (remaining interior-disjoint) so that each end of each segment either extends to infinity, or lies on another segment: if a set of k-transmitters covers the plane with respect to the extended segments then it covers the plane with respect to the original segments. With this assumption the segments partition the plane into $n + 1$ rectangular faces.

The *visibility graph* $G(S)$ has a vertex for each segment of S and an edge st if segments s and t are weakly visible, i.e. there is a point p interior to s and a point q interior to t such that the line segment pq does not cross any segment in S. Equivalently, for the case of extended segments, s and t are weakly visible if some face is incident to both of them.

Lemma 1. *If I is an independent set in $G(S)$ and T is a set of 0-transmitters that covers the whole plane with respect to $S - I$, then T is a set of 1-transmitters that covers the whole plane with respect to S.*

Proof. Suppose that a *0*-transmitter at point p covers point q with respect to $S - I$. Then the line segment from p to q does not cross any segment of $S - I$. It cannot cross two or more segments of I otherwise two such consecutive segments would be visible (and not independent). Thus a *1*-transmitter at p covers q with respect to S. □

To obtain a large independent set in $G(S)$ we will color $G(S)$ and take the largest color class. If the faces formed by S were all triangles then $G(S)$ would be planar and thus 4-colorable. Instead, we have rectangular faces, so $G(S)$ is 1-*planar* and can be colored with 6 colors. A graph is *1-planar* if it can be drawn in the plane, with points for vertices and curves for edges, in such a way that each edge crosses at most one other edge. Ringel conjectured in 1965 that 1-planar graphs are 6-colorable. This was proved in 1984 by Borodin, who gave a shorter proof in 1995 [3].

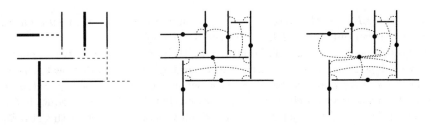

Fig. 1. (left) A set S of disjoint orthogonal segments and their extensions (dashed) with an independent set shown in bold; (middle) $G(S)$ with vertices drawn as segments and edges as dashed curves so 1-planarity is clear; (right) contracting a segment to a point to get a conventional drawing of the graph

Lemma 2. *If S is a set of extended orthogonal segments then $G(S)$ is 1-planar.*

Proof. The idea is the same as that used to show that the visibility graph of horizontal line segments is planar. If $G(S)$ is drawn in the natural way, with every vertex represented by its original segment, and every edge drawn as a straight line segment crossing a face, then it is clear that each edge crosses at most one other edge. See Figure 1. We can contract each segment to a point while maintaining this. Note that we end up with a multi-graph in case two segments are incident to more than one face. □

We now wrap up the proof of Theorem 1. Since $G(S)$ is 1-planar it has a 6-coloring by Borodin's result. The largest color class has at least $n/6$ vertices and forms an independent set I. The set $S - I$ has at most $5n/6$ segments, so by the result of Czyzowicz et al. [6], it has a set of *0*-transmitters of cardinality at most $\lceil (\frac{5n}{6} + 1)/2 \rceil = \lceil (5n + 6)/12 \rceil$ that covers the entire plane. By Lemma 1, placing *1*-transmitters at those points covers the entire plane with respect to S. □

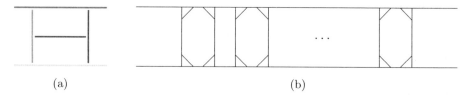

Fig. 2. (a) An arrangement of five segments whose visibility graph is complete and thus requires 5 colors. (b) A guillotine subdivision with $n = 6k + 2$ segments that requires $4k$ 0-transmitters. Each of the $4k$ triangular faces must have a 0-transmitter on its boundary and no two triangular faces share a boundary.

We note that the above proof relies on a 6-coloring of $G(S)$. An example that requires 5 colors is shown in Figure 2(a).

Theorem 2. *In order to cover the plane in the presence of n disjoint orthogonal line segments, $\lceil \frac{1}{2}((5/6)^{\log(k+1)}n + 1)\rceil$ k-transmitters are always sufficient and $\lceil (n+1)/2(k+1)\rceil$ are sometimes necessary.*

Proof. As for $k = 1$, the lower bound is realized by parallel segments. One k-transmitter can only cover $2(k + 1)$ of the $n + 1$ regions.

For the upper bound, we build on the proof technique for $k = 1$. We repeatedly remove independent sets, extending the remaining segments after each removal.

For a set of segments S, let $X(S)$ be a set of segments formed by extending those of S until they touch. It will not matter that $X(S)$ is not unique. Let R_0 be S and for $i = 1, 2, \ldots$ let S_i be a maximal independent set in the visibility graph of $X(R_{i-1})$ and let $R_i = S - (\cup_{j=1}^{i} S_j)$. Then R_i has cardinality at most $(5/6)^i n$.

Lemma 3. *If T is a set of 0-transmitters that covers the whole plane with respect to R_i, then T is a set of $(2^i - 1)$-transmitters that covers the whole plane with respect to $S = R_0$.*

Proof. We prove by induction on $j = 0, \ldots, i$ that T is a set of $(2^j - 1)$-transmitters that covers the whole plane with respect to R_{i-j}. Suppose this holds for $j - 1$. Suppose a $(2^{j-1} - 1)$-transmitter at point p sees point q in R_{i-j+1}. Then the line segment pq crosses at most $2^{j-1} - 1$ segments of R_{i-j+1}, and thus 2^{j-1} faces. Consider putting back the segments of S_{i-j+1} to obtain R_{i-j}. The segments of S_{i-j+1} are independent in R_{i-j}. Therefore the line segment pq can cross at most one segment of S_{i-j+1} in each face. The total number of segments of R_{i-j} crossed by pq is thus $2^{j-1} - 1 + 2^{j-1} = 2^j - 1$. In other words, a $(2^j - 1)$-transmitter at p in R_{i-j} covers the same area as the original $(2^{j-1} - 1)$-transmitter at p in R_{i-j+1}. □

We use this lemma to complete the proof of the theorem. Since we have the power of k-transmitters, we can continue removing independent sets until R_i, where $k = 2^i - 1$, i.e. $i = \log(k + 1)$. Then R_i has size $(5/6)^{\log(k+1)}n$, and the number of 0-transmitters needed to cover the plane with respect to R_i is $\lceil \frac{1}{2}((5/6)^{\log(k+1)}n+1) \rceil$. Applying the lemma, this is the number of k-transmitters we need to cover the plane with respect to S.   □

2.2   Guillotine Subdivisions

A *guillotine subdivision* S is obtained by inserting a sequence $s_1, \ldots, s_n$ of line segments (possibly rays or lines), such that each inserted segment s_i splits a face of the current subdivision S_{i-1} into two new faces yielding a new subdivision S_i. We start with one unbounded face S_0, which is the entire plane.

As the example in Figure 2(b) shows, a guillotine subdivision with n segments can require $2(n-2)/3$ 0-transmitters. In this section, we show that no guillotine subdivision requires more than $(n+1)/2$ 1-transmitters. We begin with a lemma:

Lemma 4. *Let F be a face in a guillotine subdivision S. If there are 1-transmitters on every face that shares an edge with F then these 1-transmitters see all of F.*

Proof. Consider the segment s_i whose insertion created the face F. Before the insertion of s_i, the subdivision S_{i-1} contained a convex face that was split by s_i into two faces F and F' (Figure 3(a)). No further segments were inserted into F, but F' may have been further subdivided, so that there are now several faces $F'_1, \ldots, F'_k$, with $F'_j \subseteq F'$ and F'_j incident on s_i for all $j \in \{1, \ldots, k\}$ (Figure 3(b)).

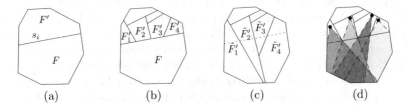

Fig. 3. The proof of Lemma 4

We claim that the 1-transmitters in $F'_1, \ldots, F'_k$ guard the interior of F. To see this, imagine removing s_i from the subdivision and instead, constructing a guillotine subdivision $\tilde{S}$ from the sequence $s_1, \ldots, s_{i-1}, s_{i+1}, \ldots, s_n$ (Figure 3(c)). In this case, each face F'_j in S becomes a larger face $\tilde{F}'_j$ in $\tilde{S}$ and together $\bigcup_{j=1}^{k} \tilde{F}'_j \supseteq F$. Finally, we observe that each 1-transmitter in S in face F'_j guards at least $\tilde{F}'_j$, so together, the 1-transmitters in $F'_1, \ldots, F'_k$ guard all of F (Figure 3(d)).   □

Theorem 3. *Any guillotine subdivision can be guarded with at most $(n+1)/2$ 1-transmitters.*

Proof. Consider the dual graph T of the subdivision. T is a triangulation with $n+1$ vertices. Let M be any maximal matching in T. Consider the unmatched vertices of T. Each such vertex is adjacent only to matched vertices (otherwise M would not be maximal). Let G be the set of 1-transmitters obtained by placing a single 1-transmitter on the primal edge associated with each edge $e \in M$. Then $|G| = |M| \le (n+1)/2$. For every face F of S, F either contains a 1-transmitter in G, or all faces that share an edge with F contain a 1-transmitter in G. In the former case, F is obviously guarded. In the latter case, Lemma 4 ensures that F is guarded. Therefore, G is a set of 1-transmitters that guards all faces of F and has size at most $(n+1)/2$. □

2.3 Nested Convex Polygons

The problems analyzed in this section are essentially two:

1. How many 2-transmitters are always sufficient (and sometimes necessary) to cover the edges of a set of nested convex polygons?
2. How many 2-transmitters are always sufficient (and sometimes necessary) to cover the plane in the presence of a set of nested convex polygons?

Henceforth, we use the *bounding box* of a polygon to refer to the smallest axis-parallel rectangle containing the polygon.

Some notation. We call a set of k convex polygons $\{P_1, P_2, \ldots, P_k\}$ *nested* if $P_1 \supseteq P_2 \supseteq \cdots \supseteq P_k$. The total number of vertices of the set of polygons $\{P_1, P_2, \ldots, P_k\}$ is n.

Given such a set, we use the term *layers* for the boundaries of the polygons and *rings* for the portions of the plane between layers, i.e., the the the i-th ring is $R_i = P_i - P_{i+1}$, for $i = 1, \ldots, k-1$. In addition, $R_0 = \mathbb{R} - P_1$ and $R_k = P_k$.

We assume that vertices on each layer have labels with indices increasing counterclockwise. Given a vertex $v_j \in P_i$, we call the positive angle $\angle v_{j-1} v_j v_{j+1}$ its *external visibility angle*. (Positive angles are measured counterclockwise, and negative angles are measured clockwise.) Its *internal visibility angle* is the negative angle $\angle v_{j-1} v_j v_{j+1}$.

Lemma 5. *Placing a 2-transmitter at every other vertex in a given layer i guarantees to completely cover layers $i-3$, $i-2$, $i-1$ and i, as well as rings $i-3$, $i-2$ and $i-1$.*

Proof. The fact that layer i is covered is obvious. As for the previous layers, notice that the convexity of P_i guarantees that the external visibility angles of any vertex pair v_j and v_{j+2} overlap, as illustrated in Figure 4(a). Since $v_j \in P_i \subseteq P_{i-1} \subseteq P_{i-2} \subseteq P_{i-3}$ and the polygons are convex, all rays from v_j within its external visibility angle traverse exactly two segments before reaching layer $i-3$. □

Lemma 6. *Placing a 2-transmitter at each vertex of a given layer i guarantees to completely cover layers $i-3$, $i-2$, $i-1$, i, $i+1$, $i+2$ and $i+3$, as well as rings $i-3$, $i-2$, $i-1$, i, $i+1$ and $i+2$.*

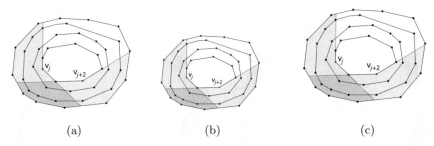

(a)   (b)   (c)

Fig. 4. (a) External visibility angles of two vertices v_j, v_{j+2} of layer i. Only layers $i-3$, $i-2$, $i-1$ and i are shown. (b) External and internal visibility from a 2-transmitter located in a vertex of layer i. Only layers $i-3$, $i-2$, $i-1$, i, $i+1$, $i+2$ and $i+3$ are shown. (c) The shaded region is not covered by the 2-transmitters located at the red vertices. Only the three involved layers are shown.

Proof. The fact that layers $i-3$, $i-2$, $i-1$, i and rings $i-3$, $i-2$ and $i-1$ are covered is a consequence of Lemma 5. As for the remaining layers and rings, notice that, in the internal visibility angle of a 2-transmitter $v_j \in P_i$, visibility is determined by the supporting lines from v_j to layers $i+1$, $i+2$ and $i+3$, as illustrated in Figure 4(b). Having a 2-transmitter on each of the vertices of layer i, combined with the fact that all polygons are convex, guarantees total covering of layers $i+1$, $i+2$ and $i+3$ and rings i, $i+1$ and $i+2$.   □

Theorem 4. $\lfloor \frac{n}{7} \rfloor + 5$ *2-transmitters are always sufficient to cover the edges of any nested set of convex polygons with a total of n vertices.*

Proof. If the number of layers is $k \in \{1, 2, 3, 4, 5, 6\}$, five 2-transmitters trivially suffice: one in the interior of P_k and the other four at the corners of the bounding box of P_1. If $k \geq 7$, from the pigeonhole principle one of $i \in \{1, 2, 3, 4, 5, 6, 7\}$ is such that the set $G = \{P_j \mid j \in \{1, \ldots, k\}, \; j \equiv i \pmod 7\}$ has no more than $\lfloor \frac{n}{7} \rfloor$ vertices. Place one 2-transmitter at each vertex of each $P_j \in G$. From Lemma 6, for a certain value of $m \in \mathbb{Z}$ all edges in the following layers are covered: $i-3, i-2, i-1$ (if they exist), $i, \ldots, i+7m$, $i+7m+1$, $i+7m+2$ and $i+7m+3$ (if they exist). In the worst case, the only layers that may remain uncovered are 1, 2 and 3, as well as $k-2$, $k-1$ and k. Because of the convexity of the polygons, four 2-transmitters conveniently located at the corners of the bounding box of P_1, and one 2-transmitter located in the interior of P_k, can take care of covering these remaining layers. The total number of 2-transmitters used is at most $\lfloor \frac{n}{7} \rfloor + 5$.   □

The transmitter placement from Theorem 4 guarantees that all edges are covered, while some rings remain uncovered.

Theorem 5. $\lfloor \frac{n}{6} \rfloor + 3$ *2-transmitters are always sufficient to cover the plane in the presence of any nested set of convex polygons with a total of n vertices.*

Proof. The proof is similar to Theorem 4, but locating the 2-transmitters at all vertices of every 6th layer (as opposed to every 7th layer in Theorem 4).   □

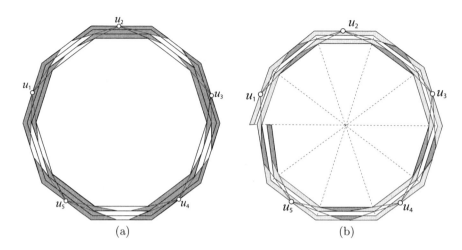

Fig. 5. (a) $\lfloor \frac{n}{8} \rfloor$ 2-transmitters are necessary to cover the edges of these four nested convex layers. (b) $\lfloor \frac{n}{8} \rfloor$ 2-transmitters are necessary to cover the edges of this spirangle polygon.

Lemma 7. $\lfloor \frac{n}{8} \rfloor$ *2-transmitters are sometimes necessary to cover the plane in the presence of any nested set of convex polygons with a total of n vertices.*

Proof. This lower bound is established by the example from Figure 5, which shows four nested regular t-gons, with t even (so $n = 4t$). Consider the set S of midpoints of alternating edges of the outermost convex layer (marked u_i in Figure 5). The gap between adjacent layers controls the size of the visibility regions of the points in S (by symmetry, all visibility regions have identical size). A small enough gap guarantees that the visibility regions of the points in S are all disjoint, as illustrated in Figure 5. This means that at least $t/2$ 2-transmitters are necessary to cover all points in S (one transmitter in the visibility region of each point). So the number of 2-transmitters necessary to cover all edges is $t/2 = n/8$. □

Lemmas 8 and 9 establish improved upper bounds for the case when all layers (convex polygons) have an even number of vertices. Due to space constraints, we omit the proofs of these lemmas.

Lemma 8. $\lfloor n/8 \rfloor + 1$ *2-transmitters are always sufficient to cover the edges of any nested set of convex polygons with a total of n vertices, if each of the polygons has an even number of vertices.*

Lemma 9. $\lfloor \frac{n}{6} \rfloor + 1$ *2-transmitters are always sufficient to cover the plane in the presence of any nested set of convex polygons with a total of n vertices, if each of the polygons has an even number of vertices.*

3   Coverage of Simple Polygons

This section addresses the problem of covering a polygonal region P with 2-transmitters placed *interior* to P. Therefore, when we talk about a *vertex* or an *edge* transmitter, the implicit assumption is that the transmitter is placed just inside the polygonal region, and so must penetrate one wall to reach the exterior. Our construction places a small (constant) number of transmitters outside P, but still within the bounding box for P.

3.1   Lower Bounds for Covering Polygons

Theorem 6. *There are simple polygons that require at least $\frac{n}{6}$ 2-transmitters to cover when transmitters are restricted to the interior of the polygon.*

Proof. Figure 6 shows the construction for a $n = 36$ vertex polygon, which generalizes to $n = 6m$, for any $m \geq 2$. It is a pinwheel whose $n/3$ arms alternate between spikes and barbs. Consider an interior point p at the tip of a barb. The locus of all interior points from which a 2-transmitter can cover p includes the spike counterclockwise from the barb, the barb containing p, and a small section of the pinwheel center. This region is shown shaded for the point p labeled in Figure 6. Observe that this shaded region is disjoint from the analogous regions associated with the other barb tips. Hence no two barb tips can be covered by the same 2-transmitter. Since there are $n/6$ barbs, the lower bound is obtained.   □

3.2   Spirangles

Two edges are *homothetic* if one edge is a scaled and translated image of the other. A *t-spirangle* is a polygonal chain $A = a_1, a_2, ..., a_m$ that spirals inward about a center point such that every t edges it completes a 2π turn, and each edge pair $a_i a_{i+1}, a_{i+t} a_{i+1+t}$ is homethetic, for $1 \leq i \leq m - t$. We assume that

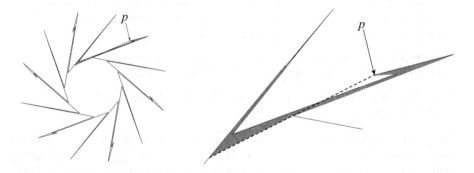

Fig. 6. A family of polygons requiring at least $n/6$ interior 2-transmitters to cover. For labeled point p located in the tip of a barb (shown magnified on the right with the arms shortened), the locus of all interior points from which a 2-transmitter can cover p is shown shaded.

the spiral direction is clockwise. A t-sided convex polygon may be thought of as generating a family of t-spirangles where the ith edge of each spirangle is parallel to the $(i \mod t)^{th}$ edge of the polygon, for $i = 0, 1, 2, \ldots$. See Figure 7(a) for a 4-spirangle example and a polygon generating it.

A *homothetic t-spirangle* polygon P is a simple polygon whose boundary consists of two nested t-spirangles $A = a_1, a_2, \ldots, a_m$ and $B = b_1, b_2, \ldots, b_m$ generated by the same t-sided convex polygon, plus two additional edges $a_1 b_1$ and $a_m b_m$ joining their endpoints. We assume that chain B is nested inside of chain A, as shown in Figure 7(b). We refer to A as the convex chain and B as the reflex chain in reference to the type of vertices found on each.

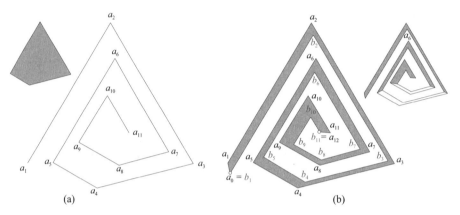

Fig. 7. Definitions (a) A 4-spirangle and corresponding convex polygon (b) Edge-homothetic spiral polygon (left) and quadrilaterals entirely visible to a_6 (right)

Property 1. Let P be a homothetic spirangle polygon, composed of a convex spirangle $A = a_1, a_2, \ldots$, and a reflex spirangle $B = b_1, b_2, \ldots$. Then a_i and b_i see each other, and the set of diagonals $\{a_i b_i \mid i = 1, 2, \ldots\}$, induces a partition of P into quadrilaterals. Furthermore, the visibility region of a_i includes six quadrilaterals: two quadrilaterals adjacent to $a_{i-t} b_{i-t}$, two adjacent to $a_i b_i$, and two adjacent to $a_{i+t} b_{i+t}$. See right of Figure 7(b).

Theorem 7. $\lfloor \frac{n}{8} \rfloor$ *2-transmitters are sufficient, and sometimes necessary, to cover a homothetic t-spirangle polygon P with n vertices.*

Proof. The algorithm that places transmitters at vertices of P to cover the interior of P is fairly simple, and is outlined in Table 1.

The proof that this algorithm covers the interior of P is fairly intuitive. Due to space constraints, we omit this proof. The fact that the $\lfloor \frac{n}{8} \rfloor$ bound is tight is established by the spirangle polygon example from Figure 5(b), and the arguments are similar to the one used in the proof of Lemma 7. The example from Figure 5(b) depicts a worst-case scenario, in which transmitters do not get the chance to use their full coverage potential, since the total turn angle of the spirangle is between 2π and 6π.   □

Table 1. Covering the interior of a homothetic spirangle polygon with 2-transmitters

Homothetic t-Spirangle Polygon Cover(P)
Let $A = a_1, a_2, \ldots a_m$ be the convex spirangle of P, with a_1 outermost. Let $B = b_1, b_2, \ldots b_m$ be the reflex spirangle of P.   1.   If $m \leq t + 2$ (or equivalently, the total turn angle of A is $\leq 2\pi$): Place one transmitter at a_m, and return (see Figure 8a).   2.   Place the first transmitter at vertex a_{t+2} (see a_7 in Figure 8b).   3.   Starting at a_{t+2}, place transmitters at every other vertex of A, up to a_{2t+1} (i.e., for a 2π turn angle of A, but excluding a_{2t+2}).   4.   Let a_j be the vertex hosting the last transmitter placed in step 3. ($j = 2t + 1$ for t odd, $j = 2t$ for t even.) Let P_1 be the subpolygon of P induced by vertices $a_1, \ldots, a_{j+t+1}$ and $b_1, \ldots, b_{j+t+1}$ (shaded left of Figure 8b.)   Recurse on $P \setminus P_1$: Homothetic t-Spirangle Polygon Cover$(P \setminus P_1)$.

3.3 Arbitrary Spirals

A spiral polygon P consists of a clockwise convex chain and a clockwise reflex chain that meet at their endpoints. A trivial $\lfloor \frac{n}{4} \rfloor$ upper bound for the number of 2-transmitters that are sufficient to cover P can be obtained as follows. Pick the chain Γ of P with fewer vertices (i.e., Γ is the reflex chain of P, if the number of reflex vertices exceeds the number of convex vertices, and the convex chain of P otherwise). Then simply place one vertex 2-transmitter at every other vertex of Γ. By definition, the visibility ray from one 2-transmitter can cross the boundary of P at most twice. Note however that, even under the restriction that transmitters be placed interior of P, the visibility ray of one transmitter can leave and re-enter P, as depicted in Fig. 9(a) for transmitter labeled a. Then arguments similar to the ones used in Lemma 5 show that the union of the external visibility angles of all these 2-transmitters cover the entire plane. So we have the following result:

Lemma 10. $\lfloor \frac{n}{4} \rfloor$ *2-transmitters placed interior to an arbitrary polygonal spiral P are sufficient to cover P (in fact, the entire plane).*

We remark on two special situations. In the case of transmitters placed at every other reflex vertex of P, 0-transmitters are sufficient to cover the interior of P, and 1-transmitters are sufficient to cover the entire plane. In the case of transmitters placed at every other convex vertex of P, 1-transmitters are sufficient to cover P, *if* they are placed *outside* of P.

An improved upper bound can be established for *non-degenerate* spirals, which we define as spirals in which each 2π-turn of each of the convex and reflex chain of P is homothetic to a convex polygon (i.e., it contains at least 3 vertices). The result (whose proof we omit due to space restrictions) is as follows.

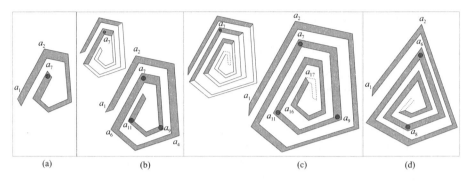

Fig. 8. Covering spirangles with 2-transmitters. (a) A t-spirangle ($t = 5$) with $2t + 4$ edges covered with one transmitter. (b) A t-spirangle ($t = 5$) with $8t$ edges. (c) A t-spirangle ($t = 5$) with $6t + 4$ edges covered with $t/2 + 1$ transmitters. (d) A t-spirangle ($t = 4$) with $6t$ edges covered with $t/2$ transmitters.

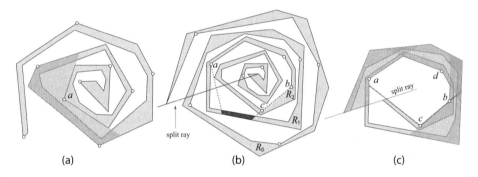

Fig. 9. Transmitters marked with small circles (a) Visibility angle of a (b) The dark area is not covered by a or b (c) P is covered

Lemma 11. *Let P be a polygonal spiral whose every 2π turn chain has at least 3 vertices. Then $\lceil \frac{2n}{9} \rceil + 1$ 2-transmitters placed interior to P are sufficient to cover the interior of P (in fact, the entire plane).*

4   Conclusion

In this paper we study the problem of covering ("guarding") a target region in the plane with k-transmitters, in the presence of obstacles. We develop lower and upper bounds for the problem instance in which the target region is the plane, and the obstacles are lines and line segments, a guillotine subdivision, or nested convex polygons. We also develop lower and upper bounds for the problem instance in which the target region is the set of rings created by nested convex polygons, or the interior of a spiral polygon. Our work leaves open two main problems: (i) closing the gap between the $\lfloor \frac{n}{8} \rfloor$ lower bound and the $\lfloor \frac{n}{6} \rfloor$ upper bound in the case of nested convex layers, and (ii) closing the gap between the

$\lfloor \frac{n}{8} \rfloor$ lower bound and the $\lfloor \frac{n}{4} \rfloor$ upper bound for spiral polygons. Investigating the k-transmitter problem for other classes of polygons (such as orthogonal polygons) also remains open.

Acknowledgement. We thank Joseph O'Rourke for the pinwheel example from Fig. 6 and for initiating this line of work.

References

1. Aichholzer, O., Aurenhammer, F., Hurtado, F., Ramos, P., Urrutia, J.: k-convex polygons. In: EuroCG, pp. 117–120 (2009)
2. Aichholzer, O., Fabila-Monroy, R., Flores-Pealoza, D., Hackl, T., Huemer, C., Urrutia, J., Vogtenhuber, B.: Modem illumination of monotone polygons. In: EuroCG (2009)
3. Borodin, O.: A new proof of the 6 color theorem. Journal of Graph Theory 19(4), 507–521 (1995)
4. Christ, T., Hoffmann, M., Okamoto, Y., Uno, T.: Improved bounds for wireless localization. In: Gudmundsson, J. (ed.) SWAT 2008. LNCS, vol. 5124, pp. 77–89. Springer, Heidelberg (2008)
5. Chvátal, V.: A combinatorial theorem in plane geometry. Journal of Combinatorial Theory Series B 18, 39–41 (1975)
6. Czyzowicz, J., Rivera-Campo, E., Santoro, N., Urrutia, J., Zaks, J.: Guarding rectangular art galleries. Discrete Applied Math. 50, 149–157 (1994)
7. Damian, M., Flatland, R., O'Rourke, J., Ramaswami, S.: A new lower bound on guard placement for wireless localization. In: Proc. of the 17th Fall Workshop on Computational Geometry, FWCG 2007, pp. 21–24 (November 2007)
8. Dean, A.M., Evans, W., Gethner, E., Laison, J., Safari, M.A., Trotter, W.T.: Bar k-visibility graphs: Bounds on the number of edges, chromatic number, and thickness. In: Healy, P., Nikolov, N.S. (eds.) GD 2005. LNCS, vol. 3843, pp. 73–82. Springer, Heidelberg (2005)
9. Eppstein, D., Goodrich, M.T., Sitchinava, N.: Guard placement for efficient point-in-polygon proofs. In: SoCG, pp. 27–36 (2007)
10. Fabila-Monroy, R., Vargas, A.R., Urrutia, J.: On modem illumination problems. In: XIII Encuentros de Geometria Computacional, Zaragoza, Spain (June 2009)
11. Felsner, S., Massow, M.: Parameters of bar k-visibility graphs. Journal of Graph Algorithms and Applications 12(1), 5–27 (2008)
12. Fulek, R., Holmsen, A.F., Pach, J.: Intersecting convex sets by rays. Discrete Comput. Geom. 42(3), 343–358 (2009)
13. Hartke, S.G., Vandenbussche, J., Wenger, P.: Further results on bar k-visibility graphs. SIAM Journal of Discrete Mathematics 21(2), 523–531 (2007)
14. Lee, D.T., Lin, A.K.: Computational complexity of art gallery problems. IEEE Trans. Inf. Theor. 32(2), 276–282 (1986)
15. O'Rourke, J.: Art gallery theorems and algorithms. Oxford University Press, Inc., New York (1987)
16. Urrutia, J.: Art gallery and illumination problems. In: Sack, J.-R., Urrutia, J. (eds.) Handbook of Computational Geometry, pp. 973–1027. North-Holland, Amsterdam (2000)

On Symbolic OBDD-Based Algorithms for the Minimum Spanning Tree Problem

Beate Bollig

LS2 Informatik, TU Dortmund,
44221 Dortmund, Germany

Abstract. The minimum spanning tree problem is one of the most fundamental algorithmic graph problems and OBDDs are a very common dynamic data structure for Boolean functions. Since in some applications graphs become larger and larger, a research branch has emerged which is concerned with the design and analysis of so-called symbolic algorithms for classical graph problems on OBDD-represented graph instances. Here, a symbolic minimum spanning tree algorithm using $O(\log^3 |V|)$ functional operations is presented, where V is the set of vertices of the input graph. Furthermore, answering an open problem posed by Sawitzki (2006) it is shown that every symbolic OBDD-based algorithm for the minimum spanning tree problem needs exponential space (with respect to the OBDD size of the input graph). This result even holds for planar input graphs.

Keywords: minimum spanning tree algorithms, ordered binary decision diagrams, symbolic algorithms.

1 Introduction

A spanning tree of a connected undirected graph G with real edge weights is a minimum spanning tree if its weights, i.e., the total weight of its edges, is minimal among all total weights of spanning trees of G. Constructing a minimum spanning tree is a well-known fundamental problem in network analysis with numerous applications. Besides the importance of the problem in its own right, the problem arises in solutions of other problems (see, e.g., [19] for a nice survey on results from the earliest known algorithm of Borůvka [8] to the invention of Fibonacci heaps and [2] for a survey and empirical study on various minimum spanning tree algorithms). Since modern applications require huge graphs, explicit representations by adjacency matrices or adjacency lists may cause conflicts with memory limitations and even polynomial time algorithms seem not to be applicable any more. As time and space resources do not suffice to consider individual vertices, one way out seems to be to deal with sets of vertices and edges represented by their characteristic functions. Ordered binary decision diagrams, denoted OBDDs, introduced by Bryant in 1986 [10], are well suited for the representation and manipulation of Boolean functions, therefore, a research branch has emerged which is concerned with the design and analysis of

W. Wu and O. Daescu (Eds.): COCOA 2010, Part II, LNCS 6509, pp. 16–30, 2010.
© Springer-Verlag Berlin Heidelberg 2010

so-called symbolic algorithms for classical graph problems on OBDD-represented graph instances (see, e.g., [17, 18], [20], [27], [28, 29], and [33]). Symbolic algorithms have to solve problems on a given graph instance by efficient functional operations offered by the OBDD data structure.

Representing graphs with regularities by means of data structures smaller than adjacency matrices or adjacency lists seems to be a natural idea. In [1, 16, 25] it has been shown that problems typically get harder when their input is implicitly represented by circuits. Since there are Boolean functions like some output bits of integer multiplication whose OBDD complexity is exponentially larger than its circuit size [3, 11], these results do not directly carry over to problems on OBDD-represented inputs. However, in [14] it has been shown that even the very basic problem of deciding whether two vertices s and t are connected in a directed graph G, the so-called graph accessibility problem GAP, is PSPACE-complete on OBDD-represented graphs. Nevertheless, OBDD-based algorithms are successful in many applications and already in [14] it has been pointed out that worst-case hardness results do not adequately capture the complexity of the problems on real-world instances. Therefore, one aim is to find precise characterizations of the special cases that can be solved efficiently and on the other hand to find simple instances that are hard to process. In [28] exponential lower bounds on the space complexity of OBDD-based algorithms for the single-source shortest paths problem, the maximum flow problem, and a restricted class of algorithms for the reachability problem have been presented. Recently, a general exponential lower bound on the space complexity of OBDD-based algorithms for the reachability problem and exponential lower bounds on the space complexity of symbolic algorithms for the maximum matching and the maxflow problem in 0-1-networks have been shown [4-6]. The results are not very astonishing but the proofs present worst-case examples which could be helpful to realize which cases are difficult to process. Due to the problem's rich area of applications the minimum spanning tree problem has received a considerable amount of attention for explicit graph representations. The best currently known upper bound on the complexity of the minimum spanning tree problem in the explicit setting was established in [12], where an algorithm that runs on input $G = (V, E)$ in time $O(|E|\alpha(|E| \cdot |V|))$ has been presented. Here, α is the inverse of the Ackermann function. In [26] an optimal algorithm has been given but nothing better than $O(|E|\alpha(|E| \cdot |V|))$ is known about the running time. An expected linear time algorithm has been shown in [22]. For restricted graph classes problems could be easier and for the explicit setting already in [13] a linear time algorithm for minimum spanning trees on planar graphs has been shown.

Here, answering an open question posed by Sawitzki (see Table 1, page 785 in [28]), we prove that OBDD-based representations of (unique) minimum spanning trees can be exponentially larger than the OBDD representation of the input graph even if the input graph is planar. Despite the exponential blow-up from input to output size in the implicit setting, it is still possible that there exists an OBDD-based algorithm that solve the minimum spanning tree problem polynomially with respect to the number of vertices of the input graph and often with sublinear space. In the paper we present a symbolic algorithm that uses a polylogarithmic number of functional operations with respect to the number of vertices of the input graph.

The paper is organized as follows. In Section 2 we define some notation and review some basics concerning OBDDs, symbolic graph representations, graphs, and the minimum spanning tree problem. Section 3 contains a symbolic minimum spanning tree algorithm that uses $O(\log^3 |V|)$ OBDD-operations, where V is the set of vertices of the input graph. Afterwards, in Section 4 we show that symbolic OBDD-based algorithms for the minimum spanning tree problem need exponential space with respect to the size of the implicit representation of the input graph even if the graph is planar. Here, we do not introduce a new lower bound method in order to prove the exponential lower bound on the implicit representation of the minimum spanning tree but the merit of the result is the presentation of a very simple input graph for which an exponential blow-up from input to output size can be shown. Finally, we finish the paper with some concluding remarks.

2   Preliminaries

In order to make the paper self-contained we briefly recall the main notions we are dealing with in this paper.

2.1   Ordered Binary Decision Diagrams

When working with Boolean functions as in circuit verification, synthesis, and model checking, ordered binary decision diagrams are one of the most often used data structures supporting all fundamental operations on Boolean functions, like binary operators, quantifications or satisfiability tests, efficiently. (For a history of results on binary decision diagrams see, e.g., the monograph of Wegener [32]).

Definition 1. *Let $X_n = \{x_1, \ldots, x_n\}$ be a set of Boolean variables. A variable ordering π on X_n is a permutation on $\{1, \ldots, n\}$ leading to the ordered list $x_{\pi(1)}, \ldots, x_{\pi(n)}$ of the variables.*

In the following a variable ordering π is sometimes identified with the corresponding ordering $x_{\pi(1)}, \ldots, x_{\pi(n)}$ of the variables if the meaning is clear from the context.

Definition 2. *A π-OBDD on X_n is a directed acyclic graph $G = (V, E)$ whose sinks are labeled by the Boolean constants 0 and 1 and whose non-sink (or decision) nodes are labeled by Boolean variables from X_n. Each decision node has two outgoing edges one labeled by 0 and the other by 1. The edges between decision nodes have to respect the variable ordering π, i.e., if an edge leads from an x_i-node to an x_j-node, then $\pi^{-1}(i) \leq \pi^{-1}(j)$ (x_i precedes x_j in $x_{\pi(1)}, \ldots, x_{\pi(n)}$). Each node v represents a Boolean function $f_v \in B_n$, i.e., $f_v : \{0,1\}^n \to \{0,1\}$, defined in the following way. In order to evaluate $f_v(b)$, $b \in \{0,1\}^n$, start at v.*

After reaching an x_i-node choose the outgoing edge with label b_i until a sink is reached. The label of this sink defines $f_v(b)$. The width of a π-OBDD is the maximum number of nodes labeled by the same variable. The size of a π-OBDD G is equal to the number of its nodes and the π-OBDD size of a function f is the size of the minimal π-OBDD representing f. The π-OBDD of minimal size for a given function f is unique up to isomorphism. A π-OBDD is called reduced, if it is the minimal π-OBDD.

Let g be a Boolean function on the variables $x_1, \ldots, x_n$. The subfunction $g_{|x_i=c}$, $1 \leq i \leq n$ and $c \in \{0,1\}$, is defined as $g(x_1, \ldots, x_{i-1}, c, x_{i+1}, \ldots, x_n)$. It is well known that the size of an OBDD representing a function f that depends essentially on n Boolean variables (a function g depends essentially on a Boolean variable z if $g_{|z=0} \neq g_{|z=1}$) may be different for different variable orderings and may vary between linear and exponential size with respect to n.

Definition 3. *The OBDD size or OBDD complexity of f is the minimum of all π-OBDD(f).*

The size of the reduced π-OBDD representing f is described by the following structure theorem [30].

Theorem 1. *The number of $x_{\pi(i)}$-nodes of the minimal π-OBDD for f is the number s_i of different subfunctions $f_{|x_{\pi(1)}=a_1,\ldots,x_{\pi(i-1)}=a_{i-1}}$, $a_1, \ldots, a_{i-1} \in \{0,1\}$, that essentially depend on $x_{\pi(i)}$.*

Theorem 1 implies the following simple observation which is helpful in order to prove lower bounds. Given an arbitrary variable ordering π the number of nodes labeled by a variable x in the reduced π-OBDD representing a given function f is not smaller than the number of x-nodes in a reduced π-OBDD representing any subfunction of f.

Now, we briefly describe a list of important operations on data structures for Boolean functions and the corresponding time and additional space requirements for OBDDs (for a detailed discussion see, e.g., [32]). In the following let f and g be Boolean functions in B_n on the variable set $X_n = \{x_1, \ldots, x_n\}$ and G_f and G_g be π-OBDDs for the representations of f and g, respectively.

- Evaluation: Given G_f and an input $b \in \{0,1\}^n$, compute $f(b)$. This can be done in time $O(n)$.
- Replacements by constants: Given G_f, an index $i \in \{1, \ldots, n\}$, and a Boolean constant $c_i \in \{0,1\}$, compute a π-OBDD for the subfunction $f_{|x_i=c_i}$. This can be done in time $O(|G_f|)$ and the π-OBDD for $f_{|x_i=c_i}$ is not larger than G_f.
- Equality test: Given G_f and G_g, decide, whether f and g are equal. This can be done in time $O(|G_f| + |G_g|)$.
- Satisfiability count: Given G_f, compute $|f^{-1}(1)|$. This can be done in time $O(|G_f|)$.

- Synthesis: Given G_f and G_g and a binary Boolean operation $\otimes \in B_2$, compute a π-OBDD G_h for the function $h \in B_n$ defined as $h := f \otimes g$. This can be done in time and space $O(|G_f| \cdot |G_g|)$ and the size of G_h is bounded above by $O(|G_f| \cdot |G_g|)$.
- Quantification: Given G_f, an index $i \in \{1, \ldots, n\}$, and a quantifier $Q \in \{\exists, \forall\}$, compute a π-OBDD G_h for the function $h \in B_n$ defined as $h := (Qx_i)f$, where $(\exists x_i)f := f_{|x_i=0} \vee f_{|x_i=1}$ and $(\forall x_i)f := f_{|x_i=0} \wedge f_{|x_i=1}$. The computation of G_h can be realized by two replacements of constants and a synthesis operation. This can be done in time and space $O(|G_f|^2)$.

In the rest of the paper quantifications over k Boolean variables $(Qx_1, \ldots, x_k)f$ are denoted by $(Qx)f$, where $x = (x_1, \ldots, x_k)$.

2.2 Symbolic OBDD-Based Graph Representations and the Minimum Spanning Tree Problem

In the following for $z = (z_{n-1}, \ldots, z_0) \in \{0,1\}^n$ let $|z| := \sum_{i=0}^{n-1} z_i 2^i$. Let $G = (V, E)$ be a graph with N vertices $v_0, \ldots v_{N-1}$. The edge set E can be represented by an OBDD for its characteristic function, where

$$\mathcal{X}_E(x, y) = 1 \Leftrightarrow (|x|, |y| < N) \wedge (v_{|x|}, v_{|y|}) \in E, x, y \in \{0,1\}^n \text{ and } n = \lceil \log N \rceil.$$

If G is a weighted graph, i.e., there exists a function $c : E \to \{1, \ldots, B\}$, where B is the maximum weight, the definition of the characteristic function of G's edge set is extended by $\mathcal{X}_E(x, y, d) = 1 \Leftrightarrow (v_{|x|}, v_{|y|}) \in E \wedge c(v_{|x|}, v_{|y|}) = |d| + 1$, $d = (d_0, \ldots, d_{\lceil \log B \rceil - 1})$. Undirected edges are represented by symmetric directed ones. In the rest of the paper we assume that B and N are powers of 2 since it has no bearing on the essence of our results. It is well known that for every variable ordering π the size of the reduced π-OBDD for a given function $f \in B_n$ is upper bounded by $(2 + o(1))2^n/n$ (see, e.g., [9]). Moreover, it is not difficult to prove that the size is also upper bounded by $O(n \cdot |f^{-1}(1)|)$. Therefore, the characteristic function $\mathcal{X}_E$ of an edge set $E \subseteq V \times V$ can be represented by OBDDs of size $O(\min(|V|^2/\log |V|, |E| \log |V|))$.

By simple counting arguments it is easy to see that almost all graphs on N vertices cannot be represented by OBDDs of polylogarithmic size with respect to N. On the other hand, it is quite obvious that very simply structured graphs, e.g., grid graphs, have a small OBDD representation. Therefore, in [23, 24] the question has been investigated whether succinct OBDD representations can be found for significant graph classes. OBDD-represented graphs on N vertices are typically only defined on $\log N$ Boolean variables in comparison to other implicit graph representations where at least $c \log N$ bits for some constant $c > 1$ are allowed [21, 31]. One of the reasons is that the number of variables of intermediate OBDDs during the computation of a symbolic algorithm can be seen as a performance parameter. Multiplying the number of variables on which a function essentially depends by a constant c enlarges the worst-case OBDD size asymptotically from S to S^c. (See, e.g., [15] for the importance to keep the number of variables as low as possible.)

A graph is called planar if it can be drawn in the plane so that its edges intersect only at their ends. A sequence of vertices $v_{i_1}, \ldots, v_{i_k}$ is said to be a path from u to w of length $k-1$ in an unweighted graph $G = (V, E)$, $u, w \in V$, if $v_{i_1} = u$, $v_{i_k} = w$, and $(v_{i_j}, v_{i_{j+1}}) \in E$, $j \in \{0, \ldots, k-1\}$. Given two vertices u and w in V, we say that u reaches w, if there exists a path from u to w in G. The distance between two vertices u and w in G is the number of vertices minus 1 on a shortest paths from u to w in G. The diameter of G is the maximum distance on all vertex pairs in G. For an edge weighted graph we define the diameter in a similar way (using the assumption that each edge in G has weight 1). A connected component CC in G is a subgraph in G in which for any two vertices there exists a path from one vertex to the other one and no more vertices or edges (from G) can be added while preserving its connectivity. In other words CC is a maximal connected subgraph. A spanning tree in G is a subgraph in G that contains all vertices in V and is a tree. A minimum spanning tree (MST) of an undirected weighted graph $G = (V, E)$ is a minimum total weight subset of E that forms a spanning tree of G. In the symbolic setting the MST problem is the following one. Given an OBDD for the characteristic function of the edge set of an undirected weighted input graph G, the output is an OBDD that represents the characteristic function of a minimum spanning tree in G. In order to obtain small size representations, we may distinguish the problem to represent only the edges or the weighted edges of a minimum spanning tree symbolically. The proof of Theorem 2 in Section 4 works for both.

3   A Symbolic Minimum Spanning Tree Algorithm

Here, we present a symbolic OBDD-based algorithm for the minimum spanning tree problem. Given an implicitly represented edge weighted graph $G = (V, E, d)$ the task is to compute an implicit representation for a minimum spanning tree in G. The idea is to use Borůvka's well-known algorithm for the computation of minimum spanning trees on explicitly defined input graphs and to adapt it to the implicit setting. Since symbolic algorithms have to deal with sets of vertices and edges in order to save time and space, we deal with a parallel variant of Borůvka's algorithm. Although we assume that the reader is quite familiar with Borůvka's algorithm, we briefly recall the method in the following.

The edges of a minimum spanning tree in G are iteratively computed. We start with an empty set of edges. Each vertex $v \in G$ can be seen as a connected component of size 1 with respect to the edges already computed for the minimum spanning tree. In each iteration for each connected component C_i the edge with the smallest weight with respect to the remaining edges incident to another different connected component C_i' is parallel computed. If such an edge is not unique, we choose for every connected component an edge in an appropriate way. The chosen edges are added to the already computed edges of the minimum spanning tree. Afterwards, the computation of the connected components with respect to the edges already computed for the minimum spanning tree is updated. The computation terminates if there is only one connected component. The correctness of this method follows directly from the correctness of Borůvka's algorithm.

Since sometimes we have to choose an edge out of a given set of edges, we define an ordering $<$ on the Boolean encoding of edges of a given graph. For $(x, y, d), (x', y', d') \in \mathcal{X}_E^{-1}(1)$ we define $(d, x, y) < (d', x', y')$ iff one of the following requirements is fulfilled:

- $|d| < |d'|$,
- $|d| = |d'|$ and $\min(|x|, |y|) < \min(|x'|, |y'|)$, or
- $|d| = |d'|$, $\min(|x|, |y|) = \min(|x'|, |y'|)$, and $\max(|x|, |y|) < \max(|x'|, |y'|)$.

The ordering of the edges can easily be described by a Boolean function P_n in the following way: $P_n((x, y, d), (x', y', d')) = 1 \Leftrightarrow (x, y, d) < (x', y', d')$. Note, that the function P_n is defined on all Boolean inputs not only on $((x, y, d), (x', y', d'))$, where $(x, y, d), (x', y', d') \in \mathcal{X}_E^{-1}(1)$. It is not difficult to show that P_n can be represented by OBDDs of constant width 9 and therefore linear size with respect to the variable ordering

$$d_{\log B - 1}, d'_{\log B - 1}, \ldots, d_0, d'_0, x_{n-1}, x'_{n-1}, y_{n-1}, y'_{n-1} \ldots, x_0, x'_0, y_0, y'_0.$$

Next, we present a well-known algorithm for the problem transitive closure, the problem to compute an OBDD representing all connected vertex pairs for a graph symbolically represented by an OBDD. The algorithm uses the method of iterative squaring.

Algorithm findTransitiveClosure$(\mathcal{X}_E(x, y, d))$

(1) $R(x, y) \leftarrow (x = y) \vee (\exists d)\mathcal{X}_E(x, y, d)$
(2) **repeat**
(3) $R'(x, y) \leftarrow R(x, y)$
(4) $R(x, y) \leftarrow (\exists z)(R'(x, z) \wedge R'(z, y))$
(5) **until** $R(x, y) = R'(x, y)$
(6) **return** $R(x, y)$

It is easy to see that the algorithm uses $O(\log^2 |V|) = O(n^2)$ functional operations. There are at most $\log |V| = n$ iterations since the diameter of each graph on $|V|$ vertices is at most $|V| - 1$ and for each iteration $O(\log |V|)$ quantifications are necessary.

Finally, we present our symbolic algorithm for the computation of a minimum spanning tree in a given input graph.

Algorithm `findMinimumSpanningTree`$(\mathcal{X}_E(x,y,d))$

(1) $MST(x,y,d) \leftarrow 0$
(2) **repeat**
(3) $R(x,y) \leftarrow$ `findTransitiveClosure`$(MST(x,y,d))$
(4) $C(x,y,d) \leftarrow \mathcal{X}_E(x,y,d) \wedge \overline{R(x,y)} \wedge$
$$(\exists y', z, d'(R(x,z) \wedge \mathcal{X}_E(z,y',d') \wedge \overline{R(z,y')} \wedge P_n((d',z,y'),(d,x,y))))$$
(5) $C(x,y,d) \leftarrow C(x,y,d) \vee C(y,x,d)$
(6) $MST'(x,y,d) \leftarrow MST(x,y,d)$
(7) $MST(x,y,d) \leftarrow MST'(x,y,d) \vee C(x,y,d)$
(8) **until** $MST(x,y,d) = MST'(x,y,d)$
(9) **return** $MST(x,y,d)$

In the explicit setting it is often a good idea to work with contracted graphs during the computation of a graph algorithm, because the running time mostly depends on the number of vertices and edges of the considered graph. In the implicit setting the situation is different. The representation size for a subgraph can be larger than the representation size for the graph as in Section 4 our worst-case instance for the maximum spanning tree problem will show. Therefore, our algorithm works in each iteration with $\mathcal{X}_E(x,y,d)$.

Lemma 1. *Given the characteristic function of the edge set of an undirected weighted graph $G = (V, E, c)$ the algorithm* `findMinimumSpanningTree` *computes the characteristic function of a minimum spanning tree in G using $O(\log^3 |V|)$ functional operations.*

Proof. At the beginning the set of already computed edges for the minimum spanning tree is empty (1). In each iteration the transitive closure on the graph of the already determined edges for the minimum spanning tree is computed, in other words the connected components are determined and $R(x,y) = 1$, iff the two vertices encoded by x and y belong to the same connected component. A new edge (u,v) is added in (3), if the two vertices u and v do not already belong to the same connected component and there exists no smaller edge according to the ordering given by P_n that connect a vertex in the connected component C of u to a vertex of another connected component C', $C \neq C'$. Since undirected edges are represented by two directed ones, the set of the new computed edges of the minimum spanning tree is updated in (4) and afterwards added to the set of the already computed edges of the minimum spanning tree. The computation stops if no new edge can be added because the minimum spanning tree is complete. Altogether the correctness of the algorithm follows from the correctness of Borůvka's algorithm.

Since in each iteration the number of connected components is at least halved, the number of iterations is at most $\log |V|$. Furthermore, in each iteration there is a constant number of synthesis, negation, and equality operations and

$O(\log |V|)$ quantifications. Moreover, there is an algorithm for the computation of the transitive closure on the graph of the already computed edges for the minimum spanning tree. Summarizing the running time for the algorithm findMinimumSpanningTree is $O(\log^3 |V|)$.   □

In [6] it has been shown that the problem transitive closure is not computable in polynomial space with respect to the size of an implicitly defined input graph. Nevertheless, the situation during the computation of the algorithm findMinimumSpanningTree is a special one, since the input graphs for findTransitiveClosure are trees and it could be that in this case the problem is easier to solve. However, we will see in the next section that there can be an exponential blow-up from input to output size for the minimum spanning tree problem in the implicit setting.

4   On the Complexity of the Minimum Spanning Tree Problem on OBDD-Represented Graphs

In this section we demonstrate that there can be an exponential blow-up from input to output size for the minimum spanning tree problem even for planar graphs in the symbolic setting.

Theorem 2. *Symbolic OBDD-based algorithms for the minimum spanning tree problem need exponential space with respect to the size of the implicit representation of the input graph even if the input graph is planar.*

Proof. Our proof structure is the following one. First, we define a planar input graph $G = (V, E, c)$ for the minimum spanning tree problem. The size of the corresponding OBDD representation for the characteristic function of the weighted edge set is polynomial with respect to the number of Boolean variables. Afterwards we prove that the symbolic OBDD representation of the unique minimum spanning tree in G needs exponential space. Therefore, every OBDD-based algorithm solving the minimum spanning tree problem needs exponential space with respect to its input length. We start with the definition of a function which is well known in the BDD literature.

Definition 4. *The hidden weighted bit function* $\mathrm{HWB}_n : \{0,1\}^n \to \{0,1\}$ *computes the bit* b_{sum} *on the input* $b = (b_1, \ldots, b_n)$, *where* $sum := \sum_{i=1}^{n} b_i$ *and* $b_0 := 0$.

Bryant [11] has introduced this function as a very simple version of a storage access function, where each variable is control and data variable. He has also already shown that the OBDD complexity of HWB_n is $\Omega(2^{(1/5-\epsilon)n})$ which has been slightly improved up to $\Omega(2^{n/5})$ in [7].

1) The definition of the input graph G:

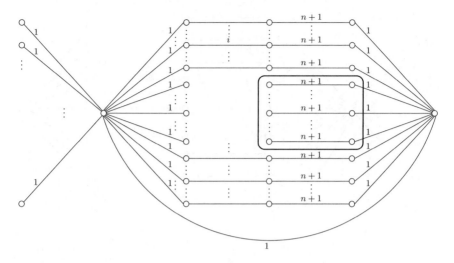

Fig. 1. The weighted input graph G. In the first two columns (on the left side) are the $v_{3,\cdot}$-vertices and the vertex $v_{3,2^n-1}$, the right vertex is $v_{3,0}$, on the left side of $v_{3,0}$ are the $v_{1,\cdot}$-vertices, next on the left side are the $v_{2,\cdot}$-vertices. The other vertices are the $v_{0,\cdot}$-vertices.

The graph $G = (V, E)$ consists of 2^{n+2} vertices $v_{i,j}$, $i \in \{0, \ldots, 3\}$, $j \in \{0, \ldots, 2^n - 1\}$. In the following let $b^\ell = (b^\ell_{n-1}, \ldots, b^\ell_0)$ be the binary representation of an integer $\ell \in \{0, \ldots, 2^n - 1\}$. There exists an edge between a vertex v_{i_1,j_1} and a vertex v_{i_2,j_2}

- with weight 1 if one of the following requirements is fulfilled:
    - $i_1 = 3$, $j_1 = 0$, and $i_2 = 1, j_2 \in \{0, \ldots, 2^n - 1\}$ (or vice versa),
    - $i_1 = 3$, $j_1 \neq 2^n - 1$, $i_2 = 3$ and $j_2 = 2^n - 1$ (or vice versa),
    - $i_1 = 3$, $j_1 = 2^n - 1$, $i_2 = 0$ and $j_2 \in \{0, \ldots, 2^n - 1\}$ (or vice versa),
- with weight $n + 1$ if $i_1 = 1$, $i_2 = 2$ and $j_1 = j_2$ (or vice versa),
- with weight i, $1 \leq i \leq n$, if $i_1 = 0$, $i_2 = 2$, $\sum_{k=0}^{n-1} b^{j_1}_k = i$, $b^{j_1}_{i-1} = 1$, and $j_1 = j_2$ (or vice versa). (Note, that if $\sum_{k=0}^{n-1} b^{j_1}_k = i$, and $b^{j_1}_{i-1} = 0$ or $j_1 \neq j_2$, there is no edge between v_{0,j_1} and v_{2,j_2}.)

Figure 1 shows the structure of the input graph G. Obviously, G is planar and the minimum spanning tree in G is unique. The important property of G is that an edge between a vertex v_{1,j_1} and a vertex v_{2,j_2} is in the minimum spanning tree if $j_1 = j_2$, and the binary representation of j_1 respectively j_2 corresponds to an input that belongs to $\mathrm{HWB}_n^{-1}(0)$. Therefore, the characteristic function of this edge set is a difficult function but in our input graph this edge set is in some sense hidden such that the characteristic function of the edge set of the input graph can be represented by OBDDs of small size. Figure 2 shows the minimum spanning tree in G. The vertices $v_{3,j}$, $j \neq 2^n - 1$, are only auxiliary vertices in order to obtain a number of vertices which is a power of 2.

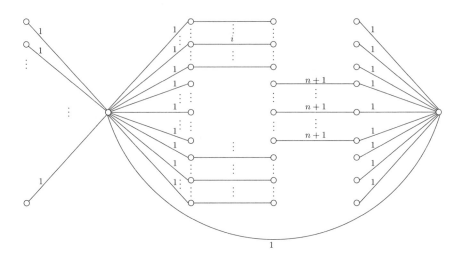

Fig. 2. The minimum spanning tree in G

2) The polynomial upper bound on the size of the OBDD representation for
$\mathcal{X}_E$:

Let $x_1^1, x_0^1, x_{n-1}^2, \ldots, x_0^2$ be the variables of the Boolean encoding of a vertex
$v_{i,j}$, where x_0^1 and x_0^2 denote the least significant bits, the x^1-variables represent
i and the x^2-variables j. In the rest of the proof we assume that $n+1$ is a
power of 2 because it has no bearing on the essence of our result. Let $d = (d_{\log(n+1)-1}, \ldots, d_0)$ be the binary representation of the edge weight $|d|$. The
characteristic function $\mathcal{X}_E$ of the edge set depends on $2n+4+\log(n+1)$ Boolean
variables $((x_1^1, x_0^1, x_{n-1}^2, \ldots, x_0^2), (y_1^1, y_0^1, y_{n-1}^2 \ldots, y_0^2), (d_{\log(n+1)-1}, \ldots, d_0))$. Our
aim is to prove that $\mathcal{X}_E$ can be represented by OBDDs of size $O(n^3)$ according
to the variable ordering

$$d_{\log(n+1)-1}, \ldots, d_0, x_1^1, y_1^1, x_0^1, y_0^1, x_{n-1}^2, y_{n-1}^2, \ldots, x_0^2, y_0^2.$$

Since there are $n+1$ different weights, the first part of the OBDD is a complete
binary tree of size $O(n)$. In the second part of the OBDD we distinguish three
different disjoint edge sets, between $v_{1,\cdot}$- and $v_{2,\cdot}$-vertices, between $v_{0,\cdot}$- and $v_{2,\cdot}$-
vertices, and the remaining edges. We prove that each of them can be represented
by OBDDs of small size. Since the different edge sets can be identified by the
assignments to the x^1- and y^1-variables which are tested at the beginning of the
OBDD, it suffices to add the OBDD sizes in order to obtain an upper bound on
the OBDD complexity of $\mathcal{X}_E$. (Note, that we can also use the well-known result
on the worst-case complexity of the synthesis operation, here the $\vee$-operation,
that the width of the OBDD for $\mathcal{X}_E$ can be asymptotically bounded above by
the product of the widths of the OBDDs for the three different edge sets. Since
for two of them the width is a constant we are done.)

- If $x_0^1 = 0, x_1^1 = 1$, $y_0^1 = 1$, and $y_1^1 = 0$ (or vice versa), it is checked whether $x_i^2 = y_i^2$. This is a simple equality check which can be done in linear size.
- If $x_1^1 = x_1^1 = 1$, $y_0^1 = 1$, and $y_1^1 = 0$, it is checked whether $|x^2| = 0$. (If $x_0^1 = 1, x_1^1 = 0$, $y_0^1 = y_1^1 = 1$, the roles of the x- and y-variables are exchanged.) If $x_0^1 = x_1^1 = 0$, $y_0^1 = y_1^1 = 1$, it is checked whether $|y^2| = 2^n - 1$. (If $x_0^1 = x_1^1 = 1$, $y_0^1 = y_1^1 = 0$, the roles of the x- and y-variables are exchanged.) If $x_0^1 = x_1^1 = y_0^1 = y_1^1 = 1$, it is checked whether $|x^2| = 2^n - 1$ and $|y^2| \neq 2^n - 1$ (or vice versa). Altogether the set of edges can be represented in linear size.
- If $x_0^1 = x_1^1 = 0$, $y_0^1 = 0$, and $y_1^1 = 1$ (or vice versa) and the edge weight is i, i.e. $|d| = i - 1$, the number of x^2-variables is counted. The function value is 1 if $\sum_{k=0}^{n-1} x_k^2 = i$, $x_{i-1}^2 = 1$, and $y_k^2 = x_k^2$, $0 \leq k \leq n - 1$. Since we only have to distinguish $i + 1$ different values for $|x^2|$, this can be done by an OBDD of width $O(n)$ if the edge weight is fixed. As there are n possible edge weights, the considered edge set can be represented by OBDDs of width $O(n^2)$ and size $O(n^3)$.

Summarizing, we have seen that $\mathcal{X}_E$ can be represented by an OBDD of size $O(n^3)$.

3) The exponential lower bound on the size of OBDDs for the characteristic function of the minimum spanning tree $\mathcal{X}_{MST}$ in G:

Here, we use some ideas presented in [5] for maximum matchings with exponential OBDD complexity (but in G there exist maximum matchings with $2^n + 2$ edges that can be represented by OBDDs of linear size).

Due to our definition of G the minimum spanning tree contains an edge between a vertex v_{1,j_1} and a vertex v_{2,j_2}, $j_1, j_2 \in \{0, \ldots, 2^n - 1\}$ iff and the binary representation of j_1 respectively j_2 corresponds to an input that belongs to $\mathrm{HWB}_n^{-1}(0)$ and $j_1 = j_2$. Our aim is to adapt the ideas for the exponential lower bound on the OBDD size of HWB_n presented in [11]. Therefore, we consider the subfunction of $\mathcal{X}_{MST}$, where all d-variables are replaced by 1, with other words the edge weight is set to $n + 1$, and $x_0^1 = 1$, $x_1^1 = 0$, $y_0^1 = 0$, and $y_1^1 = 1$. Let π be an arbitrary but fixed variable ordering. In the following our aim is to prove that the considered subfunction of $\mathcal{X}_{MST}$ has exponential π-OBDD size. As a result we can conclude that the size of any OBDD for the representation of $\mathcal{X}_{MST}$ needs exponential size. A pair (x_ℓ^2, y_ℓ^2), $\ell \in \{0, \ldots, n - 1\}$, is called (x, y)-pair and x_ℓ^2 a partner of y_ℓ^2 and vice versa. Now, we define a cut in the variable ordering after for the first time for exactly $(3/5)n$ (x, y)-pairs there exist at least one variable. T contains the variables before the cut according to π and B the remaining variables. Let P_H be the set of all pairs (x_i^2, y_i^2), $i \in \{n/2, \ldots, (9/10)n - 1\}$, and P_L be the set of all pairs (x_j^2, y_j^2), $j \in \{n/10, \ldots, n/2 - 1\}$. Obviously, T contains at least for $n/5$ pairs in P_H or at least for $n/5$ pairs in P_L at least one variable. W.l.o.g. we assume that T contains at least for $n/5$ pairs in P_L at least one variable. In the following we only consider assignments where variables that belong to the same (x, y)-pair are replaced by the same constant. We consider all assignments to the variables in T, where exactly $n/10$ pairs in

P_L are replaced by 1, all other variables in T are set to 0. There are at least $\binom{n/5}{n/10} = \Omega(n^{-1/2}2^{n/5})$ different assignments. Using Theorem 1 it is sufficient to prove that these assignments lead to different subfunctions. For this reason we consider two different assignments b and b' to the variables in T. Let (x_ℓ^2, y_ℓ^2) be an (x, y)-pair for which at least one variable is replaced differently in b and b'. W.l.o.g. $x_{\ell-1}^2$ is set to 0 in b and to 1 in b'. Now, we consider the following assignment b_r to the variables in B. The variables for which there is a partner in T are replaced by the assignment to the partner according to b. The remaining variables are replaced in such a way that there are exactly $\ell - n/10$ pairs that are set to 1. This can be done because there are $(2/5)n$ pairs for which both variables are in B and $\ell \leq n/2$. Obviously, the function value of the subfunction induced by b on b_r is 1. The function value for the subfunction induced by b' on b_r is 0 because either $|x^2| \neq |y^2|$ or $x^2 \in \mathrm{HWB}_n^{-1}(1)$.

Altogether, we have shown that the OBDD complexity of $\mathcal{X}_{MST}$ is at least $\Omega(n^{-1/2}2^{n/5})$. Since our input graph is planar we have shown that already the minimum spanning tree problem for planar graphs needs exponential space in the OBDD setting. □

Furthermore, we obtain the following result.

Corollary 1. *Symbolic OBDD-based algorithms for the single source shortest paths problem need exponential space with respect to the size of the implicit representation of the input graph.*

In [28] it has been shown that the single source shortest paths problem needs exponential space in the symbolic setting. Here, we sketch another proof for this result that leads to a slightly larger lower bound using our planar input graph G. The input is an OBDD for $\mathcal{X}_E$ defined above and an OBDD for the characteristic function of the vertex $v_{3,2^n-1}$. Let D be the set of all solution pairs $(v, \omega) \in V \times \mathbb{N}$ such that a shortest path from $v_{3,2^n-1}$ to v has weight ω. Here, the weight of a path from a vertex to another one is the total weight of the edges that belong to the considered path. The output OBDD has to represent the characteristic function $\mathcal{X}_D$. A shortest path from $v_{3,2^n-1}$ to a vertex $v_{i,j}$ in G, $0 \leq i \leq 3$ and $j \in \{0, \ldots, 2^n - 1\}$, has distance $n + 3$ iff $i = 2$ and the binary representation of j is an element in $\mathrm{HWB}_n^{-1}(0)$. Therefore, if we replace the distance variables by the binary representation of $n + 3$, $x_1^1 = 1$, and $x_0^1 = 0$ we obtain an OBDD for $\overline{\mathrm{HWB}_n}$. Since the π-OBDD size of a subfunction of a given function cannot be larger than the π-OBDD size of the function, we are done.

Concluding Remarks

One aim in the symbolic setting is to find advantageous properties of real-world instances that cause an essentially better behavior than in the worst-case. In [27] and [33] symbolic algorithms for maximum flow in 0-1 networks and topological sorting have been presented which have polylogarithmic running time with respect to the number of vertices of a given grid graph. These results rely on the very structured input graph and on restrictions on the width of occuring OBDDs during the computation. It is open whether constant input OBDD width is sufficient to guarantee polynomial space complexity for the minimum spanning tree problem.

References

1. Balcázar, J.L., Lozano, A.: The complexity of graph problems for succinctly represented graphs. In: Nagl, M. (ed.) WG 1989. LNCS, vol. 411, pp. 277–285. Springer, Heidelberg (1989)
2. Bazlamaçci, C.F., Hindi, K.S.: Minimum-weight spanning tree algorithms. A survey and empirical study. Computers & Operations Research 28, 767–785 (2001)
3. Bollig, B.: On the OBDD complexity of the most significant bit of integer multiplication. In: Agrawal, M., Du, D.-Z., Duan, Z., Li, A. (eds.) TAMC 2008. LNCS, vol. 4978, pp. 306–317. Springer, Heidelberg (2008)
4. Bollig, B.: Exponential space complexity for symbolic maximum flow algorithms in 0-1 networks. In: Hliněný, P., Kučera, A. (eds.) MFCS 2010. LNCS, vol. 6281, pp. 186–197. Springer, Heidelberg (2010)
5. Bollig, B.: On symbolic representations of maximum matchings and (un)directed graphs. In: Proc. of TCS IFIP AICT, vol. 323, pp. 263–300 (2010)
6. Bollig, B.: Symbolic OBDD-based reachability analysis needs exponential space. In: van Leeuwen, J., Muscholl, A., Peleg, D., Pokorný, J., Rumpe, B. (eds.) SOFSEM 2010. LNCS, vol. 5901, pp. 224–234. Springer, Heidelberg (2010)
7. Bollig, B., Löbbing, M., Sauerhoff, M., Wegener, I.: On the complexity of the hidden weighted bit function for various BDD models. Theoretical Informatics and Applications 33, 103–115 (1999)
8. Borůvka, O.: O jistém problému minimálním. Práce Mor. Piřírodověd. Spol. v Brně (Acta Societ. Scient. Natur. Moravicae) 3, 37–58 (1926)
9. Breitbart, Y., Hunt III, H.B., Rosenkrantz, D.J.: On the size of binary decision diagrams representing Boolean functions. Theoretical Computer Science 145, 45–69 (1995)
10. Bryant, R.E.: Graph-based algorithms for Boolean function manipulation. IEEE Trans. on Computers 35, 677–691 (1986)
11. Bryant, R.E.: On the complexity of VLSI implementations and graph representations of Boolean functions with application to integer multiplication. IEEE Trans. on Computers 40, 205–213 (1991)
12. Chazelle, B.: A minimum spanning tree algorithm with inverse-Ackermann type complexity. Journal of ACM 47(6), 1028–1047 (2000)
13. Cheriton, D., Tarjan, R.E.: Finding minimum spanning trees. SIAM Journal on Computing 5, 724–742 (1976)
14. Feigenbaum, J., Kannan, S., Vardi, M.V., Viswanathan, M.: Complexity of problems on graphs represented as OBDDs. In: Meinel, C., Morvan, M. (eds.) STACS 1998. LNCS, vol. 1373, pp. 216–226. Springer, Heidelberg (1998)
15. Fisler, K., Vardi, M.Y.: Bisimulation, minimization, and symbolic model checking. Formal Methods in System Design 21(1), 39–78 (2002)
16. Galperin, H., Wigderson, A.: Succinct representations of graphs. Information and Control 56, 183–198 (1983)
17. Gentilini, R., Piazza, C., Policriti, A.: Computing strongly connected components in a linear number of symbolic steps. In: Proc. of SODA, pp. 573–582. ACM Press, New York (2003)

18. Gentilini, R., Piazza, C., Policriti, A.: Symbolic graphs: linear solutions to connectivity related problems. Algorithmica 50, 120–158 (2008)
19. Graham, R.L., Hell, P.: On the history of the minimum spanning tree problem. Ann. Hist. Comput. 7, 43–57 (1985)
20. Hachtel, G.D., Somenzi, F.: A symbolic algorithm for maximum flow in 0–1 networks. Formal Methods in System Design 10, 207–219 (1997)
21. Kannan, S., Naor, M., Rudich, S.: Implicit representations of graphs. SIAM Journal on Discrete Mathematic 5, 596–603 (1992)
22. Karger, D.R., Klein, P.N., Tarjan, R.E.: A randomized linear-time algorithm to find minimum spanning trees. Journal of ACM 42, 321–328 (1995)
23. Meer, K., Rautenbach, D.: On the OBDD size for graphs of bounded tree- and clique-width. Discrete Mathematics 309(4), 843–851 (2009)
24. Nunkesser, R., Woelfel, P.: Representation of graphs by OBDDs. Discrete Applied Mathematics 157(2), 247–261 (2009)
25. Papadimitriou, C.H., Yannakakis, M.: A note on succinct representations of graphs. Information and Control 71, 181–185 (1986)
26. Pettie, S., Ramachandran, V.: An optimal minimum spanning tree algorithm. Journal of ACM 49, 16–34 (2002)
27. Sawitzki, D.: Implicit flow maximization by iterative squaring. In: Van Emde Boas, P., Pokorný, J., Bieliková, M., Štuller, J. (eds.) SOFSEM 2004. LNCS, vol. 2932, pp. 301–313. Springer, Heidelberg (2004)
28. Sawitzki, D.: Exponential lower bounds on the space complexity of OBDD-based graph algorithms. In: Correa, J.R., Hevia, A., Kiwi, M. (eds.) LATIN 2006. LNCS, vol. 3887, pp. 781–792. Springer, Heidelberg (2006)
29. Sawitzki, D.: The complexity of problems on implicitly represented inputs. In: Wiedermann, J., Tel, G., Pokorný, J., Bieliková, M., Štuller, J. (eds.) SOFSEM 2006. LNCS, vol. 3831, pp. 471–482. Springer, Heidelberg (2006)
30. Sieling, D., Wegener, I.: NC-algorithms for operations on binary decision diagrams. Parallel Processing Letters 48, 139–144 (1993)
31. Talamo, M., Vocca, P.: Representing graphs implicitly using almost optimal space. Discrete Applied Mathematics 108, 193–210 (2001)
32. Wegener, I.: Branching Programs and Binary Decision Diagrams - Theory and Applications. SIAM Monographs on Discrete Mathematics and Applications (2000)
33. Woelfel, P.: Symbolic topological sorting with OBDDs. Journal of Discrete Algorithms 4(1), 51–71 (2006)

Reducing the Maximum Latency of Selfish Ring Routing via Pairwise Cooperations*

Xujin Chen, Xiaodong Hu, and Weidong Ma

Institute of Applied Mathematics
Chinese Academy of Sciences, Beijing 100190, China
{xchen,xdhu}@amss.ac.cn, mawd335@163.com

Abstract. This paper studies the selfish routing game in ring networks with a load-dependent linear latency on each link. We adopt the asymmetric atomic routing model. Each player selfishly chooses a route to connect his source-destination pair, aiming at a lowest latency of his route, while the system objective is to minimize the maximum latency among all routes of players. Such a routing game always has a Nash equilibrium (NE) that is a "stable state" among all players, from which no player has the incentive to deviate unilaterally. Furthermore, 16 is the current best upper bound on its price of anarchy (PoA), the worst-case ratio between the maximum latencies in a NE and in a system optimum. In this paper we show that the PoA is at most 10.16 provided cooperations within pairs of players are allowed, where any two players could change their routes simultaneously if neither would experience a longer latency and at least one would experience a shorter latency.

Keywords: Selfish Routing, Price of Anarchy.

1 Introduction

In contrast to traditional routing in small-scale networks, routing in modern large networks often has no central control, and involves a large number of disparate participants who are not interested in any global optimization and simply seek to minimize their own cost by acting selfishly. *Selfish routing* [12] models network routing from a game-theoretic perspective, in which network users are viewed as self-interested strategic *players* participating in a competitive game. Each player, with his own pair of source and destination nodes in the network, aims to establish a communication path (between his source and destination) along which he would experience latency as low as possible, given the link congestion caused by all other players. Despite the system objective to minimize the maximum latency among all source-destination pairs, in the absence of a central authority who can impose and maintain globally efficient routing strategies on the network traffic [9], network designers are often interested in a *Nash equilibrium* (NE) that is as close to the system optimum as possible, where the NE is a "stable state" among the players, from which no player has the incentive to

* Supported in part by the NSF of China under Grant No. 10771209, 10721101, 10928102 and Chinese Academy of Sciences under Grant No. kjcx-yw-s7.

W. Wu and O. Daescu (Eds.): COCOA 2010, Part II, LNCS 6509, pp. 31–45, 2010.
© Springer-Verlag Berlin Heidelberg 2010

deviate unilaterally. The (in)efficiency of NE is predominantly quantified by the so called *price of anarchy* (PoA) [8], which is the worst-case ratio between the maximum latencies in a NE and in a system optimum. This paper focuses on selfish routing in ring networks whose links are associated with load-dependent linear latencies. In an effort to improve the efficiency of the selfish ring routing whose PoA has been upper bounded by 16 [5], we show that the upper bound reduces to 10.16 provided cooperations within pairs of players are allowed, where two players cooperate with each other only if neither would experience a longer latency and at least one would experience a shorter latency when simultaneous changing their routes in the current routing.

Related work. Among vast literature on selfish routing [9], the study on egalitarian system objective falls behind its utilitarian counterpart [3,7,11]. In particular, when the system performance is measured by the maximum latency (whose minimization is thus desirable), the PoA of atomic congestion games [10] with linear latency is 2.5 in single-commodity network, but it explodes to $\Theta(\sqrt{m})$ in m-commodity networks [7]. Analogously, selfish routing in general network can have unbounded PoA [6]. When network topology is further restricted to rings, the selfish ring routing with linear latency possesses the nice property that for any instance either every optimal routing is a NE or the PoA of the instance is at most $4 + 2\sqrt{2} < 6.83$ [6]. Recently, the PoA of this selfish ring routing has been shown to be bounded above by 16 [5].

The stable solutions investigated in this paper are similar to and stronger than the extensively studied strong equilibria and k-strong equilibria [1]. Our study on the ring topology is inspired by the fact that rings have been a fundamental topology frequently encountered in communication networks, and attract considerable attention and efforts from the research community [2,4,6,13,14].

Our contributions. In this paper, by method of allowing cooperations within pairs of self-interested players, we establish a lower PoA upper bound for selfish ring routing with linear latency with respect to the system objective of minimizing maximum latency. In addition to the theoretical improvement, our concrete example shows that this kind of cooperation does lead to shorter maximum latency of the network system. The improvement on global efficiency brought by coordination within small-sized coalitions is highly realizable in decentralized environments since players themselves are able to determine easily (say by enumeration) coalitions of size at most k (say $k \leq 2$) whose deviations can make every member "better off". This approach is particularly useful for competitive games involving a large number of players in large networks, where only small-scale communication and computation are achievable.

Organization of the paper. In Section 2, we provide the mathematical model SRLC for the selfish ring routing with cooperation/collusion. In Section 3, we prove some basic properties for the NE in SRLC. In Section 4, we establish the main result that the PoA of SRLC is at most 10.16 when cooperations within pairs of players are allowed. In Section 5, we conclude this paper with remarks on future research.

2   Model

Our model, *selfish ring latency with collusion* (SRLC), is specified by a quadruple $\mathcal{I} = (R, l, (s_i, t_i)_{i=1}^m, k)$, usually called a $SRLC_k$ *instance* or a *SRLC instance*. As illustrated in Fig. 1(a), the underlying network of $\mathcal{I}$ is a ring $R = (V, E)$, an undirected cycle, with node set $V = \{v_1, v_2, \ldots, v_n\}$ and link set $E = \{e_i = v_i v_{i+1} : i = 1, 2, \ldots, n\}$, where $v_{n+1} = v_1$. By writing $P \subseteq R$, we mean that P is a subgraph of R (possibly R itself) with node set $V(P)$ and link set $E(P)$. Each link $e \in E$ is associated with a load-dependent linear *latency (function)* $l_e(x) = a_e x + b_e$, where a_e, b_e are nonnegative constants, and x is an integer variable indicating the load on e. There are m (≥ 2) source-destination node pairs (s_i, t_i), $i = 1, 2, \ldots, m$, corresponding to m players $1, 2, \ldots, m$. Each player i $(1 \leq i \leq m)$ has a communication request for routing one unit of flow from his source $s_i \in V$ to his destination $t_i \in V - \{s_i\}$, and his strategy set consists of two internally disjoint paths P_i and $\bar{P}_i$ in ring R with ends s_i and t_i satisfying

$$V(P_i) \cap V(\bar{P}_i) = \{s_i, t_i\} \text{ and } P_i \cup \bar{P}_i = R, \ i = 1, 2, \ldots, m.$$

We set $\bar{\bar{P}}_i := P_i$ for $i = 1, 2, \ldots, m$. Different players may have the same source-destination pair, and vertices $s_i, t_i, i = 1, 2, \ldots, m$, are not necessarily distinct.

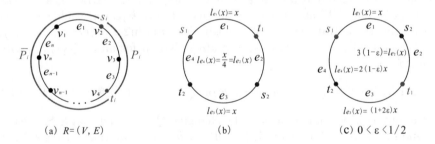

(a) $R = (V, E)$ (b)   (c) $0 < \varepsilon < 1/2$

Fig. 1. The SRLC instances

A *(feasible) routing* π for the SRLC instance $\mathcal{I}$ is a 0-1 function π on multiset $\mathcal{P} := \cup_{i=1}^m \{P_i, \bar{P}_i\}$ such that $\pi(P_i) + \pi(\bar{P}_i) = 1$ for every $i = 1, 2, \ldots, m$. In view of the correspondence between π and player strategies adopted for the SRLC instance, we abuse the notation slightly by writing $\pi = \{Q_1, Q_2, \ldots, Q_m\}$ with the understanding that, for each $i = 1, 2, \ldots, m$, the one unit of flow requested by player i is routed along path $Q_i \in \{P_i, \bar{P}_i\}$, and correspondingly $\pi(Q_i) = 1 > 0 = \pi(\bar{Q}_i)$. Also we write $Q_i \in \pi$ for $i = 1, 2, \ldots, m$. Each link $e \in E$ bears a *load* π_e with respect to π defined as

$$\pi_e := \sum_{P \in \mathcal{P} : e \in E(P)} \pi(P) = |\{Q_i : e \in E(Q_i), i = 1, 2, \ldots, m\}|$$

equal the number of paths in $\{Q_1, Q_2, \ldots, Q_m\}$ each of which goes through e. Every $P \subseteq R$ is associated with a nonnegative integer

$$l_P(\pi) := \sum_{e \in E(P)} l_e(\pi_e) = \sum_{e \in E(P)} (a_e \pi_e + b_e),$$

which indicates roughly the total latencies of links on P experienced in π. (The wording "indicates roughly" changes to "equals" when every link of P is used by some player in the routing π.) Naturally, the latency experienced by player i and the maximum latency experienced by the system are

$$L_i(\pi) := l_{Q_i}(\pi) \text{ for } i = 1, 2, \ldots, m, \text{ and } L(\pi) := \max_{i=1}^{m} L_i(\pi), \qquad (2.1)$$

respectively. We call $L_i(\pi)$ the *latency of player* i with respect to π, and $L(\pi)$ the *maximum latency of the routing* π. A routing for $\mathcal{I}$ is *optimal* if its maximum latency is minimum among all routings for $\mathcal{I}$.

Given a routing π, a coalition of players *gets a gain* by changing simultaneously its members' strategies/choices in π if the change makes no latency of any member increase and makes the latency of at least one member decrease. In $\mathcal{I} = (R, l, (s_i, t_i)_{i=1}^m, k)$, integer k lies in $[1, m]$, and a coalition of at most k players is allowed to form in order to get a gain by changing whenever possible. In response to this kind of collusion, a "stable" solution in the $SRLC_k$ is a *k-robust Nash equilibrium* (NE) at which no coalition of at most k players could get a gain by changing their strategies simultaneously. So, a routing $\pi = \{Q_1, Q_2, \ldots, Q_m\}$ is a *k-robust NE* or a *k-robust Nash routing* (NR) if for all $S \subseteq \{1, 2, \ldots, m\}$ with $|S| \leq k$,

$$\text{either } l_{Q_i}(\pi) < l_{\bar{Q}_i}(\pi') \text{ for some } i \in S \text{ or } l_{Q_i}(\pi) = l_{\bar{Q}_i}(\pi') \text{ for all } i \in S, \quad (2.2)$$

where routing π' is obtained from π by changing Q_i to $\bar{Q}_i$ for all $i \in S$. Clearly, a k-robust NE is an h-robust NE for all $1 \leq h \leq k$. Note that 1-robust NE is exactly the classical Nash equilibrium, concerning with special cases $S = \{i\}$, $i = 1, 2, \ldots, m$, for which (2.2) gives

$$l_{Q_i}(\pi) \leq \sum_{e \in E(\bar{Q}_i)} l_e(\pi_e + 1) \text{ for all } i = 1, 2, \ldots, m. \qquad (2.3)$$

The term "1-robust" will often be omitted for short. We point out that the notion of k-robust NE studied in is closely related to the so called k-strong equilibrium [1] referring to a strategy profile in which no coalition of size at most k has any joint deviation beneficial to (strictly reducing the latencies of) all members. Note that every k-robust NE is a k-strong equilibrium while the converse is necessarily true.

In the $SRLC_k$ instance $\mathcal{I}$, let π^* be an optimal routing, and π a k-robust NR with maximum $L(\pi)$. The ratio $L(\pi)/L(\pi^*)$ is called the *k-robust price of anarchy* (*k-RPoA*) of $\mathcal{I}$. For fixed k, the notion of the k-RPoA extends to the *$SRLC_k$ problem* of all $SRLC_k$ instances, whose *k-RPoA* is set to be the supremum of k-RPoA over all SRLC$_k$ instances. Take 2-player case as an example, where $k = 1$ or 2. For the SRLC$_k$ instances depicted in Fig. 1(b), the k-RPoA is 2 when $k = 1$ and 1 when $k = 2$ (Collusion of two players does help!). In contrast, the k-RPoA of the SRLC$_k$ instances in Fig. 1(c) is $\frac{5-3\varepsilon}{4-\varepsilon}$ for $k = 1$ and 2.

3   Basic Properties

In this section, we investigate Nash routings for an arbitrary SRLC$_k$ instance $\mathcal{I} = (R, l, (s_i, t_i)_{i=1}^m, k)$. For any $P \subseteq R$ and any routing π for $\mathcal{I}$, we often consider $l_P(\pi) := \sum_{e \in E(P)} l_e(\pi_e) = \sum_{e \in E(P)} (a_e \pi_e + b_e)$ as the sum of

$$l_P^a(\pi) := \sum_{e \in E(P)} a_e \pi_e \text{ and } l_P^b(\pi) := \sum_{e \in E(P)} b_e.$$

Define notations:

$$\|P\|_a := \sum_{e \in E(P)} a_e, \quad \|P\|_b := \sum_{e \in E(P)} b_e, \text{ and } \|P\| := \|P\|_a + \|P\|_b.$$

It is worth noting that the equation $l_P^b(\pi) = \|P\|_b$ always holds, though in contrast the integer $l_P^a(\pi)$ may be smaller or bigger than or equal to $\|P\|_a$. So for any routing π we particularly have

$$l_P(\pi) = l_P^a(\pi) + l_P^b(\pi) = l_P^a(\pi) + \|P\|_b. \tag{3.4}$$

When P ($\subseteq R$) is a path, complementary to it is the other path $\bar{P} \subseteq R$ whose edge-disjoint union with P forms R. We will make explicit or implicit use of the following equations in our discussion:

$$\|P\|_a + \|\bar{P}\|_a = \|R\|_a, \ \|P\|_b + \|\bar{P}\|_b = \|R\|_b, \text{ and } \|P\| + \|\bar{P}\| = \|R\|. \tag{3.5}$$

In the rest of the paper, we denote by $\pi^\nabla = \{Q_1, Q_2, \ldots, Q_k\}$ a given routing for the SRLC$_k$ instance $\mathcal{I} = (R, l, (s_i, t_i)_{i=1}^m, k)$ in which players $1, 2, \ldots, m$ are named such that for a minimum j with $1 \leq j \leq m$, we have

$$\pi^N = \{\bar{Q}_1, \ldots, \bar{Q}_j, Q_{j+1}, \ldots, Q_m\} \text{ is a NR for } \mathcal{I}, \text{ and}$$
$$\gamma := \max_{i=1}^j \frac{\|\bar{Q}_i\|_a}{\|Q_i\|_a} = \frac{\|\bar{Q}_1\|_a}{\|Q_1\|_a}; \text{ so } l_R^a(\pi^N) \leq \max\{\gamma, 1\} l_R^a(\pi^\nabla). \tag{3.6}$$

If $\bar{Q}_p = Q_q$ for some p, q with $1 \leq p \neq q \leq j$, then without loss of generality $\{p, q\} = \{j-1, j\}$; it follows that $Q_{j-1} = \bar{Q}_j \in \pi^N$, $Q_j = \bar{Q}_{j-1} \in \pi^N$, and we can express π^N as $\pi^N = \{\bar{Q}_1, \ldots, \bar{Q}_{j-2}, Q_{j-1}, \ldots, Q_m\}$, contradicting the minimality of j. Thus

$$\{\bar{Q}_1, \ldots, \bar{Q}_j\} \cap \{Q_1, \ldots, Q_j\} = \emptyset. \tag{3.7}$$

By (3.5), we see from $\|\bar{Q}_1\|_a = \gamma \|Q_1\|_a$ in (3.6) that

$$\|Q_1\|_a = \|R\|_a/(\gamma + 1). \tag{3.8}$$

Since R is the edge-disjoint union of Q_i and $\bar{Q}_i$ for every $i = 1, 2, \ldots, m$, from (3.4), with R in place of P, we derive for $i = 1, 2, \ldots, m$,

$$l_{\bar{Q}_i}(\pi^N) + l_{Q_i}(\pi^N) = l_R(\pi^N) = l_R^a(\pi^N) + l_R^b(\pi^N) = l_R^a(\pi^N) + ||R||_b. \quad (3.9)$$

Applying (2.3) to the NR $\pi^N = \{\bar{Q}_1, \ldots, \bar{Q}_j, Q_{j+1}, \ldots, Q_m\}$, we obtain

$$\begin{aligned} l_{\bar{Q}_i}(\pi^N) &\leq l_{Q_i}(\pi^N) + ||Q_i||_a \text{ for } i = 1, 2, \ldots, j; \\ l_{Q_i}(\pi^N) &\leq l_{\bar{Q}_i}(\pi^N) + ||\bar{Q}_i||_a \text{ for } i = j+1, j+2, \ldots, m. \end{aligned} \quad (3.10)$$

With the definition of $L(\pi^N)$ given by (2.1), an easy case analysis on (3.10) shows that $L(\pi^N)$ is bounded above by $(l_Q(\pi^N) + l_{\bar{Q}}(\pi^N) + \max\{||Q||_a, ||\bar{Q}||_a\})/2$ for $Q \in \pi^N$ with $l_Q(\pi^N) = L(\pi^N)$. This in combination with (3.9) gives

$$L(\pi^N) \leq \frac{l_R(\pi^N) + ||R||_a}{2} = \frac{l_R^a(\pi^N) + ||R||_a + ||R||_b}{2} = \frac{l_R^a(\pi^N) + ||R||}{2}. \quad (3.11)$$

Note from (3.9) and (3.10) that $l_R(\pi^N) = l_{\bar{Q}_i}(\pi^N) + l_{Q_i}(\pi^N) \leq 2l_{Q_i}(\pi^N) + ||Q_i||_a$ for $i = 1, 2, \ldots, j$. Thus the leftmost inequality in (3.11) implies

$$L(\pi^N) \leq l_{Q_i}(\pi^N) + \frac{||Q_i||_a}{2} + \frac{||R||_a}{2}, \text{ for } i = 1, 2, \ldots, j. \quad (3.12)$$

Lemma 1. *If positive numbers β and ρ satisfy $\beta = L(\pi^N)/L(\pi^\nabla)$, $l_R(\pi^\nabla) \leq 2\rho L(\pi^\nabla)$, and $\beta > \rho$, then the following hold:*

(i) $\beta \leq \rho \max\{\gamma, 1\} + ||R||_a/(2L(\pi^\nabla))$.
(ii) $(\beta\gamma - \beta - 2\rho) l_{Q_1}(\pi^N) \leq 2\rho(\beta\gamma - \rho)L(\pi^\nabla) + (\beta + \rho)||Q_1||_a + \rho||R||_a - (\beta - \rho)||R||_b$.

Proof. From (3.11) we have $L(\pi^N) \leq \frac{1}{2}(l_R(\pi) + ||R||_a)$, which in combination of (3.6) implies (i):

$$\beta = \frac{L(\pi^N)}{L(\pi^\nabla)} \leq \frac{\max\{\gamma, 1\} l_R(\pi^\nabla) + ||R||_a}{2L(\pi^\nabla)} \leq \max\{\gamma, 1\}\rho + \frac{||R||_a}{2L(\pi^\nabla)}.$$

To prove (ii), we deduce from (3.11) that $l_R^a(\pi^N) \geq 2L(\pi^N) - ||R|| = 2\beta L(\pi^\nabla) - ||R||$. Thus $l_R^a(\pi^N) \geq \frac{\beta}{\rho} l_R(\pi^\nabla) - ||R||$ which can be expressed using (3.4) as

$$\sum_{i=1}^{j} ||\bar{Q}_i||_a + \sum_{i=j+1}^{k} ||Q_i||_a \geq \frac{\beta}{\rho} \sum_{i=1}^{j} ||Q_i||_a + \frac{\beta}{\rho} \sum_{i=j+1}^{k} ||Q_i||_a + \frac{\beta}{\rho}||R||_b - ||R||_a - ||R||_b.$$

By applying (3.5) and substituting $||R||_a - ||\bar{Q}_i||_a$ for $||Q_i||_a$, $i = 1, 2, \ldots, j$, in the above inequality, we obtain

$$\sum_{i=1}^{j} ||\bar{Q}_i||_a \geq \frac{\beta}{\rho}\left(j \cdot ||R||_a - \sum_{i=1}^{j} ||\bar{Q}_i||_a\right) + \left(\frac{\beta}{\rho} - 1\right) \sum_{i=j+1}^{k} ||Q_i||_a + \left(\frac{\beta}{\rho} - 1\right)||R||_b - ||R||_a.$$

Rearranging terms in the above inequality yields

$$\left(\frac{\beta}{\rho}+1\right)\sum_{i=1}^{j}||\bar{Q}_i||_a \geq \left(\frac{\beta}{\rho}j-1\right)||R||_a + \left(\frac{\beta}{\rho}-1\right)\sum_{i=j+1}^{k}||Q_i||_a + \left(\frac{\beta}{\rho}-1\right)||R||_b.$$

Since $\beta/\rho > 1$, ignoring the nonnegative middle term on the right-hand side and dividing both sides by positive number $\beta/\rho + 1$, we derive from the above inequality that

$$\sum_{i=1}^{j}||\bar{Q}_i||_a \geq \frac{\beta j - \rho}{\beta + \rho}||R||_a + \frac{\beta - \rho}{\beta + \rho}||R||_b. \tag{3.13}$$

Let us now consider sum $\sum_{i=1}^{j}||\bar{Q}_i \cap Q_1||_a$, which equals the total contributions of paths $\bar{Q}_1, \bar{Q}_2, \ldots, \bar{Q}_j$ in the NR π^N to the value of $l_{Q_1}^a(\pi^N)$. Clearly, the sum of the contributions is at least

$$l_{Q_1}^a(\pi^N) - \sum_{i=j+1}^{k}||Q_i||_a \geq l_{Q_1}^a(\pi^N) - l_R^a(\pi^\nabla),$$

and thus at least $l_{Q_1}^a(\pi^N) - l_R(\pi^\nabla) + ||R||_b$ by (3.4). It follows from $l_R(\pi^\nabla) \leq 2\rho L(\pi^\nabla)$ that

$$\sum_{i=1}^{j}||\bar{Q}_i \cap Q_1||_a \geq l_{Q_1}^a(\pi^N) - 2\rho L(\pi^\nabla) + ||R||_b. \tag{3.14}$$

On the other hand, since R is the link-disjoint union of Q_1 and $\bar{Q}_1$, we have

$$l_{\bar{Q}_1}^a(\pi^N) \geq \sum_{i=1}^{j}||\bar{Q}_i \cap \bar{Q}_1||_a \geq \sum_{i=1}^{j}\left(||\bar{Q}_i||_a - ||Q_1||_a\right).$$

In turn, using (3.13) and $||R||_a = (\gamma + 1)||Q_1||_a$ in (3.8), we can lower bound $l_{\bar{Q}_1}^a(\pi^N)$ as follows:

$$\begin{aligned}
l_{\bar{Q}_1}^a(\pi^N) &\geq \sum_{i=1}^{j}\left(||\bar{Q}_i||_a - ||Q_1||_a\right) \\
&\geq \frac{\beta j - \rho}{\beta + \rho}||R||_a - j \cdot ||Q_1||_a + \frac{\beta - \rho}{\beta + \rho}||R||_b \\
&= j\left(\frac{\beta(\gamma + 1)}{\beta + \rho} - 1\right)||Q_1||_a - \frac{\rho}{\beta + \rho}||R||_a + \frac{\beta - \rho}{\beta + \rho}||R||_b \\
&\geq \frac{\beta\gamma - \rho}{\beta + \rho}\sum_{i=1}^{j}||\bar{Q}_i \cap Q_1||_a + \frac{(\beta - \rho)||R||_b - \rho||R||_a}{\beta + \rho}.
\end{aligned}$$

Furthermore, it follows from (3.14) that

$$l_{\bar{Q}_1}^a(\pi^N) \geq \frac{\beta\gamma - \rho}{\beta + \rho}\left(l_{Q_1}^a(\pi^N) - 2\rho L(\pi^\nabla) + ||R||_b\right) + \frac{(\beta - \rho)||R||_b - \rho||R||_a}{\beta + \rho}. \tag{3.15}$$

Applying (3.10) and (3.4), we have

$$l_{Q_1}(\pi^N) + ||Q_1||_a \geq l_{\bar{Q}_1}(\pi^N) = l_{\bar{Q}_1}^a(\pi^N) + ||\bar{Q}_1||_b.$$

Combining the above inequality with (3.15) and using $||R||_b = ||Q_1||_b + ||\bar{Q}_1||_b \geq ||Q_1||_b$, we deduce that

$$l_{Q_1}(\pi^N) + ||Q_1||_a$$
$$\geq \frac{\beta\gamma - \rho}{\beta + \rho}\left(l_{Q_1}^a(\pi) - 2\rho L(\pi^\nabla) + ||R||_b\right) + \frac{(\beta - \rho)||R||_b - \rho||R||_a}{\beta + \rho} + ||\bar{Q}_1||_b$$
$$\geq \frac{\beta\gamma - \rho}{\beta + \rho}\left(l_{Q_1}^a(\pi^N) - 2\rho L(\pi^\nabla) + ||Q_1||_b\right) + \frac{\beta(||Q_1||_b + ||\bar{Q}_1||_b) - \rho||R||_a - \rho||R||_b}{\beta + \rho}$$
$$\geq \frac{\beta\gamma - \rho}{\beta + \rho}\left(l_{Q_1}(\pi^N) - 2\rho L(\pi^\nabla)\right) + \frac{\beta||R||_b - \rho||R||_a - \rho||R||_b}{\beta + \rho}$$
$$= \frac{\beta\gamma - \rho}{\beta + \rho}\left(l_{Q_1}(\pi^N) - 2\rho L(\pi^\nabla)\right) + \frac{(\beta - \rho)||R||_b - \rho||R||_a}{\beta + \rho}.$$

Thus we obtain

$$(\beta + \rho)\left(l_{Q_1}(\pi^N) + ||Q_1||_a\right) \geq (\beta\gamma - \rho)\left(l_{Q_1}(\pi^N) - 2\rho L(\pi^\nabla)\right) + (\beta - \rho)||R||_b - \rho||R||_a,$$

which is equivalent to the inequality in (ii). The lemma is then proved. □

Lemma 2. Let $\beta = L(\pi^N)/L(\pi^\nabla)$. Then the following hold:

(i) $\beta \leq 10.16$ if $l_R(\pi^\nabla) \leq 5L(\pi^\nabla)$ and $||R||_a \leq 2.5L(\pi^\nabla)$.
(ii) $\beta \leq 7.05$ if $l_R(\pi^\nabla) \leq 3L(\pi^\nabla)$ and $||R||_a \leq 3L(\pi^\nabla)$.

Proof. To see (i), assume to the contrary $\beta > 10.16$. With $\rho = 2.5$, we deduce from Lemma 1 that

$$\gamma = \max\{\gamma, 1\} \geq \frac{\beta}{\rho} - \frac{||R||_a}{2\rho L(\pi^\nabla)} > \frac{10.16}{2.5} - \frac{2.5}{5} = 3.564, \qquad (3.16)$$
$$(\beta\gamma - \beta - 5)l_{Q_1}(\pi^N) \leq 5(\beta\gamma - 2.5)L(\pi^\nabla) + (\beta + 2.5)||Q_1||_a + 2.5||R||_a - (\beta - 2.5)||R||_b.$$

Note from (3.16) that $\beta\gamma - \beta - 5 > 0$, and from (3.8) that $||Q_1||_a = \frac{||R||_a}{\gamma + 1} \leq 2.5\frac{L(\pi^\nabla)}{\gamma + 1}$. With (3.12) we get

$$L(\pi^N) \leq l_{Q_1}(\pi^N) + \frac{||Q_1||_a}{2} + \frac{||R||_a}{2}$$
$$\leq \frac{5(\beta\gamma - 2.5)}{\beta\gamma - \beta - 5}L(\pi^\nabla) + \left(\frac{\beta + 2.5}{\beta\gamma - \beta - 5} + \frac{1}{2}\right)||Q_1||_a + \left(\frac{2.5}{\beta\gamma - \beta - 5} + \frac{1}{2}\right)||R||_a$$
$$\leq \frac{5(\beta\gamma - 2.5)}{\beta\gamma - \beta - 5}L(\pi^\nabla) + \frac{\beta(\gamma + 1)}{2(\beta\gamma - \beta - 5)} \cdot \frac{2.5L(\pi^\nabla)}{\gamma + 1} + \frac{\beta(\gamma - 1)}{2(\beta\gamma - \beta - 5)} \cdot 2.5L(\pi^\nabla)$$
$$= \frac{12.5\beta\gamma - 25}{2(\beta\gamma - \beta - 5)}L(\pi^\nabla).$$

As $\gamma > 0$ by (3.16), the derivative of $\frac{12.5\beta\gamma - 25}{2(\beta\gamma - \beta - 5)}$ with respect to β is negative for all $\beta > 0$. So, using $\beta > 10.16$, we obtain

$$10.16 < \beta = \frac{L(\pi^N)}{L(\pi^\nabla)} \leq \frac{12.5\beta\gamma - 25}{2(\beta\gamma - \beta - 5)} \leq \frac{12.5(10.16\gamma) - 25}{2(10.16\gamma - 10.16 - 5)} = \frac{127\gamma - 25}{20.32\gamma - 30.32}.$$

Now $\frac{127\gamma - 25}{20.32\gamma - 30.32} > 10.16$ implies $\gamma < 3.563$, a contradiction to (3.16), proving (i).

We verify (ii) similarly. By contradiction assume $\beta > 7.05$. With $\rho = 1.5$, Lemma 1 gives

$$\gamma = \max\{\gamma, 1\} \geq \frac{\beta}{\rho} - \frac{||R||_a}{2\rho L(\pi^\nabla)} > \frac{7.05}{1.5} - \frac{3}{3} = 3.7, \tag{3.17}$$

$$(\beta\gamma - \beta - 3)l_{Q_1}(\pi^N) \leq 3(\beta\gamma - 1.5)L(\pi^\nabla) + (\beta + 1.5)||Q_1||_a + 1.5||R||_a - (\beta - 1.5)||R||_b.$$

Note from (3.17) that $\beta\gamma - \beta - 3 > 0$, and from (3.8) that $||Q_1||_a = \frac{||R||_a}{\gamma + 1} \leq 3\frac{L(\pi^\nabla)}{\gamma + 1}$. With (3.12) we get

$$L(\pi^N) \leq l_{Q_1}(\pi^N) + \frac{||Q_1||_a}{2} + \frac{||R||_a}{2}$$

$$\leq \frac{3(\beta\gamma - 1.5)}{\beta\gamma - \beta - 3}L(\pi^\nabla) + \left(\frac{\beta + 1.5}{\beta\gamma - \beta - 3} + \frac{1}{2}\right) \cdot \frac{3L(\pi^\nabla)}{\gamma + 1} + \left(\frac{1.5}{\beta\gamma - \beta - 3} + \frac{1}{2}\right) \cdot 3L(\pi^\nabla)$$

$$= \frac{9\beta\gamma - 9}{2(\beta\gamma - \beta - 3)}L(\pi^\nabla).$$

As $\gamma > 0$ by (3.17), the derivative of $\frac{9\beta\gamma - 9}{2(\beta\gamma - \beta - 3)}$ with respect to β is negative for all $\beta > 0$. So, using $\beta > 7.05$, we obtain

$$7.05 < \beta = \frac{L(\pi^N)}{L(\pi^\nabla)} \leq \frac{9\beta\gamma - 9}{2(\beta\gamma - \beta - 3)} \leq \frac{9(7.05\gamma) - 9}{2(7.05\gamma - 7.05 - 3)} = \frac{63.45\gamma - 9}{14.1\gamma - 20.1},$$

implying $\gamma < 3.6909$. The contradiction to (3.17) establishes (ii).   □

4   2-Robust Nash Routings

The result established in this section provides the evidence that small-sized collusions might help decrease PoA in selfish ring routing games, which is desirable for network design. In case of $k = 1$, no collusion is allowed, and the PoA of the SRLC$_1$ problem has been shown to be bounded above by 16 [5]. In case of $k = 2$, two players may act simultaneously to gain benefit for at least one without increasing latency of the other (See Fig. 1(b)). The following theorem shows a smaller upper bound for the PoA of the SRLC$_2$ problem in case of 2-robust NE existing.

Theorem 1. *The 2-robust price of anarchy of the SRLC$_2$ problem is at most 10.16.*

Proof. Consider an arbitrary 2-robust Nash routing π^N for a SRLC$_2$ instance $\mathcal{I} = (R, l, (s_i, t_i)_{i=1}^m, 2)$. If some link e of $R = (V, E)$ has $a_e + b_e = 0$, then shrinking e gives a SRLC$_2$ instance with the same PoA. So for ease of description, we assume without loss of generality that

$$a_e + b_e > 0 \text{ for all } e \in E. \tag{4.18}$$

For any subgraphs P and Q of the ring R, by $P \cup Q$ (resp. $P \cap Q$) we mean the subgraph of R with node set $V(P) \cup V(Q)$ (resp. $V(P) \cap V(Q)$) and link set $E(P) \cup E(Q)$ (resp. $E(P) \cap E(Q)$), which consists of at most two paths.

Clearly $\mathcal{I}$ admits an optimal routing π^* that is *irredundant* in the sense that any two paths $P, Q \in \pi^*$ with $P \cup Q = R$ are link-disjoint. Set $\beta := L(\pi^N)/L(\pi^*)$. It suffices to show $\beta \leq 10.16$. To this end, we may assume $\pi^* = \pi^\nabla \neq \pi^N$ as described in Section 3, as otherwise $\beta = 1$ and we are done.

If some $\bar{Q}_g$ and $\bar{Q}_h$ with $1 \leq g < h \leq j$ are link-disjoint, then $Q_g \cup Q_h = R$, and since π^∇ is irredundant, it must be the case that $\bar{Q}_g = Q_h$ and $\bar{Q}_h = Q_g$, a contradiction to (3.7). Hence

$$E(\bar{Q}_g) \cap E(\bar{Q}_h) \neq \emptyset \text{ for all } 1 \leq g < h \leq j. \tag{4.19}$$

With (3.9), we may assume

$$l_{Q_i}(\pi^N) + l_{\bar{Q}_i}(\pi^N) = l_R(\pi^N) > 10.16 L(\pi^\nabla) \text{ for all } i = 1, 2, \ldots, m, \tag{4.20}$$

as otherwise (2.1) implies $L(\pi^N) \leq l_R(\pi^N) \leq 10.16 L(\pi^\nabla)$ giving $\beta \leq 10.16$. By definition,

$$\|Q_i\|_a \leq \|Q_i\| \leq l_{Q_i}(\pi^\nabla) \leq L(\pi^\nabla) \text{ for all } i = 1, 2, \ldots, m. \tag{4.21}$$

For the Nash routing π^N, we deduce from (3.10) that

$$l_{Q_i}(\pi^N) \geq l_{\bar{Q}_i}(\pi^N) - \|Q_i\|_a \geq l_{\bar{Q}_i}(\pi^N) - L(\pi^\nabla) \text{ for } 1 \leq i \leq j, \tag{4.22}$$

and then from (4.20) that

$$l_{Q_i}(\pi^N) \geq \frac{l_{\bar{Q}_i}(\pi^N) + l_{Q_i}(\pi^N) - L(\pi^\nabla)}{2} > 4.58 \cdot L(\pi^\nabla) \text{ for } 1 \leq i \leq j. \tag{4.23}$$

If some Q_g with $1 \leq g \leq j$ is link-disjoint from $\cup_{i=1}^j \bar{Q}_i$, then $l_{Q_g}(\pi^N) \leq l_{Q_g}(\pi^\nabla) \leq L(\pi^\nabla)$ indicates a contradiction to (4.23). So we have

$$E(Q_g) \cap \left(\cup_{i=1}^j E(\bar{Q}_i) \right) \neq \emptyset \text{ for all } 1 \leq g \leq j; \text{ in particular } j \geq 2. \tag{4.24}$$

It is not difficult to see from (4.19) and (4.24) that one of the following three cases (illustrated in Fig. 2) must be true:

Case 1: There exist p, q, and r with $1 \leq p < q < r \leq j$ such that $\bar{Q}_p \cup \bar{Q}_q \subsetneq R$, $\bar{Q}_q \cup \bar{Q}_r \subsetneq R$, $\bar{Q}_r \cup \bar{Q}_p \subsetneq R$, and $\bar{Q}_p \cup \bar{Q}_q \cup \bar{Q}_r = R$.

Case 2: There exist p and q with $1 \leq p < q \leq j$ such that $\bar{Q}_p \cup \bar{Q}_q = R$.

Case 3: There exist p and q with $1 \leq p < q \leq j$ such that $\cup_{i=1}^j \bar{Q}_i \subseteq \bar{Q}_p \cup \bar{Q}_q \subsetneq R$.

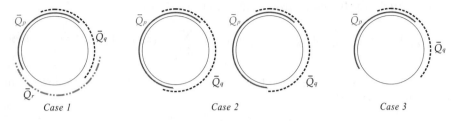

Fig. 2. Possible configurations of π^N when $l_R(\pi^N) > 10.16L(\pi^\triangledown)$

Here, no other cases are possible, since (4.19) says that each pair of paths in $\bar{Q}_1, \bar{Q}_2, \ldots, \bar{Q}_j$ must have a least one common link. Let the indices p, q and r satisfying one of the above three cases be fixed. Our case analysis goes as follows:

Case 1. It is easy to see that $Q_p \cup Q_q \cup Q_r = R$, which implies

$$\|R\|_a \leq \|R\| \leq l_R(\pi^\triangledown) \leq l_{Q_p}(\pi^\triangledown) + l_{Q_q}(\pi^\triangledown) + l_{Q_r}(\pi^\triangledown) \leq 3L(\pi^\triangledown).$$

Hence Lemma 2(ii) guarantees $\beta \leq 7.05$ as desired.

Case 2. Consider any g and h with $1 \leq g < h \leq j$ and $\bar{Q}_g \cup \bar{Q}_h = R$. Let routing π be obtained from π^N by changing $\bar{Q}_g$ and $\bar{Q}_h$ to Q_g and Q_h, respectively. Suppose by symmetry $l_{\bar{Q}_g}(\pi^N) \leq l_{\bar{Q}_h}(\pi^N)$. Since $E(\bar{Q}_g) \cap E(\bar{Q}_h) \neq \emptyset$ by (4.19), we deduce from (4.18) that $l_{\bar{Q}_g \cap \bar{Q}_g}(\pi^N) > 0$ and further that

$$l_{Q_h}(\pi) = l_{\bar{Q}_g}(\pi^N) - l_{\bar{Q}_g \cap \bar{Q}_h}(\pi^N) \leq l_{\bar{Q}_h}(\pi^N) - l_{\bar{Q}_g \cap \bar{Q}_h}(\pi^N) < l_{\bar{Q}_h}(\pi^N).$$

As π^N is a 2-robust NE, it can be seen from (2.2) that $l_{\bar{Q}_g}(\pi^N) < l_{Q_g}(\pi) = l_{Q_g}(\pi^N)$. Hence $l_{Q_g}(\pi^N) > l_{\bar{Q}_g}(\pi^N) \geq l_{Q_h}(\pi^N)$ follows from $\bar{Q}_g \supseteq Q_h$ and in turn $l_{Q_g}(\pi^N) > l_{\bar{Q}_h}(\pi^N) - L(\pi^\triangledown)$ follows from (4.22). Using the fact that $\bar{Q}_h$ is the link-disjoint union of $\bar{Q}_g \cap \bar{Q}_h$ and Q_g, we obtain

$$\begin{aligned} l_{\bar{Q}_g \cap \bar{Q}_h}(\pi^N) = l_{\bar{Q}_h}(\pi^N) - l_{Q_g}(\pi^N) < L(\pi^\triangledown) \\ \text{for all } 1 \leq g < h \leq j \text{ with } \bar{Q}_g \cup \bar{Q}_h = R. \end{aligned} \tag{4.25}$$

Recall from (4.23) that $l_{Q_i}(\pi^N) > 4.5L(\pi^\triangledown)$ for all $1 \leq i \leq j$, which in combination with (4.25) implies

$$\begin{aligned} E(Q_i) \cap E(Q_g \cup Q_h) \neq \emptyset \text{ for all } 1 \leq i \leq j \\ \text{and all } 1 \leq g < h \leq j \text{ with } \bar{Q}_g \cup \bar{Q}_h = R. \end{aligned} \tag{4.26}$$

Now we turn to the indices p, q that have been fixed in the hypothesis of Case 2. By (4.26) one can easily find (not necessarily distinct) paths $P_p^1, P_p^2, P_q^1, P_q^2$ in $\{Q_1, Q_2, \ldots, Q_j\} - \{Q_p, Q_q\}$ such that

$$E(P_g^h) \cap E(Q_g) \neq \emptyset \text{ for } h = 1, 2, g = p, q; \quad \bigcup_{i=1}^{j} Q_j \subseteq \bigcup_{g=p,q} (Q_g \cup P_g^1 \cup P_g^2); \tag{4.27}$$

and subject to (4.27), $\mathcal{Q} := \{P_p^1, P_p^2, P_q^1, P_q^2\}$ is as small as possible. $\tag{4.28}$

Note $0 \le |\mathcal{Q}| \le 4$. It follows from (4.27) and (4.21) that

$$l_{\bar{Q}_p \cap \bar{Q}_q}(\pi^\nabla) - l_{\bar{Q}_p \cap \bar{Q}_q}(\pi^N) \le l_{\cup_{P \in \mathcal{Q}} P}(\pi^\nabla) \le \sum_{P \in \mathcal{Q}} l_P(\pi^\nabla) \le |\mathcal{Q}| \cdot L(\pi^\nabla). \quad (4.29)$$

Observe that R is the link-disjoint union of Q_p, Q_q and $\bar{Q}_p \cap \bar{Q}_q$. It follows from (4.21) and (4.25) that

$$\|R\|_a \le \|Q_p\|_a + \|Q_q\|_a + \frac{1}{2} l_{\bar{Q}_p \cap \bar{Q}_q}(\pi^N) \le 2.5 L(\pi^\nabla).$$

When $l_R(\pi^\nabla) \le 5L(\pi^\nabla)$, Lemma 2(i) gives $\beta \le 10.16$ as desired. When $l_R(\pi^\nabla) > 5L(\pi^\nabla)$, we derive from (4.21) and (4.25) that

$$l_{\bar{Q}_p \cap \bar{Q}_q}(\pi^\nabla) - l_{\bar{Q}_p \cap \bar{Q}_q}(\pi^N) = l_R(\pi^\nabla) - l_{Q_p}(\pi^\nabla) - l_{Q_q}(\pi^\nabla) - l_{\bar{Q}_p \cap \bar{Q}_q}(\pi^N)$$
$$> 5L(\pi^\nabla) - 3L(\pi^\nabla) = 2L(\pi^\nabla),$$

which together with (4.29) implies that $\mathcal{Q} = \{P_p^1, P_p^2, P_q^1, P_q^2\}$ has size 3 or 4. In particular we see $\bar{Q}_p \cap \bar{Q}_q$ consists of two paths X, Y each of which has at least one edge.

In case of $|\mathcal{Q}| = 3$, the minimality in (4.28) together with symmetry allows us to assume that $\mathcal{Q} = \{P_p^1, P_p^2, P_q^1\}$ such that $E(P_p^1) \cap E(Q_q) = \emptyset$, $E(P_q^1) \cap E(Q_p) = \emptyset$, $E(P_p^2) \cap E(X) \subsetneq E(P_p^1) \cap E(X) \ne \emptyset \ne E(P_q^1) \cap E(X)$ and $E(P_p^1) \cap E(Y) \subsetneq E(P_p^2) \cap E(Y) \ne \emptyset$. If P_q^1 has an end in P_p^2, then $P_p^1 \cup P_p^2 \cup P_q^1 \cup Q_p \cup (\bar{Q}_p \cap \bar{Q}_q) = R$ and (4.27) guarantees $l_R(\pi^\nabla) \le \sum_{P \in \mathcal{Q}} l_P(\pi^\nabla) + l_{Q_p}(\pi^\nabla) + l_{\bar{Q}_p \cap \bar{Q}_q}(\pi^N)$; consequently a contradiction $l_R(\pi^\nabla) < 5L(\pi^\nabla)$ would implied by (4.21) and (4.25). Thus

$$V(P_q^1) \cap V(P_p^2) = \emptyset. \quad (4.30)$$

Similarly if $V(P_p^1) \cap V(P_p^2)$ is nonempty, then (4.30) ensures the existence of common node of P_p^1 and P_p^2 in Q_p, and $l_R(\pi^\nabla) \le \sum_{P \in \mathcal{Q}} l_P(\pi^\nabla) + l_{Q_q}(\pi^\nabla) + l_{\bar{Q}_p \cap \bar{Q}_q}(\pi^N) < 5L(\pi^\nabla)$ turns out a contradiction. Hence we have

$$V(P_p^1) \cap V(P_p^2) = \emptyset. \quad (4.31)$$

By (4.30), we see that (4.25) applies with $Q_g = P_p^1$ and $Q_g = P_p^2$, and provides $l_{\bar{P}_p^1 \cap \bar{P}_p^2}(\pi^N) < L(\pi^\nabla)$. By (4.31) we deduce that $P_q^1 \subseteq (\bar{P}_p^1 \cap \bar{P}_p^2) \cup (\bar{Q}_p \cap \bar{Q}_q)$ and therefore $l_{P_q^1}(\pi^N) \le l_{\bar{P}_p^1 \cap \bar{P}_p^2}(\pi^N) + l_{\bar{Q}_p \cap \bar{Q}_q}(\pi^N) \le 2L(\pi^\nabla)$, which is a contradiction to (4.23).

In case of $|\mathcal{Q}| = 4$, the minimality in (4.28) enforces

$$E(Q_p) \cap E(P_q^1 \cup P_q^2) = \emptyset \text{ and } E(Q_q) \cap E(P_p^1 \cup P_p^2) = \emptyset. \quad (4.32)$$

As $\mathcal{Q} \subseteq \{Q_1, Q_2, \ldots, Q_j\}$, by (4.26) we must have $\bar{P}_q^1 \cup \bar{P}_q^2 \ne R$ and $\bar{P}_p^1 \cup \bar{P}_p^2 \ne R$, which in combination with (4.32) yields $E(P_g^1) \cap E(P_g^2) \ne \emptyset$ for $g = p, q$ and $(\cup_{P \in \mathcal{Q}} P) \cup (\bar{Q}_p \cap \bar{Q}_q) = R$. Moreover it can be deduced from (4.27) that $\cup_{i=1}^j Q_i \subset$

$\cup_{P \in \mathcal{Q}} P$. Therefore $l_R(\pi^\nabla) \leq \sum_{P \in \mathcal{Q}} l_P(\pi^\nabla) + l_{\bar{Q}_p \cap \bar{Q}_q}(\pi^\nabla) \leq 4L(\pi^\nabla) + l_{\bar{Q}_p \cap \bar{Q}_q}(\pi^N)$ by (4.21), and consequently $l_R(\pi^\nabla) \leq 5L(\pi^\nabla)$ by (4.25). The contradiction completes the proof of Case 2.

Case 3. By symmetry suppose $l_{\bar{Q}_p}(\pi^N) \leq l_{\bar{Q}_q}(\pi^N)$. Let routing π be obtained from π^N by changing $\bar{Q}_p$ and $\bar{Q}_q$ to Q_p and Q_q, respectively. The hypothesis $\cup_{i=1}^j \bar{Q}_i \subseteq \bar{Q}_p \cup \bar{Q}_q$ of Case 3 implies that

$$Q_p \cap Q_q \subseteq Q_i \text{ for all } 1 \leq i \leq j, \text{ and therefore}$$
$$l_{Q_p \cap Q_q}(\pi^N) = l_{Q_p \cap Q_q}(\pi^\nabla) - j\|Q_p \cap Q_q\|_a \qquad (4.33)$$

by the relation between π^∇ and π^N specified in (3.6). It follows that

$$\cup_{i=1}^j Q_j \subseteq Q_g \cup Q_h \text{ for some } 1 \leq g, h \leq j, \text{ and} \qquad (4.34)$$
$$l_{Q_p \cap Q_a}(\pi) = l_{Q_p \cap Q_a}(\pi^N) + 2\|Q_p \cap Q_q\|_a \leq l_{Q_p \cap Q_a}(\pi^\nabla). \qquad (4.35)$$

Moreover, recalling (4.21), it is obvious from $\cup_{i=1}^j \bar{Q}_i \subseteq \bar{Q}_p \cup \bar{Q}_q$ that

$$l_{Q_p \cap Q_q}(\pi^N) + \|Q_d\|_a \leq l_{Q_d}(\pi^\nabla) \leq L(\pi^\nabla) \text{ for } d = p \text{ and } q, \qquad (4.36)$$

and from (4.34) that

$$l_R(\pi^\nabla) \leq l_{Q_g}(\pi^\nabla) + l_{Q_h}(\pi^\nabla) + l_{\bar{Q}_p \cap \bar{Q}_q}(\pi^N) \leq 2L(\pi^\nabla) + l_{\bar{Q}_p \cap \bar{Q}_q}(\pi^N). \qquad (4.37)$$

Since Q_q is the link-disjoint union of $Q_p \cap Q_q$ and a subpath of $\bar{Q}_p$, yielding $l_{\bar{Q}_p}(\pi^N) \geq l_{Q_q}(\pi^N) - l_{Q_p \cap Q_q}(\pi^N)$, we derive from (4.22) that

$$l_{Q_p}(\pi^N) \geq l_{\bar{Q}_p}(\pi^N) - \|Q_p\|_a \geq l_{Q_q}(\pi^N) - l_{Q_p \cap Q_q}(\pi^N) - \|Q_p\|_a$$
$$\geq l_{\bar{Q}_q}(\pi^N) - \|Q_q\|_a - l_{Q_p \cap Q_q}(\pi^N) - \|Q_p\|_a,$$

and further that

$$l_{\bar{Q}_p \cap \bar{Q}_q}(\pi^N) = l_{\bar{Q}_q}(\pi^N) - \left(l_{Q_p}(\pi^N) - l_{Q_p \cap Q_q}(\pi^N) \right)$$
$$\leq 2 l_{Q_p \cap Q_q}(\pi^N) + \|Q_p\|_a + \|Q_q\|_a.$$

It follows from (4.36) that $l_{\bar{Q}_p \cap \bar{Q}_q}(\pi^N) \leq 2L(\pi^\nabla)$, which in combination with (4.37) gives

$$l_R(\pi^\nabla) \leq 4L(\pi^\nabla). \qquad (4.38)$$

Consider the case where $l_{Q_p \cap Q_a}(\pi^\nabla) < l_{\bar{Q}_p \cap \bar{Q}_a}(\pi^N)$. From $l_{\bar{Q}_p}(\pi^N) \leq l_{\bar{Q}_q}(\pi^N)$ and (4.35) we deduce that

$$l_{Q_q}(\pi) = l_{\bar{Q}_p}(\pi^N) - l_{\bar{Q}_p \cap \bar{Q}_q}(\pi^N) + l_{Q_p \cap Q_a}(\pi)$$
$$\leq l_{\bar{Q}_q}(\pi^N) - l_{\bar{Q}_p \cap \bar{Q}_q}(\pi^N) + l_{Q_p \cap Q_q}(\pi^\nabla) < l_{\bar{Q}_q}(\pi^N).$$

Since π^N is a 2-robust NE, by (2.2) it must be the case that $l_{Q_p}(\pi) > l_{\bar{Q}_p}(\pi^N)$ saying $l_{\bar{Q}_q}(\pi^N) - l_{\bar{Q}_p \cap \bar{Q}_q}(\pi^N) + l_{Q_p \cap Q_q}(\pi) > l_{\bar{Q}_p}(\pi^N)$. On the one hand (4.35) shows

$$l_{\bar{Q}_q}(\pi^N) - l_{\bar{Q}_p \cap \bar{Q}_q}(\pi^N) + l_{Q_p \cap Q_q}(\pi^\nabla) > l_{\bar{Q}_p}(\pi^N). \tag{4.39}$$

On the other hand, since π^N is a Nash routing, (2.3) says

$$l_{\bar{Q}_q}(\pi^N)$$
$$\leq \sum_{e \in E(Q_q)} l_e(\pi_e^N + 1)$$
$$= \left[l_{\bar{Q}_p}(\pi^N) - l_{\bar{Q}_p \cap \bar{Q}_q}(\pi^N) \right] + [\|Q_q\|_a - \|Q_p \cap Q_q\|_a] + l_{Q_p \cap Q_q}(\pi^N) + \|Q_p \cap Q_q\|_a.$$

Using (4.33), we get

$$l_{\bar{Q}_q}(\pi^N) \leq l_{\bar{Q}_p}(\pi^N) - l_{\bar{Q}_p \cap \bar{Q}_q}(\pi^N) + \|Q_q\|_a - \|Q_p \cap Q_q\|_a + l_{Q_p \cap Q_q}(\pi^\nabla).$$

Substituting the right hand side of the above inequality for $l_{\bar{Q}_q}(\pi^N)$ in (4.39) gives $l_{\bar{Q}_p}(\pi^N) - l_{\bar{Q}_p \cap \bar{Q}_q}(\pi^N) + \|Q_q\|_a - \|Q_p \cap Q_q\|_a + l_{Q_p \cap Q_q}(\pi^\nabla) - l_{\bar{Q}_p \cap \bar{Q}_q}(\pi^N) + l_{Q_p \cap Q_q}(\pi^\nabla) > l_{\bar{Q}_p}(\pi^N)$. Rearranging and collecting terms provides $l_{\bar{Q}_p \cap \bar{Q}_q}(\pi^N) < l_{Q_p \cap Q_q}(\pi^\nabla) + \frac{1}{2}(\|Q_q\|_a - \|Q_p \cap Q_q\|_a)$. Hence we have shown that in any case

$$l_{\bar{Q}_p \cap \bar{Q}_a}(\pi^N) \leq l_{Q_p \cap Q_q}(\pi^\nabla) + \frac{1}{2}(\|Q_q\|_a - \|Q_p \cap Q_q\|_a).$$

It follows that

$$\|R\|_a = \|Q_p \cup Q_q\|_a + \|\bar{Q}_p \cap \bar{Q}_q\|_a \leq l_{Q_p \cup Q_q}(\pi^\nabla) + l_{\bar{Q}_p \cap \bar{Q}_q}(\pi^N)$$
$$= l_{Q_p}(\pi^\nabla) + l_{Q_q}(\pi^\nabla) - l_{Q_p \cap Q_q}(\pi^\nabla) + l_{\bar{Q}_p \cap \bar{Q}_q}(\pi^N)$$
$$\leq l_{Q_p}(\pi^\nabla) + l_{Q_q}(\pi^\nabla) + \frac{1}{2}\|Q_q\|_a.$$

Thus $\|R\|_a \leq 2.5L(\pi^\nabla)$ by (4.21). Since $l_R(\pi^\nabla) \leq 4L(\pi^\nabla)$ by (4.38), Lemma 2(i) ensures $\beta \leq 10.16$.

We are now able to conclude that $\beta \leq 10.16$ in all cases, which establishes Theorem 1.   □

5   Concluding Remark

In this paper we have studied the selfish ring routing with linear latency for minimizing maximum latency that allows coalitions among self-interested players (SRLC). We have proved that the 2-RPoA of SRLC is bounded above by 10.16. It deserves further efforts to obtain smaller upper bound on k-RPoA for greater k. Since improvement on global efficiency due to cooperation within small-sized coalitions is highly realizable in decentralized environments, it is interesting to see if the method could be extended to selfish routing games in general networks for minimizing maximum latency, shortening the system maximum latency via small coalitions and some other techniques. Obviously, it is a big challenge to shift study from ring networks to other networks of more complicated topologies.

References

1. Andelman, N., Feldman, M., Mansour, Y.: Srong Price of Anarchy. Games and Economic Behavior 65, 289–317 (2009)
2. Anshelevich, E., Zhang, L.: Path Decomposition under a New Cost Measure with Applications to Optical Network Design. ACM Transactions on Algorithms 4, Artical No. 15 (2008)
3. Awerbuch, B., Azar, Y., Epstein, L.: The Price of Routing Unsplittable Flow. In: 37th Annual ACM Symposium on Theory of Computing, pp. 57–66 (2005)
4. Bentza, C., Costab, M.-C., Létocartc, L., Roupin, F.: Multicuts and Integral Multiflows in Rings. European Journal of Operational Research 196, 1251–1254 (2009)
5. Chen, B., Chen, X., Hu, J., Hu, X.: Stability vs. Optimality in Selfish Ring Routing. Submitted to SIAM Journal on Discrete Mathematics
6. Chen, B., Chen, X., Hu, X.: The Price of Atomic Selfish Ring Routing. Journal of Combinatorial Optimization 19, 258–278 (2010)
7. Christodoulou, G., Koutsoupias, E.: The Price of Anarchy of Finite Congestion Games. In: 37th Annual ACM Symposium on Theory of Computing, pp. 67–73 (2005)
8. Koutsoupias, E., Papadimitriou, C.H.: Worst-case Equilibria. In: Meinel, C., Tison, S. (eds.) STACS 1999. LNCS, vol. 1563, pp. 404–413. Springer, Heidelberg (1999)
9. Nisan, N., Roughtgarden, T., Tardos, É., Vazirani, V.V. (eds.): Algorithmic Game Theory. Cambridge University Press, Cambridge (2007)
10. Rosenthal, R.W.: A Class of Games Possessing Pure-strategy Nash Equilibira. International Jouranl of Game Theory 2, 65–67 (1973)
11. Roughgarden, T.: The Price of Anarchy Is Independent of the Network Topology. Jouranl of Computer and System Sciences 67, 342–364 (2003)
12. Roughgarden, T., Tardos, É.: How Bad Is Selfish Routing? Journal of the ACM 49, 236–259 (2002)
13. Schrijver, A., Seymour, P., Winkler, P.: The Ring Loading Problem. SIAM Journal on Discrete Mathematics 11, 1–14 (1998)
14. Wang, B.F.: Linear Time Algorithms for the Ring Loading Problem with Demand Splitting. Journal of Algorithms 54, 45–57 (2005)

Constrained Surface-Level Gateway Placement for Underwater Acoustic Wireless Sensor Networks

Deying Li, Zheng Li, Wenkai Ma, and Hong Chen

Key Laboratory of Data Engineering and Knowledge Engineering, MOE,
School of Information, Renmin University of China, China

Abstract. One approach to guarantee the performance of underwater
acoustic sensor networks is to deploy multiple Surface-level Gateways
(SGs) at the surface. This paper addresses the connected (or survivable)
Constrained Surface-level Gateway Placement (C-SGP) problem for 3-D
underwater acoustic sensor networks. Given a set of candidate locations
where SGs can be placed, our objective is to place minimum number
of SGs at a subset of candidate locations such that it is connected (or
2-connected) from any USN to the base station. We propose a polynomial time approximation algorithm for the connected C-SGP problem
and survivable C-SGP problem, respectively. Simulations are conducted
to verify our algorithms' efficiency.

Keywords: Underwater sensor networks, Surface-level gateway placement, Connectivity, Survivability, Approximation algorithm.

1 Introduction and Motivations

Underwater Acoustic Wireless Sensor Networks (UA-WSNs) consist of underwater sensors that are deployed to perform collaborative monitoring tasks over
a given region [1]. Underwater sensors are prone to failures because of fouling
and corrosion. It is important that the deployed network is highly reliable, so
as to avoid failure of monitoring missions due to failure of single or multiple
sensors. Additionally, the network topology is in general a crucial factor in determining the energy consumption, the capacity and the communication delay of
a network [2]. Hence, the network topology should be carefully engineered, and
post-deployment topology optimization should be performed, when possible.

There is an architecture for 3-D UA-WSNs, consisting of resource-constrained
Underwater Sensor Nodes (USNs) floating at different depths in order to observe
a given phenomenon, some resource-rich SGs which are placed at the surface,
and BSs (onshore sink or satellite etc.). The SG is equipped with an acoustic transceiver that is able to handle multiple parallel communications with the
USNs. It is also endowed with a long range Radio Frequency (RF) transmitter to
communicate with other SGs and the Base Stations (BSs). This network architecture provides better QoS and is used to quickly forward sensing data packets

W. Wu and O. Daescu (Eds.): COCOA 2010, Part II, LNCS 6509, pp. 46–57, 2010.
© Springer-Verlag Berlin Heidelberg 2010

to the user [1,3,4]. In practice, however, there are some physical constraints on the placement of the SGs (or relay nodes). For example, there should be a minimum distance between two SGs in the network to avoid interference. Also, there may be some regions where SGs cannot be placed. In practice, there may be a forbidden regions where SGs cannot be placed [8].

In this paper, we study the Constrained Surface-level Gateway Placement (C-SGP) problem for 3-D underwater acoustic sensor networks, in which the optimization objective is to place the minimum number of SGs at a subset of candidate locations to meet 1-connectivity and Survivability (2-connectivity) requirements. We propose an approximate algorithm for the two problems respectively, and corroborate the algorithms' performance through theoretical analysis and simulations.

The rest of this paper is organized as follows. In Section II we present related works. Section III describes the network model and basic notations. The 1-connected and survivable C-SGP problems are studied in section IV. Section V presents the simulation results, and Section VI concludes this paper.

2   Related Works

The benefits of using SGs have been presented in previous research[1,3,4]. The work in [1] introduces a type of 3-D UA-WSNs architecture, consisting of USNs, SGs, and BSs (onshore sink or satellite etc.). The role of SGs is to communicate USNs with BSs. The work in [3] mainly focuses on the surface gateway placement. And the tradeoff between the number of surface gateways and the expected delay and energy consumption was analyzed. In [4], the authors propose a novel virtual sink architecture for UA-WSNs that aims to achieve robustness and energy efficiency in harsh under water channel conditions.

The majority of the existing work in relay node deployment problem is based on the 2-D network model derived from the terrestrial wireless sensor networks [5–8]. In addition, almost all of the above works study unconstrained version problem, in the sense that the relay nodes can be placed anywhere. However, in reality there are some physical constraints on the placement of the SGs (or relay nodes). Only works in [3,8] address the constrained surface-level gateway (relay node) placement problem. In this paper, we focus on the constrained surface-level gateway placement problem in 3-D networks to meet 1-connectivity and 2-connectivity, which is different from the problems in [3,8]. The authors in [3] only formulated the problem as Integer Linear Programming, but did not give any algorithm. In [8], the authors studied the constrained relay node placement problem in 2-D WSNs to meet 1-connectivity and survivability requirements. However, approaches proposed for 2-D networks can not be directly applied in 3-D networks. Therefore, some new research challenges are posed.

3   Notations and Basic Concepts

Let us consider a 3-D heterogeneous UA-WSN consisting of USNs, SGs and a BS. The USNs are pre-deployed in the sensing area and floated at different depths, each of them is equipped with an acoustic communicator which has communication range R_A. On the other hand, SGs only can be deployed on the surface, and are equipped with acoustic communicators and RF transceivers which have communication ranges R_A and R_{RF}, respectively. Compared with wireless RF links among ground-based or surface-level gateways, underwater acoustic wireless links have higher attenuation and path loss [1]. We assume that the wireless RF transceiver has longer effective distance than the acoustic modem. R_A and R_{RF} are given positive constants and $R_{RF} > R_A > 0$. We also assume that the BS is powerful enough so that its communication range is much greater than R_{RF} and R_A.

In this paper, $d_{Euc}(u, v)$ represents the Euclidean distance between u and v. Let b be the base station, S be a set of USNs, and L be a set of candidate locations where SGs can be placed. We use an undirected graph $G(V, E)$ to model the network architecture of a 3-D UA-WSN, where $V(G)=\{b\} \cup S \cup L$. The edge set $E(G)$ can be defined as follows:

- For any SG $u \in L$, and any node $v \in \{b\} \cup L$ which could be either a SG or the BS, $(u, v) \in E$ if and only if $d_{Euc}(u, v) \leq R_{RF}$.
- For any USN $w \in S$ and any node $z \in S \cup L \cup \{b\}$ which could be either a USN, a SG or the BS, $(w, z) \in E$ if and only if $d_{Euc}(w, z) \leq R_A$.

Definition 1. Suppose $G(V, E)$ is a 3-D graph to model a 3-D UA-WSN. Let H be a subgraph of G and u be a SG in H. The *USN degree* of u in H, denoted by $\delta_s(u, H)$, is the number of USNs that are neighbors of u in H. The *maximum USN degree* of H is defined as $\Delta_s(H)=\max\{\delta_s(u, H)|u \in V(H) \cap L\}$.

Definition 2. Suppose $G(V, E)$ is a graph. For $V' \subseteq V$, we denote $G[V'] = G(V', E')$ as an induced subgraph of $G(V, E)$ by V', in which, for any two nodes u and v in V', $(u, v) \in E'$ if and only if $(u, v) \in E$. For $E' \subseteq E$, we denote $G[E'] = G(V', E')$ as an induced subgraph of $G(V, E)$ by E', where V' is a set of endpoints of all edges in E'.

In this paper, we focus on the connected (or survivable) C-SGP Problem, which are formally represented as follows:

The connected (or survivable) C-SGP Problem: Given an UA-WSN (R_{RF}, R_A, $\{b\}$, S, L), the connected (or survivable) C-SGP problem is to place SGs at a subset L' of candidate locations in L such that there exists 1 routing path (or 2 node disjoint routing paths) connecting any USN in S to the BS and $|L'|$ is minimized.

The connected and survivable C-SGP problems are NP-hard since they have been proved to be NP-hard even for the scenario of 2-D network model[8]. In addition, the authors [9] proved that the 1-connected node cover placement problem (which is the special case of the connected C-SGP problem) is NP-hard where all the nodes are on regular triangular grid points.

4   Algorithms for the Constrained Surface-Level Gateway Placement Problems

In order to design approximation algorithms for the C-SGP problems, we construct a weighted graph $G(V, E, w)$. We give a weight for each edge in $G(V, E)$ as follows:

- For any edge $(u, v) \in E(G)$, we define its *weight* as $w(u, v) = |\{u, v\} \cap L|$. Let H be a subgraph of G, the weight of H is define as: $w(H) = \sum_{e \in E(H)} w(e)$.

From above definition, we know that the weight of any edge in E connecting two nodes u and v in L is assigned to 2. Similarly, any edge in E connecting a node in $\{b\} \cup S$ with a node in L is assigned weight of 1. Any edge connecting two USNs is assigned weight of 0. We have the following lemma.

Lemma 1. Let H be a subgraph of $G(V, E, w)$ such that each node's degree in $V(H) \cap L$ is at least 2 (within H). Then $w(H) \geq 2 \cdot |V(H) \cap L|$.

Proof. Initially, each node's weight in H is initialized to 0. Let (u, v) be an edge of H which is incident with two SGs. According to our definition, the weight of this edge is 2. In this case, we divide the edge weight into two equal pieces, add weight 1 to node u, add another 1 weight to node v. Let (u, v) be an edge of G where u is a SG and v is not. According to our definition, the weight of this edge is 1. In this case, we add weight 1 to node u, add weight 0 to node v. Let (u, v) be an edge of H where neither u nor v is a SG. According our definition, the weight of this edge is 0. In this case, we do not add any weight to node u and v. When all edges are executed over, we have shifted the edge weights of H to the SGs in H. Note that any SG u is getting a weight of 1 from every edge of H which is incident with u, resulting in that the weight of u is equal to the degree of u. Since each SG in H is incident to at least two edges in H, it receives at least weight 2. Therefore, $w(H) \geq 2 \cdot |V(H) \cap L|$.   ∎

4.1   An Algorithm for the Connected C-SGP Problem

In this subsection, we propose a polynomial time approximation algorithm for the connected C-SGP problem.

The algorithm includes two steps: (1) construct an edge-weighted undirected graph $G(V, E, w)$; (2) using the existing algorithm for the minimum Steiner tree problem on weight graph $G(V, E, w)$ to get a feasible solution for the connected C-SGP problem.

The algorithm is presented as Algorithm 1.

Lemma 2. Suppose Y_{opt} is an optimal solution to the connected C-SGP problem. Let T_{opt} be a Minimum Spanning Tree (MST) of $G[\{b\} \cup Y_{opt} \cup S]$ which is a induced subgraph of $G(V, E, w)$ by $\{b\} \cup Y_{opt} \cup S$. Then $\Delta_s(T_{opt}) \leq 12$.

Proof. We prove this lemma by contradiction. Assume that there is a SG u which can be connected to more than 12 USNs in T_{opt}. Without of the generality, we

---

Algorithm 1. An approximation algorithm for the connected C-SGP Problem

---

Input: An UA-WSN (R_{RF}, R_A, $\{b\}$, S, L).
Output: A feasible solution Y_A for the connected C-SGP problem.

Begin:
1: Construct an edge-weighted undirected graph $G(V, E, w)$ based on this UA-WSN, where $V = \{b\} \cup S \cup L$.
2: **if** The nodes in $\{b\} \cup S$ are not in a single connected component H of $G(V, E)$ **then**
3:    The connected C-SGP problem does not have a feasible solution. Stop.
4: **end if**
5: Apply the existing algorithm A for the Steiner Minimum Tree problem to compute a low weight Steiner Tree T_A of $G(V, E, w)$ for $\{b\} \cup S$.
6: Output $Y_A = V(T_A) \cap L$.
End.

---

assume that u is connected to 13 USNs $v_1, v_2, ..., v_{13}$. We will prove that these 13 USNs can not communicate with each other. Otherwise, we assume v_1 and v_2 can communicate with each other, i. e., (v_1, v_2) is an edge in $G(V, E, w)$. Since T_{opt} is a tree, it does not contain edge (v_1, v_2), otherwise there would be a cycle (u, v_1, v_2, u) in T_{opt}. Replacing edge (u, v_1) in T_{opt} by edge (v_1, v_2), we obtain another tree T_1 spanning all nodes in $\{b\} \cup Y_{opt} \cup S$. Since $w(u, v_1) = 1$ and $w(v_1, v_2) = 0$, we have $w(T_1) < w(T_{opt})$, contracting to the assumption that T_{opt} is a MST.

Since the acoustic communication range of any SG u is at most R_A, this assumption that SG u is connected to at least 13 USNs (which can not communicate with each other) implies that the maximum cardinality of the *MIS* (Maximal Independent Set) in u's neighbors in 3-D space is at least 13. Note that SG u and its USN neighbors all have the communication range R_A, i.e., when the Euclidean distance between u and one of its USN neighbors is less than R_A, there is a edge in G. Thus SG u and its neighbors can construct a local UBG (Unit Ball Graph). However, the authors in [11,12] had proved that the maximum cardinality of the *MIS* in a node's neighbors in 3-D space is at most 12. Therefore, this contradiction proves that SG u can not be connected to more than 12 USNs in T_{opt}, i.e., $\Delta_s(T_{opt}) \le 12$. This lemma holds.   ■

Theorem 1. Algorithm 1 can guarantee getting a feasible solution which uses no more than $(7.5 \cdot \alpha \cdot |Y_{opt}|)$ SGs, where α is an approximation ratio of algorithm A for the Steiner minimum tree problem.

Proof. Let T_{min} be the minimum weight tree of $G(V, E, w)$ which connects all nodes in $\{b\} \cup S$, and T_{opt} be a minimum weight spanning tree of $G[\{b\} \cup Y_{opt} \cup S]$ which is an induced subgraph of $G(V, E, w)$ by $\{b\} \cup Y_{opt} \cup S$. Y_A and T_A be a feasible solution and a subgraph corresponding Y_A got by Algorithm 1, respectively.

We denote T_{opt}^1 as an induced subgraph of T_{opt} by all 1-weight edges, and T_{opt}^2 as an induced subgraph of T_{opt} by all 2-weight edges. Then $w(T_{opt}) = w(T_{opt}^1) + w(T_{opt}^2)$.

From the definition of $\Delta_s(T_{opt})$ and the structure of T_{opt}^1, since each edge in T_{opt}^1 has weight 1 and can only contain a SG and a USN (or BS), and there is only one BS, we know each SG in (T_{opt}^1) is incident with at most $\Delta_s(T_{opt}) + 1$ edges. Therefore we have

$$w(T_{opt}^1) \leq (\Delta_s(T_{opt}) + 1) \cdot |Y_{opt}|. \tag{1}$$

Since T_{opt} is a tree of $G(V, E)$, it has at most $(|Y_{opt}| - 1)$ 2-weight edges. Then,

$$w(T_{opt}^2) \leq 2 \cdot (|Y_{opt}| - 1). \tag{2}$$

Therefore

$$w(T_{opt}) \leq (2 + \Delta_s(T_{opt}) + 1) \cdot |Y_{opt}| - 2. \tag{3}$$

Since T_{min} is a minimum weight tree for $\{b\} \cup S$, we have

$$w(T_{min}) \leq w(T_{opt}). \tag{4}$$

Since algorithm A's approximation ratio is α, we have

$$w(T_A) \leq \alpha \cdot w(T_{min}) \leq \alpha \cdot w(T_{opt}) \tag{5}$$

Note that T_A must satisfy the condition of Lemma 1, this is because, if there exists a node u in $V(T_A) \cap L$ such that $d_{T_A}(u) = 1$, where $d_{T_A}(u)$ represents u's degree in T_A, then we delete u from T_A and still get a feasible solution. Therefore, $|Y_A| \leq \frac{1}{2}w(T_A)$.

Combining above inequations, we have

$$|Y_A| \leq \frac{\alpha}{2} \cdot (2 + \Delta_s(T_{opt}) + 1) \cdot |Y_{opt}|. \tag{6}$$

From Lemma 2, we have $\Delta_s(T_{opt}) \leq 12$, therefore,

$$|Y_A| \leq 7.5 \cdot \alpha \cdot |Y_{opt}|. \tag{7}$$

This theorem holds. ∎

We can use $(1 + \frac{\ln 3}{2})$-approximation algorithm in [16] as algorithm A in Algorithm 1. From theorem 1, we have the following corollary:

Corollary 1. The connected C-SGP problem has a polynomial time 11.625-approximation algorithm.

4.2 An Algorithm for Survivable C-SGP Problem

In the UA-WSNs, USNs are prone to failures because of fouling and corrosion. Thus, survivability is an important requirement for topology construction or data routing. The network connectivity should be preserved even when some USNs fail or deplete their power. One way to preserve survivability is to construct 2-connected paths from any USN to base station. In this section, we present a polynomial time approximation algorithm for the survivable C-SGP problem. Our algorithm is based on polynomial time approximation algorithms for minimum weight 2-connected many-to-one routing problem. The algorithm for the survivable C-SGP problem is presented as Algorithm 2.

Algorithm 2. A approximation algorithm for the Survivable C-SGP problem.

Input: An UA-WSN $(R_{RF}, R_A, \{b\}, S, L)$.
Output: A feasible solution $Y_A \subseteq L$.
Begin:
  1: Construct an edge-weighted undirected graph $G(V, E, w)$ based on this UA-WSN, where $V = \{b\} \cup S \cup L$.
  2: **if** The nodes in $\{b\} \cup S$ are not in a single 2-connected component H of $G(V, E, w)$ **then**
  3:    The survivable C-SGP problem does not have a feasible solution. Stop.
  4: **end if**
  5: Apply the existing algorithm A for the 2-connected many-to-one routing problem to compute a low weight subgraph H_A of $G(V, E, w)$ from S to b.
  6: Output $Y_A = V(H_A) \cap L$.
End.

Lemma 3[10]. Let $G(V, E)$ be an undirected k-connected graph where $|V| \geq 3k - 2$ and $H(V, E')$ be a k-connected subgraph of H with minimum number of edges. Then $|E'| \leq k \cdot (|V| - k)$.

Lemma 4. Let $G(V, E)$ be an undirected graph and $H(V, E')$ be a many-to-one 2-connected subgraph from D to b with minimum number of edges, where D and b are a source set and a destination node, respectively. Then $|E'| \leq 2 \cdot (|V| - 1)$.

Proof. Since each node in D has two node disjoint paths to b which can construct a cycle containing b, a many-to-one 2-connected subgraph with minimum number of edges consists of some 2-node-connected components of H. Let $H_1, H_2, ..., H_m$ be these 2-connected components, where H_i has $|V_i| \geq 3$ vertices, $i = 1, 2, ..., m$. Note that these 2-connected components must contain source node b and any two 2-connected components can not share a common node in $V \setminus \{b\}$, otherwise, the two components can merge into one 2-connected component. Furthermore, each 2-connected component $H_i(V_i, E_i)$ in $H(V, E')$ is a 2-connected spanning subgraph for V_i in $H(V, E')$ with minimum number of edges. If not, we can construct another many-to-one 2-connected subgraph from D to b in G with less number of edges than H, which contradicts with the assumption that H is a many-to-one 2-connected subgraph with minimum number of edges. We apply Lemma 3 for each 2-connected component H_i with $|V_i| \geq 4$ $(i = 1, 2, ..., m)$, and since those 2-connected components with 3 nodes must be a cycle with 3 edges, therefore,

$$|E'| = \sum_{i=1}^{m} |E(H_i)| \leq \sum_{i=1}^{m} 2(|V_i| - 2) = 2(\sum_{i=1}^{m} |V_i| - 2m)$$

$$= 2(|V| + m - 1 - 2m) = 2(|V| - 1 - m) \leq 2(|V| - 1) \qquad (8)$$

Note that $|V| = \sum_{i=1}^{m} |V_i| - m + 1$. ∎

Lemma 5. Y_{opt} is an optimal solution for the Survivable C-SGP problem. Let H_{opt} be a minimum weight spanning subgraph of $G[\{b\} \cup Y_{opt} \cup S]$ which meets the 2-connected requirement from all USNs to b. Then $\Delta_s(H_{opt}) \leq 12$.

Proof. We first prove that for any SG u, if it connects to more than 3 USNs in H_{opt}, then these USNs can not communicate with each other in $G(V, E)$. We prove it by contradiction. Without loss of generality, we assume that a SG u is to connect to m USNs $v_1, v_2, ..., v_m$ in H_{opt}($m \geq 3$), and v_1 and v_2 can communicate with each other, i.e., (v_1, v_2) is an edge in $G(V, E, w)$. Since H_{opt} is 2-connected from all USNs to b, and $m \geq 3$, there are two node disjoint paths from v_1 to b and v_3 to b, respectively. Therefore, there must be a path P from v_1 to v_3 which does not go through u. If P does not go through the USN v_2, H_{opt} contains a cycle $C_1 = \{(u, v_2), (v_2, v_1), P, (v_3, u)\}$. Then we have following obversion: For a USN node x, there are two node disjoint paths from x to b in H_{opt} which can construct a cycle C_2 containing b and x. If C_2 does not contain the edge (u, v_1), replacing (u, v_1) by (v_1, v_2) in H_{opt} does not destroy the 2-connectivity from x to b. If C_2 contains edge (u, v_1), then there are at least two intersect points (u and v_1) for C_1 and C_2. Then we can find at least three node disjoint paths between u and v_1. If delete the edge (u, v_1), there also exists a cycle containing b and x, i.e., there exist two node disjoint paths from x to b. For the arbitrary choice of x, we know that replacing edge (u, v_1) by (v_1, v_2) in H_{opt} does not destroy the many-to-one 2-connectivity from S to b.

From above discussion, we know that the subgraph H_1 got by replacing edge (u, v_1) in H_{opt} by edge (v_1, v_2) also can span all nodes in $\{b\} \cup Y_{opt} \cup S$ while meeting the many-to-one 2-connected requirement. Since $w(u, v_1) = 1$ and $w(v_1, v_2) = 0$, we have $w(H_1) < w(H_{opt})$, contradicting to the assumption that H_{opt} is a minimum weight subgraph. If path P goes through the USN v_2, H_{opt} has to contain a cycle $\{(u, v_1), P, (v_3, u)\}$. Similarly, deleting the edge (u, v_2) from H_{opt} will reduce its weight and H_{opt} is also a subgraph which can meet the many-to-one 2-connected requirement. This again contradicts to the minimum weight property of H_{opt}. So, we proved that for any SG u, if it connects to more than 3 USNs in H_{opt}, then these USNs can not communicate with each other in H_{opt}.

We prove this lemma by contradiction. Assume that in H_{opt}, a SG u can connect to more than 12 USNs. From above result, these USNs with at least 13 can not communicate with each other. Therefore, this also contradicts with the conclusions in [11,12]. Similar with the proof of Lemma 2, we also can prove that a SG u cannot be connected to more than 12 USNs in H_{opt}, i.e., $\Delta_s(H_{opt}) \leq 12$. This proves this lemma. ∎

Theorem 2. Algorithm 2 can guarantee getting a feasible solution which uses no more than $(8.5 \cdot \alpha \cdot |Y_{opt}|)$ SGs, where α is an approximation ratio of algorithm A for the 2-connected Steiner Minimum Subgraph problem.

Proof. Let H_{min} be the minimum weight many-to-one 2-node connected subgraph of $G(V, E, w)$ from S to b and H_{opt} be a minimum weight many-to-one 2-node connected subgraph of $G[Y_{opt} \cup S \cup \{b\}]$. Suppose Y_A is a feasible solution got by Algorithm 2, and H_A is a subgraph corresponding to Y_A.

We denote H_{opt}^1 as an induced subgraph of H_{opt} by all 1-weight edges, and H_{opt}^2 as an induced subgraph of H_{opt} by all 2-weight edges. Then $w(H_{opt}) = w(H_{opt}^1) + w(H_{opt}^2)$.

From the definition of $\Delta_s(H_{opt})$ and the structure of H_{opt}^1, since each edge in H_{opt}^1 has weight 1 and can only contain a SG and a USN (or BS), and there is only one BS, we know each SG in (H_{opt}^1) is incident with at most $\Delta_s(H_{opt}) + 1$ edges. Therefore we have

$$w(H_{opt}^1) \leq (\Delta_s(H_{opt}) + 1) \cdot |Y_{opt}|. \tag{9}$$

$$w(H_{opt}^2) \leq 2|E(H_{opt})| \leq 2 \cdot (2|Y_{opt}| - 2). \tag{10}$$

Note that we use Lemma 4 to get the second inequation in (10) since H_{opt} satisfies the condition of Lemma 4. Therefore,

$$w(H_{opt}) \leq (4 + \Delta_s(H_{opt}) + 1) \cdot |Y_{opt}| - 4. \tag{11}$$

Since H_{min} is the minimum weight many-to-one 2-node connected subgraph of $G(V, E, w)$ from S to b, we have $w(H_{min}) \leq w(H_{opt})$. Because the approximation ratio of algorithm A is α, therefore

$$w(H_A) \leq \alpha \cdot w(H_{min}) \leq \alpha \cdot w(H_{opt}) \tag{12}$$

Note that H_A must satisfy the condition of Lemma 1, this is because, if there exists a node $u \in V(H_A) \cap L$ such that $d_{H_A}(u) = 1$, then we delete u from H_A and still get a feasible subgraph. So, $|Y_A| \leq \frac{1}{2} w(H_A)$. Combining above inequations, we have

$$|Y_A| \leq \frac{\alpha}{2} \cdot (5 + \Delta_s(H_{opt})) \cdot |H_{opt}|. \tag{13}$$

From Lemma 5, we have $\Delta_s(H_{opt}) \leq 12$, therefore

$$|Y_A| \leq 8.5 \cdot \alpha \cdot |Y_{opt}|. \tag{14}$$

This proves this theorem. ∎

Corollary 2. The survivable C-SGP problem has a polynomial time 17-approximation algorithm.

Proof. According to the algorithm in [15], there is a polynomial time 2-approximation algorithm for the many-to-one 2-connected problem. The conclusion of this corollary can be achieved by choosing the algorithm in [15] as A in Algorithm 2.   ∎

5   Performance Evaluations

In this section, we evaluate the performance of our algorithms by simulations. We implemented approximation Algorithm 1 with A being the MST based 2-approximation SMT algorithm in [13] (denoted by A1-A and simpler than the algorithm in [16]) and another algorithms in [14] (denoted by A1-B and A1-C) for Steiner Minimum Tree problem. In the simulations, we focus on comparing the approximation algorithms A1-A and heuristic algorithms A1-B and A1-C. We study how the required number of SGs is affected by two parameters varying over a wide range: the number of USNs in the space and the number of the candidate locations in the upper plane.

The simulation is conducted in a $100 \times 100 \times 30$ 3-D space. We used both regular grid points and randomly generated points as the candidate locations for the SGs, and obtained similar results. For convenience of presentation, we used regular grid points as the surface-level candidate locations for the SGs in upper plane. In this setting, the playing field consists of $K \times K$ small squares contained by the upper plane of the space, with the $(K + 1)^2$ grid points as L. We present averages of 50 separate runs for each result.

In Fig. 1 (a) and (b), we compare the number of SGs required with number of USNs varying. In both cases, the number of candidate locations was 100 (10×10). The number of USNs was varied from 10 to 100. The R_A is set to 25 for both cases and R_{RF} is set to 25 and 50 for case (a) and (b), respectively. In both cases, we can see that the required number of SGs decreases with the increment of USNs. With the increment of USNs, the USNs trend to self-connected and only few SGs are required to connectethe USNs and BSs. The algorithm A1-A always performs better than the algorithms A1-B and A1-C.

In Fig. 1 (c) and (d), we also study the relationship between the number of SGs required and number of the candidate locations. We addressed two cases for 20 and 40 USNs respectively. We set $R_{RF} = 40$ and $R_A = 25$. There is no obvious variety of the number of SGs used for A1-A in Fig. 1 (c) and (d). Since, the SGs' main function is to connect USNs and BSs, some isolated USNs have to send data to BS by the SGs nearby them. Thus the variety of number of candidate locations may change the number of required SGs a little when the total number of used SGs trend to the optimal solution. This indicates that our approximation algorithm performs well. However, for the results delivered by A1-B and A1-C, these results are worse than A1-A's, and the more choice of candidate locations, the less redundant SGs will be used. Therefore, with the increment of candidate locations, the numbers of used SGs produced by A1-B and A1-C decrease gradually.

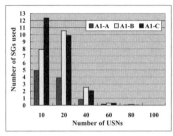
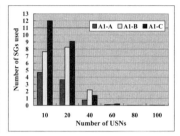

(a) Number of SGs used vs. Number of USNs.

(b) Number of SGs used vs. Number of USNs.

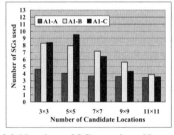
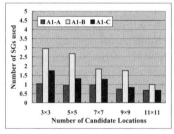

(c) Number of SGs used vs. Number of candidate locations.

(d) Number of SGs used vs. Number of candidate locations.

Fig. 1. The simulation results for the 1-connected C-SGP problem

6   Conclusions

In this paper, we studied the C-SGP problems in UA-WSNs. We mainly addressed the connected and survivable C-SGP problems, which can ensure to meet the connectivity and survivability requirements for some application environments of UA-WSNs. We discussed the computational complexity and presented an approximate algorithm for the two problems respectively, and corroborate the algorithms' performance through theoretical analysis and simulations.

Acknowledgment

This paper was supported in part by the National Natural Science Foundation of China under grant 61070191, Renmim University of China under Grants 10XNJ032 and 863 high-tech project under Grant 2008AA01Z120.

References

1. Akyildiz, I.F., Pompili, D., Melodia, T.: Underwater Acoustic Sensor Networks: Research Challenges. Elesviers Journal of Ad Hoc Networks 3(3), 257–279 (2005)
2. Pompili, D., Melodia, T., Akyildiz, I.F.: Deployment Analysis in Underwater Acoustic Wireless Sensor Networks. In: Proc. of the ACM WUWNet (2006)

3. Ibrahim, S., Cui, J.H., Ammar, R.: Surface-Level Gateway Deployment for Underwater Sensor Networks. In: Proc. of the IEEE MILCOM (2007)
4. Seah, W.K.G., Tan, H.X.: Multipath Virtual Sink Architecture for Underwater Sensor Networks. In: Proc. of the OCEANS (2006)
5. Zhang, W., Xue, G., Misra, S.: Fault-tolerant Relay Node Placement in Wireless Sensor Networks: Problem and Algorithms. In: Proc. of the IEEE INFOCOM (2007)
6. Han, X., Cao, X., Lloyd, E.L., Shen, C.C.: Fault-Tolerant Relay Node Placement in Heterogeneous Wireless Sensor Networks. In: Proc. of the IEEE INFOCOM (2007)
7. Lloyd, E., Xue, G.: Relay Node Placement in Wireless Sensor Networks. IEEE Trans. on Computers 56, 134–138 (2007)
8. Misra, S., Hong, S.D., Xue, G., Tang, J.: Constrained Relay Node Placement in Wireless Sensor Networks to Meet Connectivity and Survivability Requirements. In: Proc. of the IEEE INFCOM (2008)
9. Ke, W., Liu, B., Tsai, M.: Constructing a Wireless Sensor Network to Fully Cover Critical Grids by Deploying Minimum Sensors on Grid Points is NP-Complete. IEEE Trans. on Computers 56, 710–715 (2007)
10. Mader, W.: Uber Minimal n-fach Zusammenhangende, Unendliche Graphen Undein Extremal Problem. Arch. Math. 23, 553–560 (1972)
11. Conway, J.H., Sloane, N.J.A.: Sphere Packing, Lattices and Groups, 3rd edn. Springer, New York (1999)
12. Butenko, S., Ursulenko, O.: On Minimum Connected Dominating Set Problem in Unit-Ball Graphs. Preprint Submitted to Elervier Science (2007)
13. Kou, L.T., Markowsky, G., Berman, L.: A Fast Algorithm for Steiner Tree. Acta Informatica 15, 141–145 (1981)
14. Du, D., Hu, X.: Steiner Tree Problems in Computer Communication Networks. World Scientific Publishing Co. Pte. Ltd., Singapore (2008)
15. Fleischer, L.: A 2-approximation for Minimum Cost {0, 1, 2}-Vertex Connectivity. In: Aardal, K., Gerards, B. (eds.) IPCO 2001. LNCS, vol. 2081, p. 115. Springer, Heidelberg (2001)
16. Robins, G., Zelikovsky, A.: Tighter Bound for Graph Steiner Tree Approximation. SIAM J. on Discrete Mathmatics 19, 122–134 (2005)

Time Optimal Algorithms for Black Hole Search in Rings*

Balasingham Balamohan[1], Paola Flocchini[1], Ali Miri[2], and Nicola Santoro[3]

[1] University of Ottawa, Ottawa, Canada
{bbala078,flocchin}@site.uottawa.ca
[2] Ryerson University, Toronto, Canada
samiri@scs.ryerson.ca
[3] Carleton University, Ottawa, Canada
santoro@scs.carleton.ca

Abstract. In a network environments supporting mobile entities (called robots or agents), a *black hole* is harmful site that destroys any incoming entity without leaving any visible trace. The *black-hole search* problem is the task of a team of $k > 1$ mobile entities, starting from the same safe location and executing the same algorithm, to determine within finite time the location of the black hole. In this paper we consider the black hole search problem in asynchronous *ring* networks of n nodes, and focus on the *time complexity*.

It is known that any algorithm for black-hole search in a ring requires at least $2(n-2)$ time in the worst case. The best algorithm achieves this bound with a team of $n-1$ agents with an average time cost $2(n-2)$, equal to the worst case. In this paper we first show how the same number of agents using 2 extra time units from optimal in the worst case, can solve the problem in only $\frac{7}{4}n - O(1)$ time on the *average*. We then prove that the optimal average case complexity $\frac{3}{2}n - O(1)$ can be achieved without increasing the worst case using $2(n-1)$ agents Finally we design an algorithm that achieves asymptotically optimal both worst case and average case time complexity employing an optimal team of $k = 2$ agents, thus improving on the earlier results that required $O(n)$ agents.

1 Introduction

1.1 The Problem

Black Hole Search (BHS) is a multi-agents problem set in graph G: a team of (identical) cooperating mobile entities called agents (or robots) must determine the location in G of a *black hole* (BH): a node where any incoming agent is destroyed without leaving any detectable trace. The problem is solved if at least one agent survives and knows the location of the black hole.

A black hole can model several types of faults, both hardware and software, arising in networked systems with code mobility. For example, the crash failure of

---

* Research partially supported by NSERC.

W. Wu and O. Daescu (Eds.): COCOA 2010, Part II, LNCS 6509, pp. 58–71, 2010.
© Springer-Verlag Berlin Heidelberg 2010

a site in an asynchronous network turns such a site into a black hole; similarly, the presence at a site of a malicious process (e.g., a virus) that thrashes any incoming message (e.g., by classifying it as spam) also renders that site a black hole. Clearly, in presence of such a harmful host, the first step must be to *identify* it, if possible; i.e., to determine and report its location; following this phase, a "rescue" or "repair" activity would conceivably be initiated [15]. The black hole search problem is also theoretically interesting because it is a generalization of the classical problem of *graph exploration* (e.g., see [1, 6, 16]). In fact, it is easy to see that to locate a black hole the agents have to necessarily "explore" all safe nodes; in this exploration process some agents may disappear in the black hole. In other words, while the existing wide body of literature on exploration assumes that the graph is *safe*, BHS opens the problem of the exploration of *dangerous graphs*.

In this paper we consider the *Black Hole Search* problem in a *ring* network.

1.2 Related Work

This black hole search problem has been originally studied in *ring* networks [11] and has been extensively investigated in various settings since then (e.g., see [3, 5, 9, 7, 12, 13, 18, 24]).

In order to locate the black hole, some of the agents of the team will necessarily have to enter the dangerous site. The goal of all location algorithms studied in the literature is to minimize the *size* of the exploring team, the number of *moves* performed by the agents and the *time* spent in the search.

The main distinctions made in the literature are whether the system is synchronous or asynchronous, and whether the agents communicate through whiteboards or by using tokens.

The majority of the work focuses on the *asynchronous whiteboard model*, which is the one considered in this paper. In this model, there are no assumptions on the time required for each operation or movement other than it is finite. Each network node provides a shared memory area, the *whiteboard*, which visiting agents can access (in fair mutual exclusion) to write on and/or read from. The communication and coordination between agents takes place solely via the whiteboards. Within this model, a complete characterization has been done for the localization of a black hole in ring networks [11], providing protocols that are optimal in size, time, and asymptotically move-optimal number of moves. In [9], arbitrary topologies have been considered and asymptotically optimal location algorithms have been proposed under a variety of assumptions on the agents' knowledge (knowledge of the topology, presence of sense of direction). An improved algorithm when the topology is known has been described in [10], while optimal algorithms for common interconnection networks have been studied in [7]. In [18] the effects of knowledge of incoming link on the optimal team size has been studied and lower bounds provided. The case of black links in arbitrary networks has been studied in [19, 14], respectively for anonymous and non-anonymous nodes. Black hole search in directed graphs has been investigated for the first time in [3], where it is shown that the requirements in number of agents change considerably. A variant of dangerous node behavior has been studied in [22], where the authors introduce black holes with Byzantine behavior (they do not always destroy a passing agent) and consider the periodic ring exploration problem.

In the *asynchronous token* model, there are no whiteboards, but each agent is provided with *pebbles* that it can place on (and pick up from) a node; the communication and coordination among agents is achieved solely by placing on the nodes. This model has been investigated in [8, 12, 13, 24].

In *synchronous* networks, where movements are synchronized and it is assumed that it takes one unit of time to traverse a link, the techniques and the results are quite different. Tight bounds on the number of moves have been established for some classes of trees [5]. In the case of general networks finding the optimal strategy is shown to be NP-hard [4, 20] and approximation algorithms are given in [20,21]. The case of multiple black holes have been investigated in [2] where a lower bound on the cost and close upper bounds are given.

1.3   Main Contributions

In this paper we turn our attention to the *time complexity* of locating a black hole in a ring of n nodes using a team of $k > 1$ *asynchronous* agents communicating by means of whiteboards (shared memory available at each node). The asynchrony of the computational entities means that the algorithm must work regardless of the time required for each computation or movement, which is finite but a priori unknown (i.e., determined by an adversary); however, the time complexity of the algorithm is measured only over those executions where time delays are unitary (i.e., determined by a synchronous scheduler), as traditional in distributed computing (e.g., [11, 17, 23]).

It has been shown in [11] that any asynchronous black hole search algorithm for rings requires $T_{worst}(n, k) = 2(n - 2)$ time in the worst case regardless of the number $k > 1$ of agents. The best algorithm achieves this bound with a team of $k = n - 1$ agents with an average time cost $2(n - 2)$, equal to the worst case [11].

In this paper we first show how the same number of agents can solve the problem using on the *average* only $\frac{7}{4}n$ time, and in the worst case 2 extra time units from optimal. We also show that any asynchronous black hole search algorithm for rings requires $T_{average}(n, k) = \frac{3}{2}n$ time regardless of the number $k > 1$ of agents, and then prove that, with $2(n - 1)$ agents, the optimal average case complexity $\frac{3}{2}n - O(1)$ can be achieved *without increasing the worst case*. Finally, observing that all considered protocols achieve (worst and average) $\Theta(n)$ time using $O(n)$ agents, we prove that it is possible to locate a black hole in asymptotically optimal (worst and average) $\Theta(n)$ time with just $k = 2$ agents. In fact, we design an algorithm that uses $8n + O(1)$ time in the worst case and $\frac{15}{2}n + O(1)$ on the average, employing an optimal team of 2 agents thus improving the earlier result that employed $n - 1$ agents. These results are summarized in Table 1. The costs in terms of moves of all these algorithms is $O(n^2)$, the same as that of the algorithm in [11] they improve upon.

Table 1. Summary of results; ($\star$) indicates an optimal exact bound

Algorithm	Agents	Time Complexity	
		Average	Worst
[11]	$n-1$	$2(n-2)$	$2(n-2)$ ($\star$)
GROUP	$n-1$	$\frac{7}{4}n - O(1)$	$2(n-1)$
OPTAVGTIME	$2(n-1)$	$\frac{3}{2}n - O(1)$ ($\star$)	$2(n-2)$ ($\star$)
OPTTEAMSIZE	2 ($\star$)	$\frac{15}{2}n + O(1)$	$8n + O(1)$

2   Preliminaries

2.1   Definitions and Notations

The network environment is a ring $\mathcal{R}$ of n anonymous nodes (for simplicity indicated as $0, 1, \cdots, n-1$ in clockwise direction). Each node has two ports, labelled *left* and *right*. Without loss of generality we assume that this labeling is globally consistent and the ring is oriented (if it is not the case, orientation can be easily obtained) . Each node is equipped with a limited amount of storage, called *whiteboard*. For all our algorithms $O(\log n)$ bits of storage are sufficient.

In this network there is a set $\mathcal{A}$ of *anonymous* (i.e., identical) mobile agents, which are all initially located on the same node, called the *homebase* (w.l.g. node 0). The topology is known to the agents, as well as the number of nodes (as shown in [11] not knowing the number of nodes make the location process impossible). The agents can move from node to neighboring node in $\mathcal{R}$ and have computing capabilities and bounded storage. The agents obey the same set of behavioral rules, the protocol, and all their actions are performed asynchronously, i.e., they take a finite but unpredictable amount of time. The agents communicate by writing on and reading from the whiteboards. Access to the whiteboards is governed by fair mutual exclusion.

A *black hole* (BH) is a stationary process located at a node, which destroys any agent arriving at the node; no observable trace of such a destruction will be evident external to the node in which black hole is located. The *Black Hole Search* problem is the one of *finding the location of the black hole*. More precisely, the black hole search problem is solved if at least one agent survives, and all surviving agents know the location of the black hole within a finite amount of time.

We evaluate the efficiency of our solutions based on the following measures:

1. *Number of agents* used/needed in the protocol.
2. *Total number of moves* performed by all agents.
3. *The amount of time* between the earliest start time of the protocol by any agent and the time all the agents that started the protocol have terminated the execution of protocol. Since the system is asynchronous, when evaluating the time complexity we will employ *ideal time*; i.e., we will assume that it time delays are unitary (e.g., see [11, 17, 23]).

2.2 Cautious Walk

We first recall the *cautious walk* technique, which is central to the algorithms presented in this paper, and the existing asymptotically optimal algorithm of [11].

At any time during the search for the black hole, the ports (corresponding to the incident links) of a node can be classified as follows:

1. *unexplored*: if no agent has moved across this port.
2. *safe*: if an agent arrived via this port.
3. *active*: if an agent departed via this port, but no agent has arrived via it.

Clearly, both unexplored and active links are dangerous in that they might lead to the black hole; however, active links are being explored, so there is no need for another agent to go there unless it is declared safe. Cautious walk is defined by the following two rules:

1. when an agent moves from node u to v via an unexplored port (turning it into active), if it does not disappears (i.e., v is not the black hole) the agent immediately returns to u (making the port safe), and only then resumes its execution;
2. no agent leaves via an active port.

3 Improved Algorithm

In this section we improve on the average time complexity of [11]. We describe an algorithm that uses only $n-1$ agents, as in [11], but only $\frac{7}{4}n + O(1)$ time on the average (instead of $2(n-2)$). The worst case is $2(n-1)$ (instead of $2(n-2)$).

The idea is to determine the location of the black hole on some node by having a particular pair of agents (witnessing pair) returning successfully to the homebase after exploring a subset of nodes that does not include the black hole. The way subsets of nodes are associated to agents is complicated by the objective of reducing the number of agents entering the black hole.

We recall that node 0 indicates the homebase and the other nodes are indicated as $1, 2, \ldots, n-1$ in clockwise direction. For simplicity of description let $n = 4q + 1$. The $4q$ agents are divided into four groups: *Left*, *Right*, *Middle* and

TieBreakers. The groups *Left* and *Right* contain q agents each. The *Middle* group consists of $q + 1$ agents and the *TieBreakers* group consists of $q - 1$ agents.

The witnessing pairs are chosen in a different way depending on the potential location of the black hole. If the black hole is far from the homebase, it will be witnessed by a pair from *Left* and *Middle* or from *Right* and *Middle*. If instead, the black hole is closer to the homebase the witnessing pair will belong to *TieBreakers* and *Right* or to *TieBreakers* and *Left*.

More precisely, the idea is that the agents of the *Left*, *Right* and *Middle* groups explore each a region of size $3q$ appropriately chosen in such a way that the $2q$ nodes farthest from the homebase are endpoints of complements of explored areas of some agents. In other words, the presence of the black hole in one of those $2q$ nodes would be witnessed by a pair of agents being able to successfully return after their exploration.

The agents of the *Tiebreakers* group are instead used to pair themselves either with an agent from *Right* or with one from *Left* to locate the black hole when it is within q nodes from the homebase. The details are given in Algorithm 1.

Algorithm 1. Algorithm GROUP

---

1. The Left group consists of $left_i$ for $1 \leq i \leq q$. An agent $left_i$ in this group explores all node except the nodes $\{i, i+1, i+2, \cdots, i+q-1\}$. It moves left first, then right and then returns to homebase. (See Figure 1).

2. The Right group consists of $right_i$ for $1 \leq i \leq q$. An agent $right_i$ in this group explores all node except the nodes $\{n-i-q, n-i-q+1, \cdots, n-i\}$. It moves right first, then left and then returns to homebase.

3. The Middle group consists of $middle_i$ for $1 \leq i \leq q+1$. An agent $middle_i$ in this group explores all node except the nodes $\{q+i-1, q+i, \cdots, 2q+i-1\}$. It moves left first, then right and then returns to homebase. (See Figure 2).

4. The Tiebreaker group consists of $tiebreaker_i$ for $1 \leq i \leq q-1$. An agent $tiebreaker_i$ in this group explores all nodes except nodes $\{i, i-1, i-2, \cdots, 1\}$ at the right of the homebase starting as soon as either $right_{i+1}$ or $left_{i+1}$ passes through the homebase. (See Figure 3).

5. The black hole is located on node i iff one of the following witnessing pairs return safely: $(left_i, middle_i)$, $(right_i, middle_{q-i})$, $(tiebreaker_i, left_i)$, $(tiebreaker_i, right_i)$.

We have that:

Theorem 1. *Algorithm* GROUP *solves the black hole search problem with average time* $\frac{7n}{4} + O(1)$ *and worst case time* $2(n-1)$.

Proof. Let us first consider the correctness of the algorithm. If the black hole is one of nodes $\{q, q+1, q+2, \cdots, 2q\}$ then it is the unique node in the intersection of the excluded segments of a pair of agents $left_i$ and $middle_i$ from the *Left* and the *Middle* groups. (for example, $q+1$ is the only node not explored by the

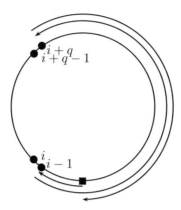

Fig. 1. Algorithm GROUP: Protocol For Left Group

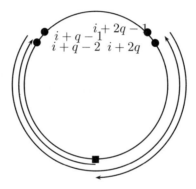

Fig. 2. Algorithm GROUP: Protocol For Middle Group

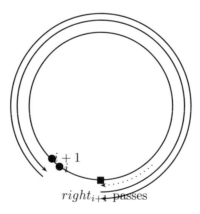

Fig. 3. Algorithm GROUP: Protocol For Tie-Breaker Group

pair $left_2$, $middle_2$). In such a case the black hole is located by the return of such a witnessing pair and the location takes $\frac{3}{2}(n-2)$ time units as both agents $left_i$ and $middle_i$ explore all but $q = \frac{1}{4}(n-1)$ agents. If the black hole is in node i with $i < q$ then agent $right_{i+1}$, after exploring the i nodes on the right of the homebase, passes back through the homebase. At this moment, by the rules of the algorithm, agent $tiebreaker_i$ starts exploring all the nodes (except node i and the $i - 1$ nodes between node i and homebase) moving to the right of the homebase. If also agent $left_i$ returns, it means that node i contains the black hole because it is the only unexplored node. Hence, the return of the pair $tiebreaker_i, left_i$ (or $tiebreaker_i, right_i$) signals the presence of the black hole in node i.

Consider now the time complexity of the algorithm. Agent $tiebreaker_i$ begins the execution after $2i$ time units, and it explores all but i nodes. Hence it returns back to the homebase $2(n-1)$ time units after the start of the execution of the first agent. So, when the black hole is one of the $q - 1$ nodes closest to the homebase, it will be located within $2(n-1)$ units. A similar argument apply when the blackhole is symmetrically placed on the other (right) half of the ring, and the worst case result follows.

As for the average time complexity, we have two situations: when the black hole is within q nodes of the homebase, the time for locating it is $2(n-1)$; otherwise, it is $\frac{3}{2}(n-2)$ time units. Hence the average ideal time complexity is:

$$\frac{2(q-1)(2(n-1)) + 2(q+1)(\frac{3(n-2)}{2})}{4q} = \frac{7(n-1)}{4} - O(1).$$

4 Optimal Average Time

In this section we show that, by using $2(n-1)$ agents, it is possible to achieve simultaneously optimal time both in the average and in the worst case, establishing a lower bound on the average time complexity of black hole search.

The idea of the algorithm is to identify pairs of agents ($left_i$, $right_i$, $i \leq 1 \leq n-1$) among the $2(n-1)$ available, and to assign each pair to "check" a node of the ring. To check node i, an agent of the pair would move to node $i-1$ clockwise (thus exploring nodes $1, 2, \ldots, i - 1$) and the other would move to node $i + 1$ counterclockwise (thus exploring nodes $i+1, i+2, \ldots n-1$). The presence of the black hole in the ring insures that only one pair will come back to the homebase intact while one agents of each of the other pairs will disappear in the black hole. Once the successful pair returns, the black hole is located.

Theorem 2. *Algorithm* OPTAVGTIME *solves the black hole location problem. in average ideal time complexity* $\frac{3}{2}n + O(1)$ *and worst case ideal time complexity* $2(n-2)$. *Both complexities are optimal.*

Proof. Correctness follows from the fact that for each node i there are two agents, namely $left_i$ and $right_i$ such that the singleton set $\{i\}$ is the intersection of the areas that they do not explore.

Algorithm 2. Algorithm OPTAVGTIME

$2(n-1)$ co-located agent $left_i, right_i, i \leq 1 \leq n-1$ at homebase node 0.

1. Agent $left_i$ explores nodes $(0, 1, 2, \cdots, i-1)$ and returns.
2. Agent $right_i$ explores nodes $(n-1, n-2 \cdots, i+2, i+1)$ and returns.
3. Let $(left_j, right_j)$ be the only full pair safely returning. The black hole is node j.

Let us now consider worst case time complexity: The time spent by $left_i$ and $right_i$ to reach it and come back is $2Max\{i-1, n-i\}$; the worst case clearly occurs when the black hole is located on node 1 (or $n-2$) and the corresponding time complexity is $2(n-2)$, which is optimal [11].

As for the average time. The presence of the black hole at a node i is witnessed by agents reaching nodes $i-1$ and $n-i-1$. Hence, the ideal time delay for the algorithm when the black hole is located at node i is $2Max\{i-1, n-1-i\}$. $2(i-1)$ is greater than or equal to $2(n-1-i)$ whenever $i \geq \frac{n}{2}$. Since all nodes other than the homebase are equally likely to contain the black hole, the average time complexity is:

$$\frac{\sum_{i=1}^{\frac{n}{2}-1} 2(n-1-i) + \sum_{i=\frac{n}{2}}^{n-1} 2(i-1)}{n-1}$$

$$= \frac{(n-1)(n-2) + \sum_{i=1}^{\frac{n}{2}-1} -2i + \sum_{\frac{n}{2}}^{n-1} 2i}{n-1}$$

$$= \frac{(n-1)(n-2) - (\frac{n}{2}-1)n + (n-1)n}{n-1} = \frac{3}{2}n + O(1)$$

Notice that nodes on either side of the black hole have necessarily to be reached by some agents and their visit reported back. Hence the time when the black hole is located at node i must be greater than or equal to both $2(i-1)$ and $2(n-1-i)$, which precisely corresponds to the time complexity of our algorithm for node i. We can then conclude that our bound is optimal.

5   Optimal Team Size

The algorithm of Dobrev et al [11] as well as the improvements presented here have optimal time complexities both in the worst and in the average case; however they all use $O(n)$ agents, which is order of magnitude larger that than the optimal team size $k = 2$. One might think that this large number of agents used by time-optimal solutions is necessary. This is however not true, as we show in this section.

In the following, we present an algorithm that allows $k = 2$ agents to locate the black hole with asymptotically optimal time in both the worst and the average case. The cost in terms of messages of this algorithm is $O(n^2)$, the same as all the others considered here.

The algorithm, called OPTTEAMSIZE, is as follows. At each point in time the nodes of the ring are partitioned into an *Explored* area and an *Unexplored* one. The explored area has been already visited by some agent and it is known to be safe, the unexplored area is still to be visited and contains the black hole. Moreover, during the algorithm, the unexplored area is partitioned between the two agents. More precisely, it is always divided into two disjoint areas of different sizes to which agents are assigned: one part containing a single node and the other containing all other unexplored nodes. In each step of the algorithm one of the agents (called *small*) is given the task to explore the area containing a single node, while the other (called *big*) has to explore the other area. The exploration proceeds with cautious walk. Since the two areas are disjoint, one of the agents will certainly succeed in its exploration. If the *big* agent succeeds, the blackhole is obviously located and the algorithm terminates. On the other hand, if the *small* agent returns successfully, it further divides the remaining unexplored area and notifies the *big* agent of the update by leaving a message on the whiteboard of the last node successfully visited by the other agent. The way the update of the unexplored area is performed is such that an agent stays *small* for *two consecutive steps* before switching role. A stage of the algorithm consists of these two consecutive steps and the algorithm is a sequence of stages which terminates when $n - 1$ nodes are known to be safe.

This division process is preceded by a preprocessing phase where the two agents divide the ring in two disjoint parts of almost equal size: only when one of the two returns to the homebase the asymmetric workload division starts to take place.

In the following when we say that an agent acts as *big* we mean that it cautiously explores *all but the last* nodes of the unexplored area. When an agent acts as *small* it cautiously explores the *first* node of the unexplored area. The location of the homebase in the various steps of the algorithms is variable and it is always the central node of the current explored area. An *update* message contains the update information about the current unexplored area and the current location of the homebase.

We now prove that the algorithm terminates correctly and we study its complexity.

Theorem 3. *Algorithm* OPTTEAMSIZE *solves the black hole search in optimal time* $\Theta(n)$ *using 2 agents and performing* $O(n^2)$ *moves.*

Proof. After the end of the preprocessing phase at least one agent, say *right*, survives and returns. Since the segments of the ring explored by each agent are always disjoint, at least one agent survives every stage. If the *big* agent survives, the algorithm terminates correctly, if the *small* agent survives the size of the unexplored area decreases by one and the algorithm correctly moves to the next stage (or it terminates if the new size is equal to one). Hence, one of the agents eventually discovers the location of the black hole.

Algorithm 3. Algorithm OPTTEAMSIZE

Two co-located agents l and r . $E = \{v_h\}$. $U = V - E$.

1. Preprocessing Phase: Agent l (resp. r) explores cautiously the leftmost (resp. rightmost) $\lfloor |U|/2 \rfloor$ nodes of the unexplored area and when finished returns to the homebase v_h.
2. Exploration Phase: One of the agents (say l) arrives at the homebase and becomes *small*. Agent l moves to the last explored node on agent r's side, it leaves an update message to r indicating to act as *big*. Agent l then moves to its side and act as *small*. Stage 1 of Phase 2 begins.
3. Stage i of Phase 2:
   (a) If the *big* agent (say r) returns to the homebase (or the *small* agent returns and the size of the unexplored area is one) then the blackhole is located and the algorithm terminates.
   (b) Otherwise, the *small* agent l returns, it moves to the last explored node on agent r's side, it leaves an update message for r indicating to maintain the same role *big*.
   (c) Agent l moves back to its side and it acts as *small*.
   (d) If r returns, then the blackhole is located and the algorithm terminates.
   (e) Otherwise agent l returns, it moves to the last explored node on agent r's side, it leaves an update message for r instructing to reverse role. Agent l then moves to the other side and it changes role acting as *big*.
   (f) If l returns the blackhole is located and the algorithm terminates.
   (g) Otherwise Agent r returns and becomes *small*; it moves to agent l's side, it leaves an update message for agent l instructing it to act as *big*. Agent r then moves back to its side and acts as *small*.
   (h) If l returns then the blackhole is located and the algorithm terminates.
   (i) Otherwise agent r returns, it changes role becoming *big*, it moves to agent l's side and it leaves a message to agent l at the last explored node updating the unexplored area and instructing to reverse roles. Agent l moves to its side and acts as *big*.
   (j) Stage $i + 1$ starts.

   To prove that the algorithm has $\Theta(n)$ time complexity, we first observe that when the exploration phase of the algorithm begins the explored area is at least of size $\frac{n-1}{2}$. While the *big* agent, say r, is exploring all but one nodes on its side, the other agent (if it did not disappear in the blackhole before) performs two steps as *small* making at least $\frac{3}{2}(n-1)$ moves (and spending the same amount of time). By that time, under the ideal time assumption, agent r would have either i) returned safely determining the location of the black hole, or ii) died in the black hole. In the first case we obviously have a time complexity of $O(n)$. In the other case, when agent l switches role becoming *big* and moves to explore all but one nodes, it necessarily completes its task locating the black hole, again with an overall time complexity of $O(n)$.

To show that the worst case move complexity is $O(n^2)$, it suffices to notice that in the worst possible asynchronous execution it is always the *small* agent that completes a step, while the big agent is slow on a link. Since the *small* agent manages to explore a single node in each step, and the size of the unexplored area when this procedure starts is $\frac{n}{2}$, $O(n)$ steps are necessary to locate the black hole. In each step however $O(n)$ moves are performed by the *small* agent to explore and report the update on the other side of the explored area, for a total of $O(n^2)$ moves.

We now show the exact average and worst case time complexities of the algorithm.

Theorem 4. *Algorithm* OPTTEAMSIZE *solves the black hole search in average ideal time* $\frac{15}{2}n + O(1)$ *and worst case* $8n + O(1)$.

Proof. By symmetry of the algorithm we may assume that the black hole is located on the right half of the ring (w.l.g let n be even). We then calculate the ideal time delay when the black hole is located at node $i \leq \frac{n}{2}$ (i.e., $\frac{n}{2}$ nodes to the right of the homebase). We consider different cases.

- *Case 1*: Node i is the border node of the partition between the right and the left agents. In this case the left agent returns after $\frac{3}{2}n + \frac{n}{2}$ time units to the homebase. In the sum, the first addend is for the cautious exploration and the second is for the time taken to return to the homebase. Now the left agents follows the path of the right agent. The right agent must have died at the last node of its partition. So in another n time units the left agent will reach the last safe node explored by the right agent and return. So, in this case the black hole is located in $3n$ total time units.
- *Case 2*: Node i is the neighbor of the border node of the partition between the right and the left agents. Similarly to the previous case, the left agent will take $3n - O(1)$ time units to return to the homebase after exploring all nodes, except the black hole and its neighbor. Now the left agent in the role of *small* explores the last safe node and return in further $n + O(1)$ time units. In this case the black hole is then located in $4n + O(1)$ time units.
- *Case 3*: Node i is the third node from the border of the partition between the right and the left agents. In this case the left agent discovers the black hole after the end of the second round as *small*. The total ideal time delay in this case is $5n + O(1)$.
- *Case 4*: Node i is the fourth node or the node further from the border of the partition between the right and the left agents. In this case the left agent (l) performs two rounds as *small* and a round as *big*. Observe that under ideal conditions the right agent would die in the black hole and would not return, so it suffices to count the time taken by the left agent. Agent l explores $(n - i) + O(1)$ nodes in total and this cautious exploration costs in total $3(n - i) + O(1)$ time units. Let us now compute the time necessary for the other movements of the left agent. Agent l takes $\frac{n}{2} + i + O(1)$ time units for reaching the last safe node explored by the right agent. Moreover, agent l

takes $4(\frac{n}{2} + i) + O(1)$ time unit for the two rounds as *small*; after becoming *big*, agent l reaches the last explored node on its side in $\frac{n}{2} + i$ time units. At this point it cautiously explores (the cost of the exploration has been already accounted for earlier). Finally, the agent returns to the homebase in $n - i + O(1)$ time units. Thus the total ideal time delay is $7n + 2i + O(1)$ time units.

Hence the average ideal time delay is :

$$\frac{12n + \sum_{i=1}^{i=\frac{n}{2}}(7n + 2i)}{\frac{n}{2}} = 7n + \frac{(\frac{n}{2})(\frac{n}{2} + 1)}{\frac{n}{2}} = \frac{15}{2}n + O(1)$$

The worst case occurs in correspondence of Case 4, when $i = n/2 - O(1)$, which yields $8n + O(1)$.

References

1. Bender, M.A., Fernández, A., Ron, D., Sahai, A., Vadhan, S.P.: The power of a pebble: Exploring and mapping directed graphs. Information and Computation 176(1), 1–21 (2002)
2. Cooper, C., Klasing, R., Radzik, T.: Searching for black-hole faults in a network using multiple agents. In: Shvartsman, M.M.A.A. (ed.) OPODIS 2006. LNCS, vol. 4305, pp. 320–332. Springer, Heidelberg (2006)
3. Czyzowicz, J., Dobrev, S., Královic, R., Miklík, S., Pardubská, D.: Black hole search in directed graphs. In: Kutten, S., Žerovnik, J. (eds.) SIROCCO 2009. LNCS, vol. 5869, pp. 182–194. Springer, Heidelberg (2009)
4. Czyzowicz, J., Kowalski, D.R., Markou, E., Pelc, A.: Complexity of searching for a black hole. Fundamenta Informaticae 71(2-3), 229–242 (2006)
5. Czyzowicz, J., Kowalski, D.R., Markou, E., Pelc, A.: Searching for a black hole in synchronous tree networks. Combinatorics, Probability & Computing 16(4), 595–619 (2007)
6. Deng, X., Papadimitriou, C.H.: Exploring an unknown graph. J. Graph Theory 32(3), 265–297 (1999)
7. Dobrev, S., Flocchini, P., Kralovic, R., Ruzicka, P., Prencipe, G., Santoro, N.: Black hole search in common interconnection networks. Networks 47(2), 61–71 (2006)
8. Dobrev, S., Flocchini, P., Kralovic, R., Santoro, N.: Exploring an unknown graph to locate a black hole using tokens. In: 5th IFIP Int. Conference on Theoretical Computer Science (TCS), pp. 131–150 (2006)
9. Dobrev, S., Flocchini, P., Prencipe, G., Santoro, N.: Searching for a black hole in arbitrary networks: Optimal mobile agents protocols. Distributed Computing 19(1), 1–19 (2006)
10. Dobrev, S., Flocchini, P., Santoro, N.: Improved bounds for optimal black hole search with a network map. In: Kralovic, R., Sýkora, O. (eds.) SIROCCO 2004. LNCS, vol. 3104, pp. 111–122. Springer, Heidelberg (2004)
11. Dobrev, S., Flocchini, P., Santoro, N.: Mobile search for a black hole in an anonymous ring. Algorithmica 48, 67–90 (2007)
12. Dobrev, S., Santoro, N., Shi, W.: Using scattered mobile agents to locate a black hole in an un-oriented ring with tokens. International Journal of Foundations of Computer Science 19(6), 1355–1372 (2008)

13. Flocchini, P., Ilcinkas, D., Santoro, N.: Ping pong in dangerous graphs: Optimal black hole search with pure tokens. In: Taubenfeld, G. (ed.) DISC 2008. LNCS, vol. 5218, pp. 227–241. Springer, Heidelberg (2008)
14. Flocchini, P., Kellett, M., Mason, P., Santoro, N.: Map construction and exploration by mobile agents scattered in a dangerous network. In: 24th IEEE International Parallel and Distributed Processing Symposium (IPDPS), pp. 1–10 (2009)
15. Flocchini, P., Santoro, N.: Distributed security algorithms for mobile agents. In: CaO, J., Das, S.K. (eds.) Mobile Agents in Networking and Distributed Computing, ch. 3. Wiley, Chichester (2009)
16. Fraigniaud, P., Ilcinkas, D., Peer, G., Pelc, A., Peleg, D.: Graph exploration by a finite automaton. Theoretical Computer Science 345(2-3), 331–344 (2005)
17. Garay, J.A., Kutten, S., Peleg, D.: A sublinear time distributed algorithm for minimum-weight spanning trees. SIAM J. Comput. 27(1), 302–316 (1998)
18. Glaus, P.: Locating a black hole without the knowledge of incoming link. In: Dolev, S. (ed.) ALGOSENSORS 2009. LNCS, vol. 5804, pp. 128–138. Springer, Heidelberg (2009)
19. Chalopin, J., Das, S., Santoro, N.: Rendezvous of mobile agents in unknown graphs with faulty links. In: Pelc, A. (ed.) DISC 2007. LNCS, vol. 4731, pp. 108–122. Springer, Heidelberg (2007)
20. Klasing, R., Markou, E., Radzik, T., Sarracco, F.: Hardness and approximation results for black hole search in arbitrary networks. Theoretical Computer Science 384(2-3), 201–221 (2007)
21. Klasing, R., Markou, E., Radzik, T., Sarracco, F.: Approximation bounds for black hole search problems. Networks 52(4), 216–226 (2008)
22. Královic, R., Miklík, S.: Periodic data retrieval problem in rings containing a malicious host. In: Patt-Shamir, B., Ekim, T. (eds.) Structural Information and Communication Complexity. LNCS, vol. 6058, pp. 157–167. Springer, Heidelberg (2010)
23. Kutten, S., Peleg, D.: Fast distributed construction of small k-dominating sets and applications. J. Algorithms 28(1), 40–66 (1998)
24. Shi, W.: Black hole search with tokens in interconnected networks. In: Guerraoui, R., Petit, F. (eds.) SSS 2009. LNCS, vol. 5873, pp. 670–682. Springer, Heidelberg (2009)

Strong Connectivity in Sensor Networks with Given Number of Directional Antennae of Bounded Angle

Stefan Dobrev[1], Evangelos Kranakis[2], Danny Krizanc[3], Jaroslav Opatrny[4], Oscar Morales Ponce[5], and Ladislav Stacho[6]

[1] Institute of Mathematics, Slovak Academy of Sciences, Bratislava, Slovak Republic. Supported in part by VEGA and APVV grants
[2] School of Computer Science, Carleton University, Ottawa, ON, K1S 5B6, Canada. Supported in part by NSERC and MITACS grants
[3] Department of Mathematics and Computer Science, Wesleyan University, Middletown CT 06459, USA
[4] Department of Computer Science, Concordia University, Montréal, QC, H3G 1M8, Canada. Supported in part by NSERC grant
[5] School of Computer Science, Carleton University, Ottawa, ON, K1S 5B6, Canada. Supported in part by CONACYT and NSERC grants
[6] Department of Mathematics, Simon Fraser University, 8888 University Drive, Burnaby, British Columbia, Canada, V5A 1S6. Supported in part by NSERC grant

Abstract. Given a set S of n sensors in the plane we consider the problem of establishing an ad hoc network from these sensors using directional antennae. We prove that for each given integer $1 \leq k \leq 5$ there is a strongly connected spanner on the set of points so that each sensor uses at most k such directional antennae whose range differs from the optimal range by a multiplicative factor of at most $2 \cdot \sin(\frac{\pi}{k+1})$. Moreover, given a minimum spanning tree on the set of points the spanner can be constructed in additional $O(n)$ time. In addition, we prove NP completeness results for $k = 2$ antennae.

Keywords: Antenna, Directional Antenna, Minimum Spanning Tree, Sensors, Spanner, Strongly Connected.

1 Introduction

The nodes of a wireless network can be connected using either omnidirectional antennae that transmit in all directions around the source or directional antennae that transmit only along a limited predefined angle. The energy usage of an antenna is proportional to its coverage area (for directional antennae, this is usually taken as the area of the sector delimited by the angle of the antenna). Therefore directional antennae can often perform more efficiently than omnidirectional ones in order to attain overall network connectivity. Given that the sensor range for a set S of sensors cannot be less than the length of the maximum edge of a minimum spanning tree on the set S, a reasonable way to lower energy consumption is by reducing the breadth (or angle or spread) of the

W. Wu and O. Daescu (Eds.): COCOA 2010, Part II, LNCS 6509, pp. 72–86, 2010.

antenna being used. However, by reducing antenna angles connectivity may be lost since communication between sensors can no longer be assumed to be bidirectional. Therefore an interesting question arising is how to maintain network connectivity when antenna angles are being reduced while at the same time the transmission range of the sensors is being kept as low as possible.

Formally, consider a set S of n sensors in the plane with identical range. Let $0 \leq \varphi \leq 2\pi$ be a given angle. Each sensor is allowed to use at most k directional antennae each of angle at most φ, for some integer value k. By directing an antenna at a sensor u towards another sensor v a directed edge (u, v) from u to v is formed provided that v is within u's range and lies inside the sector of angle φ formed by the antenna at u. By appropriately orienting such antennae at all the sensors we would like to form a strongly connected graph which spans all the sensors.

1.1   Preliminaries and Notation

Given spread φ and number of antennae k per sensor let $r_k(S, \varphi)$ denote the minimum range of directed antennae of angular spread at most φ so that if every sensor in S uses at most k such antennae then it is possible to direct them so that a strongly connected network (or spanner) on S is formed. A special case of this is to have angle $\varphi = 0$ i.e. a direct line connection, in which case we use the simpler notation $r_k(S) := r_k(S, 0)$. Let $\mathcal{D}_k(S)$ be the set of all strongly connected graphs on S which have out degree at most k. For any graph $G \in \mathcal{D}_k(S)$ let $r_k(G)$ be the length of the maximum length edge of G. It is easy to see that $r_k(S) := \min_{G \in \mathcal{D}_k(S)} r_k(G)$, i.e., $r_k(S)$ is the minimum length of a directed edge among all edges of a strongly connected graph with out degree k, for all such graphs in $\mathcal{D}_k(S)$.

It is useful to relate $r_k(S)$ to another quantity which arises from the Minimum Spanning Tree (MST) on S. Let $MST(S)$ denote the set of all MSTs on S. For $T \in MST(S)$ let $r(T)$ denote the length of longest edge of T, and let $r_{MST}(S) := \min\{r(T) : T \in MST(S)\}$. Further, for any angle $\varphi \geq 0$, it is clear that $r_{MST}(S) \leq r_k(S, \varphi)$ since every strongly connected, directed graph on S has an underlying spanning tree.

1.2   Related Work

The first paper to address the problem of converting a connected unidirectional graph consisting of omnidirectional sensors to a strongly connected graph of directional sensors having only one directional antenna each is [4]. In that paper the authors present polynomial time algorithms for the case when the sector angle of the antennae is at least $8\pi/5$. For smaller sector angles, they present algorithms that approximate the minimum radius. When the sector angle is smaller than $2\pi/3$, they show that the problem of determining the minimum radius in order to achieve strong connectivity is NP-hard. A different problem is considered in a subsequent paper [2]. Each sensor has multiple (fixed number of) directional antennae and the strong connectivity problem is considered under the assumption that the maximum (taken over all sensors) sum of angles is minimized. The authors present trade-offs between antennae range and specified sums of antennae per sensor.

When each sensor has one antenna and the angle $\varphi = 0$ then our problem is equivalent to finding a Hamiltonian cycle that minimizes the maximum length of an edge. For a set of n points $1, 2, \ldots, n$ with associated weights $c(i, j)$ satisfying the triangle inequality, the *Bottleneck Traveling Salesman Problem (BTSP)* is the min-max Hamiltonian cycle problem concerned with finding a Hamiltonian cycle for the complete graph which minimizes the maximum weight of an edge, i.e., $\min\{\max_{(i,j)\in H} c(i, j) : H$ is a Hamiltonian cycle$\}$. [10] shows that no polynomial time $(2 - \epsilon)$-approximation algorithm is possible for BTSP unless $P = NP$, and also gives a 2-approximation algorithm for this problem.

No literature is known on the connection between the MST of a set of points and strongly connected geometric spanners with given out-degree. Two papers relating somewhat these two concepts are the following. First, [5] shows that it is an NP-hard problem to decide for a given set S of n points in the Euclidean plane and a given real parameter k, whether S admits a spanning tree of maximum vertex degree four whose sum of edge lengths does not exceed k. Second, [7] gives a simple algorithm to find a spanning tree that simultaneously approximates a shortest-path tree and a minimum spanning tree.

Directional antennae are known to enhance ad hoc network capacity and performance and when replacing omnidirectional antennae can reduce the total energy consumption on the network. A theoretical model to this effect is presented in [6] showing that when n omnidirectional antennae are optimally placed and assigned optimally chosen traffic patterns the transport capacity is $\Theta(\sqrt{W/n})$, where each antenna can transmit W bits per second over the common channel(s). When both transmission and reception is directional [14] proves an $\sqrt{2\pi/\alpha\beta}$ capacity gain as well as corresponding throughput improvement factors, where α is the transmission angle and β is a parameter indicating that $\beta/2\pi$ is the average proportion of the number of receivers inside the transmission zone that will get interfered with. Additional experimental studies confirm the importance of using directional antennae in ad hoc networking (see, for example, [1,9,8,11,12,13]).

1.3   Results of the Paper

We are interested in the problem of providing an algorithm for orienting the antennae and ultimately for estimating the value of $r_k(S, \varphi)$. Without loss of generality antennae ranges will be normalized, i.e., $r_{MST}(S) = 1$. The two main results are the following.

Theorem 1. *Consider a set S of n sensors in the plane and suppose each sensor has k, $1 \leq k \leq 5$, directional antennae with any angle $\varphi \geq 0$. Then the antennae can be oriented at each sensor so that the resulting spanning graph is strongly connected and the range of each antenna is at most $2 \cdot \sin(\frac{\pi}{k+1})$ times the optimal. Moreover, given a MST on the set of points the spanner can be constructed with additional $O(n)$ overhead.*

Note that the case $k = 1$ was derived in [10].

Theorem 2. *For $k = 2$ antennae and angular sum of the antennae at most α, it is NP-hard to approximate the optimal radius to within a factor of x. where x and α are the solutions of equations $x = 2\sin(\alpha) = 1 + 2\cos(2\alpha)$.*

Using the identity $\cos(2\alpha) = 1 - 2\sin^2\alpha$ and solving the resulting quadratic equation with unknown $\sin\alpha$ we obtain numerical solutions $x \approx 1.30, \alpha \approx 0.45\pi$. Figure 9 depicts the geometric relation between α and x.

2 Upper Bound Result on Strongly Connected Spanners

The proof given in the sequel is in three parts. Due to space constraints only the proof for $k = 2$ is presented in detail in Subsection 2.1 and part of the pseudocode in Subsection 2.2, while in the full paper we prove the cases for $k = 3$ and $k = 4$ and present the remaining algorithm.

Preliminary Definitions. $D(u; r)$ is the open disk with radius r, centered at u and $C(u, r)$ is the circle with radius r and centered at u. $d(\cdot, \cdot)$ denotes the usual Euclidean distance between two points. In addition, we define the concept of *Antenna-Tree* (*A-Tree*, for short) which isolates the particular properties of a MST that we need in the course of the proof.

Definition 1. *An A-Tree is a tree T embedded in the plane satisfying the following three rules:*

1. *Its maximum degree is five.*
2. *The minimum angle among nodes with a common parent is at least $\pi/3$.*
3. *For any point u and any edge $\{u, v\}$ of T, the open disk $D(v; d(u, v))$ does not have a point $w \neq v$ which is also a neighbor of u in T.*

It is well known and easy to prove that for any set of points there is an MST on the set of points which satisfies Definition 1. We also recall that we consider normalized ranges (i.e. we assume $r(T) = 1$).

Definition 2. *For each real $r > 0$, we define the geometric r-th power of a A-Tree T, denoted by T^r, as the graph obtained from T by adding all edges between vertices of (Euclidean) distance at most r.*

For simplicity, in the sequel we slightly abuse terminology and refer to *geometric r-th power* as *r-th power*.

Definition 3. *Let G be a graph. An orientation $\overrightarrow{G}$ of G is a digraph obtained from G by orienting every edge of G in at least one direction.*

As usual, we denote with (u, v) a directed edge from u to v, whereas $\{u, v\}$ denotes an undirected edge between u and v. Let $d^+_{\overrightarrow{G}}(u)$ be the out-degree of u in $\overrightarrow{G}$ and $\Delta^+(\overrightarrow{G})$ the maximum out-degree of a vertex in $\overrightarrow{G}$.

2.1   Maximum Out-Degree 2

Theorem 3. *Given an A-Tree T, there exists a spanning subgraph $G \subseteq T^{\sqrt{3}}$ such that $\overrightarrow{G}$ is strongly connected and $\Delta^+(\overrightarrow{G}) \leq 2$. Moreover, $d^+_{\overrightarrow{G}}(u) \leq 1$ for each leaf u of T and either every edge of T which is incident to a leaf is contained in G or a leaf is connected to its two consecutive siblings in G.*

Before proving Theorem 3, we need to introduce a definition and a lemma which provides information on the proximity among the neighbors of two adjacent vertices in the tree depending on their degree. The proof of the lemma is technical and given in the full paper.

Definition 4. *We say that two consecutive neighbors of a vertex are* close *if the distance between them is at most $\sqrt{3}$. Otherwise we say that they are* far.

Lemma 1. *Let u, v and w be three consecutive siblings with parent p of an A-Tree T such that $\widehat{upv} + \widehat{vpw} \leq \pi$. If $d(v) = 3$ and the only two children of v are far, then at least one of them is close to either u or w.*

If $d(v) = 4$ and each pair of consecutive children of v are close, then at least one of them is close to either u or w.

If $d(v) = 4$, two consecutive children of v are far and all children of v are at distance at least $\sqrt{3} - 1$ of v, then one child of v is close to u and another child of v is close to w.

If $d(v) = 4$, two consecutive children of v are far and one child x of v is at distance at most $\sqrt{3} - 1$ of v, then at most one child of v different from x are far from u and w.

If $d(v) = 5$, then at least one child of v is close to either u or w.

Proof (**Theorem 3**). The proof is by induction on the diameter of the tree. Firstly, we do the base case. Let l be the diameter of T. If $l \leq 1$, let $G = T$ and the result follows trivially.

If $l = 2$, then T is an A-Tree which is a star with $2 \leq d \leq 5$ leaves, respectively. Four cases can occur:

$d = 2$. Let $G = T$ and orient every edge in both directions. This results in a strongly connected digraph which trivially satisfies the hypothesis of the theorem.

$d = 3$. Let u be the center of T. Since T is a star, two consecutive neighbors, say u_1 and u_2 are close. Let $G = T \cup \{\{u_1, u_2\}\}$ and orient edges of G as depicted in Figure 1a[1]. It is easy to check that G satisfies the hypothesis of the Theorem.

$d = 4$. Let u be the center of T and u_1, u_2, u_3, u_4 be the four neighbors of u in clockwise order around u starting at any arbitrary neighbor of u. Observe that at most two consecutive neighbors of u are far since T is a star and the angle between two nodes with a common parent is at least $\pi/3$. Assume without loss of generality that u_4 and u_1 are far. Let $G = T \cup \{\{u_1, u_2\}, \{u_3, u_4\}\}$ and orient edges of G as depicted in Figure 1b. Thus, G satisfies trivially the hypothesis of the Theorem.

---

[1] In all figures boldface arrows represent the newly added edges.

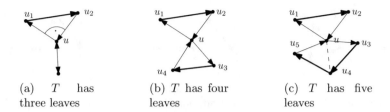

(a) T has three leaves

(b) T has four leaves

(c) T has five leaves

Fig. 1. T is a tree with diameter $l = 2$ (The angular sign with a dot depicts an angle of size at most $2\pi/3$ at vertex u and the dashed edge indicates that it exists in T but not in G.)

$d = 5$. Let u be the center of T and u_1, u_2, u_3, u_4, u_5 be the five neighbors of u in clockwise order around u starting at any arbitrary neighbor of u. Observe that all consecutive neighbors are close since T is a star and the angle between two nodes with a common parent is at least $\pi/3$. Let $G = T \setminus \{u, u_4\} \cup \{\{u_1, u_2\}, \{u_3, u_4\}\}, \{u_4, u_5\}$ and orient edges of G as depicted in Figure 1c. Observe that $\widehat{u_3 u u_5} \leq \pi$. Thus, $\overrightarrow{G}$ is strongly connected and $\Delta^+(\overrightarrow{G}) \leq 2$. Moreover, $d^{\pm}_{\overrightarrow{G}}(u) \leq 1$, all edges of T except $\{u, u_4\}$ are contained in G and $\{u_3, u_4\}$ and $\{u_4, u_5\}$ are contained in G.

Next we continue with the inductive step. Let T' be the tree obtained from T by removing all leaves. Since removal of leaves does not violate the property of being an A-Tree, T' is also an A-Tree and has diameter less than the diameter of T. Thus, by inductive hypothesis there exists $G' \subseteq T'^{\sqrt{3}}$ such that $\overrightarrow{G'}$ is strongly connected, $\Delta^+(\overrightarrow{G'}) \leq 2$. Moreover, $d^{\pm}_{\overrightarrow{G}}(u) \leq 1$ for each leaf u of T and either every edge of T which is incident to a leaf is contained in G or a leaf is connected to its two consecutive siblings in G.

Let u be a leaf of T', u_0 be the neighbor of u in T' and $u_1, \ldots, u_c$ be the c neighbors of u in $T \setminus T'$ in clockwise order around u starting from u_0. Four cases can occur:

u has one neighbor in $T \setminus T'$. Let $G = G' \cup \{\{u, u_1\}\}$ and orient it in both directions. It is easy to see that $\overrightarrow{G}$ satisfies the inductive hypothesis.

u has two neighbors in $T \setminus T'$. We consider two cases. In the first case suppose that u_1 and u_2 are close. Let $G = G' \cup \{\{u, u_1\}, \{u, u_2\}, \{u_1, u_2\}\}$ and orient edges of G as depicted in Figure 2a. In the second case, u_1 and u_2 are far. Again we need to consider two subcases:

Subcase 1 ($\{u_0, u\}$ is in G'.) Either u_0 and u_1 are close or u_2 and u_0 are close. Without loss of generality, lets assume that u_1 and u_0 are close. Let $G = \{G' \setminus \{u_0, u\}\} \cup \{\{u, u_1\}, \{u, u_2\}, \{u_0, u_1\}\}$. The orientation of G will depend on the orientation of $\{u, u_0\}$ in G'. If (u_0, u) is in $\overrightarrow{G'}$, then orient edges of G as depicted in Figure 2b. Otherwise if (u, u_0) is in $\overrightarrow{G'}$, then orient edges of G as depicted in Figure 2c. Thus, $\overrightarrow{G}$ is strongly connected and $\Delta^+(\overrightarrow{G}) \leq 2$. Moreover, the leaves u_1 and u_2 of T have degree one and the edges of T incident to them are contained in G.

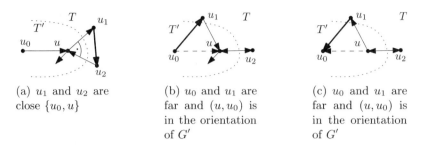

(a) u_1 and u_2 are close $\{u_0, u\}$

(b) u_0 and u_1 are far and (u, u_0) is in the orientation of G'

(c) u_0 and u_1 are far and (u, u_0) is in the orientation of G'

Fig. 2. Depicting the inductive step when u has two neighbors in $T \setminus T'$ (The dashed edge $\{u_0, u\}$ indicates that it does not exist in G but exists in G' and the dotted curve is used to separate T' from T.)

Subcase ($\{u_0, u\}$ is not in G'.) By inductive hypothesis, u is connected to its two siblings v and w in G'. Thus, by Lemma 1, either u_1 or u_2 are close to v or w. Without loss of generality, assume that u_1 and v are close. Let $G = (G' \setminus \{v, u\}) \cup \{\{u_1, u\}, \{u_2, u\}, \{v, u_1\}\}$. The orientation of G will depend on the orientation of $\{v, u\}$ in G'. If (v, u) is in $\overrightarrow{G'}$, then orient edges of G as depicted in Figure 3a. Otherwise if (u, v) is in $\overrightarrow{G'}$, then orient edges of G as depicted in Figure 3b. Thus, $\overrightarrow{G}$ is strongly connected and $\Delta^+(\overrightarrow{G}) \leq 2$. Moreover, the leaves u_1 and u_2 of T have degree one and the edges of T incident to them are contained in G.

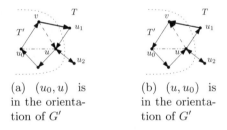

(a) (u_0, u) is in the orientation of G'

(b) (u, u_0) is in the orientation of G'

Fig. 3. Depicting the inductive step when u has two neighbors in $T \setminus T'$, u_0 and u_1 are far and $\{u_0, u\}$ is not in G' (The dashed edge $\{v, u\}$ indicates that it does not exist in G but exists in G', the dash dotted edge $\{u_0, u\}$ indicates that it exists in T' but not in G' and the dotted curve is used to separate T' from T.)

u has three neighbors in $T \setminus T'$. Two subcases can occur:

Subcase 1 ($\{u_0, u\}$ is in G'). At most two neighbors of u are far. Firstly, suppose that u_0 and u_3 are far (This case is equivalent to the case when u_1 and u_2 are far.) Let $G = \{G' \setminus \{u_0, u\}\} \cup \{\{u_1, u\}, \{u_2, u\}, \{u_3, u\}, \{u_1, u_0\}, \{u_2, u_3\}\}$. If (u_0, u) is in $\overrightarrow{G'}$, then orient edges of G as depicted in Figure 4a. Otherwise if (u, u_0) is in $\overrightarrow{G'}$, then orient edges of G as depicted in Figure 4b. Thus, $\overrightarrow{G}$ is strongly connected and $\Delta^+(\overrightarrow{G}) \leq 2$. Moreover, the leaves u_1, u_2 and u_3 of T

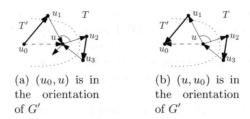

(a) (u_0, u) is in the orientation of G'

(b) (u, u_0) is in the orientation of G'

Fig. 4. Depicting the inductive step when u has three neighbors in $T \setminus T'$, u_1 and u_2 are far and $\{u_0, u\}$ is in G' (The dashed edge $\{u_0, u\}$ indicates that it does not exist in G but exists in G', the dotted curve is used to separate T' from T and the angular sign depicts an angle of size greater than $2\pi/3$ at vertex u.)

have degree one and the edges of T incident to them are contained in G. The case when u_1 and u_0 are far or u_2 and u_3 are far can be solved analogously by symmetry.

Subcase 2 ($\{u_0, u\}$ is not in G'). By inductive hypothesis u is connected to its two siblings v and w in G'. Three cases can occur.

Subcase 2.1 (u_1 is close to u_2 and u_2 is close to u_3.) By Lemma 1, either u_1 or u_3 are close to v or w. Without loss of generality, we assume that v and u_1 are close. Let $G = \{G' \setminus \{v, u\}\} \cup \{\{u_1, u\}, \{u_2, u\}, \{u_3, u\}, \{v, u_1\}, \{u_2, u_3\}\}$. The orientation of G will depend on the orientation of $\{v, u\}$ in G'. If (v, u) is in $\overrightarrow{G'}$, then orient edges of G as depicted in Figure 5a. Otherwise if (u, v) is in $\overrightarrow{G'}$, then orient edges of G as depicted in Figure 5b. Thus, $\overrightarrow{G}$ is strongly connected and $\Delta^+(\overrightarrow{G}) \leq 2$. Moreover, the leaves u_1, u_2 and u_3 of T have degree one and the edges of T incident to them are contained in G.

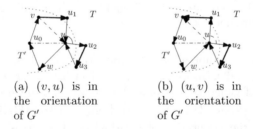

(a) (v, u) is in the orientation of G'

(b) (u, v) is in the orientation of G'

Fig. 5. Depicting the inductive step when u has three neighbors in $T \setminus T'$, u_1 and u_2 are far and $\{u_0, u\}$ is not in G' (The dashed edge $\{v, u\}$ indicates that it does not exist in G but exists in G', the dash dotted edge $\{u_0, u\}$ indicates that it exists in T' but not in G' and the dotted curve is used to separate T' from T.)

Subcase 2.2 (Either u_1 is far from u_2 or u_2 is far from u_3 and u_1, u_2 and u_3 are at distance greater than $\sqrt{3} - 1$ from u.) By Lemma 1 u_1 is close to one sibling of u, say v and u_3 is close to another sibling of u, say w. Notice that if u_1 is far from u_2, it is exactly the Case one. However, if u_2 are far from u_3, let $u'_i = u_{3-i+1}$, and Case 1 applies.

Subcase 2.3 (Either u_1 is far from u_2 or u_2 is far from u_3 and at least one child of u is at distance less than $\sqrt{3}-1$.) Without loss of generality, assume that u_1 is far from u_2. Therefore, $d(u, u_1) > \sqrt{3}-1$ and $d(u, u_3) \leq \sqrt{3}-1$. Observe that u_3 is close to u_1 and u_2. By Lemma 1 either u_1 or u_2 are close to v or w. Thus, if v is close to u_1, then apply the Case one. If w is close to u_2, then let $u_1' = u_2$, $u_2' = u_1$ and $u_3' = u_3$ and apply Case 1.

u has four neighbors in $T \setminus T'$. Two subcases can occur:

Subcase 1 ($\{u_0, u\}$ is in G'). Let

$$G = \{G' \setminus \{u_0, u\}\} \cup \{\{u_1, u\}, \{u_2, u\}, \{u_4, u\}, \{u_1, u_0\}, \{u_2, u_3\}, \{u_3, u_4\}\}.$$

The orientation of G will depend on the orientation of $\{u_0, u\}$ in G'. If (u_0, u) is in $\overrightarrow{G'}$, then orient edges of G as depicted in Figure 6a. Otherwise if (u, u_0) is in $\overrightarrow{G'}$, then orient edges of G as depicted in Figure 6b. Thus, $\overrightarrow{G}$ is strongly connected and $\Delta^+(\overrightarrow{G}) \leq 2$. Moreover, the leaves u_1, u_2, u_3 and u_4 of T have degree one, the edges of T incident to u_1, u_2 and u_4 are contained in G and u_3 is connected to u_2 and u_4 in G. Observe that $\widehat{u_2 u u_4} \leq \pi/2$.

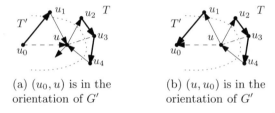

(a) (u_0, u) is in the orientation of G'

(b) (u, u_0) is in the orientation of G'

Fig. 6. Depicting the inductive step when u has four neighbors in $T \setminus T'$, $\{u_0, u\}$ is in G' (The dashed edge $\{u_0, u\}$ indicates that it does not exist in G but exists in G', the dotted curve is used to separate T' from T and the dash dotted edge $\{u, u_3\}$ indicates that it exists in T but not in G.)

Subcase 2 ($\{u_0, u\}$ is not in G'). By inductive hypothesis u is connected to its two siblings v and w in G'. By Lemma 1 either u_1 or u_4 is close to v or w. Without loss of generality, assume that u_1 and v are close. Let $G = \{G' \setminus \{v, u\}\} \cup \{\{u_1, u\}, \{u_2, u\}, \{u_4, u\}, \{v, u_1\}, \{u_2, u_3\}, \{u_3, u_4\}\}$. The orientation of G will depend on the orientation of $\{v, u\}$ in G'. If (v, u) is in $\overrightarrow{G'}$, then orient edges of G as depicted in Figure 7a. Otherwise if (u, v) is in $\overrightarrow{G'}$, then orient edges of G as depicted in Figure 7b. Thus, $\overrightarrow{G}$ is strongly connected and $\Delta^+(\overrightarrow{G}) \leq 2$. Moreover, u_1, u_2, u_3 and u_4 have degree one, the edges of T incident to u_1, u_2 and u_4 are contained in G and u_3 is connected to u_2 and u_4 in G. Observe that $\widehat{u_2 u u_4} \leq \pi/2$. This completes the proof of Theorem 3.

2.2 Algorithm

In this section we present the pseudocode for Algorithm 1 that constructs a strongly connected spanner with max out-degree $2 \leq k \leq 5$ and range bounded

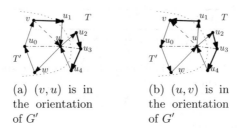

(a) (v, u) is in the orientation of G'

(b) (u, v) is in the orientation of G'

Fig. 7. Depicting the inductive step when u has four neighbors in $T \setminus T'$ and $\{u_0, u\}$ is not in G' (The dashed edge $\{v, u\}$ indicates that it does not exist in G but exists in T', the dotted curve is used to separate T' from T, the dash dotted edge $\{u_0, u\}$ idicates that it exists in T' but not in G' and the dash dotted edge $\{u, u_3\}$ indicates that it exist in T but not in G.)

by $2 \cdot \sin\left(\frac{\pi}{k+1}\right)$ times the optimal. It uses the recursive Procedure kAntennae when $3 \le k \le 5$ and the recursive Procedure TwoAntennae when $k = 2$ which is presented in the full paper. The correctness of TwoAntennae procedure is derived from Theorem 3 and the correctness of kAntennae procedure is derived in the full paper. It is not difficult to see that Algorithm 1 runs in linear time.

Algorithm 1. Strongly connected spanner with max out-degree $2 \ge k \ge 5$ and edge length bounded by $2 \cdot \sin\left(\frac{\pi}{k+1}\right)$

> **input** : T, k; where T is a MST with max length 1 and k an integer in $[2, 5]$.
> **output**: Strongly connected spanner G with max out-degree k and range
>   bounded by $2 \cdot \sin\left(\frac{\pi}{k+1}\right)$
> 1 Let u be any leaf of T and v its neighbor in T;
> 2 Let $G \leftarrow \{(v, u), (u, v)\}$;
> 3 **if** $k = 2$ **then** TwoAntennae(G, T, v, u);
> 4 **if** $3 \le k < 5$ **then** kAntennae(G, T, v, u, k);

---

3   NP Completeness

*Proof (***Theorem 2***).* By reduction from the well-known NP-hard problem for finding Hamiltonian cycles in degree three planar graphs. Take a degree three planar graph $G = (V, E)$ and replace each vertex v_i by a vertex-graph (meta-vertex) G_{v_i} shown in Figure 8a. Furthermore, replace each edge $e = \langle v_i, v_j \rangle$ of G by an edge-graph (meta-edge) G_e shown in Figure 8b.

Each meta-vertex has three parts connected in a cycle, with each part consisting of a pair of vertices (called *connecting vertices*) connected by two paths. Each meta-edge G_e has a pair of connecting vertices at each endpoint – these vertices coincide with the connecting vertices in the corresponding parts of the meta-vertices G_{v_i} and G_{v_j}. This means that after each vertex and each edge is replaced, each connecting vertex is of degree 4.

Procedure kAntennae(G, T, u, w, k)

1 Let $u_0 = w, u_1, \cdots, u_{d(u)-1}$ be the neighbors of $u \in T$ in clockwise order around u;

2 **if** $d(u) \leq k$ **then** Add to G a bidirectional arc for each u_i such that $i > 0$;

3 **else if** $d(u) = k + 1$ **then**

4   Let u_i, u_{i+1} be the consecutive neighbor of u with smallest angle;

5   **if** $i = 0$ **or** $i + 1 = 0$ **then**

6   **if** $i = 0$ **then** Let $i \leftarrow 1$;

7   **if** $(u, u_0) \in G$ **then** Let $G \leftarrow \{G \setminus \{(u, u_0)\}\} \cup \{(u, u_i), (u_i, u_0)\}$;

8   **else** Let $G \leftarrow \{G \setminus \{(u_0, u)\}\} \cup \{(u_0, u_i), (u_i, u)\}$;

9   **end**

10   **else** Let $G \leftarrow G \cup \{(u, u_i), (u_i, u_{i+1}), (u_{i+1}, u)\}$;

11   Add to G a bidirectional arc for each u_j such that $j \notin \{0, i, i+1\}$;

12 **end**

13 **else if** $d(u) = k + 2$ **then**

14   Let u_i, u_{i+1} be the consecutive neighbors of u with longest angle;

15   **if** $i = 0$ **or** $i = 2$ **or** $i = 4$ **then** Let
$G \leftarrow G \cup \{(u, u_1), (u_1, u_2), (u_2, u), (u, u_3), (u_3, u_4), (u_4, u)\}$;

16   **else**

17   **if** $(u, u_0) \in G$ **then** Let $G \leftarrow \{G \setminus \{(u, u_0)\}\} \cup \{(u, u_1), (u_1, u_0)\}$;

18   **else** Let $G \leftarrow \{G \setminus \{(u_0, u)\}\} \cup \{(u_0, u_1), (u_1, u)\}$;

19   Let $G \leftarrow G \cup \{(u, u_2), (u_2, u_3), (u_3, u), (u, u_4), (u_4, u)\}$;

20   **end**

21 **end**

22 **for** $i \leftarrow 1$ **to** $d(u) - 1$ **do if** $d(u_i) > 1$ **then** $G \leftarrow$kAntennae(G, T, u_i, u, k) ;

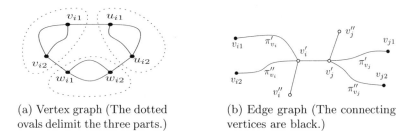

(a) Vertex graph (The dotted ovals delimit the three parts.)

(b) Edge graph (The connecting vertices are black.)

Fig. 8. Meta-vertex and meta-edge for the NP completeness proof

Take the resulting graph G' and embed it in the plane in such a way that:

- the distance (in the embedding) between neighbours in G' is at most 1,
- the distance between non-neighbours in G' is at least x, and
- the smallest angle between incident edges in G' is at least α.

Let us call the resulting embedded graph G''. Note that such an embedding always exists [3]: We have a freedom to choose the length of the paths in the meta-graphs the way we need as we can stretch the configurations apart to fit everything in without violating the embedding requirements. The only constraining

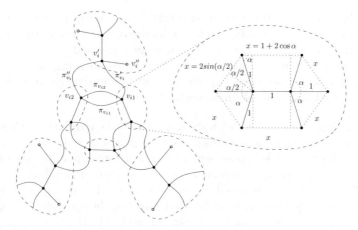

Fig. 9. Connecting meta-edges with meta-vertices (The dashed ovals show the places where embedding is constrained.)

places are the midpoints of the meta-edges and the three places in each meta-vertex where the parts are connected to each other. These can be embedded as shown in the right part of Figure 9. Note that the need to embed these parts without violating embedding requirements gives rise to the equations defining x and α (see Figure 9). This completes details of the main construction.

The proof of the Theorem is based on the following claim:

Claim. There is a Hamiltonian cycle in G if and only if there exists an assignment of two antennae with sum of angles less than α and radius less than x to the vertices of G'' such that the resulting connectivity graph is strongly connected.

Proof **(Claim).** First we show that if G has a Hamiltonian cycle then there exists the assignment of such antennae that makes the resulting connectivity graph of G'' strongly connected. Figure 10 shows antenna assignments in the meta-edges corresponding to edges used and not used by the Hamiltonian cycle, respectively. Figure 11 shows the antenna assignments in a meta-vertex. Since each vertex of G has one incoming, one outgoing and one unused incident edge, and each edge is either used in one direction, or not used at all, this provides the full description of antenna assignments in G''.

Observe that the connecting pair of vertices at the meta-vertex uses two antennae towards the meta-edge it is connected to if and only if this meta-edge is outgoing; otherwise only one antenna is used towards the meta-edge and another is used towards the next part of the meta-vertex. It is easy to verify that the resulting connectivity graph is strongly connected:

- if the edge $e = \langle v_i, v_j \rangle$ is not used in the Hamiltonian path in the direction from v_i to v_j, then the near half of the meta-edge G_e (i.e. v'_j, v''_j, π'_{v_j} and π''_{v_j}) together with the connecting part of the meta-vertex G_{v_j} form a strongly connected subgraph,

- in each meta-vertex the part corresponding to the outgoing edge is reachable from the part corresponding to the unused edge, which is in turn reachable from the part corresponding to the incoming edge, and
- all vertices of a meta-edge corresponding to an outgoing edge $\langle v_i, v_j \rangle$ are reachable from either v_{i1} or v_{i2}; furthermore the destination vertices v_{j_1} and v_{j_2} are reachable from all these vertices.

Combining these observations with the fact that the Hamiltonian cycle spans all vertices yields that the resulting graph is strongly connected.

Next we show that if it is possible to assign the antennae in G'' such that the resulting graph is strongly connected then there exists a Hamiltonian cycle in G. Recall that G'' is constructed in such a manner that no antenna of radius less than x and angle less than α can reach two neighbouring vertices, and that no antenna can reach a vertex that is not a neighbor in G''.

Assume an assignment of antennae such that the resulting graph is strongly connected. First, consider a pair of connecting vertices v_{i1} and v_{i2}. Since both path $\pi_{v_{i1}}$ and $\pi_{v_{i2}}$ are connected only to them, v_{i1} and v_{i2} must together use at least two antennae towards these two paths.

Let us call a meta-edge corresponding to edge $\langle v_i, v_j \rangle$ *directed* if in the connectivity graph there is an edge $\langle v_i', v_j' \rangle$. Without loss of generality assume the direction is from v_i' to v_j', i.e. v_i' used an antenna to reach v_j'. Since v_i'' is reachable only from v_i' (and hence v_i' used its second antenna on v_i''), this means that there is no antenna pointing from v_i' towards the paths π_{v_i}' and π_{v_i}''. Therefore, the only way for the vertices of these two paths to be reachable is to have both connecting vertices (which for simplicity we call v_{i1} and v_{i2}, respectively) use an antenna towards these paths. Since they already used two antennae to ensure reachability of $\pi_{v_i}1$ and $\pi_{v_i}2$ are reachable, they have no antenna left to connect to another part of the meta-vertex.

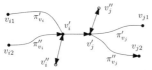
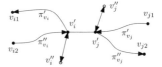

(a) Edge used in the Hamiltonian cycle.

(b) Edge not used in the Hamiltonian cycle.

Fig. 10. Antenna assignments in a meta-edges corresponding to an edge v_i to v_j

Consider now the other half of the meta-edge. Observe that since v_j' must use one antenna on v_j'', it can use at most one antenna towards the paths π_{v_j}' and π_{v_j}''. Hence, either v_{j1} or v_{j2} must use an antenna towards one of these paths. Since these vertices must use two more antennae to ensure that the paths $\pi_{v_j}1$ and $\pi_{v_j}2$ are reachable, only one antenna is left for connecting to other parts of the meta vertex. Note that this argument holds both for receiving ends of directed meta-edges, as well as for non-directed meta-edges.

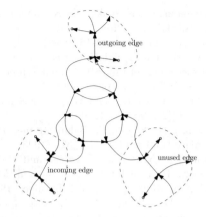

Fig. 11. Antenna assignments at the meta-vertex and incident meta-vertices

However, this means that in a meta-vertex there can be at most one outgoing directed meta-edge – otherwise there is no way to make the meta-vertex connected. Since each meta-vertex must have at least one outgoing directed meta-edge (otherwise the rest of the graph would be unreachable) and at least one incoming directed meta-edge (otherwise it would not be reachable from the rest), from the fact that the whole graph is strongly connected it follows that each meta-vertex must have exactly one undirected meta-edge, one directed incoming meta-edge and one directed outgoing meta-edge. Obviously, these correspond to unused/incoming/outgoing edges in the original graph G, with the directed edges forming the Hamiltonian cycle.

4   Conclusion

We have provided an algorithm which when given as input a set of n points (modeling sensors) in the plane and an integer $1 \leq k \leq 5$ produces a strongly connected spanner so that each sensor uses at most k directional antennae of angle 0 and range at most $2 \cdot \sin\left(\frac{\pi}{k+1}\right)$ times the optimal. Interesting open problems include looking at tradeoffs when the angle of the antennae is $\varphi > 0$ as well as deriving better lower bounds.

References

1. Bao, L., Garcia-Luna-Aceves, J.J.: Transmission scheduling in ad hoc networks with directional antennas. In: Proceedings of the 8th Annual International Conference on Mobile Computing and Networking, pp. 48–58 (2002)
2. Bhattacharya, B., Hu, Y., Kranakis, E., Krizanc, D., Shi, Q.: Sensor Network Connectivity with Multiple Directional Antennae of a Given Angular Sum. In: 23rd IEEE IPDPS 2009, May 25-29 (2009)
3. Calamoneri, T., Petreschi, R.: An Efficient Orthogonal Grid Drawing Algorithm For Cubic Graphs. In: Li, M., Du, D.-Z. (eds.) COCOON 1995. LNCS, vol. 959, pp. 31–40. Springer, Heidelberg (1995)

4. Caragiannis, I., Kaklamanis, C., Kranakis, E., Krizanc, D., Wiese, A.: Communication in Wireless Networks with Directional Antennae. In: Proceedings of 20th ACM SPAA, Munich, Germany, June 14-16 (2008)
5. Francke, A., Hoffmann, M.: The Euclidean degree-4 minimum spanning tree problem is NP-hard. In: Proceedings of the 25th Annual Symposium on Computational Geometry, pp. 179–188. ACM, New York (2009)
6. Gupta, P., Kumar, P.R.: The Capacity of Wireless Networks. IEEE Transactions on Information Theory 46(2), 388–404 (2000)
7. Khuller, S., Raghavachari, B., Young, N.: Balancing minimum spanning trees and shortest-path trees. Algorithmica 14(4), 305–321 (1995)
8. Kranakis, E., Krizanc, D., Urrutia, J.: Coverage and Connectivity in Networks with Directional Sensors. In: Danelutto, M., Vanneschi, M., Laforenza, D. (eds.) Euro-Par 2004. LNCS, vol. 3149, pp. 917–924. Springer, Heidelberg (2004)
9. Kranakis, E., Krizanc, D., Williams, E.: Directional versus omnidirectional antennas for energy consumption and k-connectivity of networks of sensors. In: Higashino, T. (ed.) OPODIS 2004. LNCS, vol. 3544, pp. 357–368. Springer, Heidelberg (2004)
10. Parker, R.G., Rardin, R.L.: Guaranteed performance heuristics for the bottleneck traveling salesman problem. Oper. Res. Lett. 2(6), 269–272 (1984)
11. Ramanathan, R.: On the Performance of Ad Hoc Networks with Beamforming Antennas. In: Proceedings of the 2nd ACM International Symposium on Mobile Ad Hoc Networking & Computing, pp. 95–105 (2001)
12. Spyropoulos, A., Raghavendra, C.S.: Energy efficient communications in ad hoc networks using directional antennas. In: Proceedings of the Twenty-First Annual Joint Conference of the IEEE Computer and Communications Societies, INFOCOM 2002, vol. 1. IEEE, Los Alamitos (2002)
13. Spyropoulos, A., Raghavendra, C.S.: Capacity Bounds for Ad-Hoc Networks Using Directional Antennas. In: IEEE International Conference on Communications, ICC 2003, vol. 1 (2003)
14. Yi, S., Pei, Y., Kalyanaraman, S.: On the capacity improvement of ad hoc wireless networks using directional antennas. In: Proceedings of the 4th ACM International Symposium on Mobile Ad hoc Networking & Computing, pp. 108–116 (2003)

A Constant-Factor Approximation Algorithm for the Link Building Problem

Martin Olsen[1], Anastasios Viglas[2], and Ilia Zvedeniouk[2]

[1] Center for Innovation and Business Development,
Institute of Business and Technology, Aarhus University,
Birk Centerpark 15, DK-7400 Herning, Denmark
martino@hih.au.dk
[2] School of Information Technologies,
University of Sydney, 1 Cleveland St, NSW 2006, Australia
taso.viglas@sydney.edu.au, izve6419@uni.sydney.edu.au

Abstract. In this work we consider the problem of maximizing the PageRank of a given target node in a graph by adding k new links. We consider the case that the new links must point to the given target node (backlinks). Previous work [7] shows that this problem has no fully polynomial time approximation schemes unless $P = NP$. We present a polynomial time algorithm yielding a PageRank value within a constant factor from the optimal. We also consider the naive algorithm where we choose backlinks from nodes with high PageRank values compared to the outdegree and show that the naive algorithm performs much worse on certain graphs compared to the constant factor approximation scheme.

1 Introduction

Search engine optimization (SEO) is a fast growing industry that deals with optimizing the ranking of web pages in search engine results. SEO is a complex task, especially since the specific details of search and ranking algorithms are often not publicly released, and also can change frequently. One of the key elements of optimizing for search engine visibility is the "external link popularity" [9], which is based on the structure of the web graph. The problem of obtaining optimal new backlinks in order to achieve good search engine rankings is known as Link Building and leading experts from the SEO industry consider Link Building to be an important aspect of SEO [9].

The PageRank algorithm is one of the popular methods of defining a ranking according to the link structure of the graph. The definition of PageRank [3] uses random walks based on the random surfer model. The random surfer walk is defined as follows: the walk can start from any node in the graph and at each step the surfer chooses a new node to visit. The surfer "usually" chooses (uniformly at random) an outgoing link from the current node, and follows it. But with a small probability at each step the surfer might choose to ignore the current node's outgoing links, and just zap to any node in the graph (chosen uniformly at random). The random surfer walk is a random walk on the graph with random

W. Wu and O. Daescu (Eds.): COCOA 2010, Part II, LNCS 6509, pp. 87–96, 2010.
© Springer-Verlag Berlin Heidelberg 2010

restarts every few steps. This random walk has a unique stationary probability distribution that assigns the probability value π_i to node i. This value is the PageRank of node i, and can be interpreted as the probability for the random surfer of being at node i at any given point during the walk. We refer to the random restart as zapping. The parameter that controls the zapping frequency is the probability of continuing the random walk at each step, $\alpha > 0$. The high level idea is that the PageRank algorithm will assign high PageRank values to nodes that would appear more often in a random surfer type of walk. In other words the nodes with high PageRank are hot-spots that will see more random surfer traffic, resulting directly from the link structure of the graph. If we add a small number of new links to the graph, the PageRank values of certain nodes can be affected very significantly. The Link Building problem arises as a natural question: given a specific target node in the graph, what is the best set of k links that will achieve the maximum increase for the PageRank of the target node?

We consider the problem of choosing the optimal set of *backlinks* for maximizing π_x, the PageRank value of some target node x. A backlink (with respect to a target node x) is a link from any node towards x. Given a graph $G(V, E)$ and an integer k, we want to identify the $k \geq 1$ links to add to node x in G in order to maximize the resulting PageRank of x, π_x. Intuitively, the new links added should redirect the random surfer walk towards the target node, as much as possible. For example adding a new link from a node of very high PageRank would usually be a good choice.

1.1   Related Work and Contribution

The PageRank algorithm [3] is based on properties of Markov chains. There are many results related to the computation of PageRank values [5,2] and recalculating PageRank values after adding a set of new links in a graph [1].

The Link Building problem that we consider in this work is known to be NP-hard [7] where it is even showed that there is no fully polynomial time approximation scheme (FPTAS) for Link Building unless NP = P and the problem is also shown to be W[1]-hard with parameter k. A related problem considers the case where a target node aims at maximizing its PageRank by adding new outlinks. Note that in this case, new outlinks can actually decrease the PageRank of the target node. This is different to the case of the Link Building problem with backlinks where the PageRank of the target can only increase [1]. For the problem of maximizing PageRank with outlinks we refer to [1,4] containing, among other things, guidelines for optimal linking structure.

In Sect. 2 we give background to the PageRank algorithm. In Sect. 3 we formally introduce the Link Building problem. In Sect. 3.1 we consider the naive and intuitively clear algorithm for Link Building where we choose backlinks from the nodes with the highest PageRank values compared to their outdegree (plus one). We show how to construct graphs where we obtain a surprisingly high approximation ratio. The approximation ratio is the value of the optimal solution divided by the value of the solution obtained by the algorithm. In

Sect. 3.2, we present a polynomial time algorithm yielding a PageRank value within a constant factor from the optimal and therefore show that the Link Building problem is in the class APX.

2 Background: The PageRank Algorithm

The PageRank algorithm was proposed by Brin, Page [3] and Brin, Page, Motwani and Winograd [8] as a webpage ranking method that captures the importance of webpages. Loosely speaking, a link pointing to a webpage is considered a vote of importance for that webpage. A link from an important webpage is better for the receiver than a link from an unimportant webpage.

We consider directed graphs $G = (V, E)$ that are unweighted and therefore we count multiple links from a node u to a node v as a single link. The graph may represent a set of webpages V with hyperlinks between them, E, or any other linked structure.

We define the following *random surfer* walk on G: at every step the random surfer will choose a new node to visit. If the random surfer is currently visiting node u then the next node is chosen as follows: (1) with probability α the surfer chooses an outlink from u, (u, v), uniformly at random and visits v. If the current node u happens to be a sink (and therefore has no outlinks) then the surfer picks any node $v \in V$ uniformly at random, (2) with probability $1 - \alpha$ the surfer visits any node $v \in V$ chosen uniformly at random– this is referred to as *zapping*. A typical value for the probability α is 0.85. The random surfer walk is therefore a random walk that usually follows a random outlink, but every few steps it essentially restarts the random walk from a random node in the graph.

Since the new node depends only on the current position in the graph, the sequence of visited pages is a Markov chain with state space V and transition probabilities that can be defined as follows. Let $P = \{p_{ij}\}$ denote a matrix derived from the adjacency matrix of the graph G, such that $p_{ij} = \frac{1}{\text{outdeg}(i)}$ if $(i, j) \in E$ and 0 otherwise (outdeg(i) denotes the outdegree of i, the number of out-going edges from node $i \in V$). If outdeg$(i) = 0$ then $p_{ij} = \frac{1}{n}$. The transition probability matrix of the Markov chain that describes the random surfer walk can therefore be written as $Q = \frac{1-\alpha}{n} \mathbb{1}_{n,n} + \alpha P$, where $\mathbb{1}_{n,n}$ is an $n \times n$ matrix with every entry equal to 1.

This Markov chain is aperiodic and irreducible and therefore has a unique stationary probability distribution π - the eigenvector associated with the dominant eigenvalue of Q. For any positive initial probability distribution x_0 over V, the iteration $x_0^T Q^l$ will converge to the stationary probability distribution π^T for large enough l. This is referred to as the *power method* [5].

The distribution $\pi = (\pi_1, \ldots, \pi_n)^T$ is defined as the PageRank vector of G. The PageRank value of a node $u \in V$ is the *expected* fraction of visits to u after i steps for large i regardless of the starting point. A node that is reachable from many other nodes in the graph via short directed paths will have a larger PageRank, for example.

3   The Link Building Problem

The k backlink (or Link Building) problem is defined as follows:

Definition 1. *The* LINK BUILDING *problem:*

- *Instance: A triple (G, x, k) where $G(V, E)$ is a directed graph, $x \in V$ and $k \in \mathbb{Z}^+$.*
- *Solution: A set $S \subseteq V \setminus \{x\}$ with $|S| = k$ maximizing π_x in $G(V, E \cup (S \times \{x\}))$.*

For fixed $k = 1$ this problem can be solved in polynomial time by simply calculating the new potential PageRanks of the target node after adding a link from each node. This requires $O(n)$ PageRank calculations. The argument is similar for any fixed k. As mentioned in Sect. 1.1, if k is part of the input then the problem becomes NP-hard.

3.1   Naive Selection of Backlinks

When choosing new incoming links in a graph, based on the definition of the PageRank algorithm, higher PageRank nodes appear to be more desirable. If we naively assume that the PageRank values will not change after inserting new links to the target node then the optimal new sources for links to the target would be the nodes with the highest PageRank values compared to outdegree plus one. This leads us to the following naive but intuitively clear algorithm:

Naive(G, x, k)
    Compute all PageRanks π_i, for all $(i \in V : (i, x) \notin E)$
    Return the k webpages with highest values of $\frac{\pi_i}{d_i+1}$, where d_i is the outdegree of page i

Fig. 1. The naive algorithm

The algorithm simply computes all initial PageRanks and chooses the k nodes with the highest value of $\frac{\pi_i}{d_i+1}$. It is well understood [7] that the naive algorithm is not always optimal. We will now show how to construct graphs with a surprisingly high approximation ratio – roughly 13.8 for $\alpha = 0.85$ – for the naive algorithm.

Lower Bound for the Approximation Ratio of the Naive Algorithm.
We define a family of input graphs ("cycle versus sink" graphs) that have the following structure: There is a cycle with k nodes, where each node has a number of incoming links from t_c other nodes (referred to as *tail nodes*). Tail nodes are used to boost the PageRanks of certain pages in the input graph and have an outdegree of 1. There are also k sink nodes (no outlinks) each one with a tail of t_s nodes pointing to them. The target node is x and it has outlinks towards all of the sinks. Figure 2 illustrates this family of graphs. Assume also that there is an isolated large clique with size t_i.

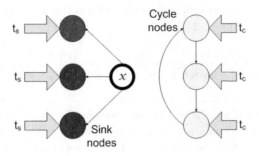

Fig. 2. A "cycle versus sink" graph for the naive algorithm

Due to symmetry all pages in the cycle will have the same PageRank π_c and the k sink pages will have the PageRank π_s. All tail nodes have no incoming links and will also have the same PageRank denoted by π_t. The PageRank of the target node is π_x and the PageRank of each node in the isolated clique is π_i.

The initial PageRanks for this kind of symmetric graph can be computed by writing a linear system of equations based on the identity $\pi^T = \pi^T Q$. The total number of nodes is $n = k \, (t_s + t_c + 2) + t_i + 1$.

$$\pi_t = \frac{1-\alpha}{n} + \frac{\alpha \, k \, \pi_s}{n}$$

$$\pi_x = \pi_t = \frac{1-\alpha}{n} + \frac{\alpha \, k \, \pi_s}{n}$$

$$\pi_s = \pi_t + \alpha \left(\frac{\pi_x}{k} + t_s \pi_t \right)$$

$$\pi_c = \pi_t + \alpha \left(\pi_c + t_c \pi_t \right)$$

$$\pi_i = \pi_t + \alpha \pi_i \ .$$

We need to add k new links towards the target node. We will pick the sizes of the tails t_c, t_s and therefore the PageRanks in the initial network so that the PageRank (divided by outdegree plus one) of the cycle nodes is slightly higher than the PageRank over degree for the sink nodes. Therefore the naive algorithm 1 will choose to add k links from the k cycle nodes. Once one link has been added, the rest of the cycle nodes are not desirable anymore, a fact that the naive algorithm fails to observe. The optimal solution is to add k links from the sink nodes.

In order to make sure cycle nodes are chosen by the naive algorithm, we need to ensure that $\frac{\pi_c}{outdeg(c)+1} > \frac{\pi_s}{outdeg(s)+1} \Leftrightarrow \frac{\pi_c}{2} > \pi_s \Leftrightarrow \pi_c/\pi_s = 2 + \delta$ for some $\delta > 0$. We then parameterize our tails:

$$t_c = u \tag{1}$$

$$t_i = u^2 \tag{2}$$

$$t_s = \frac{u}{2(1 - \lambda \alpha)} \ . \tag{3}$$

where u determines the size of the graph and λ is the solution of $\pi_c/\pi_s = 2 + \delta$, giving

$$\lambda = \frac{\left((\alpha^2 - \alpha)\,\delta + 2\alpha^2\right) ku + 2\left((\alpha - 1)\,\delta + 2\alpha - 1\right) k + 2\left(\alpha^2 - \alpha\right)\delta + 4(\alpha^2 - \alpha)}{2\,\alpha^2 ku + \left((2\,\alpha^2 - 2\,\alpha)\,\delta + 4\,\alpha^2 - 2\,\alpha\right) k + (2\,\alpha^3 - 2\,\alpha^2)\,\delta + 4\,\alpha^3 - 4\,\alpha^2}$$

We can solve for λ for any desired value of δ. Note also that we choose the tails of the clique nodes to be u^2 in order to make them asymptotically dominate all the other tails. The naive algorithm therefore will add k links from the cycle nodes which will result in the following linear system for the PageRanks:

$$\pi_t^g = \frac{1 - \alpha}{n} + \frac{\alpha k \pi_s^g}{n}$$
$$\pi_x^g = \pi_t^g + \alpha k \frac{\pi_c^g}{2}$$
$$\pi_s^g = \pi_t^g + \alpha \left(\frac{\pi_x^g}{k} + t_s \pi_t^g\right)$$
$$\pi_c^g = \pi_t^g + \alpha\left(\pi_c^g/2 + t_c \pi_t^g\right)$$
$$\pi_i^g = \pi_t^g + \alpha \pi_i^g .$$

The optimal is to choose k links from the sink nodes with a resulting PageRank vector described by the following system:

$$\pi_t^o = \frac{1 - \alpha}{n}$$
$$\pi_x^o = \pi_t^o + \alpha k \pi_s^o$$
$$\pi_s^o = \pi_t^o + \alpha \left(\frac{\pi_x^o}{k} + t_s \pi_t^o\right)$$
$$\pi_c^o = \pi_t^o + \alpha\left(\pi_c^o + t_c \pi_t^o\right)$$
$$\pi_i^o = \pi_t^o + \alpha \pi_i^o .$$

We solve these systems and calculate the approximation ratio of the naive algorithm:

$$\frac{\pi_x^o}{\pi_x^g} = \frac{\left(\alpha^3 - 2\alpha^2\right) k t_s + \left(\alpha^2 - 2\alpha\right) k + \alpha - 2}{\left(\alpha^4 - \alpha^2\right) k t_c + \left(\alpha^3 - \alpha\right) k - \alpha^3 + 2\alpha^2 + \alpha - 2} . \tag{4}$$

We now set our tails as described above in Equations 1-3 and let $u, k \to \infty$. So for large values of the tail sizes we get the following limit:

$$\lim_{u,k \to \infty} \frac{\pi_x^o}{\pi_x^g} = \frac{2 - \alpha}{\left(\alpha^3 - \alpha^2 - \alpha + 1\right)\delta + 2\alpha^3 - 2\alpha^2 - 2\alpha + 2} . \tag{5}$$

Now letting $\delta \to 0$ (as any positive value serves our purpose) we get the following theorem.

Theorem 1. *Consider the Link Building problem with target node x. Let $G = (V, E)$ be some directed graph. Let π_x^o denote the highest possible PageRank that the target node can achieve after adding k links, and π_x^g denote the PageRank* after adding the links returned by the naive algorithm from Fig. 1. Then for any $\epsilon > 0$ there exist infinitely many different graphs G where

$$\frac{\pi_x^o}{\pi_x^g} > \frac{2 - \alpha}{2\,(1 - \alpha)\,(1 - \alpha^2)} - \epsilon \ . \tag{6}$$

Note that ϵ can be written as function of u, δ, k and α. As $u, k \to \infty$, $\epsilon \to 0$ giving the asymptotic lower bound. For $\alpha = 0.85$ the lower bound is about 13.8.

3.2   Link Building Is in APX

In this section we present a greedy polynomial time algorithm for Link Building; computing a set of k new backlinks to target node x with a corresponding value of π_x^G within a constant factor from the optimal value. In other words we prove that Link Building is a member of the complexity class APX. We also introduce z_{ij} as the expected number of visits of node j starting at node i without zapping within the random surfer model. These values can be computed in polynomial time [1].

Proof of APX Membership.   Now consider the algorithm consisting of k steps where we at each step add a backlink to node x producing the maximum increase in $\frac{\pi_x}{z_{xx}}$ – the pseudo code of the algorithm is shown in Fig. 3. This algorithm runs in polynomial time, producing a solution to the Link Building problem within a constant factor from the optimal value as stated by the following theorem. So, Link Building is a member of the complexity class APX.

> **r-Greedy**(G, x, k)
> $S := \emptyset$
>   **repeat** k **times**
>   Let u be a node which maximizes the value of $\frac{\pi_x}{z_{xx}}$ in $G(V, E \cup \{(u, x)\})$
> $S := S \cup \{u\}$
> $E := E \cup \{(u, x)\}$
>   Report S as the solution

Fig. 3. Pseudo code for the greedy approach

Theorem 2.   *We let π_x^G and z_{xx}^G denote the values obtained by the r-Greedy algorithm in Fig. 3. Denoting the optimal value bye π_x^o, we have the following*

$$\pi_x^G \geq \pi_x^o \frac{z_{xx}^G}{z_{xx}^o}\left(1 - \frac{1}{e}\right) \geq \pi_x^o(1 - \alpha^2)\left(1 - \frac{1}{e}\right) \ .$$

where $e = 2.71828\ldots$ and z_{xx}^o is the value of z_{xx} corresponding to π_x^o.

Proof. Proposition 2.1 in [1] by Avrachenkov and Litvak states the following

$$\pi_x = \frac{1 - \alpha}{n} z_{xx}\left(1 + \sum_{i \neq x} r_{ix}\right) \ . \tag{7}$$

where r_{ix} is the probability that a random surfer starting at i reaches x before zapping. This means that the algorithm in Fig. 3 greedily adds backlinks to x in an attempt to maximize the probability of reaching node x before zapping, for a surfer dropped at a node chosen uniformly at random. We show in Lemma 1 below that r_{ix} in the graph obtained by adding links from $X \subseteq V$ to x is a *submodular* function of X – informally this means that adding the link (u, x) early in the process produces a higher increase of r_{ix} compared to adding the link later. We also show in Lemma 2 below that r_{ix} is not decreasing after adding (u, x), which is intuitively clear. We now conclude from (7) that $\frac{\pi_x}{z_{xx}}$ is a submodular and nondecreasing function since $\frac{\pi_x}{z_{xx}}$ is a sum of submodular and nondecreasing terms.

When we greedily maximize a nonnegative nondecreasing submodular function we will always obtain a solution within a fraction $1 - \frac{1}{e}$ from the optimal according to [6] by Nemhauser *et al.* We now have that:

$$\frac{\pi_x^G}{z_{xx}^G} \geq \frac{\pi_x^o}{z_{xx}^o}\left(1 - \frac{1}{e}\right) .$$

Finally, we use the fact that z_{xx}^G and z_{xx}^o are numbers between 1 and $\frac{1}{1-\alpha^2}$. □

For $\alpha = 0.85$ this gives an upper bound of $\frac{\pi_x^o}{\pi_x^G}$ of approximately 5.7 *It must be stressed that this upper bound is considerably smaller if z_{xx} is close to the optimal value prior to the modification – if z_{xx} cannot be improved then the upper bound is $\frac{e}{e-1} = 1.58$.* It may be the case that we obtain a bigger value of π_x by greedily maximizing π_x instead of $\frac{\pi_x}{z_{xx}}$, but π_x (the PageRank of the target node throughout the Link Building process) is *not* a submodular function of X so we cannot use the approach above to analyze this situation. To see that π_x is not submodular we just have to observe that adding a backlink from a sink node creating a short cycle late in the process will produce a higher increase in π_x compared to adding the link early in the process.

Proof of Submodularity and Monotonicity of r_{ix}. Let $f_i(X)$ denote the value of r_{ix} in $G(V, E \cup (X \times \{x\}))$ – the graph obtained after adding links from all nodes in X to x.

Lemma 1. f_i *is submodular for every* $i \in V$.

Proof. Let $f_i^r(X)$ denote the probability of reaching x from i without zapping, in r steps or less, in $G(V, E \cup (X \times \{x\}))$. We will show by induction in r that f_i^r is submodular. We will show the following for arbitrary $A \subset B$ and $y \notin B$:

$$f_i^r(B \cup \{y\}) - f_i^r(B) \leq f_i^r(A \cup \{y\}) - f_i^r(A) . \tag{8}$$

We start with the induction basis $r = 1$. It is not hard to show that the two sides of (8) are equal for $r = 1$. For the induction step; if you want to reach x in $r + 1$ steps or less you have to follow one of the links to your neighbors and reach x in r steps or less from the neighbor:

$$f_i^{r+1}(X) = \frac{\alpha}{outdeg(i)} \sum_{j:i \to j} f_j^r(X) \ . \tag{9}$$

where $j : i \to j$ denotes the nodes that i links to – this set includes x if $i \in X$. The outdegree of i is also dependent on X. If i is a sink in $G(V, E \cup (X \times \{x\}))$ then we can use (9) with $outdeg(i) = n$ and $j : i \to j = V$ – as explained in Sect. 2, the sinks can be thought of as linking to all nodes in the graph. Please also note that $f_x^r(X) = 1$.

We will now show that the following holds for every $i \in V$ assuming that (8) holds for every $i \in V$:

$$f_i^{r+1}(B \cup \{y\}) - f_i^{r+1}(B) \le f_i^{r+1}(A \cup \{y\}) - f_i^{r+1}(A) \ . \tag{10}$$

1. $i \in A$: The set $j : i \to j$ and $outdeg(i)$ are the same for all four terms in (10). We use (9) and the induction hypothesis to see that (10) holds.
2. $i \in B \setminus A$:
   (a) i is a sink in $G(V, E)$: The left hand side of (10) is 0 while the right hand side is positive or 0 according to Lemma 2 below.
   (b) i is not a sink in $G(V, E)$: In this case $j : i \to j$ includes x on the left hand side of (10) but not on the right hand side – the only difference between the two sets – and $outdeg(i)$ is one bigger on the left hand side. We now use (9), the induction hypothesis and $\forall X : f_x^r(X) = 1$.
3. $i = y$: We rearrange (10) such that the two terms including y are the only terms on the left hand side. We now use the same approach as for the case $i \in B \setminus A$.
4. $i \in V \setminus (B \cup \{y\})$: As the case $i \in A$.

Finally, we use $\lim_{r \to \infty} f_i^r(X) = f_i(X)$ to prove that (8) holds for f_i. □

Lemma 2. f_i is nondecreasing for every $i \in V$.

Proof. We shall prove the following by induction in r for $y \notin B$:

$$f_i^r(B \cup \{y\}) \ge f_i^r(B) \ . \tag{11}$$

We start with the induction basis $r = 1$.

1. $i = y$: The left hand side is $\frac{\alpha}{outdeg(y)}$ where $outdeg(y)$ is the new outdegree of y and the right hand side is at most $\frac{\alpha}{n}$ (if y is a sink in $G(V, E)$).
2. $i \ne y$: The two sides are the same.

For the induction step; assume that (11) holds for r and all $i \in V$. We will show that the following holds:

$$f_i^{r+1}(B \cup \{y\}) \ge f_i^{r+1}(B) \ . \tag{12}$$

1. $i = y$:
   (a) i is a sink in $G(V, E)$: The left hand side of (12) is α and the right hand side is smaller than α.

   (b) i is not a sink in $G(V, E)$: We use (9) in (12) and obtain simple averages on both sides with bigger numbers on the left hand side due to the induction hypothesis.
2. $i \neq y$: Again we can obtain averages where the numbers are bigger on the left hand side due to the induction hypothesis.

Again we use $\lim_{r \to \infty} f_i^r(X) = f_i(X)$ to conclude that (11) holds for f_i. $\qquad\square$

4 Discussion and Open Problems

We have presented a constant-factor approximation polynomial time algorithm for Link Building. We also presented a lower bound for the approximation ratio achieved by a perhaps more intuitive and simpler greedy algorithm.

The problem of developing a polynomial time approximation scheme (PTAS) for Link Building remains open.

References

1. Avrachenkov, K., Litvak, N.: The effect of new links on Google PageRank. Stochastic Models 22(2), 319–331 (2006)
2. Bianchini, M., Gori, M., Scarselli, F.: Inside pagerank. ACM Transactions on Internet Technology 5(1), 92–128 (2005)
3. Brin, S., Page, L.: The anatomy of a large-scale hypertextual Web search engine. Computer Networks and ISDN Systems 30(1-7), 107–117 (1998)
4. Dekerchove, C., Ninove, L., Vandooren, P.: Maximizing PageRank via outlinks. Linear Algebra and its Applications 429(5-6), 1254–1276 (2008)
5. Langville, A., Meyer, C.: Deeper inside pagerank. Internet Mathematics 1(3), 335–380 (2004)
6. Nemhauser, G., Wolsey, L., Fisher, M.: An analysis of approximations for maximizing submodular set functionsI. Mathematical Programming 14(1), 265–294 (1978)
7. Olsen, M.: Maximizing PageRank with New Backlinks. In: Calamoneri, T., Diaz, J. (eds.) Algorithms and Complexity. LNCS, vol. 6078, pp. 37–48. Springer, Heidelberg (2010)
8. Page, L., Brin, S., Motwani, R., Winograd, T.: The PageRank citation ranking: Bringing order to the web. Technical Report 1999-66, Stanford InfoLab (November 1999), http://ilpubs.stanford.edu:8090/422/
9. SEOmoz: Search engine 2009 ranking factors (2009), http://www.seomoz.org/

XML Reconstruction View Selection in XML Databases: Complexity Analysis and Approximation Scheme*

Artem Chebotko and Bin Fu

Department of Computer Science, University of Texas-Pan American,
Edinburg, TX 78539, USA
{artem,binfu}@cs.panam.edu

Abstract. Query evaluation in an XML database requires reconstructing XML subtrees rooted at nodes found by an XML query. Since XML subtree reconstruction can be expensive, one approach to improve query response time is to use reconstruction views - materialized XML subtrees of an XML document, whose nodes are frequently accessed by XML queries. For this approach to be efficient, the principal requirement is a framework for view selection. In this work, we are the first to formalize and study the problem of XML reconstruction view selection. The input is a tree T, in which every node i has a size c_i and profit p_i, and the size limitation C. The target is to find a subset of subtrees rooted at nodes $i_1, \cdots, i_k$ respectively such that $c_{i_1} + \cdots + c_{i_k} \leq C$, and $p_{i_1} + \cdots + p_{i_k}$ is maximal. Furthermore, there is no overlap between any two subtrees selected in the solution. We prove that this problem is NP-hard and present a fully polynomial-time approximation scheme (FPTAS) as a solution.

1 Introduction

With XML[1] [1] being the de facto standard for business and Web data representation and exchange, storage and querying of large XML data collections is recognized as an important and challenging research problem. A number of XML databases [13, 27, 6, 21, 8, 4, 7, 2, 26, 18, 17] have been developed to serve as a solution to this problem. While XML databases can employ various storage models, such as relational model or native XML tree model, they support standard XML query languages, called XPath[2] and XQuery[3]. In general, an XML query specifies which nodes in an XML tree need to be retrieved. Once an XML tree is stored into an XML database, a query over this tree usually requires two steps: (1) finding the specified nodes, if any, in the XML tree and (2) reconstructing and returning XML subtrees rooted at found nodes as a query result. The second step is called XML subtree reconstruction [9,10] and may have a significant impact on query response time. One approach to minimize XML subtree reconstruction time is to cache XML subtrees rooted at frequently accessed nodes as illustrated in the following example.

---

* This research is supported in part by the National Science Foundation Early Career Award 0845376.
[1] http://www.w3.org/XML
[2] http://www.w3.org/TR/xpath
[3] http://www.w3.org/XML/Query

W. Wu and O. Daescu (Eds.): COCOA 2010, Part II, LNCS 6509, pp. 97–106, 2010.

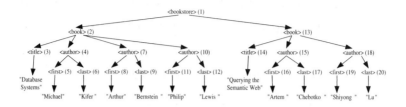

Fig. 1. XML tree

$$\mathbf{r}_{edge}$$

ID	parentID	name	content
1	NULL	bookstore	NULL
2	1	book	NULL
3	2	title	Database Systems
4	2	author	NULL
5	4	first	Michael
6	4	last	Kifer
7	2	author	NULL
8	7	first	Arthur
9	7	last	Bernstein
10	2	author	NULL
11	10	first	Philip
12	10	last	Lewis
13	1	book	NULL
14	13	title	Querying the Semantic Web
...	...	...	...

Fig. 2. Edge table

Consider an XML tree in Figure 1 that describes a sample bookstore inventory. The tree nodes correspond to XML elements, e.g., *bookstore* and *book*, and data values, e.g., *"Arthur"* and *"Bernstein"*, and the edges represent parent-child relationships among nodes, e.g., all the *book* elements are children of *bookstore*. In addition, each element node is assigned a unique identifier that is shown next to the node in the figure. As an example, in Figure 2, we show how this XML tree can be stored into a single table in an RDBMS using the edge approach [13]. The edge table r_{edge} stores each XML element as a separate tuple that includes the element ID, ID of its parent, element name, and element data content. A sample query over this XML tree that retrieves books with title "Database Systems" can be expressed in XPath as:

```
/bookstore/book[title="Database Systems"]
```

This query can be translated into relational algebra or SQL over the edge table to retrieve IDs of the *book* elements that satisfy the condition:

$\pi_{r_2.ID}$ (

$\qquad r_1 \bowtie_{r_1.ID=r_2.parentID \wedge r_1.name='bookstore' \wedge}$

$\qquad\qquad r_1.parentID \ is \ NULL \wedge r_2.name='book'$

$\qquad r_2 \bowtie_{r_2.ID=r_3.parentID \wedge r_3.name='title' \wedge}$

$\qquad\qquad r_3.content='DatabaseSystems'$

$\qquad r_3$)

where r_1, r_2, and r_3 are aliases of table r_{edge}. For the edge table in Figure 2, the relational algebra query returns ID "2", that uniquely identifies the first *book* element in the tree. However, to retrieve useful information about the book, the query evaluator must further retrieve all the descendants of the *book* node and reconstruct their parent-child relationships into an XML subtree rooted at this node; this requires additional self-joins of the edge table and a reconstruction algorithm, such as the one proposed in [9]. Instead, to avoid expensive XML subtree reconstruction, the subtree can be explicitly stored in the database as an XML reconstruction view (see Figure 3). This materialized view can be used for the above XPath query or any other query that needs to reconstruct and return the *book* node (with ID "2") or its descendant.

```
<book>
<title> Database Systems
</title>
<author>
    <first> Michael </first>
    <last> Kifer </last>
</author>
<author>
    <first> Arthur </first>
    <last> Bernstein </last>
</author>
<author>
    <first> Philip </first>
    <last> Lewis </last>
</author>
</book>
```

Fig. 3. XML reconstruction view

In this work, we study the problem of selecting XML reconstruction views to materialize: given a set of XML elements D from an XML database, their access frequencies a_i (aka workload), a set of ancestor-descendant relationships AD among these elements, and a storage capacity δ, find a set of elements M from D, whose XML subtrees should be materialized as reconstruction views, such that their combined size is no larger than δ. To our best knowledge, our solution to this problem is the first one proposed in the literature. Our main contributions and the paper organization are as follows. In Section 3, we formally define the XML reconstruction view selection problem. In Sections 4 and 5, we prove that the problem is NP-hard and describe a fully polynomial-time approximation scheme (FPTAS) for the problem.

2   Related Work

We studied the XML subtree reconstruction problem in the context of a relational storage of XML documents in [9, 10], where several algorithms have been proposed. Given an XML element returned by an XML query, our algorithms retrieve all its descendants from a database and reconstruct their relationships into an XML subtree that is returned as the query result. To our best knowledge, there have been no previous work on materializing reconstruction views or XML reconstruction view selection.

Materialized views [3, 23, 30, 22, 12, 28] have been successfully used for query optimization in XML databases. These research works rewrite an XML query, such that it can be answered either using only available materialized views, if possible, or accessing both the database and materialized views. View maintenance in XML databases has been studied in [25, 24]. There have been only one recent work [29] on materialized view selection in the context of XML databases. In [29], the problem is defined as: find views over XML data, given XML databases, storage space, and a set of queries, such that the combined view size does not exceed the storage space. The proposed solution produces minimal XML views as candidates for the given query workload, organizes them into a graph, and uses two view selection strategies to choose views to materialize. This approach makes an assumption that views are used to answer XML queries completely (not partially) without accessing an underlying XML database. The XML reconstruction view problem studied in our work focuses on a different aspect of XML query processing: it finds views to materialize based on how frequently an XML element needs to be reconstructed. However, XML reconstruction views can be complimentarily used for query answering, if desired.

Finally, the materialized view selection problem have been extensively studied in data warehouses [5, 31, 15, 16, 19, 11] and distributed databases [20]. These research results are hardly applicable to XML tree structures and in particular to subtree reconstruction, which is not required for data warehouses or relational databases.

3   XML Reconstruction View Selection Problem

In this section, we formally define the XML reconstruction view selection problem addressed in our work.

Problem formulation. Given n XML elements, $D = \{D_1, D_2, \cdots, D_n\}$, and an ancestor-descendant relationship AD over D such that if $(D_j, D_i) \in AD$, then D_j is an ancestor of D_i, let $COST_R(D_i)$ be the access cost of accessing unmaterialized D_i, and let $COST_A(D_i)$ be the access cost of accessing materialized D_i. We have $COST_A(D_i) < COST_R(D_i)$ since reconstruction of D_i takes time. We use $size(D_i)$ to denote the memory capacity required to store a materialized XML element, $size(D_i) > 0$ and $size(D_i) < size(D_j)$ for any $(D_j, D_i) \in AD$. Given a workload that is characterized by $a_i (i = 1, 2, \ldots, n)$ representing the access frequency of D_i. The *XML reconstruction view selection problem* is to select a set of elements M from D to be materialized to minimize the total access cost $\tau(D, M) = \sum_{i=1}^{n} a_i \times COST(D_i)$, under the disk capacity constraint $\sum_{D_i \in M} size(D_i) \leq \delta$, where $COST(D_i) = COST_A(D_i)$ if $D_i \in M$ or for some ancestor D_j of D_i, $D_j \in M$, otherwise $COST(D_i) = COST_R(D_i)$. δ denotes the available memory capacity, $\delta \geq 0$.

Next, let $\triangledown COST(D_i) = COST_R(D_i) - COST_A(D_i)$ means the cost saving by materialization, then one can show that function τ is minimized if and only if the following function λ is maximized $\lambda(D, M) = \sum_{D_i \in M^+} a_i \times \triangledown COST(D_i)$ under the disk capacity constraint $\sum_{D_i \in M} size(D_i) \leq \delta$, where M^+ represents all the materialized XML elements and their descendant elements in D, it is defined as $M^+ = \{D_i \mid D_i \in M \text{ or } \exists D_j.(D_j, D_i) \in AD \land D_j \in M\}$.

4 NP-Completeness

In this section, we prove that the XML reconstruction view selection problem is NP-hard. First, the maximization problem is changed into the equivalent decision problem.

Equivalent decision problem. Given D, AD, $size(D_i)$, a_i, $\triangledown COST(D_i)$ and δ as defined in Section 3, let K denotes the cost saving goal, $K \geq 0$. Is there a subset $M \subseteq D$ such that

$$\sum_{D_i \in M^+} a_i \times \triangledown COST(D_i) \geq K$$

and

$$\sum_{D_i \in M} size(D_i) \leq \delta$$

M^+ represents all the materialized XML elements and their descendant elements in D, it is defined as $M^+ = \{D_i \mid D_i \in M \text{ or } \exists D_j.(D_j, D_i) \in AD \land D_j \in M\}$.

In order to study this problem in a convenient model, we have the following simplified version.

The input is a tree T, in which every node i has a size c_i and profit p_i, and the size limitation C. The target is to find a subset of subtrees rooted at nodes $i_1, \cdots, i_k$ such that $c_{i_1} + \cdots + c_{i_k} \leq C$, and $p_{i_1} + \cdots + p_{i_k}$ is maximal. Furthermore, there is no overlap between any two subtrees selected in the solution.

We prove that the decision problem of the XML reconstruction view selection is an NP-hard. A polynomial time reduction from *KNAPSACK* [14] to it is constructed.

Theorem 1. *The decision problem of the XML reconstruction view selection is NP-complete.*

Proof. It is straightforward to verify that the problem is in NP. Restrict the problem to the well-known NP-complete problem *KNAPSACK* [14] by allowing only problem instances in which:

Assume that a Knapsack problem has input $(p_1, c_1), \cdot, (p_n, c_n)$, and parameters K and C. We need to determine a subset $S \subseteq \{1, \cdots, n\}$ such that $\sum_{i \in S} c_i \leq C$ and $\sum_{i \in S} p_i \geq K$.

Build a binary tree T with exactly leaves. Let leaf i have profit p_i and size c_i. Furthermore, each internal node, which is not leaf, has size ∞ and profit ∞.

Clearly, any solution cannot contain any internal due to the size limitation. We can only select a subset of leaves. This is equivalent to the Knapsack problem.   □

5   Fully Polynomial-Time Approximation Scheme

We assume that each parameter is an integer. The input is n XML elements, $D = \{D_1, D_2, \cdots, D_n\}$ which will be represented by an AD tree J, where each edge in J shows a relationship between a pair of parent and child nodes.

We have a divide and conquer approach to develop a fully approximation scheme. Given an AD tree J with root r, it has subtrees $J_1, \cdots, J_k$ derived from the children $r_1, \cdots, r_k$ of r. We find a set of approximate solutions among $J_1, \cdots, J_{k/2}$ and another set of approximate solutions among $J_{k/2+1}, \cdots, J_k$.

We merge the two sets of approximate solutions to obtain the solution for the union of subtrees $J_1, \cdots, J_k$. Add one more solution that is derived by selecting the root r of J. Group those solutions into parts such that each part contains all solutions P with similar $\lambda(D, P)$. Prune those solution by selecting the one from each part with the least size. This can reduce the number of solution to be bounded by a polynomial.

We will use a list P to represent the selection of elements from D.

For a list of elements P, define $\lambda(P) = \sum_{D_i \in P} a_i \times \nabla COST(D_i)$, and $\mu(P) = \sum_{D_i \in P} size(D_i)$. Define $\chi(J)$ be the largest product of the node degrees along a path from root to a leaf in the AD tree J.

Assume that ϵ is a small constant with $1 > \epsilon > 0$. We need an $(1+\epsilon)$ approximation. We maintain a list of solutions $P_1, P_2, \cdots$, where P_i is a list of elements in D.

Let $f = (1 + \frac{\epsilon}{z})$ with $z = 2\log\chi(J)$. Let $w = \sum_{i=1}^n a_i COST_R(D_i)$ and $s = \sum_{i=1}^n size(D_i)$.

Partition the interval $[0, w]$ into $I_1, I_2, \cdots, I_{t_1}$ such that $I_1 = [0,1]$ and $I_k = (b_{k-1}, b_k]$ with $b_k = f \cdot b_{k-1}$ for $k < t$, and $I_{t_1} = (b_{t-1}, w]$, where $b_{t-1} < w \le f b_{t-1}$.

Two lists P_i and P_j, are in the same region if there exist I_k such that both $\lambda(P_i)$ and $\lambda(P_j)$ are I_k.

For two lists of partial solutions $P_i = D_{i_1} \cdots D_{i_{m_1}}$ and $P_j = D_{j_1} \cdots D_{j_{m_2}}$, their link $P_i \circ P_j = D_{i_1} \cdots D_{i_{m_1}} D_{j_1} \cdots D_{j_{m_2}}$.

Prune (L)
    Input: L is a list of partial solutions $P_1, P_2, \cdots, P_m$;
    Partition L into parts $U_1, \cdots, U_v$ such that two lists P_i and P_j are in the same part if P_i and P_j are in the same region.
    For each U_i, select P_j such that $\mu(P_j)$ is the least among all P_j in U_i;
End of Prune

Merge (L_1, L_2)
    Input: L_1 and L_2 are two lists of solutions.
    Let $L = \emptyset$;
    For each $P_i \in L_1$ and each $P_j \in L_2$
        append their link $P_i \circ P_j$ to L;
    Return L;
End of Merge

Union $(L_1, L_2, \cdots, L_k)$

    Input: $L_1, \cdots, L_k$ are lists of solutions.

    If $k = 1$ then return L_1;

    Return Prune(Merge(Union($L_1, \cdots, L_{k/2}$),

                         Union($L_{k/2+1}, \cdots, L_k$)));

End of Union

Sketch (J)

Input: J is a set of n elements according to their AD.

    If J only contains one element D_i, return the list $L = D_i, \emptyset$ with two solutions.

    Partition the list J into subtrees $J_1, \cdots, J_k$ according to its k children.

    Let L_0 be the list that only contains solution J.

    for $i = 1$ to k let L_i=Sketch(J_i);

    Return Union($L_0, L_1, \cdots, L_k$);

End of Sketch

For a list of elements P and an AD tree J, $P[J]$ is the list of elements in both P and J. If $J_1, \cdots, J_k$ are disjoint subtrees of an AD tree, $P[J_1, \cdots, J_k]$ is $P[J_1] \circ \cdots \circ P[J_k]$.

A tree J is *normalized* if each node has degree 2^k for some integer $k \geq 0$. In order to make it convenient, we make the tree J normalized by adding some useless nodes D_i with $COST_R(D) = COST_A(D) = 0$. The size of tree is at most doubled after normalization. In the rest of the section, we always assume J is normalized.

Lemma 1. *Assume that L_i is a list of solutions for the problem with AD tree J_i for $i = 1, \cdots, k$. Let $P_i \in L_i$ for $i = 1, \cdots, k$. Then there exists $P \in L =$Union$(L_1, \cdots, L_k)$ such that $\lambda(P) \leq f^{\log k} \cdot \lambda(P_1 \circ \cdots \circ P_k)$ and $\mu(P) \leq \mu(P_1 \circ \cdots \circ P_k))$.*

Proof. We prove by induction. It is trivial when $k = 1$. Assume that the lemma is true for cases less than k.

Let $M_1 =$ Union$(L_1, \cdots, L_{k/2})$ and $M_2 =$ Union$(L_{k/2+1}, \cdots, L_k)$.

Assume that M_1 contains Q_1 such that $\lambda(Q_1) \leq f^{\log(k/2)} \lambda(P_1 \circ \cdots \circ P_{k/2})$ and $\mu(Q_1) \leq \mu(P_1 \circ \cdots \circ P_{k/2})$.

Assume that M_2 contains Q_2 such that $\lambda(Q_2) \leq f^{\log(k/2)} \lambda(P_{k/2+1} \circ \cdots \circ P_k])$ and $\mu(Q_2) \leq \mu(P_{k/2+1} \circ \cdots \circ P_k)$.

Let $Q = Q_1 \circ Q_2$. Let Q^* be the solution in the same region with Q and has the least $\mu(Q^*)$. Therefore, $\lambda(Q^*) \leq f\lambda(Q) \leq f^{\log(k/2)} f\lambda(P_1 \circ \cdots \circ P_k) \leq f^{\log k} \lambda(P_1 \circ \cdots \circ P_k)$.

Since $\mu(Q_1) \leq \mu(P_1 \circ \cdots \circ P_{k/2})$ and $\mu(Q_2) \leq \mu(P_{k/2+1} \circ \cdots \circ P_k)$, we also have $\mu(Q^*) \leq \mu(Q) \leq \mu(Q_1) + \mu(Q_2) \leq \mu(P_1 \circ \cdots \circ P_{k/2}) + \mu(P_{k/2+1} \circ \cdots \circ P_k) = \mu(P_1 \circ \cdots \circ P_k) = \mu(P)$. □

Lemma 2. *Assume that P is an arbitrary solution for the problem with AD tree J. For L=Sketch(J), there exists a solution P' in the list L such that $\lambda(P') \leq f^{\log \chi(J)} \cdot \lambda(P)$ and $\mu(P') \leq \mu(P)$.*

Proof. We prove by induction. The basis at $|J| \leq 1$ is trivial. We assume that the claim is true for all $|J| < m$. Now assume that $|J| = m$ and J has k children which induce k subtrees $J_1, \cdots, J_k$.

Let $L_i = P[J_i]$ for $i = 1, \cdots, k$. By our hypothesis, for each i with $1 \leq i \leq k$, there exists $Q_i \in L_i$ such that $\lambda(Q_i) \leq f^{\log \chi(J_i)} \cdot \lambda(P[J_i])$ and $\mu(Q_i) \leq \mu(P[J_i]))$.

Let $M=\text{Union}(L_1, \cdots, L_k)$. By Lemma 1, there exists $P' \in M$ such that $\lambda(P') \leq f^{\log(k)}\lambda(Q_1 \circ \cdots \circ Q_k) \leq f^{\max\{\log \chi(J_1), \cdots, \log \chi(J_k)\}} f^{\log(k)}\lambda(P[J_1, \cdots, J_k]) \leq f^{\log \chi(J)}\lambda(P[J_1, \cdots, J_k]) = f^{\log \chi(J)}\lambda(P)$, and $\mu(P') \leq \mu(Q_1 \circ \cdots \circ Q_k) \leq \mu(P[J_1, \cdots, J_k]) = \mu(P)$. $\square$

Lemma 3. *Assume that $\mu(D, J) \leq a(n)$. Then the computational time for Sketch(J) is $O(|J|(\frac{(\log \chi(J))(\log a(n))}{\epsilon})^2)$, where $|J|$ is the number of nodes in J.*

Proof. The number of intervals is $O(\frac{(\log \chi(J))(\log a(n))}{\epsilon})$. Therefore the list of each $L_i=\text{Prune}(J_i)$ is of length $O(\frac{(\log \chi(J))(\log a(n))}{\epsilon})$.

Let $F(k)$ be the time for Union($L_1, \cdots, L_k$). It satisfies the recursion $F(k) = 2F(k/2) + O((\frac{(\log \chi(J))(\log a(n))}{\epsilon})^2)$. This brings solution $F(k) = O(k(\frac{(\log \chi(J))(\log a(n))}{\epsilon})^2)$.

Let $T(J)$ be the computational time for Prune(J). Denote $E(J)$ to be the number of edges in J. We prove by induction that $T(J) \leq cE(J)(\frac{(\log \chi(J))(\log a(n))}{\epsilon})^2$ for some constant $c > 0$. We select constant c enough so that merging two lists takes $c(n \log a(n))^2)$ steps. We have that $T(J) \leq T(J_1) + \cdots + T(J_k) + F(k) \leq cE(J_1)(\frac{(\log \chi(J))(\log a(n))}{\epsilon})^2 + \cdots + cE(J_k)(\frac{(\log \chi(J))(\log a(n))}{\epsilon})^2 + ck(\frac{(\log \chi(J))(\log a(n))}{\epsilon})^2 \leq cE(J)(\frac{(\log \chi(J))(\log a(n))}{\epsilon})^2 \leq c|J|(\frac{(\log \chi(J))(\log a(n))}{\epsilon})^2$. $\square$

Algorithm
Approximate(J, ϵ)
Input: J is an AD tree with elements $D_1, \cdots, D_n$ and ϵ is a small constant with $1 > \epsilon > 0$;
    Let $L =\text{Sketch}(J)$;
    Select P_i from the list L that P_i has the optimal cost;
End of the Algorithm

Theorem 2. *For any instance of J of an AD tree with n elements, there exists an $O(n(\frac{(\log \chi(J))(\log a(n))}{\epsilon})^2)$ time approximation scheme, where $\sum_{i=1}^{n} a_i COST_R(D_i) \leq a(n)$.*

Proof. Assume that P is the optimal solution for input J. Let $L=\text{Prune}(J)$. By Lemma 2, we have $P^* \in L$ that satisfies the condition of Lemma 2. We have $\lambda(P^*) \leq f^{\log \chi(J)}\lambda(P) = (1+\frac{\epsilon}{2z})^{\log \chi(J)}\lambda(P) \leq e^{\frac{\epsilon}{2}} \cdot \lambda(P) = (1+\frac{\epsilon}{2}+(\frac{\epsilon}{2})^2) \cdot \lambda(P) < (1 + \epsilon) \cdot \lambda(P)$. Furthermore, $\mu(P^*) \leq \mu(P)$. The computational time follows from Lemma 3. $\square$

It is easy to see that $\chi(J) \leq 2^{|J|}$. We have the following corollary.

Corollary 1. *For any instance of J of an AD tree with n elements, there exists an $O(n^3(\frac{\log a(n)}{\epsilon})^2)$ time approximation scheme, where $\sum_{i=1}^{n} a_i COST_R(D_i) \leq a(n)$.*

We show an approximation scheme for the problem with an input of multiple trees. The input is a series of trees $J_1, \cdots, J_k$.

Theorem 3. *For any instance of J of an AD tree with n elements, there exists an* $O(n(\frac{(\log \chi(J_0))(\log a_0(n))}{\epsilon})^2)$ *time approximation scheme, where* $\sum_j (p_j + c_j) \leq a_0(n)$ *and J_0 is a tree via connecting all $J_1, \cdots, J_k$ into a single tree under a common root* r_0.

Proof. Build a new tree with a new node r_0 such that $J_1, \cdots, J_k$ are the subtrees under r_0. Apply the algorithm in in Theorem 2.   □

References

1. Abiteboul, S., Buneman, P., Suciu, D.: Data on the Web: From Relations to Semistructured Data and XML. Morgan Kaufmann, San Francisco (1999)
2. Atay, M., Chebotko, A., Liu, D., Lu, S., Fotouhi, F.: Efficient schema-based XML-to-relational data mapping. Inf. Syst. 32(3), 458–476 (2007)
3. Balmin, A., Özcan, F., Beyer, K.S., Cochrane, R., Pirahesh, H.: A framework for using materialized XPath views in XML query processing. In: VLDB, pp. 60–71 (2004)
4. Balmin, A., Papakonstantinou, Y.: Storing and querying XML data using denormalized relational databases. VLDB J. 14(1), 30–49 (2005)
5. Baralis, E., Paraboschi, S., Teniente, E.: Materialized views selection in a multidimensional database. In: VLDB, pp. 156–165 (1997)
6. Bohannon, P., Freire, J., Roy, P., Siméon, J.: From XML schema to relations: A cost-based approach to XML storage. In: ICDE, pp. 64–75 (2002)
7. Boncz, P.A., Grust, T., van Keulen, M., Manegold, S., Rittinger, J., Teubner, J.: Mon-etDB/XQuery: a fast XQuery processor powered by a relational engine. In: SIGMOD Conference, pp. 479–490 (2006)
8. Chaudhuri, S., Chen, Z., Shim, K., Wu, Y.: Storing XML (with XSD) in SQL databases: Interplay of logical and physical designs. IEEE Trans. Knowl. Data Eng. 17(12), 1595–1609 (2005)
9. Chebotko, A., Atay, M., Lu, S., Fotouhi, F.: XML subtree reconstruction from relational storage of XML documents. Data Knowl. Eng. 62(2), 199–218 (2007)
10. Chebotko, A., Liu, D., Atay, M., Lu, S., Fotouhi, F.: Reconstructing XML subtrees from relational storage of XML documents. In: ICDE Workshops, p. 1282 (2005)
11. Chirkova, R., Halevy, A.Y., Suciu, D.: A formal perspective on the view selection problem. VLDB J. 11(3), 216–237 (2002)
12. Fan, W., Geerts, F., Jia, X., Kementsietsidis, A.: Rewriting regular XPath queries on XML views. In: ICDE, pp. 666–675 (2007)
13. Florescu, D., Kossmann, D.: Storing and querying XML data using an RDMBS. IEEE Data Eng. Bull. 22(3), 27–34 (1999)
14. Garey, M.R., Johnson, D.S.: Computer and Intractability: A Guide to the Theory of NP-Completeness. W.H. Freeman, New York (1979)
15. Gupta, H.: Selection of views to materialize in a data warehouse. In: Afrati, F.N., Kolaitis, P.G. (eds.) ICDT 1997. LNCS, vol. 1186, pp. 98–112. Springer, Heidelberg (1997)
16. Gupta, H., Mumick, I.S.: Selection of views to materialize under a maintenance cost constraint. In: Beeri, C., Bruneman, P. (eds.) ICDT 1999. LNCS, vol. 1540, pp. 453–470. Springer, Heidelberg (1999)

17. Hündling, J., Sievers, J., Weske, M.: NaXDB - realizing pipelined XQuery processing in a native XML database system. In: XIME-P (2005)

18. Jagadish, H.V., Al-Khalifa, S., Chapman, A., Lakshmanan, L.V.S., Nierman, A., Paparizos, S., Patel, J.M., Srivastava, D., Wiwatwattana, N., Wu, Y., Yu, C.: TIMBER: A native XML database. VLDB J. 11(4), 274–291 (2002)

19. Karloff, H.J., Mihail, M.: On the complexity of the view-selection problem. In: PODS, pp. 167–173 (1999)

20. Kossmann, D.: The state of the art in distributed query processing. ACM Comput. Surv. 32(4), 422–469 (2000)

21. Krishnamurthy, R., Chakaravarthy, V.T., Kaushik, R., Naughton, J.F.: Recursive XML schemas, recursive XML queries, and relational storage: XML-to-SQL query translation. In: ICDE, pp. 42–53 (2004)

22. Lakshmanan, L.V.S., Wang, H., Zhao, Z.J.: Answering tree pattern queries using views. In: VLDB, pp. 571–582 (2006)

23. Mandhani, B., Suciu, D.: Query caching and view selection for XML databases. In: VLDB, pp. 469–480 (2005)

24. Sawires, A., Tatemura, J., Po, O., Agrawal, D., Abbadi, A.E., Candan, K.S.: Maintaining XPath views in loosely coupled systems. In: VLDB, pp. 583–594 (2006)

25. Sawires, A., Tatemura, J., Po, O., Agrawal, D., Candan, K.S.: Incremental maintenance of path expression views. In: SIGMOD Conference, pp. 443–454 (2005)

26. Schöning, H.: Tamino - a DBMS designed for XML. In: ICDE, pp. 149–154 (2001)

27. Shanmugasundaram, J., Tufte, K., Zhang, C., He, G., DeWitt, D.J., Naughton, J.F.: Relational databases for querying XML documents: Limitations and opportunities. In: VLDB, pp. 302–314 (1999)

28. Tang, N., Yu, J.X., Özsu, M.T., Choi, B., Wong, K.-F.: Multiple materialized view selection for XPath query rewriting. In: ICDE, pp. 873–882 (2008)

29. Tang, N., Yu, J.X., Tang, H., Özsu, M.T., Boncz, P.A.: Materialized view selection in XML databases. In: DASFAA, pp. 616–630 (2009)

30. Xu, W., Özsoyoglu, Z.M.: Rewriting XPath queries using materialized views. In: VLDB, pp. 121–132 (2005)

31. Yang, J., Karlapalem, K., Li, Q.: Algorithms for materialized view design in data warehousing environment. In: VLDB, pp. 136–145 (1997)

Computational Study for Planar Connected Dominating Set Problem

Marjan Marzban[1], Qian-Ping Gu[1], and Xiaohua Jia[2]

[1] School of Computing Science, Simon Fraser University, Burnaby BC Canada
{mmarzba,qgu}@cs.sfu.ca
[2] Department of Computer Science, City University of Hong Kong
csjia@cityu.edu.hk

Abstract. The connected dominating set (CDS) problem is a well studied NP-hard problem with many important applications. Dorn et al. [ESA2005, LNCS3669,pp95-106] introduce a new technique to generate $2^{O(\sqrt{n})}$ time and fixed-parameter algorithms for a number of non-local hard problems, including the CDS problem in planar graphs. The practical performance of this algorithm is yet to be evaluated. We perform a computational study for such an evaluation. The results show that the size of instances can be solved by the algorithm mainly depends on the branchwidth of the instances, coinciding with the theoretical result. For graphs with small or moderate branchwidth, the CDS problem instances with size up to a few thousands edges can be solved in a practical time and memory space. This suggests that the branch-decomposition based algorithms can be practical for the planar CDS problem.

Keywords: Branch-decomposition based algorithms, CDS problem, planar graphs, fixed-parameter algorithms, computational study.

1 Introduction

In this paper, graphs are undirected, simple and finite unless otherwise stated. Let G be a graph with vertex set $V(G)$ and edge set $E(G)$. A dominating set D of G is a subset of $V(G)$ such that for every vertex $u \in V(G)$, $u \in D$ or u is incident to a vertex $v \in D$. The *dominating number* of G, denoted by $\gamma(G)$, is the minimum size of a dominating set of G. The dominating set problem is to decide if $\gamma(G) \leq k$ for a given G and integer k. The dominating set problem is a core NP-complete problem in combinatorial optimization [15]. The rich literature of algorithms and complexity of dominating set problem can be found in [20,19].

A subset D of $V(G)$ is a connected dominating set (CDS) of G if D is a dominating set of G and the subgraph $G[D]$ induced by D is connected. The *connected dominating number* of G, denoted by $\gamma_c(G)$, is the minimum size of a CDS of G. The CDS problem is to decide if $\gamma_c(G) \leq k$ for a given G and integer k. The optimization version of the CDS problem is to find a minimum CDS of an input graph. The CDS problem is an important variant of the dominating set problem and has wide practical applications in wireless ad hoc or sensor

W. Wu and O. Daescu (Eds.): COCOA 2010, Part II, LNCS 6509, pp. 107–116, 2010.

networks such as virtual backbone construction [6], energy efficient routing and broadcasting [5]. Notice that $\gamma(G) \leq \gamma_c(G) \leq 3\gamma(G) - 2$.

The CDS problem is NP-complete [15]. Approximation algorithms and exact fixed parameter algorithms have been main topics in the algorithmic research for the CDS problem. For arbitrary graphs, there are $2(1 + \ln \Delta)$- and $(\ln \Delta + 3)$-approximation algorithms for the CDS problem, where Δ is the maximum vertex degree of the input graph [18]; the CDS problem is not approximable within a factor of $(1 - \epsilon) \ln \Delta$ for any $\epsilon > 0$ unless $NP \subseteq DTIME(n^{\log \log n})$ [18]; and the CDS problem is fixed-parameter intractable unless the parameterized complexity classes collapse [11,12]. The CDS problem remains NP-complete if the input graphs are restricted to planar [15]. However, the planar CDS problem admits a PTAS [7], and is fixed parameter tractable [9,10].

Recently, significant progresses have been made on the fixed-parameter algorithms for the planar dominating set problem [13,8] and practical performance of those algorithms have been reported [21]. The notions of tree/branch-decompositions introduced by Robertson and Seymour [23,24,25] play a central role in those algorithms. Although the dominating set problem and the CDS problem are closely related, they have different properties from the tree/branch-decomposition based algorithm point of view. In particular, the techniques used to solve the dominating set problem do not seem to work for the CDS problem. One of the main reasons of such discrepancy is that *connectivity* is a *non-local property* (see Section 3 for more details).

Along the lines to clear the hurdles caused by the non-local property, Dorn et al. [9,10] propose a new technique to design sub-exponential time exact algorithms for the non-local problems in planar graphs. This new technique is based on the geometric properties of branch-decomposition of graphs with a planar embedding in a sphere and the properties of non-crossing partitions in the embedding. Based on this new technique, they show that many non-local problems in planar graphs can be solved in $2^{O(\sqrt{n})}$ time [9,10]. Especially, they give an algorithm (called DPBF Algorithm in what follows) which solves the planar CDS problem in $O(2^{O(\mathrm{bw}(G))}n + n^3)$ and $O(2^{9.822\sqrt{n}}n + n^3)$ time [10]. The constant in $O(\mathrm{bw}(G))$ is not explicitly given in [9,10]. By a more careful analysis, it can be shown that DPBF Algorithm solves the planar CDS problem in $O(2^{4.618\mathrm{bw}(G)}n + n^3)$ and $O(2^{9.8\sqrt{n}}n + n^3)$ time. The running time can be further improved to $O(2^{3.812\mathrm{bw}(G)}n + n^3)$ and $O(2^{8.088\sqrt{n}}n + n^3)$ if the fast distance matrix multiplication is applied [8]. It is known that $\mathrm{bw}(G) \leq 3\sqrt{4.5\gamma(G)}$ for planar graph G [14,13]. Since $\gamma(G) \leq \gamma_c(G)$, the planar CDS problem admits an $O(2^{24.257\sqrt{\gamma_c(G)}}n + n^3)$ time fixed-parameter algorithm.

Because of the applications of the planar CDS problem in wireless networks, practically efficient exact algorithms for the planar CDS problem are of great interests for those applications. DPBF Algorithm suggests theoretically an efficient exact approach for the CDS problem in planar graphs of small branchwidth. However, the practical performance of the algorithm is yet to be evaluated. In this paper, we perform a computational study to evaluate DPBF Algorithm.

We also apply the recent result on the kernelization for the planar CDS problem in our study. A linear size kernel of a graph G for the CDS problem is a subgraph H of G with $O(\gamma_c(G))$ vertices and $\gamma_c(H) \leq \gamma_c(G)$ such that a minimum CDS of G can be produced efficiently from a minimum CDS of H. It is known that the planar CDS problem admits a linear size kernel and such a kernel can be computed in $O(n^3)$ time [16]. Applying the algorithm of [16] to shrink the input graph G into a linear size kernel H, we get an $O(2^{24.257\sqrt{\gamma_c(G)}}\gamma_c(G) + n^3)$ time algorithm for the planar CDS problem.

The computational study is performed on several classes of planar graphs that cover a wide range of planar graphs. The results show that the conventional version of DPBF Algorithm is more efficient than the version of using fast distance matrix multiplication even though the latter has a better theoretical running time because the fast distance matrix multiplication itself is not practical. The size of instances that can be solved in a practical time and memory space mainly depends on the branchwidth of the kernels of the instances. This coincides with the theoretical running time of DPBF Algorithm.

The computational study gives a concrete example on using the branch-decomposition based algorithms for solving important non-local problems in planar graphs and shows that the planar CDS problem can be solved in practice for a wide range of graphs. This work provides a tool for computing the optimal CDS of planar graphs and may bring the sphere-cut decomposition and noncrossing partitions based approach closer to practice.

In the rest of the paper, Section 2 provides necessary definitions. We describe DPBF Algorithm and our implementation of it in Section 3. Computational results are reported in Section 4 and the final section concludes the paper.

2   Preliminaries

A graph $G(V, E)$ consists of a set $V(G)$ of vertices and a set $E(G)$ of edges. A graph H is a subgraph of G if $V(H) \subseteq V(G)$ and $E(H) \subseteq E(G)$. For a subset $A \subseteq E(G)$ ($U \subseteq V(G)$), we denote by $G[A]$ ($G[U]$) the subgraph of G induced by A (U).

For a graph G and a subset $A \subseteq E(G)$ of edges, we denote $E(G) \setminus A$ by $\overline{A}$ when G is clear from the context. A separation of graph G is a pair $(A, \overline{A})$ of subsets of $E(G)$. For each $A \subseteq E(G)$, we denote by $\partial(A)$ the vertex set $V(A) \cap V(\overline{A})$. The *order* of separation $(A, \overline{A})$ is $|\partial(A)| = |\partial(\overline{A})|$.

A *branch-decomposition* of graph G [25] is a pair (ϕ, T) where T is a tree each internal node of which has degree 3 and ϕ is a bijection from the set of leaves of T to $E(G)$. Consider a link e of T and let L_1 and L_2 denote the sets of leaves of T in the two respective subtrees of T obtained by removing e. We say that the separation $(\phi(L_1), \phi(L_2))$ is induced by this link e of T. We define the width of the branch-decomposition (ϕ, T) to be the largest order of the separations induced by links of T. The *branchwidth* of G, denoted by $\mathrm{bw}(G)$, is the minimum width of all branch-decompositions of G. In the rest of this paper, we identify a branch-decomposition (ϕ, T) with the tree T, leaving the bijection implicit and regarding each leaf of T as an edge of G.

Let Σ be a fixed sphere. A set P of points in Σ is a *topological segment* of Σ if it is homeomorphic to an open segment $\{(x, 0)|0 < x < 1\}$ in the plane. For a topological segment P, we denote by $\overline{P}$ the closure of P and $\mathrm{bd}(P) = \overline{P} \setminus P$ the two *end points* of P. A planar embedding of a graph G is a mapping $\rho : V(G) \cup E(G) \to \Sigma \cup 2^{\Sigma}$ satisfying the following properties.

– For $u \in V(G)$, $\rho(u)$ is a point of Σ, and for distinct $u, v \in V(G)$, $\rho(u) \neq \rho(v)$.
– For each edge $e = \{u, v\}$ of $E(G)$, $\rho(e)$ is a topological segment with two end points $\rho(u)$ and $\rho(v)$.
– For distinct $e_1, e_2 \in E(G)$, $\overline{\rho(e_1)} \cap \overline{\rho(e_2)} = \{\rho(u)|u \in e_1 \cap e_2\}$.

A graph is planar if it has a planar embedding. A plane graph is a pair (G, ρ), where ρ is a planar embedding of G. We may simply use G to denote the plane graph (G, ρ), leaving the embedding ρ implicit. We do not distinguish a vertex v (resp. an edge e) from its embedding $\rho(v)$ (resp. $\rho(e)$) when there is no confusion.

Let G be a plane graph. We say that a curve μ on the sphere Σ is *normal* if μ does not intersect with itself or any edge of G. A *noose* of G is a closed normal curve on Σ. Let ν be a noose of G and let R_1 and R_2 be the two open regions of the sphere separated by ν. Then, ν induces a separation $(A, \overline{A})$ of G, with $A = \{e \in E(G) \mid \rho(e) \subseteq R_1\}$ and $\overline{A} = \{e \in E(G) \mid \rho(e) \subseteq R_2\}$. We also say that noose ν induces edge-subset A of G if ν induces a separation $(A, \overline{A})$ having A on one side. We call a separation or an edge-subset *noose-induced* if it is induced by some noose. A branch-decomposition T of G is a *sphere-cut decomposition* if every separation induced by a link of T is noose-induced [9,10]. It is known that every plane graph G has an optimal branch-decomposition (of width $\mathrm{bw}(G)$) that is a sphere-cut decomposition and such a decomposition can be found in $O(n^3)$ time [26,17].

3 Algorithm for Planar CDS Problem

DPBF Algorithm uses the branch-decomposition based approach which has two major steps: (1) compute a branch-decomposition T of the input graph, and (2) apply dynamic programming method based on T to solve the problem. A link e of T is called a *leaf link* if e contains a leaf node of T, otherwise called an *internal link*. To solve an optimization problem P in Step (2), T is first converted to a rooted binary tree by replacing a link $\{x, y\}$ of T with three links $\{x, z\}, \{y, z\}, \{z, r\}$, where z and r are new nodes to T, r is the root, and $\{z, r\}$ is an internal link. A link e' (resp. a node x) is called a *descendant link* (resp. *descendant node*) of link e if e is in the path from e' (resp. x) to the root r of T. For a link e of T, let $(A_e, \overline{A_e})$ be the separation induced by e with A_e the set of leaf nodes of T (set of edges of G) that are descendant nodes of e. For a leaf link e, all possible partial solutions of P in the subgraph $G[A_e]$ can be computed by enumeration. For an internal link e of T, e has two child links e_1 and e_2. Notice that $A_e = A_{e_1} \cup A_{e_2}$. All possible partial solutions in the subgraph $G[A_e]$ are computed by merging the partial solutions in $G[A_{e_1}]$ and those in $G[A_{e_2}]$.

A problem P is known having a *local structure*, if a partial solution of P in $G[A_e]$ can be identified by a fixed number of states of each vertex in $\partial(A_e)$, and all partial solutions in $G[A_e]$ can be computed from the states of the vertices in $\partial(A_{e_1})$ and those of the vertices in $\partial(A_{e_2})$. The local structure is a key condition for the branch-decomposition based algorithm to solve P in $O(2^{O(\mathrm{bw}(G))}n^{O(1)})$ time. However for the CDS problem, the connectivity information in a partial solution in $G[A_e]$ may not be expressed by a fixed number of states of each vertex of $\partial(A_e)$. In the merge step, the structures of the partial solutions in the entire subgraphs $G[A_{e_1}]$ and $G[A_{e_2}]$ may have to be checked. Because of this, the CDS problem is known having a *non-local structure*.

Dorn et al. give a new technique which makes the branch-decomposition based approach applicable to many problems with the non-local structure in planar graphs [9,10]. This new technique is based on two observations. One is the geometric property of the sphere-cut decomposition T of plane graph G: For any link e of T and the separation $(A_e, \overline{A_e})$ induced by e, there is a noose ν_e such that ν_e induces $(A_e, \overline{A_e})$, ν_e partitions the sphere Σ into two regions, all edges of A_e are in one region, and all edges of $\overline{A_e}$ are in the other region. Notice that ν_e intersects all vertices of $\partial(A_e)$. The other observation is known as the non-crossing partitions: Let $P_1, ..., P_r$ be the subsets of A_e such that $G[P_i]$ is connected for each $1 \leq i \leq r$ and $G[P_i \cup P_j]$ is not connected for every pair of $1 \leq i \neq j \leq r$. We call $P_1, ..., P_r$ *disjoint components*. Two components P_i and P_j are called *crossing* if there are $u, u' \in V(P_i) \cap \partial(A_e)$ and $v, v' \in V(P_j) \cap \partial(A_e)$ such that the four vertices appear on ν_e in the orders u, v, u', v', otherwise *non-crossing*. Notice that if P_i and P_j are crossing then $G[P_i \cup P_j]$ is connected because $G[A_e]$ is a plane graph. So, any pair of disjoint components are non-crossing. The sphere-cut decomposition and the non-crossing partitions make it possible to compute the partial solutions in $G[A_e]$ by only looking at the local structures of partial solutions in $G[A_{e_1}]$ at $\partial(A_{e_1})$ and those in $G[A_{e_2}]$ at $\partial(A_{e_2})$.

For a minimum CDS D of G, the subgraph $G[D \cap V(A_e)]$ of $G[A_e]$ induced by D consists of disjoint components $P_1, ..., P_r$ with $|V(P_i) \cap \partial(A_e)| \geq 1$ for every $1 \leq i \leq r$. We assume the vertices of $\partial(A_e)$ are indexed as $u_1, u_2, ..., u_k$ in the clockwise order as they appear in the noose ν_e. If $|V(P_i) \cap \partial(A_e)| \geq 2$, we call the vertex of $V(P_i) \cap \partial(A_e)$ with the smallest index the *small end*, the vertex of $V(P_i) \cap \partial(A_e)$ with the largest index the *large end* and other vertices of $V(P_i) \cap \partial(A_e)$ the *middle vertices* of P_i. In DPBF Algorithm, each vertex $u \in \partial(A_e)$ is given one of the following six colors.

- Color 0, u does not appear in any P_i and is dominated by some vertex of $D \cap V(A_e)$.
- Color $\hat{0}$, u does not appear in any P_i and is not dominated by any vertex of $D \cap V(A_e)$.
- Color $1_[$, u is the small end of some P_i.
- Color $1_]$, u is the large end of some P_i.
- Color 1^*, u is a middle vertex of some P_i.
- Color $\hat{1}$, u is the only vertex of some $V(P_i) \cap \partial(A_e)$.

From the geometric property of sphere-cut decomposition and the non-crossing partitions, each partial solution $P_1, ..., P_r$ can be identified by a coloring of $\{0, \hat{0}, 1_[, 1_], 1^*, \hat{1}\}^{|\partial(A_e)|}$.

We implemented DPBF Algorithm together with a pre-processing step which reduces the input graph to a linear size kernel. There are three major steps in our implementation. Let G be a plane graph of n vertices.

Step I: Compute a kernel H of G with $|V(H)| = O(\gamma_c(G))$. This can be done in $O(n^3)$ time [16].

Step II: Compute a sphere-cut decomposition T of H with width bw(H). This can be done in $O((\gamma_c(H))^3)$ time [26,17].

Step III: Compute a minimum CDS D of H using the dynamic programming method based on T and compute a minimum CDS of G from D.

We use 1 to express the numerical value of $1_], 1_[, 1^*, \hat{1}$. Let $b = |\partial(A_e)|$. We call $0, \hat{0}, 1$ *basic colors* and a $\lambda \in \{0, \hat{0}, 1\}^b$ a *basic-coloring* of $\partial(A_e)$.

For a coloring $\eta \in \{0, \hat{0}, 1_[, 1_], 1^*, \hat{1}\}^b$, we denote by $D_e(\eta)$ the partial solution identified by η with the minimum number of black vertices. In the merge step for the link $e = \{z, r\}$ incident to the root node r, we check the connectivity of $H[D_e(\eta)]$. A $D_e(\eta)$ with the minimum cardinality and $H[D_e(\eta)]$ connected is a minimum CDS of H. For $\eta \in \{0, \hat{0}, 1_[, 1_], 1^*, \hat{1}\}^b$, we define $a_e(\eta) = |D_e(\eta)|$ if η identifies a partial solution, otherwise $a_e(\eta) = +\infty$. For a leaf link e of T, $D_e(\eta)$ is computed for every $\eta \in \{0, \hat{0}, 1_[, 1_], 1^*, \hat{1}\}^b$ by enumeration. For an internal link e of T, e has two child links e_1 and e_2. Let $b_1 = |\partial(A_{e_1})|$ and $b_2 = |\partial(A_{e_2})|$. The sets $D_e(\eta)$ are computed by combining the sets of $D_{e_1}(\eta_1)$ and the sets of $D_{e_2}(\eta_2)$, where η_1 is a coloring of $\{0, \hat{0}, 1_[, 1_], 1^*, \hat{1}\}^{b_1}$ and η_2 is a coloring of $\{0, \hat{0}, 1_[, 1_], 1^*, \hat{1}\}^{b_2}$.

Let $X_1 = \partial(A_e) \setminus \partial(A_{e_2})$, $X_2 = \partial(A_e) \setminus \partial(A_{e_1})$, $X_3 = \partial(A_e) \cap \partial(A_{e_1}) \cap \partial(A_{e_3})$, and $X_4 = (\partial(A_{e_1}) \cup \partial(A_{e_2})) \setminus \partial(A_e)$. Then $\partial(A_e) = X_1 \cup X_2 \cup X_3$, $\partial(A_{e_1}) = X_1 \cup X_3 \cup X_4$, and $\partial(A_{e_2}) = X_2 \cup X_3 \cup X_4$. A basic-coloring λ of $\partial(A_e)$ is formed from basic-colorings λ_1 of $\partial(A_{e_1})$ and basic colorings λ_2 of $\partial(A_{e_1})$ if:

1. For $u \in X_1$, $\lambda(u) = \lambda_1(u)$.
2. For $u \in X_2$, $\lambda(u) = \lambda_2(u)$.
3. For $u \in X_3$, if $\lambda_1(u) = \lambda_2(u) = 1$ then $\lambda(u) = 1$; if $\lambda_1(u) = \lambda_2(u) = \hat{0}$ then $\lambda(u) = \hat{0}$; and if $\lambda_1(u) = 0$ and $\lambda_2(u) = \hat{0}$, or $\lambda_1(u) = \hat{0}$ and $\lambda_2(u) = 0$ then $\lambda(u) = 0$.
4. For $u \in X_4$, $\lambda_1(u) = \lambda_2(u) = 1$, or $\lambda_1(u) = 0$ and $\lambda_2(u) = \hat{0}$, or $\lambda_1(u) = \hat{0}$ and $\lambda_2(u) = 0$.

For a basic-coloring λ which is formed by two basic-colorings λ_1 and λ_2, we compute the disjoint components $P_1, ..., P_r$ of $H[D_{e_1}(\eta_1) \cup D_{e_2}(\eta_2)]$, where for $i = 1, 2$ $\eta_i(u) = \lambda_i(u)$ if $\lambda_i(u) \in \{0, \hat{0}\}$ and $\eta_i(u) \in \{1_[, 1_], 1^*, \hat{1}\}$ if $\lambda_i(u) = 1$. $D_{e_1}(\eta_1) \cup D_{e_2}(\eta_2)$ is called a *candidate* for $D_e(\eta)$ if each P_i has at least one vertex in $\partial(A_e)$. If $D_{e_1}(\eta_1) \cup D_{e_2}(\eta_2)$ is a candidate, we convert the color of u with $\lambda(u) = 1$ into one color of $\{1_[, 1_], 1^*, \hat{1}\}$ according to if u is the small end, the large end, a middle vertex, or the only vertex of $V(P_i) \cap \partial(A_e)$, respectively, to get a coloring $\eta \in \{0, \hat{0}, 1_{[}, 1_{]}, 1^*, \hat{1}\}^b$. Finally, $D_e(\eta)$ is a candidate $D_{e_1}(\eta_1) \cup D_{e_2}(\eta_2)$ with the minimum cardinality.

The colorings $\{0, \hat{0}, 1_{[}, 1_{]}, 1^*, \hat{1}\}^b$ and the corresponding partial solutions can be kept in a table of size $O(6^b)$. A bijection from $\{0, \hat{0}, 1_{[}, 1_{]}, 1^*, \hat{1}\}^b$ to $\{1, 2, ..., 6^b\}$ gives an index method to access the table. The colorings $\{0, \hat{0}, 1_{[}, 1_{]}, 1^*, \hat{1}\}^{b_1}$ and $\{0, \hat{0}, 1_{[}, 1_{]}, 1^*, \hat{1}\}^{b_2}$ are handled similarly. The index method of DPBF Algorithm described above solves the CDS problem for a plane G of n vertices in $O(2^{4.67\mathrm{bw}(G)}\gamma_c(G) + n^3)$ time and $O(6^{\mathrm{bw}(G)}\gamma_c(G))$ memory space. The running time of the index method can be improved to $O(2^{4.618\mathrm{bw}(G)}\gamma_c(G) + n^3)$ by a more complex analysis for Step III. In Step III, merging colorings can be done by the distance matrix multiplication. If the conventional $O(n^3)$ time distance matrix multiplication is used, this gives the same running time as that of the index method. If the fast distance matrix multiplication is used, the running time of DPBF Algorithm can be further improved to $O(2^{9.8\sqrt{n}}\gamma_c(G) + n^3)$ and $O(2^{24.257\sqrt{\gamma_c(G)}}\gamma_c(G) + n^3)$. We omit the analysis of the above running times due to the limit of space.

4   Computational Results

We tested the performance of DPBF Algorithm on the following classes of planar graphs. Class (1) is a set of random maximal planar graphs and their subgraphs generated by LEDA [1,3]. Class(2) includes the Delaunay triangulations of point sets taken from TSPLIB [22]. The instances of Classes (3) and (4) are the triangulations and intersection graphs generated by LEDA, respectively. The instances of Class (5) are Gabriel graphs generated using the points uniformly distributed in a two-dimensional plane. The instances of Class (6) are random planar graphs generated by the PIGALE library [2].

We use the reduction rules of [16] to compute the kernels of input instances in Step I and the $O(n^3)$ time algorithm of [4] to compute optimal sphere-cut decompositions of kernels in Step II. For Step III, we use an index method to access the tables. To save memory, we compute the colorings of links of T in the postorder manner. Once the colorings of a link e are computed for a link e, the solutions for the children links of e are discarded. Because the fast distance matrix multiplication is not practical [21], applying this technique does not improve the practical performance of the algorithm.

The computer used for testing has an AMD Athlon(tm) 64 X2 Dual Core Processor 4600+ (2.4GHz) and 3GByte of internal memory. The operating system is SUSE Linux 10.2 and the programming language used is C++.

Table 1 shows the computational results of the simple version of DPBF Algorithm. H is the kernel of an instance computed in Step I. In Step II, an optimal sphere cut decomposition of H is computed and we report $|E(H)|$, the size of H, the branchwidth $\mathrm{bw}(H)$ of H, and the running time of this step. For Step III, we give $\gamma_c(G)$ obtained, the running time of the step and the required memory in Gigabytes (GB). All times in the table are in seconds.

Table 1. Computational results (time in seconds) of DPBF Algorithm. For the instances marked with "*", the 3GByte memory is not enough for computing a minimum connected dominating set.

Class	Graph G	$|E(G)|$	$bw(G)$	Step I time	Step II $|E(H)|$	$bw(H)$	time	Step III $\gamma_c(G)$	time	total time	maximum memory
(1)	max1000	2912	4	19.4	704	4	2.3	131	4	25.7	
	max2000	5978	4	63	1133	4	6.0	252	9.9	78.9	
	max3000	8510	4	359	2531	4	37.6	417	94	491	
	max4000	10759	4	836	3965	4	145	614	458	1439	
	max5000	14311	4	848	3873	4	160	650	383	1392	
	max5000	16206	4	1702	5989	4	325	907	1769	3796	
(2)	eil51	140	8	0.1	140	8	0.2	14	253	254	0.03
	lin105	292	8	0.3	275	8	3	27	810	813	0.03
	pr144	393	9	1	347	7	0.5	25	18.1	19.7	0.06
	kroB150	436	10	1	436	10	0.8	36	133856	133858	1.05
	pr226	586	7	1.3	399	6	1.7	24	5.1	8.1	0.04
	ch130	377	10	0.3	377	10	0.6	34	38562	38563	0.74
(3)	tri100	288	7	0.7	258	6	0.6	20	7.1	8.4	0.05
	tri500	1470	7	10.1	1438	6	37.2	91	62.6	110	0.07
	tri800	2374	8	18	2279	7	86.4	149	289	393	0.13
	tri2000	5977	8	109	5751	8	603	369	5643	6355	0.48
	tri4000	11969	9	547	11236	9	3690	753	42323	46560	0.57
(4)	rand100	121	5	0.1	73	3	0.1	40	0.1	0.3	0.03
	rand500	709	7	1.7	545	6	0.4	216	10.8	12.9	0.05
	rand700	1037	7	2.9	836	6	1	301	17.8	21.8	0.07
	rand1000	1512	8	4.5	1242	7	2.5	421	422.8	429.8	0.25
	rand2000	3247	8	17.5	2852	8	17.8	839	10179	10214	0.38
	rand3000*	4943	10	-	-	10	-	-	-	-	-
(5)	Gab50	88	4	0.1	88	4	0.1	22	0.2	0.4	0.03
	Gab100	182	7	0.1	179	7	0.3	41	66.7	67.1	0.11
	Gab200	366	8	0.7	362	8	1.5	81	2290	2293	0.13
	Gab300	552	10	1.4	545	10	1.6	121	12 days	12 days	2.53
(6)	P206	269	4	0.6	163	4	0.3	78	0.3	1.2	0.02
	P495	852	5	3.2	765	5	8.4	167	11.9	23.5	0.02
	P855	1434	6	7.9	1280	6	15.1	289	77.9	101	0.06
	P1000	1325	5	4.4	777	5	2.5	378	7.3	14.2	0.07
	P2000	2619	6	24.5	1527	6	12.3	738	58.0	94.8	0.11
	P4206	7101	6	256	6377	6	1816	1423	2411	4482	0.43

Now we go over the details of our results. It is shown in [21] that the branchwidth of the instances of class (1) is at most four. Our results show that reduction rules are very effective on these graphs and that the size of the kernels is much smaller than the size of the original graphs. Thus, Step III is fast and the minimum CDS of some instances with 16000 edges can be computed in about one hour on our platform.

However, the branchwidth increases very fast in the size of the graph for the instances of Classes (2) and (5). In addition, the reduction rules do not reduce the size of the original graphs very much, and the size and branchwidth of generated kernels are the same as those of the original graphs. The running time of Step III increases significantly with the branchwidth of instances (e.g., see the running time of instances pr144 and kroB150). For instances with the same branchwidth the running time of this step depends on the size of the kernel. (see instances eil51 and lin105). For these classes of planar graphs DPBF Algorithm is time consuming and can solve the CDS problem on instances of size up to a few hundreds edges in a practical time. The branchwidth of instances of Classes (3) and (4) grows relatively slow in the instance sizes. Furthermore, data reduction rules are effective on the instances of Class (4). The branchwidth of graph instances in Class (6) does not grow in the instance size thus, DPBF Algorithm is efficient for this class.

The memory space required by DPBF Algorithm in Step III is a bottleneck for solving instances with large branchwidth. Experimental results show that DPBF Algorithm can compute a minimum CDS for instances with the branchwidth of kernels at most 10 ($\mathrm{bw}(H) \leq 10$) using 3GBytes of memory space.

5   Concluding Remarks

We evaluated the performance of DPBF Algorithm for the CDS problem on a wide range of planar graphs. The computational results coincide with the theoretical analysis of the algorithm, it is efficient for graphs with small branchwidth but may not be practical for graphs with large branchwidth. Using a computer with a CPU of 2.4GHz and 3GMBytes memory space, it is possible to find a minimum CDS for graphs with the branchwidth of their kernels at most 10 in a few hours. Since the branchwidth of a planar graph can be computed in $O(n^2 \log n)$ time by the $O(n^2)$ time rat-catching algorithm [26] and a binary search, one may first get the branchwidth of the input graph and then decide if DPBF Algorithm is applicable using the results of this paper as a guideline.

Because DPBF Algorithm runs and requires memory space exponentially in the branchwidth $\mathrm{bw}(H)$ of a kernel H for a given graph, it is worth to develop more powerful data reduction rules to reduce $\mathrm{bw}(H)$. It is known that the planar CDS problem admits PTAS [7]. The approach for the PTAS is to partition an input graph into subgraphs of fixed branchwidth, find a minimum CDS for each subgraph and combining the solutions of subgraphs into a solution of the input graph. It is interesting to apply DPBF Algorithm to develop PTAS which is efficient in practice for the planar CDS problem in graphs with large branchwidth.

References

1. Library of Efficient Data Types and Algorithms, Version 5.2 (2008), http://www.algorithmic-solutions.com/enleda.htm
2. Public Implementation of a Graph Algorithm Library and Editor (2008), http://pigale.sourceforge.net/
3. The LEDA User Manual, Algorithmic Solutions, Version 4.2.1 (2008), http://www.mpi-inf.mpg.de/LEDA/MANUAL/MANUAL.html
4. Bian, Z., Gu, Q.: Computing branch decompositions of large planar graphs. In: McGeoch, C.C. (ed.) WEA 2008. LNCS, vol. 5038, pp. 87–100. Springer, Heidelberg (2008)
5. Blum, J., Ding, M., Thaeler, A., Cheng, X.: Connected dominating set in sensor networks and MANETs. In: Du, D.-Z., Pardalos, P. (eds.) Handbooks of Combinatorial Optimization, Suppl., vol. B, pp. 329–369. Springer, Heidelberg (2004)
6. Cheng, X., Ding, M., Du, H., Jia, X.: Virtual backbone construction in multihop Ad Hoc wireless networks. Wireless Communications and Mobile Computing 6(2), 183–190 (2006)

7. Demaine, E.D., Hajiaghayi, M.: Bidimensionality: new connections between FPT algorithms and PTAS. In: Proc. of the 2005 ACM/SIAM Symposium on Discrete Algorithms (SODA 2005), pp. 590–601 (2005)
8. Dorn, F.: Dynamic programming and fast matrix multiplication. In: Azar, Y., Erlebach, T. (eds.) ESA 2006. LNCS, vol. 4168, pp. 280–291. Springer, Heidelberg (2006)
9. Dorn, F., Penninkx, E., Bodlaender, H.L., Fomin, F.V.: Efficient exact algorithms on planar graphs: exploiting sphere cut branch decompositions. In: Brodal, G.S., Leonardi, S. (eds.) ESA 2005. LNCS, vol. 3669, pp. 95–106. Springer, Heidelberg (2005)
10. Dorn, F., Penninkx, E., Bodlaender, H.L., Fomin, F.V.: Efficient exact algorithms on planar graphs: exploiting sphere cut decompositions. Technical report, UU-CS-2006-006, Department of Information and Computing Sciences (2006)
11. Downey, R.G., Fellow, M.R.: Parameterized complexity. In: Monographs in Computer Science. Springer, Heidelberg (1999)
12. Downey, R.G., Fellows, M.R.: Fixed parameter tractability and completeness. Cong. Num. 87, 161–187 (1992)
13. Fomin, F.V., Thilikos, D.M.: Dominating sets in planar graphs: branch-width and exponential speed-up. SIAM Journal on Computing 36(2), 281–309 (2006)
14. Fomin, F.V., Thilikos, D.M.: New upper bounds on the decomposability of planar graphs. Journal of Graph Theory 51(1), 53–81 (2006)
15. Garey, M.R., Johnson, D.S.: Computers and Intractability, a Guide to the Theory of NP-Completeness. Freeman, New York (1979)
16. Gu, Q., Imani, N.: Connectivity is not a limit for kernelization: planar connected dominating set. In: López-Ortiz, A. (ed.) LATIN 2010. LNCS, vol. 6034, pp. 26–37. Springer, Heidelberg (2010)
17. Gu, Q., Tamaki, H.: Optimal branch-decomposition of planar graphs in $O(n^3)$ time. ACM Transactions on Algorithms 4(3), 30:1–30:13 (2008)
18. Guha, S., Khuller, S.: Approximation algorithms for connected dominating sets. Algorithmca 20, 374–387 (1998)
19. Haynes, T.W., Hedetniemi, S.T., Slater, P.J.: Domination in graphs. In: Monographs and Textbooks in Pure and Applied Mathematics, vol. 209. Marcel Dekker, New York (1998)
20. Haynes, T.W., Hedetniemi, S.T., Slater, P.J.: Fundamentals of domination in graphs. In: Monographs and Textbooks in Pure and Applied Mathematics, vol. 208. Marcel Dekker, New York (1998)
21. Marzban, M., Gu, Q., Jia, X.: Computational study on planar dominating set problem. Theoretical Computer Science 410(52), 5455–5466 (2009)
22. Reinelt, G.: TSPLIB-A traveling salesman library. ORSA J. on Computing 3, 376–384 (1991)
23. Robertson, N., Seymour, P.D.: Graph minors I. Excluding a forest. Journal of Combinatorial Theory, Series B 35, 39–61 (1983)
24. Robertson, N., Seymour, P.D.: Graph minors II. Algorithmic aspects of tree-width. Journal of Algorithms 7, 309–322 (1986)
25. Robertson, N., Seymour, P.D.: Graph minors X. Obstructions to tree decomposition. J. of Combinatorial Theory, Series B 52, 153–190 (1991)
26. Seymour, P.D., Thomas, R.: Call routing and the ratcatcher. Combinatorica 14(2), 217–241 (1994)

Bounds for Nonadaptive Group Tests to Estimate the Amount of Defectives

Peter Damaschke and Azam Sheikh Muhammad

Department of Computer Science and Engineering
Chalmers University, 41296 Göteborg, Sweden
{ptr,azams}@chalmers.se

Abstract. The classical and well-studied group testing problem is to find d defectives in a set of n elements by group tests, which tell us for any chosen subset whether it contains defectives or not. Strategies are preferred that use both a small number of tests close to the information-theoretic lower bound $d \log n$, and a small constant number of stages, where tests in every stage are done in parallel, in order to save time. They should even work if d is completely unknown in advance. An essential ingredient of such competitive and minimal-adaptive group testing strategies is an estimate of d within a constant factor. More precisely, d shall be underestimated only with some given error probability, and overestimated only by a constant factor, called the competitive ratio. The latter problem is also interesting in its own right. It can be solved with $O(\log n)$ randomized group tests of a certain type. In this paper we prove that $\Omega(\log n)$ tests are really needed. The proof is based on an analysis of the influence of tests on the searcher's ability to distinguish between any two candidate numbers with a constant ratio. Once we know this lower bound, the next challenge is to get optimal constant factors in the $O(\log n)$ test number, depending on the desired error probability and competitive ratio. We give a method to derive upper bounds and conjecture that our particular strategy is already optimal.

Keywords: algorithm, learning by queries, competitive group testing, nonadaptive strategy, randomized strategy, lower bound.

1 Introduction

Suppose that, in a set of n elements, d unknown elements are defective, and a searcher can do group tests which work as follows. She can take any subset of elements, called a pool, and ask whether the pool contains some defective. That is, the result of a group test is binary: 0 means that no defective is in the pool, and 1 means the presence of at least one defective. The combinatorial group testing problem asks to determine at most d defectives using a minimum number of tests; we also refer to them as queries. Group testing with its variants is a classical problem in combinatorial search, with a history dating back to year 1943 [8], and it has various applications in chemical testing, bioinformatics, communication networks, information gathering, compression, streaming algorithms, etc., see for instance [3, 4, 7, 9, 11–13].

W. Wu and O. Daescu (Eds.): COCOA 2010, Part II, LNCS 6509, pp. 117–130, 2010.

By the trivial information-theoretic lower bound, essentially $d \log_2 n$ queries are necessary for combinatorial group testing. A group testing strategy using $O(d \log n)$ queries despite ignorance of d before the testing process is called competitive, and the "hidden" constant factor is the competitive ratio. The currently best competitive ratio is 1.5 when queries are asked sequentially [14]. However, group testing strategies with minimal adaptivity are preferable for applications where the tests are time-consuming. Such strategies work in a few stages, where queries in a stage are prepared prior to the stage and then asked in parallel. For 1-stage group testing, at least $\Omega((d^2 / \log d) \log n)$ queries are needed even in the case of a known d; see [1]. Clearly, 1-stage competitive group testing is impossible. As opposed to this, already 2 stages are enough to enable an $O(d \log n)$ test strategy, also the competitive ratio has been improved in several steps [2, 6, 10]. Still d must be known in advance or, to say it more accurately, d is some assumed upper bound on the true number of defectives. Apparently we were the first to study group testing strategies that are both minimal adaptive and competitive, i.e., they are suitable even when nothing about the magnitude of d is known beforehand [5]. Unfortunately, any efficient deterministic competitive group testing strategy needs $\Omega(\log d / \log \log d)$ stages (and $O(\log d)$ stages are sufficient). The picture changes when randomization is applied. If we can estimate an upper bound on the unknown d within a constant factor, using a logarithmic number of nonadaptive randomized queries, then we can subsequently apply any 2-stage $O(d \log n)$ strategy for known d, and thus obtain a randomized 3-stage competitive strategy. If we, instead, append a randomized 1-stage strategy with $O(d \log n)$ queries [2], we obtain a competitive group testing strategy that needs only 2 stages. Determining d exactly is as hard as combinatorial group testing itself [5], thus it would require $\Omega((d^2 / \log d) \log n)$ nonadaptive queries. But an estimate of d within a constant factor is sufficient (and also necessary) for minimal adaptive competitive group testing. We call the expected ratio of our estimate and the true d a competitive ratio as well; it is always clear from context which competitive ratio is meant.

It is not hard to come up with such a nonadaptive estimator of d [5]. More precisely, using $O(\log n)$ queries we can output a number which is smaller than d only with some prescribed error probability ϵ but has an expectation $O(d)$. (If the alleged d was too small, the subsequent stages will notice the failure, and we try again from scratch, thus solving the combinatorial group testing problem in $O(1)$ expected stages.) To this end we prepare pools as follows. We fix some probability q and put every element in the pool independently with probability $1 - q$. Clearly, the group test gives the result 0 and 1 with probability q^d and $1 - q^d$, respectively. We prepare $O(\log n)$ of these pools such that the values $1 / \log_2(1/q)$ form an exponential sequence of numbers between 1 to n. Note that these values are the defective numbers d for which $q^d = 1/2$. Then, the position in the sequence of pools where test results 0 switch to 1 hint to the value of d, subject to some constant factor and with some constant error probability. (Of course, the details have to be specified and proved [5].) Note that the expected competitive ratio of 2-stage or 3-stage group testing is determined by three quantities: the competitive ratio of the group testing strategy used, and both the query number and competitive ratio of the randomized d estimator. The currently best 2-stage group testing strategy [2] uses $(1.44 + o(1))d \log n$ queries (asymptotically for $n \to \infty$). in this paper we focus on the estimator which requires methods completely different from the combinatorial group testing part that actually finds the defectives. Estimating d is also an interesting problem in its own right, as in some group testing applications we may only be interested in the amount of defectives rather than their identities.

An open question so far was whether $O(\log n)$ tests are really needed to estimate d, in the above sense. Intuitively this should be expected, based on the following heuristic argument. "Remote" queries with $1/\log_2(1/q)$ far from d will almost surely have a fixed result (0 or 1) and contribute little information about the precise location of d. Therefore we must have queries with values $1/\log_2(1/q)$ within some constant ratio of every possible d, which would imply an $\Omega(\log n)$ bound. However, the searcher may use the accumulated information from all queries, and even though "unexpected" results of the remote queries have low probabilities, a few such events might reveal enough useful information about d. Apparently, in order to turn the intuition into a proof we must somehow quantify the influence of remote queries and show that they actually provide too little information. To see the challenge, we first remark that the simple information-theoretic argument falls short. Imagine that we divide the interval from 1 to n into exponentially growing segments. Then the problem of estimating d up to a constant factor is in principle (don't care about technicalities) equivalent to guessing the segment where d is located, or a neighbored segment. The number of possible outcomes is some $\log n$, thus we need $\Omega(\log \log n)$ queries, which is a very weak lower bound. The next idea that comes into mind is to take the very different probabilities of binary answers into account. The entropy of the distribution of result strings is low, however it is not easy to see how to translate entropy into a measure suited to our problem.

A main result of the present paper is a proof of the $\Omega(\log n)$ query bound, for any fixed competitive ratio and any fixed error probability. A key ingredient is a suitable influence measure for queries. The proof is based on a simpler auxiliary problem that may deserve independent interest: deciding on one of two hypotheses about which we got only probabilistic information, thereby respecting a pair of error bounds. It has to be noticed that our result does not yet prove the non-existence of a randomized $o(\log n)$ query strategy in general. The result only refers to randomized pools constructed in the aforementioned simple way: adding every element to a pool independently with some fixed probability $1-q$. However, the result gives strong support to the conjecture that $\Omega(\log n)$ is also a lower bound for any other randomized pooling design. Intuitively, randomized pools that treat all elements symmetrically and make independent decisions destroy all possibilities for a malicious adversary to mislead the searcher by some clever placement of defectives. Therefore it is hard to imagine that other constructions could have benefits.

The rest of the paper is organized as follows. In Section 2 we give a formal problem statement and some useful notation. In Section 3 we study a probabilistic inference problem on two hypotheses, and we define the influence of the random bit contributed by any query. This is used in Section 4 to prove the logarithmic lower bound for estimating the defective number by group tests. In Section 5 we derive a particular $O(\log n)$ query strategy for estimating the defectives, and we have reason to conjecture that its hidden constant factor is already optimal, for every input parameter. Section 6 concludes the paper.

2   Preliminaries

Motivated by competitive group testing we study the following abstract problem.

Problem 1: Given are positive integers n and L, some positive error probability $\epsilon < 1$, and some $c > 1$ that we call the competitive ratio. Furthermore, an "invisible" number $x \in [1, n]$ is given. A searcher can prepare L nonadaptive queries to an oracle as follows. A query specifies a number $q \in (0, 1)$, and the oracle answers 0 with probability q^x, and 1 with probability $1 - q^x$. Based on the string s of these L binary answers the searcher is supposed to output some number x' such that $\mathbf{Pr}[x' < x] \leq \epsilon$ and $\mathbf{E}[x'/x] \leq c$ holds for every x.

The actual problem is to place the L queries, and to compute an x' from s, in such a way that the demands are fulfilled. The optimization version asks to minimize c given the other input parameters. We will prove that $L = \Omega(\log n)$ queries are needed, for any fixed ϵ and c. Note that randomness is not only in the oracle answers but possibly also in the rule that decides on x' based on s, and even in the choice of queries.

Symbols $\mathbf{Pr}$ and $\mathbf{E}$ in the definition refer to the resulting probability distribution of x' given x. Note that no distribution of x is assumed, rather, the conditions shall be fulfilled for any fixed x. We might, of course, define similar problem versions, e.g., with two-sided errors or with worst-case (rather than expected) competitive ratio and tail probabilities. However we stick to the above problem formulation, as it came up in this form in competitive 2-stage and 3-stage group testing, and other conceivable variations would behave similarly. In the group testing context, an oracle query obviously represents a randomized pool where every element is selected independently with probability $1 - q$, and x is the unknown number of defectives. However we will treat x as a real-valued variable. Asymptotically this does not change anything, but it simplifies several technical issues.

It turns out that some coordinate transformations reflect the geometry of the problem better than the variables originating from the group testing application. We will look at x on the logarithmic axis and reserve symbol y for $y = \ln x$. Note that $y \in [0, \ln n]$. Furthermore, we relate every q to that value y which would make $q^x = q^{e^y}$ some constant "medium" probability, such as $1/e$, the inverse of Euler's constant. (The choice of this constant is arbitrary, but it will simplify some expressions.) We denote this y value by t, in other words, we want $q^{e^t} = 1/e$, which means $q = e^{-e^{-t}}$ and $\ln(1/q) = e^{-t}$. Symbol t is reserved for this transformed q. We refer to t as a query point.

3 Probabilistic Inference of One-Out-of-Two Hypotheses

Buridan's ass could not decide on either a stack of hay or a pail of water and thus suffered from both hunger and thirst. The following problem demands a decision between two alternatives either of which could be wrong, but it also offers a clear rationale for the decision. As usual in inference problems, the term "target" refers to the true hypothesis.

Problem 2: The following items are given: two hypotheses g and h; two nonnegative real numbers $\epsilon, \delta < 1$; furthermore N possible observations that we simply denote by indices $s = 1, \ldots, N$; probabilities p_s to observe s if g is the target, and similarly, probabilities q_s to observe s if h is the target. Clearly, $\sum_{s=1}^{N} p_s = 1$ and $\sum_{s=1}^{N} q_s = 1$. Based on the observed s, the searcher can infer g with some probability x_s, and h with probability $1 - x_s$. The searcher's goal is to choose her x_s for all s, so as to limit to ϵ the probability of wrongly inferring h when g is the target, and to limit to δ the probability of wrongly inferring g when h is the target.

We rename the observations so that $p_1/q_1 \leq \ldots \leq p_N/q_N$.

In the optimization version of Problem 2, only one error probability, say ϵ, is fixed, and the searcher wants to determine $x_1, \ldots, x_N$ so as to minimize δ. We denote the optimum by $\delta(\epsilon)$. Problem 2 is easily solved in a greedy fashion:

Lemma 1. *A complete scheme of optimal strategies (one for every ϵ) for Problem 2 is described as follows. Determine u such that $p_1 + \ldots + p_{u-1} \leq \epsilon < p_1 + \ldots + p_u$, and let $f := (\epsilon - p_{u-1})/(p_u - p_{u-1})$. Infer h if $s < u$, infer h with probability f in case $s = u$, and otherwise infer g. Consequently, $\delta(\epsilon) = (1 - f)q_u + q_{u+1} + \ldots + q_N$.*

Proof. We only have to prove optimality. In any given strategy, let us change two consecutive "strategy values" simultaneously by $x_s := x_s - \Delta \cdot p_{s+1}$ and $x_{s+1} := x_{s+1} + \Delta \cdot p_s$, for some $\Delta > 0$. If the target is g, this manipulation changes the probability to wrongly infer h by $p_s\Delta \cdot p_{s+1} - p_{s+1}\Delta \cdot p_s = 0$. If the target is h, this manipulation changes the probability to wrongly infer g by $-q_s\Delta \cdot p_{s+1} + q_{s+1}\Delta \cdot p_s = -(q_sp_{s+1} - q_{s+1}p_s)\Delta \leq 0$, since $q_sp_{s+1} \geq q_{s+1}p_s$. Thus it can only improve the strategy. The manipulation is impossible only if some index u exists with $x_s = 0$ for all $s < u$, and $x_s = 1$ for all $s > u$. Now the lemma follows easily. $\qquad\square$

Lemma 1 also implies:

Corollary 1. $\delta(\epsilon)$ *is a monotone decreasing and convex (i.e., sub-additive), piecewise linear function with $\delta(0) = 1$ and $\delta(1) = 0$.* $\qquad\square$

The following technical lemma shows that certain small additive changes in the probability sequences do not change the error function much (which is quite intuitive). In order to avoid heavy notation we give the proof in a geometric language, referring to a coordinate system with abscissa ϵ and ordinate δ.

Lemma 2. *Consider the following type of rearrangement of a given instance of Problem 2. Replace every p_s with $p_s - \rho_s$, where $\sum_s \rho_s = \rho$. Similarly, replace every q_s with $q_s - \tau_s$, where $\tau_s = \rho_s q_s / p_s$ and $\sum_s \tau_s = \tau$. Then add the removed probability masses, in total ρ and τ, arbitrarily to existing pairs (p_s, q_s) or create new pairs (p_s, q_s), but in such a way that $\sum_s p_s = 1$ and $\sum_s q_s = 1$ are recovered. If such a rearrangement reduces $\delta(\epsilon)$, then the decrease is at most τ.*

Proof. By Corollary 1, the curve of function $\delta(\epsilon)$ is a chain of straight line segments whose slopes $-\delta'(\epsilon)$ get smaller from left to right, and these slopes are the ratios q_s / p_s. The rearrangement has the following effect on the curve: Pieces of the segments are cut out, whose horizontal and vertical projections have total length ρ and τ, respectively. Then their horizontal and vertical lengths may increase again by re-insertions (and all these actions may change the slopes of existing segments), and possibly new segments are created. Finally all segments are assembled to a new chain connecting the points $\delta(0) = 1$ and $\delta(1) = 0$, and having a monotone sequence of slopes again.

Consider a fixed ϵ. Let ρ_0 and τ_0 be the total horizontal and vertical length, respectively, of the pieces cut out to the left of ϵ. Let ρ_1 and τ_1 be defined similarly for the pieces to the right of ϵ. The largest possible reduction of $\delta(\epsilon)$ appears if: (a) some new vertical piece of length τ_1 forms the left end, and (b) some new horizontal piece of length ρ_0 and forms the right end of the modified curve. Note that pieces in (a) were originally located below $\delta(\epsilon)$, and pieces in (b) were originally located to the left of ϵ. This moves the remainder of the original curve (a) down by τ_1 length units, and (b) to the left by ρ_0 length units. The vertical move (a) reduces $\delta(\epsilon)$ by τ_1. The horizontal move (b) causes that the new function value at ϵ is the old function value at $\epsilon + \rho_0$. Since the slopes decrease from left to right, the slope at our fixed ϵ (and to the right of it) can be at most τ_0 / ρ_0. Thus, move (b) reduces $\delta(\epsilon)$ by at most $\rho_0 \tau_0 / \rho_0 = \tau_0$. Finally note that $\tau_1 + \tau_0 = \tau$. $\qquad\square$

One should not be confused that ρ does not appear in the decrease bound: As we have chosen to consider δ as a function of ϵ, the setting is not symmetric.

We are particularly interested in the special case of Problem 2 where the $N = 2^L$ observations s are strings of L independent bits.

Problem 3: The following items are given: two hypotheses g and h; two non-negative real numbers $\epsilon, \delta \le 1$; and 2^L possible observations described by binary strings $s = s_1 \ldots s_L$. Furthermore, for $k = 1, \ldots, L$, we are given the probability a_k to observe $s_k = 0$ if the target is g, and the probability b_k to observe $s_k = 0$ if the target is h. The s_k are independent. The rest of the problem specification is as in Problem 2.

Clearly, our p_s and q_s evaluate to $p_s = \prod_{k=1}^{L}((1 - s_k)a_k + s_k(1 - a_k))$ and $q_s = \prod_{k=1}^{L}((1 - s_k)b_k + s_k(1 - b_k))$. Since the greedy algorithm in Lemma 1 applies also to Problem 3, a complete set of optimal strategies is described as follows: Infer h for p_s/q_s below some threshold, infer g for p_s/q_s above that threshold, and infer g or h randomized (with some prescribed probability) for p_s/q_s equal to that threshold.

Remark: Since the p_s and q_s are just products of certain probabilities a_k or $1 - a_k$, and b_k or $1 - b_k$, respectively, taking the logarithm reveals a nice and simple geometric structure of the optimal strategies from Lemma 1: Note that $\log(p_s/q_s) = \sum_{k=1}^{L}((1 - s_k)(\log a_k - \log b_k) + s_k(\log(1 - a_k) - \log(1 - b_k)))$. Since log is a monotone function, comparing the p_s/q_s with some threshold is equivalent to comparing the $\log(p_s/q_s)$ with some threshold. In other words, the decision for g or h is merely a linear threshold predicate. We will not need this remark in our lower-bound proof, still it might be interesting to notice.

In the following we consider any fixed $\epsilon > 0$, and all notations are understood with respect to this fixed error bound. Now think of our L independent bits as $L - 1$ bits plus a distinguished one, say the kth bit. We define the *influence* of this kth bit as the decrease of $\delta(\epsilon)$, that is, the difference to the $\delta(\epsilon)$ value accomplished by an optimal strategy when the kth bit is ignored. Trivially, $\delta(\epsilon)$ can only decrease when more information is available.

Lemma 3. *With the above notations for Problem 3, the influence of the kth bit is at most* $\min(\max(a_k, b_k), \max(1 - a_k, 1 - b_k))$.

Proof. The kth bit splits every old observation s, consisting of the $L - 1$ other bits and generated with probabilities p_s, q_s depending on the target, in two new observations. Their new probability pairs are obviously $(p_s a_k, q_s b_k)$ for $s_k = 0$, and $(p_s(1 - a_k), q_s(1 - b_k))$ for $s_k = 1$. In order to apply Lemma 2 we can view this splitting of observations as cutting out pieces from the segment of slope q_s/p_s of the $\delta(\epsilon)$ curve in the following way. If $q_s/p_s \leq b_k/a_k$, a piece of vertical length $q_s b_k$ is cut out. If $q_s/p_s > b_k/a_k$, a piece of horizontal length $p_s a_k$ is cut out, corresponding to a piece of vertical length $p_s a_k q_s/p_s = q_s a_k$. (Note that we must first "cut out enough length" in both directions, therefore this case distinction is needed.) This is done for all old s. Since, of course, the old q_s sum up to 1, we have $\tau \leq \max(a_k, b_k)$. The same reasoning applies to $1 - a_k, 1 - b_k$, thus we have $\tau \leq \max(1 - a_k, 1 - b_k)$ as well. □

Note that the influence bound in Lemma 3 is expressed only in terms of the probabilities of the respective bit being 0/1, conditional on the hypothesis. Hence we can independently apply Lemma 3 to each of the bits, no matter in which order they are considered, and simply add the influence bounds of several bits (similarly to a union bound of probabilities).

4   The Logarithmic Lower Bound

We further narrow down our one-out-of-two inference problem to a special case of Problem 3. (Below we reuse symbol q, without risk of confusion.)

Problem 4: The following items are given: two hypotheses r and 1. where $r > 1$ is a fixed real number; two nonnegative real numbers $\epsilon, \delta \leq 1$, furthermore 2^L possible observations described by binary strings $s = s_1 \ldots s_L$. For $k = 1, \ldots, L$, let q_k^x be the probability to observe $s_k = 0$ if the target is x. We also speak of a "query at q_k". The s_k are independent. The rest of the problem specification is as before. In particular, let ϵ be the probability of wrongly inferring 1 although r is the target, and let δ be the probability of wrongly inferring r although 1 is the target.

Note that the hypothesis $x = r$ generates the string s with probability $\prod_{k=1}^{L}((1 - s_k)q_k^r + s_k(1 - q_k^r))$, and the hypothesis $x = 1$ generates s with probability $\prod_{k=1}^{L}((1 - s_k)q_k + s_k(1 - q_k))$, in other words, $a_k = q_k^r$ and $b_k = q_k$. As earlier we fix some error bound ϵ. From Lemma 3 we get immediately:

Lemma 4. *With the above notations for Problem 4, the influence of a query at q is at most $\min(q, 1 - q^r)$.* □

Problem 4 was stated, without loss of generality, for hypotheses r and 1. Similarly we may formulate it for hypotheses rx and x (for any positive x), which merely involves a coordinate transformation. We speak of the "influence of q on x" when we mean the influence of a query at q, with respect to Problem 4 for hypotheses rx and x. Clearly, the influence of q on x equals the influence of q^x on 1. Therefore Lemma 4 generalizes immediately to:

The influence of q on x is at most $\min(q^x, 1 - q^{rx})$.

Remember $y := \ln x$ from Section 2. By a slight abuse of notation, the phrase "influence of q on y" refers to the logarithmic coordinates, and Lemma 4 gets this form:

The influence of q on y is at most $\min(q^{e^y}, 1 - q^{re^y})$.

While q^{e^y} obviously decreases doubly exponentially with growing $y > 0$, it is also useful to have a simple upper bound for $1 - q^{re^y}$ when $y < 0$. Since $1 - e^{-z} \leq z$ for any variable z, we take z with $e^{-z} = q^{re^y}$ to obtain $1 - q^{re^y} \leq z = -\ln q^{re^y} = \ln(1/q)re^y$. Now we have:

The influence of q on y is at most $\min(q^{e^y}, \ln(1/q)re^y)$.

Finally we also transform q into t as introduced in Section 2, and we speak of the "influence of t on y", denoted $I_t(y)$. With $q = e^{-e^{-t}}$ and $\ln(1/q) = e^{-t}$, our influence lemma is in its final shape:

Lemma 5. $I_t(y) \leq \min(e^{-e^{y-t}}, re^{y-t})$. □

From this bound we get:

Lemma 6. *For every fixed t we have $\int_0^{\ln n} I_t(y)\, dy = \Theta(\ln r)$.*

Proof. For simplicity we bound the integral over the entire real axis. (Since $I_t(y)$ decreases rapidly on both sides of t, this is not even too generous.) The advantage is that we can assume $t = 0$ without loss of generality. We split the integral in two parts, at $y = -\ln r$. As $I_t(y)$ is a minimum of two functions, we can take either of them as an upper bound. Specifically we get $\int_{-\infty}^{\infty} I_t(y)\,dy < \int_{-\infty}^{-\ln r} re^y\,dy + \int_{-\ln r}^{\infty} e^{-e^y}\,dy = \int_{-\ln r}^{\infty} re^{-y}\,d(-y) + \int_{-\ln r}^{\infty} e^{-e^y}\,dy = re^{-\ln r} + \Theta(\ln r) = 1 + \Theta(\ln r)$. The second integral is $\Theta(\ln r)$ since both $e^{-e^{-\ln r}} = e^{-1/r}$ and (for instance) $e^{-e^0} = e^{-1}$ are between some positive constants, the function is monotone decreasing, and $\int_0^{\infty} e^{-e^y}\,dy = \Theta(1)$. □

The next lemma connects our "bipolar" number guessing problem to the problem we started from.

Lemma 7. *For every $r > 1$ and $0 < \delta < 1$ we have: Any strategy solving Problem 1 with error probability ϵ and competitive ratio $c := 1 + (r-1)\delta$ yields a strategy solving Problem 4 with hypotheses rx and x, for every $x \le n/r$, with error probabilities ϵ and δ.*

Proof. Imagine a searcher wants to solve an instance of Problem 1, and an adversary tells her that the target is either rx or x. Despite this strong help, in case that rx is the target, the searcher must still guess rx subject to an error probability ϵ, due to the definition of Problem 1. In the other case when the target is x, error probability δ means a competitive ratio of $(1-\delta) + r\delta = 1 + (r-1)\delta$. □

We are ready to state the main result of this section:

Theorem 1. *Any strategy for Problem 1, with fixed error probability ϵ and competitive ratio c, needs $\Omega(\ln n/\ln r)$ queries, where the constant factor may depend on ϵ.*

Proof. Fix some $r > c$ and $\delta = (c-1)/(r-1)$, hence $c = 1 + (r-1)\delta$. We choose $r = \Theta(c)$ large enough so that $D := 1 - \epsilon - \delta$ is positive. Due to Lemma 7, the set of queries must be powerful enough to solve Problem 4 with hypotheses rx and x, for every $x \le n/r$, with error probabilities ϵ and δ. In the case of no queries, the error tradeoff at every x would be simply $\delta(\epsilon) = 1 - \epsilon$. Since we need to reduce $\delta(\epsilon)$ down to our fixed δ, all queries together must have an influence at least $1 - \epsilon - \delta$ on x. In transformed coordinates this means $\sum_t I_t(y) \ge D$ for all $0 \le y \le \ln n - \ln r$, where the sum is taken over all t in our query set (multiple occurrences counted). Hence $\int_0^{\ln n - \ln r} \sum_t I_t(y)\,dy \ge D(\ln n - \ln r)$. Since Lemma 6 states $\int_0^{\ln n - \ln r} I_t(y)\,dy = \Theta(\ln r)$ regardless of t, the number of queries is at least $(\ln n - \ln r)D/\Theta(\ln r) = \Omega(\ln n/\ln r)$. □

Note that this integration argument also applies if the queries themselves are located according to some probability distribution, that is, Theorem 1 also holds for "fully randomized" strategies.

Theorem 1 only shows that the query number is logarithmic, for any fixed parameter values. But the proof method is not suited for deriving also good lower bounds on the hidden constant factor. For instance, this factor should increase to infinity when ϵ tends to 0. To reflect this behaviour in the lower bound, apparently the previous proof must be combined with some reduction between problem instances with different ϵ. We leave this topic here. Anyways, in practice one would apply some reasonable standard value like $\epsilon = 0.05$ rather than trading much more queries for smaller failure probabilities. A more relevant question, addressed in the next section, is which upper bounds we can accomplish.

5   Translation-Invariant Strategies and Upper Bounds

Theorem 1 states that $L/\ln n$ in Problem 1 must be at least some constant, depending on ϵ and c. In order to get upper bounds on $L/\ln n$ we consider the following "infinite extension" of Problem 1. This has merely formal reasons that will be explained below.

Problem 5: Given are some positive error probability $\epsilon < 1$, some $c > 1$ that we call the competitive ratio, and an "invisible" number x which can be any real number. A searcher can prepare countably infinitely many nonadaptive queries to an oracle as follows. A query specifies a number $q \in (0, 1)$, and the oracle gives answer 0 with probability q^x and answer 1 with probability $1 - q^x$. Based on the infinite string s of the binary answers, the searcher is supposed to output some number x' such that $\mathbf{Pr}[x' < x] \leq \epsilon$ and $\mathbf{E}[x'/x] \leq c$ holds for every x.

For Problem 5 we naturally consider the *density* of queries, i.e., the number of queries per length unit on the logarithmic axis, corresponding to $L/\ln n$ in Problem 1. We withhold a precise formal definition of density, because for our upper bound we will only study a particular strategy for which the notion of density is straightforward:

Remember that $y = \ln x$, and every query, with probability q of responding with 0, is matched to a query point t on the logarithmic axis through $q = e^{-e^{-t}}$. If y is the unknown target value (in logarithmic coordinates), the probability of answer 0 to a query at point t is $q^x = e^{-e^{y-t}}$. The logarithmic lower bound in Theorem 1 and the influence argument in its proof suggests that query points t should be spread evenly over the logarithmic axis. More specifically, we consider strategies where the query points t are placed equidistantly, with space u between neighbored points. We place our queries at points $t = ju + v$, where u is fixed, j loops over all integers, and v is a random shift being uniformly distributed, with $0 \leq v < u$. For every two-sided infinite binary sequence s of answers we also specify an y_s such that the output $y' = \ln x'$ is located y_s length units to the right of the point of the leftmost answer 0 in s (see details below). We call such strategies *translation-invariant* with density u^{-1} because, obviously, all translations of the y-axis are automorphisms. One should not worry about the uncountably infinitely many s; in practice we "cut out a finite segment" of this infinite strategy according to:

Lemma 8. *Any translation-invariant strategy for Problem 5 with bounds ϵ and c and density u^{-1} yields a strategy for the original Problem 1 that has asymptotically, i.e., for $n \to \infty$, the same characteristics as the given strategy: error probability ϵ, competitive ratio c, and $u^{-1} \ln n$ queries.*

Proof. (sketch) We simply take the query points in the interval from 0 to $\ln n$, plus some margins on both sides, whose lengths grow with n but slower than $\ln n$. Since even the total influence of the (infinitely many!) ignored queries on any point y, $0 \leq y \leq \ln n$, decreases exponentially with the margin length, the resulting finite strategy performs as the original strategy for Problem 5, subject to vanishing terms.   □

The reason for replacing Problem 1 with Problem 5 is its greater formal beauty. This way we skip some artificial treatment of the interval ends and obtain "clean" translation-invariance. In particular, in the calculations we can assume without loss of generality that $y = 0$, and the searcher does not know the shift of the coordinates (while in reality the searcher knows the coordinate system but not y). This will simplify the expressions a lot. Furthermore, the random shift v that we used to make our strategy translation-invariant does not sacrifice optimality: If, in any optimal strategy for Problem 5, the query points are first shifted randomly, the strategy remains optimal. To see this, simply note that the resulting strategy still respects the bounds ϵ and c at every y, if the original strategy did.

Next we show how to obtain the optimal values y_s for our specific strategy. For a given error probability and query density we want to minimize the competitive ratio. We need to consider only those two-sided infinite strings s that have a leftmost 0 and a rightmost 1. We call the segment bounded by these positions the significant segment. Clearly, all other response strings appear with total probability 0. We (arbitrarily) index the bits in each s so that $s_0 = 0$ is the leftmost 0, that is, $s_k = 1$ for all $k < 0$. The point on the y-axis where the leftmost query t with answer 0 is located is called the reference point.

The probability density of the event that string s appears, and its reference point is $ju + v$ (j integer, $0 \leq v < u$), is given by

$$f_s(ju + v) := u^{-1} \prod_k \left((1 - s_k)e^{-e^{-(k+j)u-v}} + s_k(1 - e^{-e^{-(k+j)u-v}}) \right)$$

where k loops over all integers, and the s_k are the bits of s as specified above.

Since, for each s, our strategy returns the point located y_s units to the right of the reference point t, the contribution of string s to the error probability (of having output $y' < 0$) amounts to $\int_{-\infty}^{-y_s} f_s(t)\, dt$. Hence our goal is to minimize $\sum_s \int_{-\infty}^{+\infty} e^{t+y_s} f_s(t)\, dt$ under the constraint $\sum_s \int_{-\infty}^{-y_s} f_s(t)\, dt \leq \epsilon$. To summarize:

Proposition 1. *For any fixed u, the solution to the problem of minimizing $\sum_s \int_{-\infty}^{+\infty} e^{t+y_s} f_s(t)\, dt$ under the constraint $\sum_s \int_{-\infty}^{-y_s} f_s(t)\, dt \leq \epsilon$ yields an upper bound on the competitive ratio c for Problem 1 when $u^{-1} \ln 2 \cdot \log_2 n$ queries are used.*   □

Now these bounds can be calculated by standard nonlinear constraint optimization problem solvers. It suffices to consider some finite set of the most likely strings s whose sum of probabilities is close enough to 1. We implemented the method using the Matlab features `fmincon` for optimization and `quadgk` for numerical integration. As a little illustration, Table 1 displays the competitive ratios for $\epsilon = 0.01, \ldots 0.05$ and $g \log_2 n$ pools, for $g = 0.5$ and $g = 1$.

Table 1. Some competitive ratios c

g	e 0.01	e 0.02	e 0.03	e 0.04	e 0.05
0.5	11.87	9.83	8.67	7.89	7.28
1.0	5.31	4.56	4.13	3.86	3.61

Of course, the optimizer also outputs the strategy variables y_s, here we do not show them due to limited space. For larger g it becomes harder to run the method in this form on a usual laptop computer. The denser the query points are, the more strings s have non-negligible probabilities, and the resulting large number of variables leads to slow convergence. However, these technical issues can be resolved by more computational power. One should also bear in mind that a strategy needs to be computed only once, for any given pair of input parameters g and ϵ, thus long waiting times might be acceptable. The only thing needed to apply the computed strategy is a look-up table of the y_s. Anyways, some optimality criterion for the problem could enable us to find the optimal strategies more efficiently than by this "naive" direct use of an optimizer.

For the original problem (of estimating the number d of defectives in an n-element set by group tests) we have also found and implemented an LP formulation. Clearly, the competitive ratios grow with n and should tend to the results for Problem 5 when $n \to \infty$. This behaviour is confirmed by our empirical results. Since our methods guarantee optimal competitive ratios for translation-invariant pooling designs, they improve upon the ad-hoc strategies in [5] where the problem was studied for the first time.

6   Open Questions

We studied the problem of estimating the number d of defective elements in a population of size n by randomized nonadaptive group tests, to within a constant factor c, and with a prescribed probability ϵ of underestimating d. A main result is that $\Omega(\log n)$ queries are needed, if the single pools are formed in a natural way by independent random choices. While this bound is intuitive, it has not been proved before, and quite some technical efforts were needed. It remains open how to show this lower bound also for arbitrary pools. A combination of our influence argument with Yao's lower bound technique may lead to an answer. The logarithmic lower bound also suggests that query points should be placed translation-invariant on the logarithmic axis; see details above. We gave such a strategy which allows numerical calculation of the output and competitive ratios, for any given query density and ϵ. One could also think of other translation-invariant strategies, for instance, query points may be chosen by a Poisson process, however this seems worse because then the density of actual query points can accidentally be low around the target value. In summary we conjecture that our strategy in Section 5 is already optimal, with respect to the constant factors and parameters, among all possible randomized strategies. But a proof (if it is true) would apparently require a different mathematical machinery. Disproving the conjecture would give interesting insights as well. Finally, the method proposed in Section 5 is a numerical one. A challenging question is whether the dependency of optimal competitive ratio, error probability and query number can be characterized in a closed analytical form.

Acknowledgments

This work has been supported by the Swedish Research Council (Vetenskapsrådet), grant no. 2007-6437, "Combinatorial inference algorithms – parameterization and clustering". We thank the referees for some helpful editorial remarks.

References

1. Chen, H.B., Hwang, F.K.: Exploring the Missing Link Among d-Separable, $\bar{d}$-Separable and d-Disjunct Matrices. Discr. Appl. Math. 155, 662–664 (2007)
2. Cheng, Y., Du, D.Z.: New Constructions of One- and Two-Stage Pooling Designs. J. Comp. Biol. 15, 195–205 (2008)
3. Clementi, A.E.F., Monti, A., Silvestri, R.: Selective Families, Superimposed Codes, and Broadcasting on Unknown Radio Networks. In: SODA 2001, pp. 709–718. ACM/SIAM (2001)
4. Cormode, G., Muthukrishnan, S.: What's Hot and What's Not: Tracking Most Frequent Items Dynamically. ACM Trans. Database Systems 30, 249–278 (2005)
5. Damaschke, P., Sheikh Muhammad, A.: Competitive Group Testing and Learning Hidden Vertex Covers with Minimum Adaptivity. In: Kutylowski, M., Gebala, M., Charatonik, W. (eds.) FCT 2009. LNCS, vol. 5699, pp. 84–95. Springer, Heidelberg (2009); Extended version to appear in Discr. Math. Algor. Appl.
6. De Bonis, A., Gasieniec, L., Vaccaro, U.: Optimal Two-Stage Algorithms for Group Testing Problems. SIAM J. Comp. 34, 1253–1270 (2005)
7. De Bonis, A., Vaccaro, U.: Constructions of Generalized Superimposed Codes with Applications to Group Testing and Conflict Resolution in Multiple Access Channels. Theor. Comp. Sc. 306, 223–243 (2003)
8. Dorfman, R.: The Detection of Defective Members of Large Populations. The Annals of Math. Stat. 14, 436–440 (1943)
9. Du, D.Z., Hwang, F.K.: Pooling Designs and Nonadaptive Group Testing. World Scientific, Singapore (2006)
10. Eppstein, D., Goodrich, M.T., Hirschberg, D.S.: Improved Combinatorial Group Testing Algorithms for Real-World Problem Sizes. SIAM J. Comp. 36, 1360–1375 (2007)

11. Gilbert, A.C., Iwen, M.A., Strauss, M.J.: Group Testing and Sparse Signal Recovery. In: Asilomar Conf. on Signals, Systems, and Computers 2008, pp. 1059–1063 (2008)
12. Goodrich, M.T., Hirschberg, D.S.: Improved Adaptive Group Testing Algorithms with Applications to Multiple Access Channels and Dead Sensor Diagnosis. J. Comb. Optim. 15, 95–121 (2008)
13. Kahng, A.B., Reda, S.: New and Improved BIST Diagnosis Methods from Combinatorial Group Testing Theory. IEEE Trans. CAD of Integr. Circuits and Systems 25, 533–543 (2006)
14. Schlaghoff, J., Triesch, E.: Improved Results for Competitive Group Testing. Comb. Prob. and Comp. 14, 191–202 (2005)

A Search-Based Approach to the Railway Rolling Stock Allocation Problem

Tomoshi Otsuki, Hideyuki Aisu, and Toshiaki Tanaka

Toshiba Corporation, 1, Komukai-Toshiba-cho,
Saiwai-ku, Kawasaki, 212-8582, Japan
tomoshi1.otsuki@toshiba.co.jp

Abstract. Experts working for railway operators still have to devote much time and effort to creating plans for rolling stock allocation. In this paper, we formulate the railway rolling stock allocation problem as a set partitioning multi-commodity flow (SPMCF) problem and we propose a search-based heuristic approach for SPMCF. We show that our approach can obtain an approximate solution near the optimum in shorter time than CPLEX for real-life problems. Since our approach deals with a wide variety of constraint expressions, it would be applicable for developing practical plans automatically for many railway operators.

1 Introduction

Railway rolling stock allocation for a given diagram is a problem that has been studied for a long time. Some simplified classes of the problem such as those with only periodic constraints can be solved easily. But realistic objectives and constraints are complex, with a variety of constraints derived from circumstances of trains or railway depots and from non-periodic maintenance schedules. Therefore, at many railway operators, experts still have to devote much time and effort to creating plans for rolling stock allocation manually in practice. Additionally, in the case that diagrams must be changed owing to traffic accidents or other eventualities, experts must modify original plans rapidly, which is frequently necessary at peak hours in urban areas in Japan.

In this paper, we formulate the rolling stock allocation problem as a set partitioning multi-commodity flow(SPMCF) problem, and we propose a search-based approach for SPMCF. We show that the approach can obtain an approximate solution near the optimum in a much shorter time than CPLEX for real-life problems.

The remainder of this paper is organized as follows. Section 2 provides a detailed description of the rolling stock allocation problem and its formulation as an SPMCF problem, and refers to related works. Section 3 proposes a search-based approach for SPMCF. In Sect. 4, computational results are reported and compared with those of CPLEX. Finally, Sect. 5 is devoted to the conclusion.

W. Wu and O. Daescu (Eds.): COCOA 2010, Part II, LNCS 6509, pp. 131–143, 2010.

2   Railway Rolling Stock Allocation Problem

We begin with some basic definitions. Let d_s be an *initial day* of the scheduling, and d_f be a *final day* of the scheduling. That is, a *planning period* is from the d_s th day to the d_f th day. Since we ignore coupling/decoupling of rolling stock in this paper, a *train* is considered to be the minimum unit of scheduling. And let $\mathcal{H} = \{h_1, h_2, \ldots\}$ be a set of trains available in the planning period. A *place* is where a train can be left for a while such as a train depot, and let $\mathcal{R} = \{r_1, r_2, \ldots\}$ be a set of places available in the planning period. A *route* is an in-service/out-of-service trip to be covered with a train between two given places with given departure and arrival times. Let $\mathcal{U} = \{u_1, u_2, \ldots\}$ be a set of routes available in the planning period. And for every $u \in \mathcal{U}$, a *departure place* $\mathbf{DEP}(u) \in \mathcal{R}$, an *arrival place* $\mathbf{ARV}(u) \in \mathcal{R}$, a *departure time* $\mathbf{dept}(u)$, and an *arrival time* $\mathbf{arvt}(u)$ are given. For our purpose, we do not need to consider more details about a route such as intermediate stations or duration time.

Figure 1 illustrates an instance of routes with solid lines on a railway diagram whose vertical axis is a series of places and horizontal axis is a time scale. Route 1 departs from the train depot α at 0615 and after some tracks it arrives at the train depot β at 0810. And route 2 departs from the depot γ at 0620 and arrives at the train depot α at 0755.

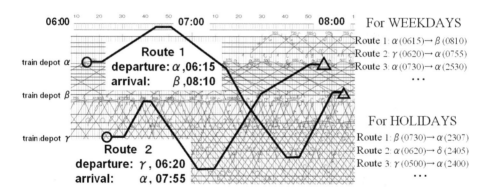

Fig. 1. An instance of routes on a diagram

The goal of the *rolling stock allocation problem* is to determine an assignment of given routes to given trains. A *route sequence* $P_h = (p_{h,1}, p_{h,2}, \ldots, p_{h,N_h})$ is a sequence of routes assigned to a train h in increasing order with respect to the arrival time $\mathbf{arvt}(\cdot)$, where N_h is the number of routes assigned to h. Supposing each train must depart from *first departure place* $s_h \in \mathcal{R}$ on the initial day d_s, we add an *origin route* $p_{h,0}$ at the beginning of P_h for convenience, where $\mathbf{ARV}(p_{h,0}) = s_h$ and $\mathbf{arvt}(p_{h,0}) = -\infty$. We also add a *final route* N^- as p_{h+1} at the end of P_h, if necessary.

We suppose trains cannot transfer by means other than routes; thus, they have to meet *time and place connection constraints* represented as $\mathbf{ARV}(p_{h,k}) = \mathbf{DEP}(p_{h,k+1})$ and $\mathbf{arvt}(p_{h,k}) < \mathbf{dept}(p_{h,k+1})$ for $0 \leq k \leq N_h - 1$. If these conditions permit, we may assign multiple routes to a single train on a single day, or may assign no routes to a single train over several days.

Each train must have several types of maintenance, such as interior/exterior cleaning, some repairs, and some checkups, regularly but not periodically. Maintenance is done in a specific train depot and takes a specific amount of time; thus, the maintenance is done between routes or on a day for which no routes are assigned. Consequently, the routes assignable to the train on the day when maintenance is done are limited.

Supposing the maintenance schedule is fixed, we can represent maintenance constraints through *hard constraints*, the conditions trains have to satisfy, and *soft constraints*, the conditions trains should satisfy as long as they can. Here, assuming that these violations can be divided into those of respective trains, and that we can sum up each train's violations to get the total number of violations, we formulate the rolling stock allocation problem as that of minimizing $\sum_h \mathbf{Soft}(P_h)$ on the condition that $\mathbf{Hard}(P_h) = 0$, where $\mathbf{Hard}(P_h)$ and $\mathbf{Soft}(P_h)$ are the total number of hard and soft constraint violations of train h, respectively.

Most of our discussion below in this paper will be independent of the particular choice of $\mathbf{Soft}(P_h)$ and $\mathbf{Hard}(P_h)$. Indeed, our approach is tractable for a wide range of other constraints such as circumstances of train depots or the train itself.

2.1 SPMCF Problem Formulation

If we regard each route as a node, each route sequence as a flow, and each train as a commodity, we can formulate the rolling stock allocation problem as the set partitioning multi-commodity flow formulation, denoted SPMCF, as follows.

SPMCF is defined over the network $G(V, E)$ comprising node set V and arc set E. SPMCF contains binary decision variable f, where f_{uv}^h equals 1 if the quantity of commodity h is assigned to arc (u, v). Supposing node u has supply of commodity h, denoted b_u^h, equal to 1 if u is the destination node for h, equal to -1 if u is the origin node for h, and equal to 0 otherwise.

The node-arc SPMCF formulation is:

$$\min \quad \sum_{h \in \mathcal{H}} \mathbf{Soft}(P_h) \tag{1}$$

$$\text{s.t.} \quad \sum_{v:(v,u) \in \mathbf{E}} f_{vu}^h - \sum_{v:(u,v) \in \mathbf{E}} f_{uv}^h = b_u^h \quad \forall u \in V, \forall h \in \mathcal{H}, \tag{2}$$

$$\sum_{h \in \mathcal{H}} \left(\sum_{v:(u,v) \in \mathbf{E}} f_{uv}^h \right) = 1 \quad \forall u \in V, \tag{3}$$

$$\mathbf{Hard}(P_h) = 0, \tag{4}$$

$$f_{uv}^h \in \{0, 1\} \quad \forall u, v \in V, \forall h \in \mathcal{H}. \tag{5}$$

Equation (1) is an objective function that minimizes the total number of violations of soft constraints. Then, (2) is the flow conservation equation. Since

$\sum_{v:(u,v)\in \mathbf{E}} f_{uv}^h$ represents the flow of commodity h out of a node u, the left-hand side of (3) denotes the total flow through the node u. Thus, (3) represents the constraints of set partitioning that means every node is covered just once by the flow of any commodity h. Finally, (4) represents that all of the hard constraints must be satisfied.

2.2 Related Works

Studies of rolling stock allocation are summarized in [3], most of them focus not on routing but on efficient freight scheduling or combination of locomotives and different classes of cars. Problems of this kind depend on the circumstances of railway network topology or typical formation of trains.

There are two main models of rolling stock allocation problems: one uses multi-commodity flow formulation and the other uses circulation formulation.

For multi-commodity flow formulation, Benders' decomposition [4] and the path enumeration approach [6] have been applied. Column generation [7], which is a similar path generation-based approach, and its extended method branch-and-price are well known. But since these path generation-based methods depend on particular properties of the problems, it would be difficult to apply these methods directly for set partitioning problems such as SPMCF.

On the other hand, [1], [2], and [5] discusses cyclic solutions of rolling stock allocation problems, supposing that all the trains are identical and most of the constraints are periodic. However, since in practical situations there are many non-periodic constraints such as emergent events or human-caused circumstances, it is difficult to apply them widely.

3   Solution Approach

In this section, we provide the solution approach of the SPMCF problem that has two principal processes: a greedy construction process for obtaining an initial solution and a backtrack search process for improving the solution.

3.1 Some Definitions

We begin with some additional definitions.

First, an *assignment* denotes the whole set of route sequence P_h for all trains $h \in \mathcal{H}$ in which each P_h satisfies time and place connection constraints and every route $u \in \mathcal{U}$ is assigned to any P_h just once. Thus, an assignment corresponds to a feasible solution of SPMCF without hard constraints represented by (4). Figure 2 illustrates an instance of assignment for G(V,E).

Next, we define $\mathbf{origin}_A$, $\mathbf{next}_A$, $\mathbf{next}_A^{-1}$ on an assignment A as follows:

$$\begin{aligned} \mathbf{origin}_A(h) &= p_{h,0}, \\ \mathbf{next}_A(p_{h,k}) &= p_{h,k+1} \quad \mathbf{for}(k=0,1,\ldots N_h), \\ \mathbf{next}_A^{-1}(p_{h,k+1}) &= p_{h,k} \quad \mathbf{for}(k=0,1,\ldots N_h). \end{aligned} \quad (6)$$

where $P_h = (p_{h,0}, p_{h,1}, \ldots, p_{h,N_h}, N^-)$ is a route sequence of train h on A.

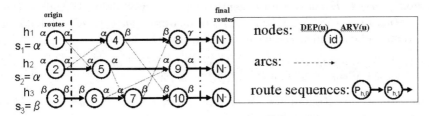

Fig. 2. An instance of an assignment for G(V,E)

And on an assignment A, $\mathbf{Bad}_A(h)$ is the total number of constraint violations of P_h, and a train that is $\mathbf{Bad}_A(h) > 0$ is called a *bad train*. Additionally, $\mathbf{penalty}(A)$ is the total number of constraint violations in A. Thus, $\mathbf{penalty}(A) = \sum_{h \in \mathcal{H}} \mathbf{Bad}_A(h)$ holds.

In addition, a *swap* $\sigma(u, v)$ on an assignment A is a transform operation into another assignment A' that satisfies $\mathbf{next}(u) = v'$ and $\mathbf{next}(v) = u'$ when $u, v \in V$, $u' = \mathbf{next}_A(u)$, $v' = \mathbf{next}_A(v)$, $(u, v') \in E$, and $(v, u') \in E$ are all satisfied. (If all these constraints are satisfied, we call the transform *well-defined*) And the relation between A and A' is represented as $A' = \sigma(u, v) \cdot A$.

Figure 3 illustrates an instance of a swap operation. Though $P_1 = (1, 3, 5)$ and $P_2 = (2, 4, 6)$ hold on an assignment A on the left, on an assignment $A' = \sigma(1, 2) \cdot A$ on the right, route sequences are changed into $P_1 = (1, 4, 6)$ and $P_2 = (2, 3, 5)$ by the swap $\sigma(1, 2)$.

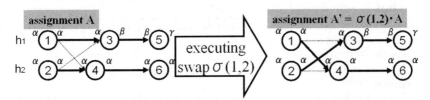

Fig. 3. An instance of a swap operation

Our solution approach is based on iteratively applying this swap operation on an initial assignment. Effectiveness of this approach is supported by the following lemma.

Lemma 1. *For any assignment pairs A_0 and A_g, A_g can be obtained by applying swaps on A_0 several times.*

Proof. By induction on the size of $\#\{u | \mathbf{next}_A(u) \neq \mathbf{next}_{A_g}(u)\}$, the case of size 0 is obvious. Now let $\#\{u | \mathbf{next}_A(u) \neq \mathbf{next}_{A_g}(u)\} = K$ and let u have maximum $\mathbf{arvt}(u)$ of all routes that satisfy $\mathbf{next}_A(u) \neq \mathbf{next}_{A_g}(u)$. Here, let $v \equiv \mathbf{next}_A^{-1}(\mathbf{next}_{A_g}(u))$. Then $\mathbf{arvt}(v) \leq \mathbf{arvt}(u)$ holds because of the maximality of $\mathbf{arvt}(u)$. Thus, $\mathbf{arvt}(v) \leq \mathbf{arvt}(u) < \mathbf{dept}(\mathbf{next}_{A_g}(u))$ holds, so that

$(v, \mathbf{next}_{A_g}(u)) \in E$ holds. Consequently, $\sigma(u, v)$ is well defined. And between $A' = \sigma(u, v) \cdot A$ and A_g, $\#\{u | \mathbf{next}_{A'}(u) \neq \mathbf{next}_{A_g}(u)\} \leq K - 1$ holds by induction, and consequently we can derive lemma 1.

Now that the swap operation has given us the metric from a basis assignment to every assignment, we define **Reach**(A, k) as follows:

Definition 1. *Let* **Reach**(A, k) *be the set of assignments that are reached by k times swaps from an assignment A.*

And we derive the following lemma for the discussion below.

Lemma 2. *Let* $i, j, k, l \in V$ *be distinct, and if* $\sigma(k, l) \cdot \sigma(i, j) \cdot A$ *is well-defined, then* $\sigma(k, l) \cdot \sigma(i, j) \cdot A = \sigma(i, j) \cdot \sigma(k, l) \cdot A$ *holds.*

Proof. Since i, j, k, l are distinct, $\sigma(i, j)$ and $\sigma(k, l)$ doesn't change the connection $k \rightarrow \mathbf{next}_A(k)$, $l \rightarrow \mathbf{next}_A(l)$, and $i \rightarrow \mathbf{next}_{\sigma(i,j) \cdot A}(i)$, $j \rightarrow \mathbf{next}_{\sigma(i,j) \cdot A}(j)$, respectively. Thus, $\sigma(k, l)$ and $\sigma(i, j)$ are well-defined for A and $\sigma(k, l) \cdot A$, respectively.

And each of $\sigma(k, l) \cdot \sigma(i, j) \cdot A$ and $\sigma(i, j) \cdot \sigma(k, l) \cdot A$ is the assignment that differs from A in that $i \rightarrow \mathbf{next}_A(j)$, $j \rightarrow \mathbf{next}_A(i)$, $k \rightarrow \mathbf{next}_A(l)$, and $l \rightarrow \mathbf{next}_A(k)$ hold, so that we can derive lemma 2.

3.2   Framework of Solution Approach

The framework of solution approach is as follows:

Algorithm 1. schedule(d_s, d_f)

Require: d_s:initial day, d_f: final day, N:upper limit of #paths
1: **for** $d = d_s$ to d_f **do**
2:     #paths $\leftarrow 0$
3:     $A \leftarrow$ the result of the greedy construction process by d th day
4:     **repeat**
5:         $A \leftarrow$ **backtrack**(A, N) (supposing **penalty**$(A) \equiv$ **Soft**(A)+c·**Hard**(A))
6:     **until** (**Soft**$(A) +$ **Hard**$(A) == 0$ || #paths $> N$)
7:     **repeat**
8:         $A \leftarrow$ **backtrack**(A, ∞) (supposing **penalty**$(A) \equiv$ **Hard**(A))
9:     **until** (**Hard**$(A) == 0$))
10: **end for**

Each day's process consists of a greedy construction process for obtaining an initial solution and a two-step backtrack search process for improving the solution. The *greedy construction process* (details are discussed below) is a process for obtaining an initial assignment for given routes, and the *backtrack search process* **backtrack**(A, N) (details are discussed below) is a process for reducing the total number of constraint violations, where N is the limit of the number of paths searched, denoted #paths. In the first backtrack search process, we consider both hard and soft constraints and reduce the weighted sum,

penalty$(A) \equiv$ **Soft**$(A) + c \cdot$ **Hard**(A), until #path exceeds N, where **Soft**(A) and **Hard**(A) are the total number of soft and hard constraint violations, respectively, and where $c > 1.0$. And at the second backtrack search process, we consider only hard constraints and reduce **penalty**$(A) \equiv$ **Hard**(A) until we obtain the assignment A that satisfies **Hard**$(A) = 0$.

Since $c > 1.0$, the first backtrack search process put more weight on hard constraints than on soft constraints. In total, we obtain the solution that minimizes **Soft**(A) on the condition that **Hard**$(A) = 0$. (We fixed $c = 10.0$ in computational experiments in this paper)

3.3 Greedy Construction Process for Obtaining Initial Solution

In the greedy construction process, we repeat the following steps. First, we select the unassigned route u that has minimum **dept**(u), and next we assign u to a train in order to satisfy time and place connection constraints.

As a result of this process, if we succeed in assigning all routes, we obtain an assignment. And inversely the following fact holds.

Lemma 3. *If an assignment exists, the greedy construction process finds one of the assignments.*

Proof. Let v be a first route that is failed for any train by the greedy construction process. Since routes are assigned in the ascending order of their departure time, v has the maximum **dept**(v). And let **PreSET**$(v) = \{u|(u,v) \in E\}$, and $u_1, u_2, \ldots u_k$ are distinct nodes of **PreSET**(v), where k is the size of **PreSET**(v). Since we failed to assign v, **next**(u_k) is predefined. Let $v_k =$ **next**(u_k). Here, **PreSET**$(v_k) \subseteq$ **PreSET**(v) holds, because of the maximality of **dept**(v). So in the greedy construction process, we have to determine the one-to-one correspondence from K routes $u_1, u_2, \ldots u_k$ to $K + 1$ routes $v, v_1, v_2, \ldots v_k$. But it's impossible and Lemma 3 follows.

3.4 Backtracking Search Process for Improve Solution

Now that we have obtained an assignment A_0 by the greedy construction process, we need to improve the assignment for a better solution because generally **penalty**$(A_0) > 0$. Thus, we propose the backtracking-based hill climbing algorithm for finding the assignment A that satisfies **penalty**$(A) <$ **penalty**(A_0), denoted an **SMB***(Swap-path Multiple Backtracking) algorithm*.

In the backtrack search process, we adopt the iterative deepening technique [8] as follows, using the property that the SMB algorithm with K swaps searches assignments in **Reach**(A_0, K).

In the codes below, we add a swap threshold k_0 by 1, and for each threshold k_0, we search assignments in **Reach**(A_0, k_0) by **SMB**(k_0, h, A_0) described below, where h is one of the bad trains on A_0. And if we succeed in finding improved solution A that satisfies **penalty**$(A) <$ **penalty**(A_0), we get back $k_0 = 1$ and search for the improved solution again. This parameter k_0 plays the important role of widening search space gradually, and, as a result, reducing total calculation cost.

Algorithm 2. backtrack(A_0, N)

Require: A_0:initial assignment, N:upper limit of #paths, $k_{\mathbf{max}}$:upper limit of swaps
1: **for** $k_0 = 1$ to $k_{\mathbf{max}}$ **do**
2:     **for all** $h \in \{h' | \mathbf{Bad}_{A_0}(h') > 0\}$ **do**
3:         **if** ($\mathbf{SMB}(k_0, h, A_0) == \mathbf{TRUE} \; \| \; \#\text{paths} > N$) **then**
4:             **return** A_g
5:         **end if**
6:     **end for**
7: **end for**

Details of SMB Algorithm Here, we explain the detail of **SMB** algorithm.

First, let a *swap path* be a path $Q = \{q_1, q_2, \ldots\}$ on $G(V, E)$, each of whose arcs (q_k, q_{k+1}) $(k = 1, 2, \ldots)$ satisfies $\mathbf{next}(q_k) = q_{k+1}$ or satisfies $(\mathbf{next}_A^{-1}(q_{k+1}),$ $\mathbf{next}_A(q_k)) \in E$.

Under this assumption, we obtain swaps $\sigma(q_k, \mathbf{next}_A^{-1}(q_{k+1}))$ from corresponding arcs (q_k, q_{k+1})s that satisfy $\mathbf{next}(q_k) \neq q_{k+1}$, on a swap path $Q = \{q_1, q_2, \ldots\}$ on an assignment A, as in the top chart of Fig.4. And we obtain a new assignment A' from these swaps.

In the bottom chart of Fig.4, $Q = (1, 5, 7, 10)$ is an instance of a swap path, since $\sigma(1, 2)$ and $\sigma(5, 6)$ are well-defined, respectively, since $(2, 4) \in E$ and $(6, 9) \in E$ hold, respectively. Consequently, we obtain a new assignment $A' = \sigma(5, 6) \cdot \sigma(1, 2) \cdot A$ from this swap path Q.

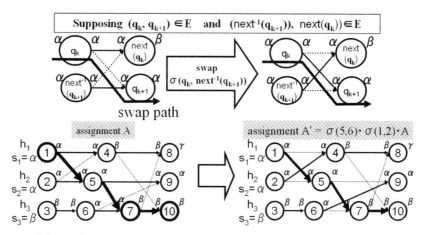

Fig. 4. An instance of a swap path and obtaining a new assignment

Using this correspondence between swap paths on A and new assignments A', we construct the **SMB** algorithm.

We explain the pseudo codes of **SMB** algorithm as follows. The pseudo codes below consist of a path generation part and a leaf evaluation part.

First, in the path generation part, we generate swap paths of a base train h that originates from the node $\mathbf{origin}_A(h)$ within K swaps, for example in the depth-first order. And executing swap operations corresponding to the swap path Q on an assignment A, we obtain a new assignment A'.

Next, in the leaf evaluation part, first we judge if $\mathbf{Bad}_{A'}(h) = 0$ holds or not. And if A' satisfies $\mathbf{Bad}_{A'}(h) = 0$ and $\mathbf{penalty}(A') < \mathbf{penalty}(A_0)$, $\mathbf{SMB}(K, h, A)$ returns $\mathbf{TRUE}$, where A_0 is an initial assignment. And if $\mathbf{penalty}(A') \geq \mathbf{penalty}(A_0)$ holds and the number of residual swaps $K - K'$ satisfies $K - K' > 0$, we call $\mathbf{SMB}$ recursively with setting $h \equiv h_{new}$, where h_{new} is a new bad train that satisfies $P_{h_{new}}(A) \neq P_{h_{new}}(A_0)$, where $P_h(A)$ represents a route consequence of train h on A.

Algorithm 3. SMB(K, h, A)

Require: K:swap threshold, h:base train, A:current assignment, A_0:initial assignment
Ensure: if return value is $\mathbf{TRUE}$, A_g satisfies $\mathbf{penalty}(A_g) < \mathbf{penalty}(A_0)$
1: **for all** Q: swap paths originating from $\mathbf{origin}_A(h)$ on A within K swaps **do**
2:     #paths $\leftarrow$ #paths $+1$
3:     $A' \leftarrow$ the assignment obtained by applying swaps corresponding to Q on A
4:     $k' \leftarrow$ the number of swaps corresponding to Q on A
5:     **if** $(\mathbf{Bad}_{A'}(h) == 0)$ **then**
6:         **if** $(\mathbf{penalty}(A') < \mathbf{penalty}(A_0))$ **then**
7:             $A_g \leftarrow A'$
8:             **return TRUE**
9:         **end if**
10:        $h_{new} \leftarrow$ one of the $h \in \{h' | \in \mathbf{Bad}_A(h') > 0 \ \&\& \ P_h(A) \neq P_h(A')\}$
11:        **return** $\mathbf{SMB}(K - k', h_{new}, A')$
12:    **end if**
13: **end for**
14: **return FALSE**

Regarding the search space of $\mathbf{SMB}$ algorithm, the theorem holds as follows:

Theorem 1. *Let A_g be an assignment that satisfies $\mathbf{penalty}(A_g) = 0$, and A_g is obtained by K times swaps from A_0 and the relation*

$$A_g = \sigma_K(u_K, v_K) \cdots \sigma_2(u_2, v_2) \cdot \sigma_1(u_1, v_1) \cdot A_0 \qquad (7)$$

holds. Here if all of $u_1, u_2, \ldots, u_K$ and $v_1, v_2, \ldots, v_K$ are distinct, then $\mathbf{SMB}(K, h_0, A_0)$ returns $\mathbf{TRUE}$, where h_0 is one of the bad trains on an assignment A_0.

Proof. By induction on the number of swaps, the case of the number of swaps 0 is obvious. Here, since all of $u_1, u_2, \ldots, u_K$ and $v_1, v_2, \ldots, v_K$ are distinct, the swaps $\sigma_1(u_1, v_1), \sigma_2(u_2, v_2), \ldots, \sigma_K(u_K, v_K)$ are all commutable from lemma 2. Thus, without a loss of generality, $k' > 0$ exists such that all $u_1 \ldots u_{k'}$ are on $P_{h_0}(A)$ and any other u_ks and v_ks are not on $P_{h_0}(A)$. In this case, $\mathbf{SMB}(K, h_0, A_0)$ enumerates a swap path Q that corresponds to swaps $\sigma_1 \ldots \sigma_k$, otherwise returning $\mathbf{TRUE}$. And if $A' = \sigma_K(u'_k, v'_k) \cdots \sigma_2(u_2, v_2) \cdot \sigma_1(u_1, v_1) \cdot A$ satisfies $\mathbf{penalty}(A') < \mathbf{penalty}(A_0)$, it returns $\mathbf{TRUE}$.

Thus, the residual case is that $\mathbf{penalty}(A') \geq \mathbf{penalty}(A_0)$. In this case, for any bad train h' that satisfies $P_{h'}(A') \neq P_{h'}(A_0)$, $P_h(A')$ includes at least one of $u_{k'+1} \ldots u_K$ or $v_{k'+1} \ldots v_K$. Consequently, whether $\mathbf{SMB}(K, h_0, A_0)$ returns $\mathbf{TRUE}$ or not, is equivalent to whether $\mathbf{SMB}(K - k', h', A')$ returns $\mathbf{TRUE}$ or not. By induction, the latter is $\mathbf{TRUE}$, so that theorem 1 holds.

This theorem suggests that $\mathbf{SMB}(K, h_0, A_0)$ searches the space within $\mathbf{Reach}(A_0, K)$ and assures that we can find one of the solutions in $\mathbf{Reach}(A_0, K)$, under the condition described in the theorem. From the numerical experiments, we observe few cases in which we can't find the solution in the $\mathbf{Reach}(A_0, K)$ by $\mathbf{SMB}(K, h_0, A_0)$. Moreover, we can use a multiple-start search technique just in case, for example, using pseudo-random number sequence in the greedy construction process in order to generate different initial assignments.

Improvement Heuristic. Lastly, we describe the improvement heuristic of the SMB algorithm.

Let A' be an assignment obtained after executing $K_0 - K$ times swaps from the assignment A. Here, if the probability of obtaining the assignment A_g that satisfies $\mathbf{penalty}(A_g) < \mathbf{penalty}(A_0)$ with residual K swaps is very small, the following pruning technique is expected to be effective.

If A' satisfies the following pruning condition:

$$\mathbf{penalty}(A') - \mathbf{penalty}(A_0) > K \cdot E(\delta \mathbf{Bad}), \qquad (8)$$

we revise so as not to generate additional swap paths from then on, where $E(\delta \mathbf{Bad})$ is an expected value of change of $\mathbf{penalty}(A)$ per swap.

Moreover, if the prune condition holds for $A' = \sigma_k, \ldots, \sigma_2 \cdot \sigma_1 \cdot A$, we revise so as not to generate all the swap paths whose initial corresponding swaps are the same as $\sigma_1, \sigma_2, \ldots, \sigma_k$. Thus, if we generate paths in the depth-first order, this pruning technique is more effective.

4   Computational Experiments

To measure the performance of the proposed approach described in Sect. 3, computational experiments were performed on practical data from a railway company. We first describe the test instance used. Secondly, we compare the solution time and quality of the proposed approach by comparison with CPLEX. The experiments were performed on a computer with Intel Xeon E5345 2.33GHz Processor and 16GB RAM.

4.1   Description of Datasets

Datasets used here are based on a monthly rolling stock allocation schedule that is a practical schedule in use, filled with routes and maintenance for all the trains, which keeps all the hard constraints.

In this schedule, the number of trains $|\mathcal{H}|$ is 47, the numbers of daily routes on weekdays and holidays are 54 and 25, respectively, and the number of places

$|\mathcal{R}|$ is 13. And the number of types of maintenance is 14, and the total number of times maintenance is done is 276 for all the trains and the entire planning period.

From these datasets, changing the pair of d_s and d_f as follows, we generate $168(= 7 \times 24)$ test cases. First we change d_s from 1 to 7, and then we change d_f in order for the planning period length to be from 1 to 24. For example, in the case that the planning period length is D, we set $d_f = d_s + D - 1$.

For each test case, we use $\mathcal{R}$, $\mathcal{H}$, $\mathcal{U}$ defined in the planning period from d_s to d_f and we create the schedule from d_s to d_f. And we set s_h as $d_s - 1$ th day's arrival place of train h in the practical schedule. Under these conditions, the original practical schedule is certainly a solution, and the problem for each test case is feasible.

Additionally, 3 hard constraint types and 3 soft constraint types exist in the test set as follows, where w, t or t_2, h, and r are specific practical values of maintenance, times, trains and places, respectively.

- Hard Constraints
  - Routes assignable to the train h on the day when maintenance w is done are limited.
  - Routes assignable to the train h are limited.
  - Every train whose earliest route is from train depot r, must depart by t on the day.
- Soft Constraints
  - Every train that arrives at a depot by t on a day should depart after t_2 on the next day.
  - Every train on which maintenance w is done on a day should be assigned more than one route on the next day.
  - Every train on which maintenance w is done on a day should depart after t on the next day.

4.2   Comparison with SMB and CPLEX

For these test cases, we compare the solution time and quality of the proposed approach with CPLEX. Here, high quality means that the value of soft constraint violates **Soft**(A) is small.

In the proposed approach, increasing the value of N, which denotes the search paths per day, resulted in a high-quality solution, but more solution time is required. So we try multiple Ns. Thus, the experiments were performed using the following 4 approaches.

- SMB algorithm-based approach when $N = 4,000$(abbreviated as SMB4K)
- SMB algorithm-based approach when $N = 16,000$(abbreviated as SMB16K)
- SMB algorithm-based approach when $N = 160,000$(abbreviated as SMB160K)
- Solution of (1)-(5) solved by ILOG CPLEX 11.2.1 (abbreviated as CPLEX)

Note that in CPLEX, we have used the SPMCF formulation itself basically but we have revised to eliminate intermediate flow variables, considering that routes are defined daily in our test cases. As a result, the number of 0-1 variables is about 100,000 even when the planning period length equals 24.

4.3   Computational Results

Figure 5 illustrates the results of the approaches of SMB and CPLEX. The left-hand chart indicates the relation between solution time and the planning period length, and the right-hand chart indicates the relation between the penalty value and the planning period length. And the plotted value for each planning period length is the average of the 7 test case results by changing the initial day d_s from 1 to 7.

CPLEX certainly obtains the optimal solution, whereas the SMB-based solution is a heuristic that does not always find the optimal solution. Thus, in SMB, we average the 5 trials, using pseudo-random number sequence in the greedy construction process.

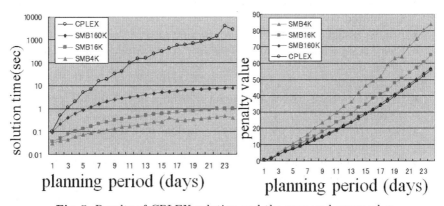

Fig. 5. Results of CPLEX solution and the proposed approaches

We first observe from the left-hand chart that the solution time of CPLEX drastically increases as planning period length increases and is more than 1000 seconds when planning period length is over 20 days. On the other hand, we observe that the solution time of SMB is less than 10 seconds in the case of SMB160K, and especially in the case of SMB4K, it is less than 1 second. That is, the solution time of proposed approach is about 1/10 to 1/100 of CPLEX's, and only for obtaining feasible solutions, it is less than 1 second.

We next observe from the right-hand chart that the solution quality of SMB tends to approach the curve of CPLEX as N increases, and the approximation ratio comes to about 104 (%) in the case of SMB160K.

Consequently, SMB160K attains a solution near the optimum in the solution time that is much shorter than CPLEX's. And our rolling stock allocation schedule is of actual use with a little expert adjustment. Consequently, we succeed in creating practical plans.

5 Conclusion

In this paper, we formulate the railway rolling stock allocation problem as the SPMCF formulation that minimizes the violation of soft constraints on the condition that all the hard constraints hold. And we propose a search-based heuristic approach for SPMCF. Our approach can obtain an approximate solution near the optimum in much shorter time than CPLEX for real-life problems. This high-speed capability of proposed approach would be effective for wide applications, for example, for creating the simultaneous scheduling of routes and maintenance. Since our approach is independent of the particular choice of soft constraints and hard constraints, it would be applicable for developing practical plans automatically for many railway operators.

References

1. Abbink, E., Berg, V.D., Kroon, L., Salomon, M.: Allocation of Railway Rolling Stock for Passenger Trains. Transportation Science 38(1), 33–41 (2004)
2. Alfieri, A., Groot, R., Kroon, L., Schrijver, A.: Efficient circulation of railway rolling stock. Transportation Science 40(3), 378–391 (2006)
3. Cordeau, J., Toth, P., Vigo, D.: A Survey of Optimization Models for Train Routing and Scheduling. Transportation Science 32(4), 380–404 (1998)
4. Cordeau, J., Soumis, F., Desrosiers, J.: A Benders Decomposition Appoach for the Locomotive and Car Assignment Problem. Transportation Science 34(2), 133–149 (2000)
5. Erlebach, T., et al.: On the Complexity of Train Assignment Problems. In: Eades, P., Takaoka, T. (eds.) ISAAC 2001. LNCS, vol. 2223, pp. 390–402. Springer, Heidelberg (2001)
6. Krishna, C., et al.: New Approaches for Solving the Block-to-train Assignment Problem. Networks 51(1), 48–62 (2007)
7. Lettovsky, L., Johnson, E.L., Nemhauser, G.L.: Airline crew recovery. Transportation Science 34(4), 337–348 (2000)
8. Russell, S., Norvig, P.: Artificial Intelligence: A Modern Approach, 2nd edn. Prentice Hall, Englewood Cliffs (2003)

Approximation Algorithm for the Minimum Directed Tree Cover

Viet Hung Nguyen

LIP6, Université Pierre et Marie Curie Paris 6, 4 place Jussieu, Paris, France

Abstract. Given a directed graph G with non negative cost on the arcs, a directed tree cover of G is a rooted directed tree such that either head or tail (or both of them) of every arc in G is touched by T. The minimum directed tree cover problem (DTCP) is to find a directed tree cover of minimum cost. The problem is known to be NP-hard. In this paper, we show that the weighted Set Cover Problem (SCP) is a special case of DTCP. Hence, one can expect at best to approximate DTCP with the same ratio as for SCP. We show that this expectation can be satisfied in some way by designing a purely combinatorial approximation algorithm for the DTCP and proving that the approximation ratio of the algorithm is $\max\{2, \ln(D^+)\}$ with D^+ is the maximum outgoing degree of the nodes in G.

1 Introduction

Let $G = (V, A)$ be a directed graph with a (non negative) cost function $c : A \Rightarrow \mathbb{Q}_+$ defined on the arcs. Let $c(u, v)$ denote the cost of the arc $(u, v) \in A$. A *directed tree* cover is a weakly connected subgraph $T = (U, F)$ such that

1. for every $e \in A$, F contains an arc f intersecting e, i.e. f and e have an end-node in common.
2. T is a rooted branching.

The *minimum directed tree cover problem* (DTCP) is to find a directed tree cover of minimum cost. Several related problems to DTCP have been investigated, in particular:

- its undirected counterpart, the minimum tree cover problem (TCP) and
- the tour cover problem in which T is a tour (not necessarily simple) instead of a tree. This problem has also two versions: undirected (ToCP) and directed (DToCP).

We discuss first about TCP which has been intensively studied in recent years. The TCP is introduced in a paper by Arkin et al. [1] where they were motivated by a problem of locating tree-shaped facilities on a graph such that all the nodes are dominated by chosen facilities. They proved the NP-hardness of TCP by observing that the unweighted case of TCP is equivalent to the *connected vertex cover* problem, which in fact is known to be as hard (to approximate)

W. Wu and O. Daescu (Eds.): COCOA 2010, Part II, LNCS 6509, pp. 144–159, 2010.
© Springer-Verlag Berlin Heidelberg 2010

as the vertex cover problem [10]. Consequently, DTCP is also NP-hard since the TCP can be easily transformed to an instance of DTCP by replacing every edge by the two arcs of opposite direction between the two end-nodes of the edge. In their paper, Arkin et al. presented a 2-approximation algorithm for the unweighted case of TCP, as well as 3.5-approximation algorithm for general costs. Later, Konemann et al. [11] and Fujito [8] independently designed a 3-approximation algorithm for TCP using a bidirected formulation. They solved a linear program (of exponential size) to find a vertex cover U and then they found a Steiner tree with U as the set of terminals. Recently, Fujito [9] and Nguyen [13] propose separately two different approximation algorithms achieving 2 the currently best approximation ratio. Actually, the algorithm in [13] is expressed for the TCP when costs satisfy the triangle inequality but one can suppose this for the general case without loss generality. The algorithm in [9] is very interesting in term of complexity since it is a primal-dual based algorithm and thus purely combinatorial. In the prospective section of [11] and [9], the authors presented DTCP as a wide open problem for further research on the topic. In particular, Fujito [9] pointed out that his approach for TCP can be extended to give a 2-approximation algorithm for the unweighted case of DTCP but falls short once arbitrary costs are allowed.

For ToCP, a 3-approximation algorithm has been developed in [11]. The principle of this algorithm is similar as for TCP, i.e. it solved a linear program (of exponential size) to find a vertex cover U and then found a traveling salesman tour over the subgraph induced by U. Recently, Nguyen [14] considered DToCP and extended the approach in [11] to obtain a $2\log_2(n)$-approximation algorithm for DToCP. We can similarly adapt the method in [11] for TCP to DTCP but we will have to find a directed Steiner tree with U a vertex cover as the terminal set. Using the best known approximation algorithm by Charikar et al. [4] for the minimum Steiner directed tree problem, we obtain a ratio of $(1 + \sqrt{|U|}^{2/3} log^{1/3}(|U|))$ for DTCP which is worse than a logarithmic ratio.

In this paper, we improve this ratio by giving a logarithmic ratio approximation algorithm for DTCP. In particular, we show that the weighted Set Cover Problem (SCP) is a special case of DTCP and the transformation is approximation preserving. Based on the known complexity results for SCP, we can only expect a logarithmic ratio for the approximation of DTCP. Let D^+ be the maximum outgoing degree of the nodes in G, we design a primal-dual $\max\{2, \ln(D^+)\}$-approximation algorithm for DTCP which is thus somewhat best possible.

The paper is organized as follows. In the remaining of this section, we will define the notations that will be used in the papers. In Section 2, we present an integer formulation and state a primal-dual algorithm for DTCP. Finally, we prove the validity of the algorithm and its approximation ratio.

Let us introduce the notations that will be used in the paper. Let $G = (V, A)$ be a digraph with vertex set V and arc set A. Let $n = |V|$ and $m = |A|$. If $x \in \mathbb{Q}^{|A|}$ is a vector indexed by the arc set A and $F \subseteq E$ is a subset of arcs, we use $x(F)$ to denote the sum of values of x on the arcs in F, $x(F) = \sum_{e \in F} x_e$. Similarly, for a vector $y \in \mathbb{Q}^{|V|}$ indexed by the nodes and $S \subseteq V$ is a subset of nodes, let $y(S)$ denote the sum of values of y on the nodes in the set S. For a subset of nodes $S \subseteq V$, let $A(S)$ denote the set of the arcs having both end-nodes in S. Let $\delta^+(S)$(respectively $\delta^-(S)$) denote the set of the arcs having only the tail (respectively head) in S. We will call $\delta^+(S)$ the *outgoing cut* associated to S, $\delta^-(S)$ the *ingoing cut* associated to S. For two subset $U, W \subset V$ such that $U \cap W = \emptyset$, let $(U : W)$ be the set of the arcs having the tail in U and the head in W. For $u \in V$, we say v an *outneighbor* (respectively *inneighbor*) of u if $(u, v) \in A$ (respectively $(v, u) \in A$). For the sake of simplicity, in clear contexts, the singleton $\{u\}$ will be denoted simply by u.

For an arc subset F of A, let $V(F)$ denote the set of end-nodes of all the arcs in F. We say F *covers* a vertex subset S if $F \cap \delta^-(S) \neq \emptyset$. We say F is a cover for the graph G if for all arc $(u, v) \in A$, we have $F \cap \delta^-(\{u, v\}) \neq \emptyset$.

When we work on more than one graph, we specify the graph in the index of the notation, e.g. $\delta_G^+(S)$ will denote $\delta^+(S)$ in the graph G. By default, the notations without indication of the the graph in the index are applied on G.

2   Minimum r-Branching Cover Problem

Suppose that T is a directed tree cover of G rooted in $r \in V$, i.e. T is a branching, $V(T)$ is a vertex cover in G and there is a directed path in T from r to any other node in $V(T)$. In this case, we call T, a *r-branching cover*. Thus, DTCP can be divided into n subproblems in which we find a minimum r-branching cover for all $r \in V$. By this observation, in this paper, we will focus on approximating the minimum r-branching cover for a specific vertex $r \in V$. An approximation algorithm for DTCP is then simply resulted from applying n times the algorithm for the minimum r-branching cover for each $r \in V$.

2.1   Weighted Set Cover Problem as a Special Case

Let us consider any instance $\mathcal{A}$ of the weighted Set Cover Problem (SCP) with a set $E = \{e_1, e_2, \ldots, e_p\}$ of ground elements, and a collection of subsets $S_1, S_2, \ldots, S_q \subseteq E$ with corresponding non-negative weights $w_1, w_2, \ldots, w_q$. The objective is to find a set $I \subseteq \{1, 2, \ldots, q\}$ that minimizes $\sum_{i \in I} w_i$, such that $\bigcup_{i \in I} S_i = E$.
We transform this instance to an instance of the minimum r-branching cover problem in some graph G_1 as follows. We create a node r, q nodes $S_1, S_2, \ldots, S_q$ and q arcs (r, S_i) with weight w_i. We then add $2p$ new nodes $e_1, \ldots, e_p$ and $e_1', \ldots, e_p'$. If $e_k \in S_i$ for some $1 \leq k \leq p$ and $1 \leq i \leq q$, we create an arc (S_i, e_k) with weight 0 (or a very insignificant positive weight). At last, we add an arc (e_k, e_k') of weight 0 (or a very insignificant positive weight) for each $1 \leq k \leq p$.

Lemma 1. *Any r-branching cover in G_1 correspond to a set cover in $\mathcal{A}$ of the same weight and vice versa.*

Proof. Let us consider any r-branching cover T in G_1. Since T should cover all the arcs (e_k, e_k') for $1 \leq k \leq n$, T contains the nodes e_k. By the construction of

G_1, these nodes are connected to r uniquely through the nodes $S_1, \ldots S_q$ with the corresponding cost $w_1, \ldots, w_q$. Clearly, the nodes S_i in T constitute a set cover in $\mathcal{A}$ of the same weight as T. It is then easy to see that any set cover in $\mathcal{A}$ correspond to r-branching cover in G_1 of the same weight.

Let D_r^+ be the maximum outgoing degree of the nodes (except r) in G_1. We can see that $D_r^+ = p$, the number of ground elements in $\mathcal{A}$. Hence, we have

Corollary 1. *Any $f(D_r^+)$-approximation algorithm for the minimum r-branching cover problem is also an $f(p)$-approximation algorithm for SCP where f is a function from $\mathbb{N}$ to $\mathbb{R}$.*

Note that the converse is not true. As a corollary of this corollary, we have the same complexity results for the minimum r-branching cover problem as known results for SCP [12,7,15,2]. Precisely,

Corollary 2

- If there exists a $c\ln(D_r^+)$-approximation algorithm for the minimum r-branching cover problem where $c < 1$ then $NP \subseteq DTIME(n^{\{O(\ln^k(D_r^+))\}})$.
- There exists some $0 < c < 1$ such that if there exists a $c\log(D_r^+)$-approximation algorithm for the minimum r-branching cover problem, then $P = NP$.

Note that this result does not contradict the Fujito's result about an approximation ratio 2 for the unweighted DTCP because in our transformation we use arcs of weight 0 (or a very insignificant fractional positive weight) which are not involved in an instance of unweighted DTCP.

Hence in some sense, the $\max\{2, \ln(D_r^+)\}$ approximation algorithm that we are going to describe in the next sections seems to be best possible for the general weighted DTCP.

3   Integer Programming Formulation for Minimum r-Branching Cover

We use a formulation inspired from the one in [11] designed originally for the TCP. The formulation is as follows: for a fixed root r, define $\mathcal{F}$ to be the set of all subsets S of $V \setminus \{r\}$ such that S induces at least one arc of A,

$$\mathcal{F} = \{S \subseteq V \setminus \{r\} \mid A(S) \neq \emptyset\}.$$

Let T be the arc set of a directed tree cover of G containing r, T is thus a branching rooted at r. Now for every $S \in \mathcal{F}$, at least one node, saying v, in S should belong to $V(T)$. By definition of directed tree cover there is a path from r to v in T and as $r \notin S$, this path should contain at least one arc in $\delta^-(S)$. This allows us to derive the following *cut* constraint which is valid for the DTCP:

$$\sum_{e \in \delta^-(S)} x_e \geq 1 \text{ for all } S \in \mathcal{F}$$

This leads to the following IP formulation for the minimum r-branching cover.

$$\min \sum_{e \in A} c(e) x_e$$

$$\sum_{e \in \delta^-(S)} x_e \geq 1 \text{ for all } S \in \mathcal{F}$$

$$x \in \{0,1\}^A.$$

A trivial case for which this formulation has no constraint is when G is a r-rooted star but in this case the optimal solution is trivially the central node r with cost 0.

Replacing the integrity constraints by

$$x \geq 0,$$

we obtain the linear programming relaxation. We use the DTC(G) to denote the convex hull of all vectors x satisfying the constraints above (with integrity constraints replaced by $x \geq 0$). We express below the dual of $DTC(G)$:

$$\max \sum_{S \in \mathcal{F}} y_S$$

$$\sum_{S \in \mathcal{F} \text{ s.t. } e \in \delta^-(S)} y_S \leq c(e) \text{ for all } e \in A$$

$$y_S \geq 0 \text{ for all } S \in \mathcal{F}$$

4   Approximating the Minimum r-Branching Cover

4.1   Preliminary Observations and Algorithm Overview

Preliminary observations. As we can see, the minimum r-branching cover is closely related to the well-known minimum r-arborescence problem which finds a minimum r-branching spanning all the nodes in G. Edmonds [6] gave a linear programming formulation for this problem which consists of the cut constraints for all the subsets $S \subseteq V \setminus \{r\}$ (not limited to $S \in \mathcal{F}$). He designed then a primal-dual algorithm (also described in [5]) which repeatedly keeps and updates a set A_0 of zero reduced cost and the subgraph G_0 induced by A_0 and at each iteration, tries to cover a chosen strongly connected component in G_0 by augmenting (as much as possible with respect to the current reduced cost) the corresponding dual variable. The algorithm ends when all the nodes are reachable from r in G_0. The crucial point in the Edmonds' algorithm is that when there still exist nodes not reachable from r in G_0, there always exists in G_0 a strongly connected component to be covered because we can choose trivial strongly connected components which are singletons. We can not do such a thing for minimum r-branching cover because a node can be or not belonging to a r-branching cover. But we shall see that if G_0 satisfies a certain conditions, we can use an Edmonds-style primal-dual algorithm to find a r-branching cover and to obtain a G_0 satisfying such conditions, we should pay a ratio of $\max\{2, \ln(n)\}$. Let us see what could be these conditions. A node j is said *connected to* to another node i (resp. a connected subgraph B) if there is a path from i (resp. a node in B) to j. Suppose that we have found a vertex cover U and a graph G_0, we define an *Edmonds connected subgraph* as a non-trivial connected (not necessarily strongly) subgraph B not containing r of G_0 such that given any node $i \in B$ and for all $v \in B \cap U$, v is connected to i in G_0. Note that any strongly connected subgraph not containing r in G_0 which contains at least a node in U is an Edmonds connected subgraph. As in the definition, for an Edmonds connected subgraph B, we will also use abusively B to denote its vertex set.

Theorem 1. *If for any node $v \in U$ not reachable from r in G_0, we have*

- *either v belongs to an Edmonds connected subgraph of G_0,*
- *or v is connected to an Edmonds connected subgraph of G_0.*

then we can apply an Edmonds-style primal-dual algorithm completing G_0 to get a r-branching cover spanning U without paying any additional ratio.

Proof. We will prove that if there still exist nodes in U not reachable from r in G_0, then there always exists an Edmonds connected subgraph, say B, uncovered, i.e. $\delta^-_{G_0}(B) = \emptyset$. Choosing any node $v_1 \in U$ not reachable from r in G_0, we can see that in both cases, the Edmonds connected subgraph, say B_1, of G_0 containing v_1 or to which v_1 is connected, is not reachable from r. In this sense we suppose that B_1 is maximal. If B_1 is uncovered, we have done. If B_1 is covered then it should be covered by an arc from a node $v_2 \in U$ not reachable from r because if $v_2 \notin U$ then $B_1 \cup \{v_2\}$ induces an Edmonds connected subgraph which contradicts the fact that B_1 is maximal. Similarly, we should have $v_2 \neq v_1$ because otherwise $B_1 \cup \{v_1\}$ induces an Edmonds connected subgraph. We continue this reasoning with v_2, if this process does not stop, we will meet another node $v_3 \in U \setminus \{v_1, v_2\}$ not reachable from r and so on .... As $|U| \leq n-1$, this process should end with an Edmonds connected subgraph B_k uncovered.

We can then apply a primal-dual Edmonds-style algorithm (with respect to the reduced cost modified by the determination of U and G_0 before) which repeatedly cover in each iteration an uncovered Edmonds connected subgraph in G_0 until every node in U is reachable from r. By definition of Edmonds connected subgraphs, in the output r-branching cover, we can choose only one arc entering the chosen Edmonds connected subgraph and it is enough to cover the nodes belonging to U in this subgraph.

Algorithm overview. Based on the above observations on $DTC(G)$ and its dual, we design an algorithm which is a composition of 3 phases. Phases I and II determine G_0 and a vertex cover U satisfying the conditions stated in Theorem 1. The details of each phase is as follows:

- Phase I is of a primal-dual style which tries to cover the sets $S \in \mathcal{F}$ such that $|S| = 2$. We keep a set A_0 of zero reduced cost and the subgraph G_0 induced by A_0. A_0 is a cover but does not necessarily contain a r-branching cover. We determine after this phase a vertex cover (i.e. a node cover) set U of G. Phase I outputs a partial solution T_0^1 which is a directed tree rooted in r spanning the nodes in U reachable from r in G_0. It outputs also a dual feasible solution y.

- Phase II is executed only if A_0 does not contain a r-branching cover, i.e. there are nodes in U determined in Phase I which are not reachable from r in G_0. Phase II works with the reduced costs issued from Phase I and tries to make the nodes in U not reachable from r in G_0, either reachable from r in G_0, or belong or be connected to an Edmonds connected subgraph in G_0. Phase II transforms this problem to a kind of Set Cover Problem and solve it by a greedy algorithm. Phase II outputs a set of arcs T_0^2 and grows the dual solution y issued from Phase I (by growing only the zero value components of y).
- Phase III is executed only if $T_0^1 \cup T_0^2$ is not a r-branching cover. Phase III applies a primal-dual Edmonds-style algorithm (with respect to the reduced cost issued from Phases I and II) which repeatedly cover in each iteration an uncovered Edmonds connected subgraph in G_0 until every node in U is reachable from r.

4.2   Initialization

Set $\mathcal{B}$ to be the collection of the vertex set of all the arcs in A which do not have r as an end vertex. In other words, $\mathcal{B}$ contains all the sets of cardinality 2 in $\mathcal{F}$, i.e. $\mathcal{B} = \{S \mid S \in \mathcal{F} \text{ and } |S| = 2\}$. Set the dual variable to zero, i.e. $y \leftarrow 0$ and set the reduced cost $\bar{c}$ to c, i.e. $\bar{c} \leftarrow c$. Set $A_0 \leftarrow \{e \in A \mid \bar{c}(e) = 0\}$. Let $G_0 = (V_0, A_0)$ be the subgraph of G induced by A_0.

During the algorithm, we will keep and update constantly a subset of $T_0 \subseteq A_0$. At this stage of initialization, we set $T_0 \leftarrow \emptyset$.

During Phase I, we also keep updating a dual feasible solution y that is initialized at 0 (i.e. all the components of y are equal to 0). The dual solution y is not necessary in the construction of a r-branching cover but we will need it in the proof for the performance guarantee of the algorithm.

4.3   Phase I

In this phase, we will progressively expand A_0 so that it covers all the sets in $\mathcal{B}$. In the mean time, during the expansion of A_0, we add the vertex set of newly created strongly connected components of G_0 to $\mathcal{B}$.

Phase I repeatedly do the followings until $\mathcal{B}$ becomes empty.

1. select a set $S \in \mathcal{B}$ which is not covered by A_0.
2. select the cheapest (reduced cost) arc(s) in $\delta^-(S)$ and add it (them) to A_0. A_0 covers then S. Let α denote the reduced cost of the cheapest arc(s) chosen above, then we modify the reduced cost of the arcs in $\delta^-(S)$ by subtracting α from them. Set $y_S \leftarrow \alpha$.
3. Remove S from $\mathcal{B}$ and if we detect a strongly connected component K in G_0 due to the addition of new arcs in A_0, in the original graph G, we add the set $V(K)$ to $\mathcal{B}$.

Proposition 1. *After Phase I, A_0 is a cover.*

Proof. As we can see, Phase I terminates when $\mathcal{B}$ becomes empty. That means the node sets of the arcs, which do not have r as an end-node, are all covered by A_0. Also all the strongly connected components in G_0 are covered.   □

At this stage, if for any node v there is a path from r to v in G_0, we say that v is *reachable* from r. Set T_0 to be a directed tree (rooted in r) in G_0 spanning the nodes reachable from r. T_0 is chosen such that for each strongly connected component K added to $\mathcal{B}$ in Phase I, there is exactly one arc in T_0 entering K, i.e. $|\delta^-(K) \cap T_0| = 1$. If the nodes reachable from r in G_0 form a vertex cover, then T_0 is a r-branching cover and the algorithm stops. Otherwise, it goes to Phase II.

4.4   Phase II

Let us consider the nodes which are not reachable from r in G_0. We divide them into three following categories:

- The nodes i such that $|\delta^-_{G_0}(i)| = 0$, i.e. there is no arc in A_0 entering i. Let us call these nodes *source nodes*.
- The nodes i such that $|\delta^-_{G_0}(i)| = 1$, i.e. there is exactly one arc in A_0 entering i. Let us call these nodes *sink nodes*.
- The nodes i such that $|\delta^-_{G_0}(i)| \geq 2$, i.e. there is at least two arcs in A_0 entering i. Let us call these nodes *critical nodes*.

Proposition 2. *The set of the source nodes is a stable set.*

Proof. Suppose that the converse is true, then there is an arc (i, j) with i, j are both source nodes. As $\delta^-_{G_0}(i) = \delta^-_{G_0}(j) = \emptyset$, we have $\delta^-_{G_0}(\{i, j\}) = \emptyset$. Hence, (i, j) is not covered by A_0. Contradiction.

Corollary 3. *The set U containing the nodes reachable from r in G_0 after Phase I, the sink nodes and the critical nodes is a vertex cover (i.e. a node cover) of G.*

Proposition 3. *For any sink node j, there is at least one critical node i such that j is connected to i in G_0.*

Proof. Let the unique arc in $\delta^-_{G_0}(j)$ be (i_1, j). Since this arc should be covered by A_0, $\delta^-_{G_0}(i_1) \neq \emptyset$. If $|\delta^-_{G_0}(i_1)| \geq 2$ then i_1 is a critical node and we have done. Otherwise, i.e. $|\delta^-_{G_0}(i_1)| = 1$ and i_1 is a sink node. Let (i_2, i_1) be the unique arc in $\delta^-_{G_0}(i_1)$, we repeat then the same reasoning for (i_2, i_1) and for i_2. If this process does not end with a critical node, it should meet each time a new sink node not visited before (It is not possible that a directed cycle is created since then this directed cycle (strongly connected component) should be covered in Phase I and hence at least one of the nodes on the cycle has two arcs entering it, and is therefore critical). As the number of sink nodes is at most $n - 1$, the process can not continue infinitely and should end at a stage k $(k < n)$ with i_k is a critical node. By construction, the path $i_k, i_{k-1}, \ldots, i_1, j$ is a path in G_0 from i_k to j.

A critical node v is said to be *covered* if there is at least one arc $(w, v) \in A_0$ such that w is not a source node, i.e. w can be a sink node or a critical node or a node reachable from r. Otherwise, we say v is *uncovered*.

Proposition 4. *If all critical nodes are covered then for any critical node v, one of the followings is verified:*

- *either v belongs to an Edmonds connected subgraph of G_0 or v is connected to an Edmonds connected subgraph of G_0,*
- *there is a path from r to v in G_0, i.e. v is reachable from r in G_0.*

Proof. If v is covered by a node reachable from r, we have done. Otherwise, v is covered by sink node or by another critical node. From Proposition 3 we derive that in the both cases, v will be connected to a critical node w, i.e. there is a path from w to v in G_0. Continue this reasoning with w and so on, we should end with a node reachable from r or a critical node visited before. In the first case v is reachable from r. In the second case, v belongs to a directed cycle in G_0 if we have revisited v, otherwise v is connected to a directed cycle in G_0. The directed cycle in the both cases is an Edmonds connected subgraph (because it is strongly connected) and it can be included in a greater Edmonds connected subgraph.

Lemma 2. *If all critical nodes are covered then for any node $v \in U$ not reachable from r in G_0,*

- *either v belongs to an Edmonds connected subgraph of G_0,*
- *or v is connected to an Edmonds connected subgraph of G_0.*

Proof. The lemma is a direct consequence of Propositions 3 and 4.

The aim of Phase II is to cover all the uncovered critical nodes. Let us see how to convert this problem into a weighted SCP and to solve the latter by adapting the well-known greedy algorithm for weighted SCP.

A source node s is *zero connecting* a critical node v (reciprocally v is *zero connected from s*) if $(s, v) \in A_0$. If $(s, v) \notin A_0$ but $(s, v) \in A$ then s is *positively connecting* v (reciprocally v is *positively connected from s*).

Suppose that at the end of Phase I, there are k uncovered critical nodes v_1, $v_2, \ldots v_k$ and p source nodes $s_1, s_2, \ldots s_p$. Let $S = \{s_1, s_2, \ldots, s_p\}$ denote the set of the source nodes.

Remark 1. An uncovered critical node v can be only covered:

- by directly an arc from a sink node or another crtitical node to v,
- or via a source node s connecting (zero or positively) v, i.e. by two arcs: an arc in $\delta^-(s)$ and the arc (s, v).

Remark 1 suggests us that we can consider every critical node v as a ground element to be covered in a Set Cover instance and the subsets containing v could be the singleton $\{v\}$ and any subset containing v of the set of the critical nodes connecting (positive or zero) from s. The cost of the the singleton $\{v\}$ is the minimum reduced cost of the arcs from a sink node or another crtitical node to v. The cost of a subset T containing v of the set of the critical nodes connecting from s is the minimum reduced cost of the arcs in $\delta^-(s)$ plus the sum of the reduced cost of the arcs (s, w) for all $w \in T$.

Precisely, in Phase II, we proceed to cover all the uncovered critical nodes by solving by the greedy algorithm the following instance of the Set Cover Problem:

- The ground set contains k elements which are the critical nodes $v_1, v_2, \ldots, v_k$.
- The subsets are

  Type I For each source node s_i for $i = 1, \ldots, p$, let $\mathcal{C}(s_i)$ be the set of all the critical nodes connected (positively or zero) from s_i. The subsets of Type I associated to s_i are the subsets of $\mathcal{C}(s_i)$ ($\mathcal{C}(s_i)$ included). To define their cost, we define

  $$\bar{c}(s_i) = \begin{cases} \min\{\bar{c}(e) \mid e \in \delta^-(s_i)\} & \text{if } \delta^-(s_i) \neq \emptyset, \\ +\infty & \text{otherwise} \end{cases}$$

  Let us choose an arc $e_{s_i} = \mathrm{argmin}\{\bar{c}(e) \mid e \in \delta^-(s_i)\}$ which denotes an arc entering s_i of minimum reduced cost. Let T be any subset of type I associated to s_i, we define $\bar{c}(T)$ the cost of T as $\bar{c}(T) = \bar{c}(s_i) + \sum_{v \in T} \bar{c}(s_i, v)$.
  Let us call the arc subset containing the arc e_{s_i} and the arcs (s_i, v) for all $v \in T$ uncovered, *the covering arc subset of* T.

  Type II the singletons $\{v_1\}, \{v_2\}, \ldots, \{v_k\}$. We define the cost of the singleton $\{v_i\}$,

  $$\bar{c}(v_i) = \begin{cases} \min\{\bar{c}(w, v_i) \mid \text{where } w \text{ is not a source node, i.e. } w \in V \setminus S\} & \text{if } (V \setminus S : \{v_i\}) \neq \emptyset, \\ +\infty & \text{otherwise} \end{cases}$$

  Let us choose an arc $e_{v_i} = \mathrm{argmin}\{\bar{c}(w, v_i) \mid \text{where } w \text{ is not a source node,} i.e. w \in V \setminus S\}$, denotes an arc entering v_i from a non source node of minimum reduced cost. Let the singleton $\{e_{v_i}\}$ be *the covering arc subset of* $\{v_i\}$.

We will show that we can adapt the greedy algorithm solving this set cover problem to our primal-dual scheme. In particular, we will specify how to update dual variables et the sets A_0 and T_0 in each iteration of the greedy algorithm. The sketch of the algorithm is explained in Algorithm 1.

Note that in Phase II, contrary to Phase I, the reduced costs $\bar{c}$ are not to be modified and all the computations are based on the reduced costs $\bar{c}$ issued from Phase I. In the sequel, we will specify how to compute the most efficient subset Δ and update the dual variables.

For $1 \leq i \leq p$ let us call $\mathcal{S}_i$ the collection of all the subsets of type I associated to s_i. Let $\mathcal{S}$ be the collection of all the subsets of type I and II.

Algorithm 1. Greedy algorithm for Phase II

---

1 **while** *there exist uncovered critical nodes* **do**
2   Compute the most efficient subset Δ ;
3   Update the dual variables and the sets A_0 and T_0;
4   Change the status of the uncovered critical nodes in Δ to covered ;
5 **end**

Computing the most efficient subset. Given a source node s_i, while the number of subsets in $\mathcal{S}_i$ can be exponential, we will show in the following that computing the most efficient subset in $\mathcal{S}_i$ is can be done in polynomial time. Let us suppose that there are i_q critical nodes denoted by $v_{s_i}^{i_1}, v_{s_i}^{i_2}, \ldots, v_{s_i}^{i_q}$ which are connected (positively or zero) from s_i. In addition, we suppose without loss of generality that $\bar{c}(s_i, v_{s_i}^{i_1}) \leq \bar{c}(s_i, v_{s_i}^{i_2}) \leq \ldots \leq \bar{c}(s_i, v_{s_i}^{i_q})$. We compute f_i and S_i which denote respectively the best efficiency and the most effecicient set in $\mathcal{S}_i$ by the following algorithm.

Step 1. Suppose that $v_{s_i}^{i_h}$ is the first uncovered critical node met when we scan
      the critical nodes $v_{s_i}^{i_1}, v_{s_i}^{i_2}, \ldots, v_{s_i}^{i_q}$ in this order.
      Set $S_i \leftarrow \{v_{s_i}^{i_h}\}$. Set $\bar{c}(S_i) \leftarrow \bar{c}(s_i) + \bar{c}(s_i, v_{s_i}^{i_h})$.
      Set $d_i \leftarrow 1$. Set $f_i \leftarrow \frac{\bar{c}(S_i)}{d_i}$ and $\Delta_i \leftarrow S_i$.

Step 2. We add progressively uncovered critical nodes $v_{s_i}^{i_j}$ for $j = h + 1, \ldots, i_q$
      to S_i while this allows to increase the efficiency of S_i:
      For $j = h + 1$ to i_q, if $v_{s_i}^{i_j}$ is uncovered and $f_i > \frac{\bar{c}(S_i) + \bar{c}(s_i, v^{i_j})}{d_i + 1}$ then $f_i \leftarrow$
      $\frac{\bar{c}(S_i) + \bar{c}(s_i, v_{s_i}^{i_j})}{d_i + 1}$, $d_i \leftarrow d_i + 1$ and $S_i \leftarrow S_i \cup \{v_{s_i}^{i_j}\}$.

Set $i_{min} \leftarrow \operatorname{argmin}\{f_i \,|\, s_i \text{ is a source node}\}$.

Choose the most efficient subset among $S_{i_{min}}$ and the singletons of type II for which the computation of efficiency is straightforward. Set Δ to be most efficient subset and set $d \leftarrow |\Delta|$ the number of the uncovered critical nodes in Δ.

Updating the dual variables and the sets A_0 and T_0
Let $g = \max\{|T| \mid T \in \mathcal{S}\}$ and let $H_g = 1 + \frac{1}{2} + \frac{1}{3} + \ldots + \frac{1}{g}$.

Remark 2. $g \leq D_r^+$.

Given a critical node v, let p_v denote the number of source nodes connecting v. Let $s_1^v, s_2^v, \ldots, s_{p_v}^v$ be these source nodes such that $\bar{c}(s_1^v, v) \leq \bar{c}(s_2^v, v) \leq \ldots \leq \bar{c}(s_{p_v}^v, v)$. We define $S_v^j = \{v, s_1^v, \ldots, s_v^j\}$ for $j = 1, \ldots, p_v$. We can see that for $j = 1, \ldots, p_v$, $S_v^j \in \mathcal{F}$. Let $y_{S_v^j}$ be the dual variable associated to the cut constraints $x(\delta^-(S_v^j)) \geq 1$. The dual variables will be updated as follows. For each critical node v uncovered in Δ, we update the value of $y_{S_v^j}$ for $j = 1, \ldots, p_v$ for that $\sum_{j=1}^{p_v} y_{S_v^j} = \frac{\bar{c}(\Delta)}{H_g \times d}$. This updating process saturates progressively the arcs (s_v^j, v) for $j = 1, \ldots, p_v$. Details are given in Algorithm 2. We add to A_0 and to T_0 the arcs in the covering arc subset of Δ.

Algorithm 2. Updating the dual variables

1 $j \leftarrow 1$;
2 **while** $(j < p_v)$ and $(\bar{c}(s_v^{j+1}, v) < \frac{\bar{c}(\Delta)}{H_g \times d})$ **do**
3      $y_{S_v^j} \leftarrow \bar{c}(s_v^{j+1}, v) - \bar{c}(s_v^j, v)$;
4      $j \leftarrow j + 1$;
5 **end**
6 **if** $\bar{c}(s_v^{p_v}, v) < \frac{\bar{c}(\Delta)}{H_g \times d}$ **then**
7      $y_{S_v^{p_v}} \leftarrow \frac{\bar{c}(\Delta)}{H_g \times d} - \bar{c}(s_v^{p_v}, v)$;
8 **end**

Let us define $\mathcal{T}$ as the set of the subsets T such that y_T is made positive in Phase II.

Lemma 3. *The dual variables which were made positive in Phase II respect the reduced cost issued from Phase I.*

Proof. For every $T \in \mathcal{T}$, the arcs in $\delta^-(T)$ can only be either an arc in $\delta^-(s_i)$ with s_i is a source node or an arc in $\delta^-(v)$ with v is a critical node. Hence, we should show that for every arc (u', u) with u is either a critical node or a source node, we have

$$\sum_{T \in \mathcal{T} \text{ s.t. } u \in T} y_T \leq \bar{c}(u', u)$$

– u is a critical node v and u' is the source node s_j^v. The possible subsets $T \in \mathcal{T}$ such that $(s_j^v, v) \in \delta^-(T)$ are the sets $S_v^1, \ldots, S_v^{j-1}$. By Algorithm 2, we can see that

$$\sum_{k=1}^{j-1} y_{S_v^k} \leq \bar{c}(s_j^v, v).$$

– u is a critical node v and $u' \in V \setminus S$. By definition of $\bar{c}(v)$, we have $\bar{c}(u', u) \geq \bar{c}(v)$. By analogy with the Set Cover problem, the dual variables made positive in Phase II respect the cost of the singleton $\{v\}$. Hence

$$\sum_{T \in \mathcal{T} \text{ s.t. } v \in T} y_T \leq \bar{c}(v) \leq \bar{c}(u', u)$$

– u is source node and $u' \in V \setminus S$. For each critical node w such that $(u, w) \in A$, we suppose that $u = s_w^{i(u,w)}$ where $1 \leq i(u, w) \leq p_w$. Let

$$T_u = \{w \mid w \text{ is a critical node, } (u, w) \in A \text{ and } y_{S_w^{i(u,w)}} > 0\}$$

We can see that $T_u \in \mathcal{S}$ and $\bar{c}(T_u) = \bar{c}(u) + \sum_{w \in T_u} \bar{c}(u, w)$. Suppose that l is the total number of iterations in Phase II. We should show that

$$\sum_{k=1}^{l} \sum_{w \in T_u \cap \Delta_k} \left(\frac{\bar{c}(\Delta_k)}{H_g \times d_k} - \bar{c}(u, w) \right) \leq \bar{c}(u) \tag{1}$$

where Δ_k is the subset which has been chosen in k^{th} iteration. Let a_k be the number of uncovered critical nodes in T_u at the beginning of the kth iteration. We have then $a_1 = |T_u|$ and $a_{l+1} = 0$. Let A_k be the set of previously uncovered critical nodes of T_u covered in the kth iteration. We immediately find that $|A_k| = a_k - a_{k+1}$. By Algorithm 1, we can see that at the kth iteration $\frac{\bar{c}(\Delta_k)}{H_g \times d_k} \leq \frac{\bar{c}(T_u)}{H_g \times a_k}$. Since $|A_k| = a_k - a_{k+1}$ then

$$\sum_{w \in T_u \cap \Delta_k} \left(\frac{\bar{c}(\Delta_k)}{H_g \times d_k}\right) - \sum_{w \in T_u \cap \Delta_k} \bar{c}(u, w) \leq \frac{\bar{c}(T_u)}{H_g} \times \frac{a_k - a_{k+1}}{a_k} - \sum_{w \in T_u \cap \Delta_k} \bar{c}(u, w)$$

Hence,

$$\sum_{k=1}^{l} \sum_{w \in T_u \cap \Delta_k} \left(\frac{\bar{c}(\Delta_k)}{H_g \times d_k} - \bar{c}(u, w)\right) \leq \frac{\bar{c}(T_u)}{H_g} \sum_{k=1}^{l} \frac{a_k - a_{k+1}}{a_k} - \sum_{k=1}^{l} \sum_{w \in T_u \cap \Delta_k} \bar{c}(u, w)$$

$$\leq \frac{\bar{c}(T_u)}{H_g} \sum_{k=1}^{l} \left(\frac{1}{a_k} + \frac{1}{a_k - 1} + \ldots + \frac{1}{a_{k+1} - 1}\right)$$

$$- \sum_{k=1}^{l} \sum_{w \in T_u \cap \Delta_k} \bar{c}(u, w)$$

$$\leq \frac{\bar{c}(T_u)}{H_g} \sum_{i=1}^{a_1} \frac{1}{i} - \sum_{k=1}^{l} \sum_{w \in T_u \cap \Delta_k} \bar{c}(u, w)$$

$$\leq \bar{c}(T_u) - \sum_{k=1}^{l} \sum_{w \in T_u \cap \Delta_k} \bar{c}(u, w) = \bar{c}(u).$$

Let $T_0^2 \subset T_0$ the set of the arcs added to T_0 in Phase II. For each $e \in T_0^2$, let $c_2(e)$ be the part of the cost $c(e)$ used in Phase II.

Theorem 2

$$c_2(T_0) = \sum_{e \in T_0^2} c_2(e) \leq H_g \sum_{T \in \mathcal{T}} y_T \leq \ln(D_r^+) \sum_{T \in \mathcal{T}} y_T$$

Proof. By Algorithm 2, at the kth iteration, a subset Δ_k is chosen and we add the arcs in the covering arc subset of Δ_k to T_0^2 for all $v \in \Delta_k$. Let T_0^{2k} be covering arc subset of Δ_k. We can see that $c_2(T_0^{2k}) = \sum_{e \in T_0^{2k}} \bar{c}_e = \bar{c}(\Delta_k)$. In this iteration, we update the dual variables in such a way that for each critical node $v \in \Delta_k$, $\sum_{j=1}^{p_v} y_{S_v^j} = \frac{\bar{c}(\Delta_k)}{H_g \times d_k}$ with $d_k = |\Delta_k|$. Together with the fact that $\bar{c}(\Delta_k) = \bar{c}(w_k) + \sum_{v \in \Delta_k} \bar{c}(w_k, v)$ we have $\sum_{v \in \Delta_k} \sum_{j=1}^{p_v} y_{S_v^j} = \frac{\bar{c}(\Delta_k)}{H_g} = \frac{c_2(T_0^{2k})}{H_g}$. By summing over l be the number of iterations in Phase II, we obtain

$$\sum_{T \in \mathcal{T}} y_T = \sum_{k=1}^{l} \sum_{v \in \Delta_k} \sum_{j=1}^{p_v} y_{S_v^j} = \sum_{k=1}^{l} \frac{\bar{c}(\Delta_k)}{H_g} = \sum_{k=1}^{l} \frac{c_2(T_0^{2k})}{H_g} = \frac{c_2(T_0)}{H_g}$$

which proves that $c_2(T_0) = H_g \sum_{T \in \mathcal{T}} y_T$. By Remark 2, we have $g \leq D_r^+$ and $H_g \approx \ln g$, hence $c_2(T_0) \leq \ln(D_r^+) \sum_{T \in \mathcal{T}} y_T$. $\qquad\square$

4.5   Phase III

We perform Phase III if after Phase II, there exist nodes in U not reachable from r in G_0. By Lemma 2, they belong or are connected to some Edmonds connected subgraphs of G_0. By Theorem 1, we can apply an Edmonds-style primal-dual algorithm which tries to cover uncovered Edmonds connected subgraphs of G_0 until all nodes in U reachable from r. The algorithm repeatedly choosing un-covered Edmonds connected subgraph and adding to A_0 the cheapest (reduced cost) arc(s) entering it . As the reduced costs have not been modified during Phase II, we update first the reduced cost $\bar{c}$ with respect to the dual variables made positive in Phase II.

Algorithm 3. Algorithm for Phase III

---

1 Update the reduced cost $\bar{c}$ with respect to the dual variables made positive in Phase II;
2 **repeat**
3   Choose B an uncovered Edmonds connected subgraph ;
4   Let y_B be the associated dual variable to B;
5   Set $\bar{c}(B) \leftarrow \min\{\bar{c}_e \mid e \in \delta^-(B)\}$; Set $y_B \leftarrow \bar{c}(B)$;
6   **foreach** $e \in \delta^-(B)$ **do**
7          $\bar{c}_e \leftarrow \bar{c}_e - \bar{c}(B)$;
8   **end**
9   Update A_0, G_0 and T_0 (see below);
10 **until** *every nodes in U reachable from r* ;

For updating A_0, at each iteration, we add all the saturated arcs belonging to $\delta^-(B)$ to A_0. Among these arcs, we choose only one arc (u,v) with $v \in B$ to add to T_0 with a preference for a u connected from r in G_0. In the other hand, we delete the arc (x,v) with $x \in B$ from T_0. We then add to T_0 an directed tree rooted in v in G_0 spanning B. If there are sink nodes directly connected to B, i.e. the path from a critical node $w \in B$ to these nodes contains only sink nodes except w. We also add all such paths to T_0.

Lemma 4. *After Phase III, T_0 is a r-branching cover.*

Proof. We can see that after Phase III, for any critical node or a sink node v, there is a path containing only the arcs in T_0 from r to v and there is exactly one arc in $\delta^-(v) \cap T_0$.

4.6   Performance Guarantee

We state now a theorem about the performance guarantee of the algorithm.

Theorem 3. *The cost of T_0 is at most $\max\{2, \ln(D_r^+)\}$ times the cost of an optimal r-branching cover.*

Proof. Suppose that T^* is an optimal r-branching cover of G with respect to the cost c. First, we can see that the solution y built in the algorithm is feasible dual solution. Hence $c^T y \leq c(T^*)$. Let $\mathcal{B}$ be the set of all the subsets B in Phase I and Phase III (B is either a subset of cardinality 2 in $\mathcal{F}$ or a subset such that the induced subgraph is a strongly connected component or an Edmonds connected subgraph in G_0 at some stage of the algorithm). Recall that we have defined $\mathcal{T}$ as the set of the subsets T such that y_T is made positive in Phase II. We have then $c^T y = \sum_{B \in \mathcal{B}} y_B + \sum_{T \in \mathcal{T}} y_T$. For any arc e in T_0, let us divide the cost $c(e)$ into two parts: $c_1(e)$ the part saturated by the dual variables y_B with $B \in \mathcal{B}$ and $c_2(e)$ the part saturated by the dual variables y_T with $B \in \mathcal{T}$. Hence $c(T_0) = c_1(T_0) + c_2(T_0)$. By Theorem 2, we have $c_2(T_0) \leq \ln(D_r^+) \sum_{T \in \mathcal{T}} y_T$ (note that the replacing in Phase III of an arc (x, v) by another arc (u, v) with $v \in B_i$ do not change the cost $c_2(T_0)$). Let us consider any set $B \in \mathcal{B}$ by the algorithm, B is the one of the followings:

- $|B| = 2$. As T_0 is a branching so that for all vertex $v \in V$, we have $|\delta^-(v) \cap T_0| \leq 1$. Hence, $|\delta^-(B) \cap T_0| \leq 2$.
- B is a vertex set of a strongly connected component or an Edmonds connected subgraph in G_0. We can see obviously that by the algorithm $|\delta^-(B) \cap T_0| = 1$.

These observations lead to the conclusion that $c_1(T_0) \leq 2 \sum_{B \in \mathcal{B}} y_B$. Hence

$$c(T_0) = c_1(T_0) + c_2(T_0) \leq 2 \sum_{B \in \mathcal{B}} y_B + \ln(D_r^+) \sum_{T \in \mathcal{T}} y_T$$

$$\leq \max\{2, \ln(D_r^+)\} c^T y \leq \max\{2, \ln(D_r^+)\} c(T^*).$$

Corollary 4. *We can approximate the DTCP within a $\max\{2, \ln(D^+)\}$ ratio.*

5   Final Remarks

The paper has shown that the weighted Set Cover Problem is a special case of the Directed Tree Cover Problem and the latter can be approximated with a ratio of $\max\{2, \ln(D^+)\}$ (where D^+ is the maximum outgoing degree of the nodes in G) by a primal-dual algorithm. Based on known complexity results for weighted Set Cover, in one direction, this approximation seems to be best possible.

In our opinion, an interesting question is whether the same techniques can be applied to design a combinatorial approximation algorithm for Directed Tour Cover. As we have seen in Introduction section, a $2 \log_2(n)$-approximation algorithm for Directed Tour Cover has been given in [14], but this algorithm is not combinatorial.

References

1. Arkin, E.M., Halldórsson, M.M., Hassin, R.: Approximating the tree and tour covers of a graph. Information Processing Letters 47, 275–282 (1993)
2. Arora, S., Sudan, M.: Improved Low-Degree Testing and Its Applications. In: Proceedings of STOC 1997, pp. 485–495 (1997)
3. Bock, F.: An algorithm to construct a minimum spanning tree in a directed network. In: Developments in Operations Research, pp. 29–44. Gordon and Breach, NY (1971)
4. Charikar, M., Chekuri, C., Cheung, T., Dai, Z., Goel, A., Guha, S., Li, M.: Approximation Algorithms for Directed Steiner Problems. Journal of Algorithms 33, 73–91 (1999)
5. Chu, Y.J., Liu, T.H.: On the shortest arborescence of a directed graph. Science Sinica 14, 1396–1400 (1965)
6. Edmonds, J.: Optimum branchings. J. Research of the National Bureau of Standards 71B, 233–240 (1967)
7. Feige, U.: A threshold of $\ln n$ for approximating set cover. Journal of the ACM 45, 634–652 (1998)
8. Fujito, T.: On approximability of the independent/connected edge dominating set problems. Information Processing Letters 79, 261–266 (2001)
9. Fujito, T.: How to Trim an MST: A 2-Approximation Algorithm for Minimum Cost Tree Cover. In: Bugliesi, M., Preneel, B., Sassone, V., Wegener, I. (eds.) ICALP 2006. LNCS, vol. 4051, pp. 431–442. Springer, Heidelberg (2006)
10. Garey, M.R., Johnson, D.S.: The rectilinear Steiner-tree problem is NP complete. SIAM J. Appl. Math. 32, 826–834 (1977)
11. Könemann, J., Konjevod, G., Parekh, O., Sinha, A.: Improved Approximations for Tour and Tree Covers. Algorithmica 38, 441–449 (2003)
12. Lund, C., Yannakakis, M.: On the hardness of approximating minimization problems. Journal of the ACM 41, 960–981 (1994)
13. Nguyen, V.H.: Approximation algorithms for metric tree cover and generalized tour and tree covers. RAIRO Operations Research 41(3), 305–315 (2007)
14. Nguyen, V.H.: A $2 \log_2(n)$-Approximation Algorithm for Directed Tour Cover. In: Proceedings of COCOA 2009. LNCS, vol. 5573, pp. 208–218. Springer, Heidelberg (2009)
15. Raz, R., Safra, R.: A sub-constant error-probability low-degree test, and a sub-constant error-probability PCP characterization of NP. In: Proceedings of STOC 1997, pp. 475–484 (1997)

An Improved Approximation Algorithm for Spanning Star Forest in Dense Graphs

Jing He and Hongyu Liang

Institute for Theoretical Computer Science,
Tsinghua University, Beijing, China
{hejing2929,hongyuliang86}@gmail.com

Abstract. A spanning subgraph of a given graph G is called a *spanning star forest* of G if it is a collection of node-disjoint trees of depth at most 1 (such trees are called *stars*). The *size* of a spanning star forest is the number of leaves in all its components. The goal of the *spanning star forest problem* [12] is to find the maximum-size spanning star forest of a given graph.

In this paper, we study this problem in c-dense graphs, where for $c \in (0, 1)$, a graph of n vertices is called c-*dense* if it contains at least $cn^2/2$ edges [2]. We design a $(\alpha + (1-\alpha)\sqrt{c} - \epsilon)$-approximation algorithm for spanning star forest in c-dense graphs for any $\epsilon > 0$, where $\alpha = \frac{193}{240}$ is the best known approximation ratio of the spanning star forest problem in general graphs [3]. Thus, our approximation ratio outperforms the best known bound for this problem when dealing with c-dense graphs. We also prove that for any $c \in (0, 1)$, there is a constant $\epsilon = \epsilon(c) > 0$ such that approximating spanning star forest in c-dense graphs within a factor of $1 - \epsilon$ is NP-hard. We then demonstrate that for weighted versions (both node- and edge- weighted) of this problem, we cannot get *any* approximation algorithm with strictly better performance guarantee in c-dense graphs than that of the best possible approximation algorithm for general graphs. Finally, we give strong hardness-of-approximation results for a closely related problem, the minimum dominating set problem, in c-dense graphs.

Keywords: spanning star forest, approximation algorithm, dense graphs.

1 Introduction

We consider the spanning star forest problem. A graph is called a *star* if it can be regarded as a tree of depth at most 1, or equivalently, there is one vertex (called the *center*) adjacent to all other vertices (called *leaves*) in the graph. A single node is by definition also a star. A *star forest* is a forest whose connected components are all stars. The *size* of a star forest is the number of its leaves. A *spanning star forest* of a graph G is a spanning subgraph of G that is also a star forest. The *spanning star forest problem* (SSF for short), introduced in [12], is the problem of finding a spanning star forest of maximum size in a given graph. This problem has found applications in various areas. Nguyen et al. [12] use it

W. Wu and O. Daescu (Eds.): COCOA 2010, Part II, LNCS 6509, pp. 160–169, 2010.
© Springer-Verlag Berlin Heidelberg 2010

as a subroutine to design an algorithm for aligning multiple genomic sequences, which is an important bioinformatics problem in comparative genomics. This model has also been applied to the comparison of phylogenetic trees [4] and the diversity problem in the automobile industry [1].

It is not hard to see that there is a one-one correspondence between spanning star forests and dominating sets of a given graph. A dominating set of a graph G is a subset of vertices D such that every vertex not in D is adjacent to at least one vertex in D. The *minimum dominating set problem* is to find a smallest dominating set of a given graph. Given a spanning star forest of G, it is easy to argue that the collection of all centers of it is a dominating set of G, and the size of the spanning star forest is equal to the number of vertices in G minus the size of the corresponding dominating set. On the other hand, given a dominating set of G, we can construct a spanning star forest of G whose centers are exactly those vertices in the dominating set. Thus, the two problems are equivalent in finding the optimum solution. We also call one problem the *complement* of another, following the notion used before [3,7].

However, the two problems appear totally different when the approximability is considered. By Feige's famous result [9], the dominating set problem cannot be approximated within $(1-\epsilon)\ln n$ for any $\epsilon > 0$ unless $NP \subseteq DTIME(n^{O(\log \log n)})$. In contrast, a fairly simple algorithm with the idea of dividing a spanning tree into alternating levels gives a 0.5-approximation to the spanning star forest problem. Nguyen et al. [12] proposed a 0.6-approximation algorithm using the fact that every graph of n vertices of minimum degree 2 has a dominating set of size at most $\frac{2}{5}n$ except for very few special cases which can be enumerated. In addition, they prove that it is NP-hard to approximate the problem to any factor larger than $\frac{259}{260}$. They also introduce the edge-weighted version of this problem, whose objective is to find a spanning star forest in which the total weight of edges is maximized, and show a 0.5-approximation algorithm for this variant. Later on, the approximation ratio for unweighted SSF is improved to 0.71 by Chen et al. [7] based on solving a natural linear programming relaxation combined with a randomized rounding stage. They also consider another generalization of SSF where each node has a non-negative weight and the objective is to find a spanning star forest in which the total weight of all leaves is maximized. Note that node-weighted SSF is just the complement of the weighted dominating set problem where each vertex has a weight and the goal is to find a minimum-weight dominating set of the given graph. For this version, they show that a similar algorithm achieves an approximation factor of 0.64. Athanassopoulos et al. [3] realize that the unweighted spanning star forest problem is actually a special case of the complementary set cover problem, and design a 0.804-approximation for it (also for complementary set cover) using the idea of semi-local search for k-set cover [8]. Regarding the hardness results, it is proved by Chakrabarty and Goelin [6] that edge-weighted SSF and node-weighted SSF cannot be approximated to $\frac{10}{11} + \epsilon$ and $\frac{13}{14} + \epsilon$ respectively, unless $P = NP$.

1.1   Our Contributions

We study variants of the spanning star forest problem in c-dense graphs. A graph on n vertices is called c-*dense*, for some $c \in (0, 1)$, if it contains at least $cn^2/2$ edges [2]. One can show by a simple probabilistic argument that almost all graphs are dense. Thus, it captures many real-world models. In fact, this setting has received extensive studies for various combinatorial problems like vertex cover, max-cut, Steiner tree, minimum maximal matching, etc. (see [2,5,10,11,14]). To our knowledge, ours is the first study of the spanning star forest problem in the class of c-dense graphs.

We first design an approximation algorithm for (unweighted) spanning star forest in c-dense graphs with an approximation ratio better than the previously best known ratio of this problem in general graphs, for any $c \in (0, 1)$. More precisely, denoting by $\alpha = 193/240 (\approx 0.804)$ the best known approximation ratio for spanning star forest [3], our algorithm achieves an approximation factor of $\alpha + (1 - \alpha)\sqrt{c} - \epsilon$, for any $\epsilon > 0$. Note that this factor is larger than 0.9 whenever $c \geq 0.25$, and is larger than 0.96 when $c \geq 0.64$. Thus, it is a quite strong performance guarantee. Our algorithm consists of two stages. The first stage is actually a greedy procedure that chooses the vertex covering the largest number of uncovered vertices, and adds it to a maintained dominating set of the input graph. It stops when the number of uncovered vertices is smaller than some prespecified threshold, and goes to the second stage. In this stage, we find a set of vertices dominating the uncovered ones by reducing it to a problem called *complementary partial dominating set*, which will be formally defined in Section 2.2. We will show in Section 2.2 that this problem can be approximated as well as the complementary set cover problem considered in [3]. Combining the two stages, we find a dominating set of the graph of relatively small size, and then construct a spanning star forest in the standard way, which can be proved to be a good approximation to the problem.

We then prove that the spanning star forest problem in c-dense graphs does not admit a polynomial-time approximation scheme (PTAS) assuming $P \neq NP$. Specifically, we prove that for any $c \in (0, 1)$, there exists $\epsilon = \epsilon(c) > 0$ such that approximating SSF in c-dense graphs to within a factor of $1 - \epsilon$ is NP-hard. Thus, the technique developed by Arora et al. [2] for designing PTAS for combinatorial problems in dense instances cannot be applied to our problem.

Next we consider the weighted versions (both node- and edge-weighted) of this problem. A little surprisingly, we show that *any* approximation algorithm for weighted spanning star forest in c-dense graphs cannot guarantee an approximation ratio strictly larger than that of the best possible appproximation for weighted SSF in general graphs. This is proved by an (almost) approximation-preserving reduction from general instances of this problem to c-dense instances.

Finally, we show that the dominating set problem in c-dense graphs shares the same inapproximability result with dominating set in general graphs. Thus, the $(1 + \ln n)$-approximation achieved by a greedy approach is nearly the best we can hope for. This again shows that the spanning star forest problem and the dominating set problem are very different regarding the approximability, although they are equivalent in exact optimization.

1.2 Notation Used for Approximation Algorithms

For $\beta \in (0,1)$ (resp. $\beta > 1$) and a maximization (resp. minimization) problem Π, an algorithm is called a β-*approximation algorithm* for Π if given an instance $\mathcal{I}$ of Π, it runs in polynomial time and produces a solution with objective value at least (resp. at most) $\beta \cdot OPT(\Pi, \mathcal{I})$, where $OPT(\Pi, \mathcal{I})$ denotes the objective value of the optimum solution to the instance $\mathcal{I}$ of the problem Π. The value β is also called the *approximation ratio*, *approximation factor*, or *performance guarantee* of the algorithm for the problem Π. Moreover, β can be a function of the input size or some parameters in the input. For standard definitions and notations not given here, we refer the readers to [15].

2   Complementary Partial Dominating Set

In this section, we introduce the *complementary partial dominating set problem*, which is useful for designing our algorithm for spanning star forest in dense graphs. Before presenting its formal definition, we need to mention another related problem called the *complementary set cover problem*.

2.1   Complementary Set Cover

We briefly review the complementary set cover problem (CSC for short) [3], since some results of it will be used later. The input of CSC is a pair $(\mathcal{S}, U)$, which consists of a ground set U of elements and a set $\mathcal{S}$ containing some subsets of U. The set $\mathcal{S}$ is guaranteed to be close under subsets, that is, for any $S \in \mathcal{S}$ and $S' \subseteq S$, we have $S' \in \mathcal{S}$. The goal is to find a collection of pairwise-disjoint subsets $S_1, S_2, \ldots, S_k \in \mathcal{S}$ whose union is U, such that $|U| - k$ is maximized. It is shown in [3] that CSC has a $\frac{193}{240}$-approximation algorithm, which only selects subsets of size at most 6.

2.2   Complementary Partial Dominating Set

Let $G = (V, E)$ be a simple undirected graph. For any vertex $v \in V$, let $N[v] = \{u \in V : (u, v) \in E\} \cup \{v\}$ be the neighborhood of v when regarding v as a neighbor of itself. Let $N[U] = \bigcup_{v \in U} N[v]$ for $U \subseteq V$. For two subsets $U_1, U_2 \subseteq V$, we say U_1 *dominates* U_2, or U_1 is a *dominating set* of U_2, if $U_2 \subseteq N[U_1]$. The *complementary partial dominating set problem* (CPDS for short) is defined as follows.

Input: A graph $G = (V, E)$ and a subset of vertices $V' \subseteq V$.
Output: A set $U \subseteq V$ that dominates V' such that $|V'| - |U|$ is maximized.

Although the objective we use seems to be equivalent to finding the minimum-size dominating set of V', they are totally different when considering the approximability. It is easy to see that the minimization version of CPDS generalizes the dominating set problem and thus cannot be approximated to within $O(\log n)$ unless $P = NP$ [9,13], while as is shown below, CPDS allows a constant factor approximation algorithm.

Theorem 1. *There is a $\frac{193}{240}$-approximation algorithm for CPDS.*

Proof. Given an instance $\mathcal{I} = (G, V')$ of CPDS, we regard it as an instance $\mathcal{I}' = (\mathcal{S}, U)$ of CSC in the following way. The ground set U is just V', and $\mathcal{S}$ contains all subsets of V' that is dominated by some vertex in V, i.e. $\mathcal{S} = \{W \subseteq V' : \exists v \in V \text{ s.t. } W \subseteq N[v]\}$. It is easy to see that $\mathcal{S}$ is close under subsets. (Note that $\mathcal{S}$ may have exponential size; we will come back to this point later.) Now, given a solution to the instance $\mathcal{I}$ of CPDS with objective value s, we can easily construct a solution to the instance $\mathcal{I}'$ of CSC with no smaller objective value, and vice versa. Therefore, the two instances have the same optimal objective value, and we can apply the $\frac{193}{240}$-approximation algorithm for CSC on $\mathcal{I}'$ to obtain a solution to $\mathcal{I}$ with the same approximation ratio. However, the instance $\mathcal{I}'$ may have exponential size since it may contain all subsets of V'. To overcome this, we just note that the $\frac{193}{240}$-approximation algorithm for CSC only deals with sets in $\mathcal{S}$ of size at most 6, and all subsets of V' of size at most 6 can surely be enumerated in polynomial time.   □

3   Algorithm Description and Analysis

In this section, we give an approximation algorithm for the spanning star forest problem in dense graphs. Fix $c \in (0, 1)$. Let $\alpha = \frac{193}{240}$ be the best known approximation ratio for CPDS. Let ϵ be any constant such that $0 < \epsilon < \sqrt{c}$. Let $\delta = 1 - \sqrt{c} + \epsilon, M = 2/(c - (\sqrt{c} - \epsilon)^2)$, and $N_0 = M/(\epsilon(1 - \delta))$. Note that δ, M and N_0 are all positive constants only depending on c and ϵ.

We present our algorithm for SSF in c-dense graphs as Algorithm 1. Note that at the beginning (and the end) of every execution of the WHILE loop, A, B and C form a partition of V. To show that the obtained star forest is large, we bound the cardinality of A and S respectively.

Lemma 1. *At the end of Stage 1, it holds that $|A| \leq M$.*

Proof. Consider the moment right before some vertex v is added to A. Due to the loop condition, we have $|C| \geq \delta n$, and $|A \cup B| = n - |C| \leq (1 - \delta)n$. Thus, the number of edges in E with both endpoints in $A \cup B$ is at most $((1 - \delta)n)^2/2$. Since $|E| \geq cn^2/2$, the number of edges in E with at least one endpoint in C is at least $cn^2/2 - ((1 - \delta)n)^2/2 = n^2/M$. Let E_1 be the set of edges with one endpoint in B and another in C, and E_2 be the set of edges with both endpoints in C. Note that the previous statement is equivalent to $|E_1| + |E_2| \geq n^2/M$, since by definition there are no edges between A and C.

For any vertex $v \in B \cup C$, let $D(v) = N[v] \cap C$ be the set of vertices in C dominated by v. Consider $D = \sum_{v \in B \cup C} |D(v)|$. It is easy to see that every edge in E_1 contributes 1 to this sum, while each edge in E_2 contributes 2. Hence, $D = |E_1| + 2|E_2| \geq n^2/M$, from which we know that there exists a vertex

---

Algorithm 1. Approximate SSF in c-dense graphs

Input: A c-dense graph $G = (V, E)$.
Output: A spanning star forest of G.

If $n \leq N_0$ we perform the exhaustive search to get the optimal solution. In the following we assume $n > N_0$.
$A \leftarrow \emptyset$, $B \leftarrow \emptyset$, $C \leftarrow V$.
Stage 1:
while $|C| \geq \delta n$ **do**
     Find the vertex $v \in B \cup C$ that dominates the largest number of vertices in C.
     Set $A \leftarrow A \cup \{v\}$, $B \leftarrow N[A] \setminus A$, and $C \leftarrow V \setminus N[A]$.
end while
Stage 2:
Construct an instance $\mathcal{I} = (G', V')$ of CPDS, where G' is the subgraph of G induced on the vertex set $B \cup C$, and $V' = C$. Run the α-approximation algorithm for CPDS on $\mathcal{I}$ to get a dominating set of C, denoted by S.
return a spanning star forest rooted on $A \cup S$.

$v^* \in B \cup C$ such that $|D(v^*)| \geq n/M$. Note that the greedy step in the algorithm is just to pick the vertex v with the largest $|D(v)|$. Therefore, after adding v to A and updating B and C correspondingly, the size of $A \cup B$ increases by at least n/M. Since there are only n vertices, we can add at most M of them to A, completing the proof of Lemma 1. $\square$

Lemma 2. $|S| \leq \delta(1 - \alpha)n + \alpha k$, where k is the size of the smallest subset $U \subseteq B \cup C$ that dominates C.

Proof. By the definition of CPDS, we know that the value of the optimum solution to its instance $\mathcal{I}$ defined in Algorithm 1 is precisely $|C| - k$. As the solution S is obtained by applying the α-approximation algorithm for CPDS, we have $|C| - |S| \geq \alpha(|C| - k)$. Rearranging terms gives $|S| \leq (1 - \alpha)|C| + \alpha k \leq \delta(1 - \alpha)n + \alpha k$, where the second inequality follows from the fact that $|C| \leq \delta n$ at the end of Stage 1. $\square$

We are ready to prove our main theorem.

Theorem 2. *Algorithm 1 is a $(\alpha + (1 - \alpha)\sqrt{c} - 2\epsilon)$-approximation algorithm for the spanning star forest problem in c-dense graphs.*

Proof. Clearly Algorithm 1 runs in polynomial time. Furthermore, it finds the optimal spanning star forest of G when $n \leq N_0$, and produces a solution of size $n - |A| - |S| \geq (1 - \delta(1 - \alpha))n - \alpha k - M$ when $n > N_0$, by Lemmas 1 and 2. The size of the optimal solution is $n - k^*$, where k^* is the size of the smallest dominating set of G. It is easy to see that k^* is not smaller than the size of the smallest subset of V that dominates C. Since no edges exist between A and C, the latter quantity is equal to k, the size of the smallest subset of $B \cup C$ that dominates C. Therefore, we have $n - k^* \leq n - k$. We also note that $k \leq |C| \leq \delta n$ since C dominates itself. The approximation ratio of Algorithm 1 can thus be bounded from below by

$$
\frac{(1 - \delta(1 - \alpha))n - \alpha k - M}{n - k^*}
$$
$$
\geq \frac{(1 - \delta(1 - \alpha))n - \alpha k - M}{n - k}
$$
$$
= \alpha + \frac{(1 - \alpha)(1 - \delta)n}{n - k} - \frac{M}{n - k}
$$
$$
\geq \alpha + (1 - \alpha)(1 - \delta) - \frac{M}{n - \delta n}
$$
$$
\geq \alpha + (1 - \alpha)(\sqrt{c} - \epsilon) - \frac{M}{(1 - \delta)N_0}
$$
$$
\geq \alpha + (1 - \alpha)\sqrt{c} - 2\epsilon,
$$

which concludes the proof of Theorem 2.   □

4   Hardness Results

We now show that for every $0 < c < 1$, SSF in c-dense graphs does not admit a polynomial-time approximation scheme, unless $P = NP$. Thus, the technique developed by Arora et al. [2] for designing PTAS for combinatorial problems in dense instances cannot be applied to this problem.

Theorem 3. *For any $c \in (0, 1)$, there exists a constant $\epsilon = \epsilon(c) > 0$, such that it is NP-hard to approximate the spanning star forest problem in c-dense graphs to a factor of $1 - \epsilon$.*

Proof. We reduce the general SSF problem to SSF in c-dense graphs. Let $G = (V, E)$ be an input to general SSF. Let $n = |V|$, $k = \lceil 2\sqrt{c}/(1 - \sqrt{c}) \rceil$, and let OPT denote the size of the largest spanning star forest of G. It is easy to verify that $k > \sqrt{c}(k + 1)$. We assume w.l.o.g. that $n \geq k/(k^2 - c(k + 1)^2) > 0$, since otherwise we can just do a brute-force search for the constant-size (note that k and c are both constants) input graph. We also assume that G is connected, since connected and disconnected versions of general SSF share the same hardness-of-approximation result. We thus have $OPT \geq n/2$, since any connected graph on n vertices has a dominating set of size at most $n/2$. Let H be a complete graph on a vertex set of size kn which is disjoint from V, and let $G' = G \cup H$.

We verify that G' is c-dense. As G' has $n' = (k + 1)n$ vertices and at least $kn(kn - 1)/2$ edges, it suffices to show that $kn(kn - 1)/2 \geq c(k + 1)^2 n^2/2$, or $n \geq k/(k^2 - c(k+1)^2)$, which is exactly our assumption on n. Since G' consists of two disjoint components, it is clear that $OPT' = OPT + kn - 1$, OPT' denoting the size of the largest spanning star forest of G'. Moreover, given a spanning star forest of G' of size s', we can easily construct a spanning star forest of

G of size at least $s' - (kn - 1)$. Thus, given any β-approximation algorithm for SSF in c-dense graphs, we can obtain a spanning star forest of G of size $\beta(OPT + kn - 1) - (kn - 1)$. On the other hand, we know that there is a constant $\gamma > 0$ such that approximating general SSF within $1 - \gamma$ is NP-hard [12]. Therefore, there exists G such that $\beta(OPT+kn-1)-(kn-1) \leq (1-\gamma)OPT$, from which we derive that

$$\beta \leq \frac{(1-\gamma)OPT + kn - 1}{OPT + kn - 1}$$
$$= 1 - \gamma + \frac{\gamma(kn - 1)}{OPT + kn - 1}$$
$$\leq 1 - \gamma + \frac{\gamma(kn - 1)}{n/2 + kn - 1}$$
$$< 1 - \gamma + \frac{\gamma k}{k + 1/2}.$$

The proof is completed by choosing $\epsilon = \gamma/(2k + 1)$.   □

We have designed an algorithm for SSF in c-dense graphs whose approximation ratio outperforms the best known bound for general SSF, for every $0 < c < 1$. A natural question is whether we can generalize our technique to weighted versions of SSF. A little surprisingly, we show in the following that this is not the case: We cannot design *any* approximation algorithm for node- (resp. edge-)weighted SSF in c-dense graphs with a strictly larger performance guarantee than that of the best approximation algorithm for general node- (resp. edge-)weighted SSF.

Theorem 4. *For any $0 < c < 1$ and any $\beta, \epsilon > 0$, the existence of a β-approximation algorithm for node- (resp. edge-)weighted SSF in c-dense graphs implies that of a $(\beta - \epsilon)$- (resp. β-)approximation algorithm for node- (resp. edge-)weighted SSF in general graphs.*

Proof. The edge-weighted case is easy since we can regard every edge-weighted graph as a complete graph (which is c-dense for any $c < 1$ and large enough n) with some edges having weight 0. Thus, in the following we consider the node-weighted version of SSF. Fix c, ϵ and β. Let $G = (V, E)$ be an input graph to node-weighted SSF, and $w : V \to \mathbb{Q}^+ \cup \{0\}$ be the weight function on its nodes. Let $n = |V|$ and OPT denote the maximum weight of a spanning star forest of G. We assume that $OPT > 0$ since the case $OPT = 0$ is easily detectable. Let $w^* = \min\{w(v) : v \in V \text{ and } w(v) > 0\}$. Clearly $OPT \geq w^*$. We apply a reduction similar to that used in the proof of Theorem 3 to get a c-dense graph $G' = G \cup H$, with the only difference that we set the weights of all vertices in H to 1, and multiply the weights of all vertices in G by a factor of $\Delta = (1 - \beta)(kn - 1)/(\epsilon w^*)$ (recall that k is the constant defined in the proof of Theorem 3). Now we have $OPT' = \Delta \cdot OPT + kn - 1$ where OPT' denotes the maximum weight of a spanning star forest of G', and a spanning star forest of G' of weight s' can be easily transformed to a spanning star forest of G of weight at least $(s' - (kn - 1))/\Delta$. Thus, given a β-approximation to node-weighted SSF in c-dense graphs, we can design an approximation algorithm for node-weighted SSF in general graphs with an approximation ratio of

$$\beta' \geq \frac{(\beta(\Delta \cdot OPT + kn - 1) - (kn - 1))/\Delta}{OPT}$$
$$= \beta - \frac{(1 - \beta)(kn - 1)}{\Delta \cdot OPT}$$
$$\geq \beta - \frac{(1 - \beta)(kn - 1)}{\Delta \cdot w^*} = \beta - \epsilon,$$

concluding the proof of Theorem 4.   □

Finally, we show that the dominating set problem, as the complement of SSF, remains hard to approximate even in dense graphs.

Theorem 5. *For any $c \in (0, 1)$ and any $\epsilon > 0$, there is no $(1 - \epsilon) \ln n$-approximation algorithm for dominating set in c-dense graphs, where n is the number of vertices in the input graph, unless $NP \subseteq DTIME(n^{O(\log \log n)})$.*

Proof. We show how to use a $(1 - \epsilon) \ln n$-approximation for dominating set in c-dense graphs to design a $(1 - \epsilon') \ln n$-approximation for dominating set in general graphs, thus proving the theorem since by [9] this implies $NP \subseteq DTIME(n^{O(\log \log n)})$. Given a graph $G = (V, E)$, we first exhaustively check if the optimal dominating set has size at most $\lceil 1/\epsilon \rceil$. If so, we can find it in polynomial time. Otherwise, we apply a reduction similar to that used in the proof of Theorem 3 to obtain a c-dense graph G'. Denoting by OPT and OPT' the size of the minimum dominating set of G and G' respectively, it is clear that $OPT' = OPT + 1$, and a dominating set of G' of size s can be easily converted to one of G of size at most $s - 1$. Therefore, given a $(1 - \epsilon) \ln n$-approximation for dominating set on c-dense graphs, we can obtain an approximation algorithm for it on general graphs with approximation ratio at most $((1 - \epsilon) \ln n(OPT + 1) - 1)/OPT < (1 - \epsilon) \ln n(1 + 1/OPT) \leq (1 - \epsilon^2) \ln n$, since $OPT \geq \lceil 1/\epsilon \rceil$. This finishes the proof of Theorem 5.

5   Conclusion

In this paper, we explored the spanning star forest problem in c-dense graphs, and devised an algorithm whose approximation ratio is better than the previously best known bound for this problem in general graphs. We also showed that this problem does not admit a PTAS unless $P = NP$, thus ruling out the possibility of applying the general technique developed by Arora et al. to this problem. We then showed hardness results for its weighted versions as well as its complementary problem, the dominating set problem in dense graphs.

An interesting question is to bridge the gap between algorithmic and hardness results for this problem, since the inapproximability factor derived by our reduction is very close to 1. It is also interesting to see whether we can design approximation algorithms for spanning star forest in dense graphs based on any approximation algorithm for spanning star forest in general graphs, instead of using that for the CPDS problem.

Acknowledgements

This work was supported in part by the National Natural Science Foundation of China Grant 60553001, 61073174, 61033001 and the National Basic Research Program of China Grant 2007CB807900, 2007CB807901. Part of this work was done while the authors were visiting Cornell University. The authors would like to thank the anonymous referees for their helpful comments on improving the presentation of this paper.

References

1. Agra, A., Cardoso, D., Cerfeira, O., Rocha, E.: A spanning star forest model for the diversity problem in automobile industry. In: Proc. of ECCO XVII (2005)
2. Arora, S., Karger, D., Karpinski, M.: Polynomial time approximation schemes for dense instances of NP-hard problems. Journal of Computer and System Sciences 58(1), 193–210 (1999)
3. Athanassopoulos, S., Caragiannis, I., Kaklamanis, C., Kuropoulou, M.: An improved approximation bound for spanning star forest and color saving. In: Královič, R., Niwiński, D. (eds.) MFCS 2009. LNCS, vol. 5734, pp. 90–101. Springer, Heidelberg (2009)
4. Berry, V., Guillemot, S., Nicholas, F., Paul, C.: On the approximation of computing evolutionary trees. In: Wang, L. (ed.) COCOON 2005. LNCS, vol. 3595, pp. 115–125. Springer, Heidelberg (2005)
5. Cardinal, J., Langerman, S., Levy, E.: Improved approximation bounds for edge dominating set in dense graphs. Theoretical Computer Science 410, 949–957 (2009)
6. Chakrabarty, D., Goel, G.: On the approximability of budgeted allocations and improved lower bounds for submodular welfare maximization and GAP. In: Proc. of FOCS 2008, pp. 687–696 (2008)
7. Chen, N., Engelberg, R., Nguyen, C.T., Raghavendra, P., Rudra, A., Singh, G.: Improved approximation algorithms for the spanning star forest problem. In: Charikar, M., Jansen, K., Reingold, O., Rolim, J.D.P. (eds.) APPROX 2007. LNCS, vol. 4627, pp. 44–58. Springer, Heidelberg (2007)
8. Duh, R., Furer, M.: Approximation of k-set cover by semi local optimization. In: Proc. of STOC 1997, pp. 256–264 (1997)
9. Feige, U.: A threshold of lnn for aproximating set cover. Journal of the ACM 45(4), 634–652 (1998)
10. Gaspers, S., Kratsch, D., Liedloff, M., Todinca, I.: Exponential time algorithms for the minimum dominating set problem on some graph classes. ACM Transactions on Algorithms 6(1), No. 9 (2009)
11. Imamura, T., Iwama, K.: Approximating vertex cover on dense graphs. In: Proc. of SODA 2005, pp. 582–589 (2005)
12. Nguyen, C.T., Shen, J., Hou, M., Sheng, L., Miller, W., Zhang, L.: Approximating the spanning star forest problem and its applications to genomic sequence alignment. SIAM Journal on Computing 38(3), 946–962 (2008)
13. Raz, R., Safra, S.: A sub-constant error-probability low-degree test, and sub-constant error-probability PCP characterization of NP. In: Proc. of STOC 1997, pp. 475–484 (1997)
14. Schiermeyer, I.: Problems remaining NP-complete for sparse or dense graphs. Discuss. Math. Graph. Theory 15, 33–41 (1995)
15. Vazirani, V.: Approximation Algorithms. Springer, Heidelberg (2001)

A New Result on $[k, k+1]$-Factors Containing Given Hamiltonian Cycles*

Guizhen Liu[1], Xuejun Pan[1], and Jonathan Z. Sun[2,3]

[1] School of Mathematical and System Sciences, Shandong University
Jinan 250100, Shandong, China
[2] School of Computer Science and Technology, Shandong University of Technology
Zibo 255049, Shandong, China
[3] School of Computing, University of Southern Mississippi
Hattiesburg, MS 39406, USA
jonathan.sun@usm.edu

Abstract. We give a sufficient condition, which guarantees that for arbitrary Hamiltonian cycle C, there exists a $[k, k+1]$-factor containing C. This improves a previous result of Cai, Li, and Kano [7].

1 Introduction

A graph is defined by its vertex set and edge set. The number of edges incident to a vertex is called the *degree* of this vertex, and a graph is *k-regular* if all vertices have the same degree k. A spanning subgraph H of G is called a *k-factor* if the degree of every vertex of H is k, and the process of partitioning a graph into (edge-disjoint) factors is called *graph factorization*. For example, in particular, a 1-factor is a perfect matching, a 2-factor is a cycle cover, and a connected 2-factor is a Hamiltonian cycle. Originated from the efforts of finding matchings in graphs, the study of factors and factorizations has yielded abundant results in graph theory and combinatorics as well as broad impact in computer science. For example, the latest elegant work on using graph expanders to construct erasure-resilient code [2,3] in trusted distributed computing infrastructures [10] is based on Hamiltonian factorization.

Generalizations of the concept of regular factors include: 1) *[a, b]-factors*, where the degrees of vertices are bounded in range $[a, b]$ between two constants a and b; 2) *f-factors*, where the degrees of vertices are determined by a function f defined on vertex set; and 3) *(g, f)-factors*, where the above two generalizations are combined so that two functions f and g bound the degrees of vertices in the factors. Specifically, a $[k, k+1]$-*factor* (following the concept of $[a, b]$-factor) is a natural relaxation of a k-factor and its' existence conditions were extensively studied in graph theory community.

Connected factors such as Hamiltonian cycles or Hamiltonian factors (factors containing a Hamiltonian cycle) are especially useful, such as in building erasure-resilient code. Unfortunately, such factors are hard to find (see [9]). Only a few

---

* This work is partially supported by Shandong Provincial Tai-Shan Scholar Award (2010-2015) and NASA Mississippi Space Grant Consortium No. NNG05GJ72H.

W. Wu and O. Daescu (Eds.): COCOA 2010, Part II, LNCS 6509, pp. 170–180, 2010.

non-trivial sufficient conditions have been discovered. Since the relaxation from k to $[k, k+1]$ resulted in the existence of such factors in significantly larger classes of graphs, we expect to see applications of such factors in building erasure-resilient code with more flexibility in network topology or with enhanced robustness that tolerate more failures with less redundancy. This paper provides a new sufficient condition for the existence of Hamiltonian $[k, k + 1]$-factors containing any given Hamiltonian cycle, which covers a previous result of Cai, Li, and Kano [7].

1.1 Related Work

Here are some necessary terminologies before we can present related work and our result. We only consider simple graphs with no multiple edges. In a graph $G = (V(G), E(G))$ of vertex set $V(G)$ and edge set $E(G)$, we denote by $d_G(v)$ the degree of a vertex v, and by $\delta(G)$ the minimum degree of all vertices. For any $S \subseteq V(G)$, the resulting subgraph of G obtained by deleting all the vertices in S and all the edges incident with the vertices in S is denoted by $G - S$. If $S = \{v\}, v \in V(G)$, we usually denote $G - v = G - \{v\}$. For any $X \subseteq E(G)$, the resulting subgraph of G obtained by deleting all the edges of X is denoted by $G - X$. If $X = \{e\}, e \in E(G)$, we usually denote $G - e = G - \{e\}$.

A graph is Hamiltonian if it contains a Hamiltonian cycle. The following two milestones by Ore [15] and Fan [8] (although only sufficient but not necessary) in recognizing Hamiltonian graphs are most related to our work.

Theorem 1. *(Ore's condition) Let G be a simple graph with $n(\geq 3)$ vertices. If for any pair of non-adjacent vertices u, v in G,*

$$d_G(u) + d_G(v) \geq n,$$

then G is a Hamiltonian graph.

Theorem 2. *(Fan's condition) Let G be a 2-connected graph of n vertices. If for any pair of vertices u, v of distance $d_G(u, v) = 2$,*

$$\max\{d_G(u), d_G(v)\} \geq \frac{n}{2},$$

then G is a Hamiltonian graph. Here $d_G(u, v)$ is the length of the shortest path from u to v.

Cai, Li, and Kano [7] established the following existence condition of Hamiltonian $[k, k + 1]$-factors containing any Hamiltonian cycles.

Theorem 3. *Let $k \geq 2$ be an integer and G be a graph of $n(\geq 3)$ vertices with $\delta(G) \geq k$. Assume $n \geq 8k - 16$ for even n and $n \geq 6k - 13$ for odd n. If for any pair of non-adjacent vertices u, v in G,*

$$d_G(u) + d_G(v) \geq n,$$

then for any Hamiltonian cycle C, G has a $[k, k + 1]$-factor containing C.

This is the best previously known non-trivial condition for the existence of such factors. The degree condition, $n \geq 8k - 16$ for even n and $n \geq 6k - 13$ for odd n, is tight.

1.2   Our Result

We improve the above result of Cai, Li and Kano [7] by weakening the Ore's condition in Theorem 3 into the following form, which is close to Fan's condition.

Theorem 4. *Let $k \geq 2$ be an integer and G be a graph of $n(\geq 3)$ vertices. G has no cut edge and $\delta(G) \geq k$. Assume $n \geq 8k - 16$ for even n and $n \geq 6k - 13$ for odd n. If for any pair of non-adjacent vertices u, v in G,*

$$\max\{d_G(u), d_G(v)\} \geq \frac{n}{2},$$

then for any Hamiltonian cycle C, G has a $[k, k+1]$-factor containing C.

1.3   Lovász's (g, f)-Factor Theorem

L. Lovász [14] gave the following necessary and sufficient condition for a graph to have (g, f)-factors in 1970.

Theorem 5. *Let G be a graph and $g, f : V(G) \to Z$ such that $g(x) \leq f(x)$ for all $x \in V(G)$. Then G has a (g, f)-factor if and only if for all disjoint subsets S and T of $V(G)$,*

$$\sum_{x \in S} f(x) + \sum_{x \in T} (d_G(x) - g(x)) - e_G(S, T) - q(S, T) \geq 0.$$

Here $e_G(S, T)$ is the number of edges between S and T, and $q(S, T)$ denotes the number of odd components, namely, the components C of $G - (S \cup T)$ such that

$$g(x) = f(x)$$

for all $x \in V(C)$ and

$$f(x) + e_G(C, T) \equiv 1 \ (mod \ 2).$$

This theorem is the foundation of many known sufficient conditions for the existence of factors. Particularly, we will use a special case of this theorem in the proof of our main result.

2   Proof of the Main Theorem

Proof. G is Hamiltonian by Theorem 2. When $k = 2$, any Hamiltonian cycle by itself is a Hamiltonian $[k, k+1]$-factor. So we assume $k \geq 3$, Write the condition in main theorem as (1): For any pair of non-adjacent vertices u, v,

$$\max\{d_G(u), d_G(v)\} \geq \frac{n}{2}. \tag{1}$$

Given a Hamiltonian cycle C, denote $H = G - C$ and $\rho = k - 2$. Clearly, $V(H) = V(G)$ and $\rho \geq 1$. Let

$$U = \{v : v \in V(G), d_G(v) \geq \frac{n}{2}\},$$

and denote

$$L = V(G) - U.$$

For any vertex $v \in V(G)$, it is obvious that $d_H(v) = d_G(v) - 2 \geq \rho$, and

$$n \geq \begin{cases} 8\rho & \text{when } n \text{ is even;} \\ 6\rho - 1 & \text{when } n \text{ is odd} \end{cases} \tag{2}$$

G has the desired $[k, k+1]$-factor if and only if H has a $[\rho, \rho + 1]$-factor. We will prove the latter by contradictions. Suppose H has no $[\rho, \rho + 1]$-factor, then by Lovász[14] (g, f)-factor Theorem, there are two disjoint subsets S, T of $V(G)$ such that

$$\sigma(S, T) := -(\rho + 1)s + \rho t - \sum_{v \in T} d_{H-S}(v) \geq 1. \tag{3}$$

Here $s = |S|$ and $t = |T|$ are the cardinalities of S and T. (Note that, since $\rho \neq \rho + 1$, there is no odd component in this case so that $q(S, T) \equiv 0$ for any S and T.)

Furthermore, we can choose such S and T so that for any $v \in T$,

$$d_{H-S}(v) \leq \rho - 1. \tag{4}$$

Otherwise there exists a vertex u in T such that $d_{H-S}(u) \geq \rho$ so that

$$\sigma(S, T\backslash\{u\}) = -(\rho + 1)s + \rho(t - 1) - \sum_{v \in T\backslash\{u\}} d_{H-S}(v)$$

$$= -(\rho + 1)s + \rho t - \sum_{v \in T} d_{H-S}(v) + (d_{H-S}(u) - \rho)$$

$$= \sigma(S, T) + (d_{H-S}(u) - \rho)$$

$$\geq \sigma(S, T).$$

Therefore, if we remove u from T, $\sigma(S, T\backslash\{u\})$ still fulfills (3).

Next, we will study the structures of the graph under such selection of S and T to get Claims 1 - 6, by introducing a contradiction in each case. Therefore we would have proved that S and T fulfilling 3 don't exist, i.e., H has a $[\rho, \rho + 1]$-factor.

Claim 1. $G(L)$ is a complete graph.

This is obvious. Given any two vertices u, v in L, by definition of L, $d_G(u)$ and $d_G(v)$ are both less than $\frac{n}{2}$. Then by (1), u and v must be adjacent in G.

Claim 2. $s \geq 1$.

Otherwise $s = 0$ so that

$$\sigma(S, T) = \rho t - \sum_{v \in T} d_H(v) \leq 0,$$

contradicting (3).

Claim 3. $t \geq \rho + 2$.

Otherwise, $t \leq \rho + 1$. Then

$$\sigma(S, T) \leq -(\rho + 1)s + \rho t - \sum_{v \in T}(d_H(v) - s)$$
$$\leq -(\rho + 1)s + \rho t - t(\rho - s)$$
$$= s(t - \rho - 1) \leq 0,$$

another contradiction to (3).

Claim 4. $T \cap U \neq \emptyset$.

Otherwise $T \subseteq L$. By Claim 1, $G(T)$ is also a complete graph, so

$$E_G[T] = \frac{t(t-1)}{2}.$$

We also have

$$|E_G[T] \cap C| \leq t - 1,$$

since C is a Hamiltonian cycle of G. Therefore

$$\sum_{v \in T} d_{H-S}(v) \geq 2|E_G[T] \setminus C| \geq t(t-1) - 2(t-1)$$
$$= (t-1)(t-2),$$

so that $\quad \sigma(S, T) \leq -(\rho + 1)s + \rho t - (t-1)(t-2)$
$$\leq -(\rho + 1)s + \rho t - (t-1)\rho \qquad \text{(by Claim 3)}$$
$$= -(\rho + 1)s + \rho < 0, \qquad \text{(by Claim 2)}$$

another contradiction to (3).

Claim 5. $s \leq \lceil \frac{n}{2} \rceil - 3$.

Denote by $\bar{S}$ the vertices of G not in S. Let

$$X = \bar{S} \cap U$$

and

$$Y = \bar{S} \cap L(= \bar{S} \setminus X).$$

Obviously,

$$
\begin{cases}
d_G(v) \geq \frac{n}{2} & , \text{ if } v \in X; \\
d_G(v) < \frac{n}{2} & , \text{ if } v \in Y.
\end{cases}
\tag{5}
$$

We then consider two cases according to the parity of n.

For even n, assume

$$
s \geq \frac{n}{2} - 2.
$$

Then let

$$
q = s - \frac{n}{2} + 2 (\geq 0)
$$

and

$$
r = n - s - t (\geq 0).
$$

Then

$$
\begin{aligned}
\sigma(S, T) &= -(\rho + 1)s + \rho(n - s - r) - \sum_{v \in T} d_{H-S}(v) \\
&= -(2\rho + 1)s + \rho(n - r) - \sum_{v \in T} d_{H-S}(v) \\
&= -(2\rho + 1)(\frac{n}{2} - 2 + q) + \rho(n - r) - \sum_{v \in T} d_{H-S}(v) \\
&= 4\rho + 2 - \frac{n}{2} - (\rho + 1)q - \rho(r + q) - \sum_{v \in T} d_{H-S}(v) \\
&\leq 0,
\end{aligned}
$$

unless

$$
q = 0, \text{ and } \begin{cases}
r = 0, \sum_{v \in T} d_{H-S}(v) \leq 1; \text{ or} \\
r = 1, \sum_{v \in T} d_{H-S}(v) = 0.
\end{cases}
$$

Furthermore, when $r = 0$,

$$
\sum_{v \in T} d_{H-S}(v) = 2|E_H[T]| \equiv 0 \ (\mathrm{mod} 2)
$$

so that

$$
\sum_{v \in T} d_{H-S}(v) = 0.
$$

Therefore, in order to make (3) true, we must have $q = \sum_{v \in T} d_{H-S}(v) = 0$ and $r \leq 1$.

Next, by $q = 0$, there must be

$$
s = \frac{n}{2} - 2;
$$

and by $\sum_{v \in T} d_{H-S}(v) = 0$ and $r \leq 1$, we have

$$E_G[\bar{S}] \subseteq C. \tag{6}$$

Since

$$d_G(v) \leq d_{H-S}(v) + s + 2 = \frac{n}{2}$$

for any $v \in \bar{S}$, we have, for any $v \in X$,

$$d_G(v) = \frac{n}{2}. \tag{7}$$

From $s = \frac{n}{2} - 2$ we know that all edges in C with one end in X must be contained in $E_G(\bar{S})$. Therefore

$$
\begin{aligned}
|X| + |Y| - 1 = |\bar{S}| - 1 &\geq |E_G[\bar{S}] \cap C| \quad \text{(by (6))} \\
&= (|E_G[X]| + |E_G(X,Y)|) + |E_G[Y]| \\
&\geq (|X| + 1) + |E_G[Y]| \\
&= |X| + 1 + \frac{|Y|(|Y| - 1)}{2}, \quad \text{(by Claim 1)}
\end{aligned}
$$

so that

$$|Y| \geq 2 + \frac{|Y|(|Y| - 1)}{2}.$$

This gives another contradiction. That is, $\sigma(S,T) \leq 0$ while n is even and $s \geq \frac{n}{2} - 2$, contradicting (3) again.

For odd n, assume

$$s \geq \frac{n-3}{2}.$$

Similar to the case of even n, let

$$q = s - \frac{n-3}{2} (\geq 0),$$

$$r = n - s - t (\geq 0).$$

Then

$$
\begin{aligned}
\sigma(S,T) &= -(\rho + 1)s + \rho(n - s - r) - \sum_{v \in T} d_{H-S}(v) \\
&= 3\rho + \frac{3}{2} - \frac{n}{2} - (\rho + 1)q - \rho(r + q) - \sum_{v \in T} d_{H-S}(v) \\
&\leq 0,
\end{aligned}
$$

unless $q = \sum_{v \in T} d_{H-S}(v) = 0$ and $r \leq 1$. Again similar to the case of even n, we can get

$$E_G[\bar{S}] \subseteq C$$

and

$$d_G(v) = \frac{n+1}{2}$$

for any vertex v in X, so that all edges in C with one end in X are contained in $E_G(\bar{S})$. Then by the same argument, we get

$$|Y| \geq 2 + \frac{|Y|(|Y|-1)}{2},$$

still a contradiction.

Claim 6. $T \cap L \neq \emptyset$.

Otherwise $T \subseteq U$. Then by definition of U and (4), for any $v \in T$,

$$\lceil \frac{n}{2} \rceil \leq d_G(v) \leq d_{H-S}(v) + s + 2 \leq \rho + s + 1.$$

That is,

$$d_{H-S}(v) \geq \lceil \frac{n}{2} \rceil - s - 2$$

and

$$\rho + s + 2 - \lceil \frac{n}{2} \rceil \geq 1.$$

Consequently,

$$\sigma(S, T) \leq -(\rho+1)s + \rho t - t(\lceil \frac{n}{2} \rceil - s - 2)$$
$$= t(\rho + s + 2 - \lceil \frac{n}{2} \rceil) - (\rho+1)s$$
$$\leq (n-s)(\rho + s + 2 - \lceil \frac{n}{2} \rceil) - (\rho+1)s.$$

Let

$$f(s) = (n-s)(\rho + s + 2 - \lceil \frac{n}{2} \rceil) - (\rho+1)s.$$

Then take differential of $f(s)$:

$$\frac{df(s)}{ds} = -2\rho - 3 + n + \lceil \frac{n}{2} \rceil - 2s$$
$$\geq -2\rho - 3 + n + \lceil \frac{n}{2} \rceil - 2\lceil \frac{n}{2} \rceil + 6 \quad \text{(by Claim 5)}$$
$$= -2\rho + 3 + \lfloor \frac{n}{2} \rfloor \geq 0. \qquad \text{(by (2))}$$

Therefore,

$$\sigma(S, T) \leq f(s) \leq f(\lceil \frac{n}{2} \rceil - 3)$$
$$= (\lfloor \frac{n}{2} \rfloor + 3)(\rho - 1) - (\rho+1)(\lceil \frac{n}{2} \rceil - 3)$$
$$= \rho(\lfloor \frac{n}{2} \rfloor - \lceil \frac{n}{2} \rceil + 6) - n \leq 0, \qquad \text{(by (2))}$$

another contradiction to (3).

Now we have finished the proofs of Claims 1 - 6.

Let

$$T_1 = T \cap U \; ;$$
$$T_2 = T \cap L \; ;$$
$$t_1 = |T_1| \; ;$$
$$t_2 = |T_2|.$$

Obviously, $t_1 \geq 1$; $t_2 \geq 1$; and for any $v \in T$,

$$d_{H-S}(v) \geq d_G(v) - s - 2.$$

Consequently, for any $v \in T_1$,

$$d_{H-S}(v) \geq \begin{cases} \frac{n}{2} - s - 2 & \text{when } n \text{ is even} \\ \frac{n}{2} - s - \frac{3}{2} & \text{when } n \text{ is odd} \end{cases} \qquad (8)$$

Together with (4), we get

$$\begin{cases} \rho + s + 2 - \frac{n}{2} \geq 1 & \text{when } n \text{ is even} \\ \rho + s + \frac{3}{2} - \frac{n}{2} \geq 1 & \text{when } n \text{ is odd} \end{cases} \qquad (9)$$

Together with Claim 5, we further get

$$\rho \geq 2. \qquad (10)$$

Then by claim 1,

$$d_{H-S}(v) \geq t_2 - 3$$

for any $v \in T_2$. Together with (4), we have

$$t_2 \leq \rho + 2. \qquad (11)$$

We then consider two cases according to the parity of n.

For even n, by (8) and (11),

$$\begin{aligned} \sigma(S, T) &\leq -(\rho + 1)s + \rho t - t_1(\frac{n}{2} - s - 2) \\ &= t_1(\rho + s + 2 - \frac{n}{2}) - (\rho + 1)s + \rho t_2 \\ &\leq (n - s - t_2)(\rho + s + 2 - \frac{n}{2}) - (\rho + 1)s + \rho t_2 \\ &= -(s - \frac{n}{2} + 3)^2 + (s - \frac{n}{2} + 3)(\frac{n}{2} + 3 - 2\rho - t_2) + 6\rho + t_2 - n \\ &\leq -2\rho + t_2 \leq 0, \end{aligned}$$

contradicting (3).

For odd n, let

$$r = n - s - t(\geq 0).$$

Then obviously,

$$\sum_{v \in T_2} d_{H-S}(v) \geq 2|E_G[T_2] \setminus C| \geq (t_2 - 1)(t_2 - 2). \tag{12}$$

By (8), (10) and (11), we have

$$\sigma(S, T) \leq -(\rho + 1)s + \rho t - t_1 \left(\frac{n}{2} - s - \frac{3}{2}\right) - (t_2 - 1)(t_2 - 2)$$

$$= t_1 \left(\rho + s + \frac{3}{2} - \frac{n}{2}\right) - (\rho + 1)s + \rho t_2 - (t_2 - 1)(t_2 - 2)$$

$$= (n - s - t_2 - r)\left(\rho + s + \frac{3}{2} - \frac{n}{2}\right) - (\rho + 1)s + \rho t_2$$

$$\quad -(t_2 - 1)(t_2 - 2)$$

$$= -\left(s - \frac{n}{2} + \frac{5}{2}\right)^2 + \left(s - \frac{n}{2} + \frac{5}{2}\right)\left(\frac{n}{2} + \frac{5}{2} - 2\rho - t_2\right)$$

$$\quad + 5\rho + t_2 - n - (t_2 - 1)(t_2 - 2) - r\left(\rho + s + \frac{3}{2} - \frac{n}{2}\right)$$

$$\leq 0,$$

unless $s = \frac{n}{2} - \frac{5}{2}$, $t_2 = 2$, $r = 0$, $\rho = 2$, and all equalities hold in (12). However, when all equalities hold in (12), we have

$$|E_G[T_2] \setminus C| = t_2 - 1 = 1.$$

Consider $s = \frac{n}{2} - \frac{5}{2}$ and $\rho = 2$, together with (4) and (8), we will have

$$d_{H-S}(v) = 1 \text{ and } d_G(v) = \frac{n+1}{2}$$

for any $v \in T_1$. This implies that all edges in C with one end in T_1 are contained in $E_G[T] \setminus E[T_2]$, and there are at least $t_1 + 1$ such edges. That is,

$$|E_G[T] \cap C| \geq (t_1 + 1) + 1 = t,$$

contradicting the fact that C is a Hamiltonian cycle. Therefore, for odd n, we again must have $\sigma(S, T) \leq 0$, the last contradiction to (3).

Following the above argument, H must have a $[\rho, \rho + 1]$-factor.   □

3   Future Work

The condition in our result is weaker than Ore's condition but stronger than Fan's condition. It remains open whether the "non-adjacent vertices" in Theorem 4 can be further weakened into "vertices of distance two", so that the condition follows exactly the form of Fan's condition.

Acknowledgements.   Authors would like to thank an anonymous reviewer for thoroughly reading the paper and providing detailed comments to help improve the presentation.

References

1. Akiyama, J., Kano, M.: Factors and Factorizations of Graphs (2007) (online manuscript)
2. Alon, N., Bruck, J., Naor, J., Naor, M., Roth, R.: Construction of asymptotically good, low-rate error-correcting codes through pseudo-random graphs. IEEE Transactions on Information Theory 38, 509–516 (1992)
3. Alon, N., Luby, M.: A linear time erasure-resilient code with nearly optimal recovery. IEEE Transactions on Information Theory 42 (1996)
4. Bondy, J.A., Murty, U.S.R.: Graph Theory with Applications. MacMillan, London (1976)
5. Cai, M., Fang, Q., Li, Y.: Hamiltonian $[k, k+1]$-factor. Advances in Mathematics 32, 722–726 (2003)
6. Cai, M., Fang, Q., Li, Y.: Existence of Hamiltonial k-factors. J. Sys. Sci. Complexity 17(4), 464–471 (2004)
7. Cai, M., Li, Y., Kano, M.: A $[k, k + 1]$-factor containing given Hamiltonian cycle. Science in China Ser. A 41, 933–938 (1998)
8. Fan, G.: New sufficient conditions for cycles in graphs. J. Comb. Theory Ser. B 37, 221–227 (1984)
9. Garey, M.R., Johnson, D.S.: Computers and Intractability: A Guide to the Theory of NP-Completeness. Books in Mathematical Sciences. W.H. Freeman, New York (1979)
10. Goodrich, M., Nelson, M., Sun, J.: The rainbow skip graph: A fault-tolerant constant-degree distributed data structure. In: Proceedings of The 17th Annual ACM-SIAM Symposium on Discrete Algorithms, SODA 2006 (2006)
11. Kano, M.: Some current results and problems on factors of graphs. In: Proc. 3rd China-USA Internet Coof. on Combinatorics, Graph Theory, Algorithm and Applications, Beijing, China (1993)
12. Liu, G.: On covered (g, f)-covered graphs. Acta Math. Scientia 8, 181–184 (1988)
13. Liu, G.: (g, f)-factors and factorizations in graphs. Acta Math. Sinica 37, 230–237 (1994)
14. Lovász, L.: Subgraphs with prescribed valencies. J. Comb. Theory Ser. B 9, 391–416 (1970)
15. Ore, O.: Note on Hamilton circuits. Amer. Math. Monthly 67, 55 (1960)
16. Tutte, W.T.: The factorization of linear graphs. J. London Math. Soc. 22, 107–111 (1947)
17. Tutte, W.T.: The factors of graphs. Can. J. Math. 4, 314–328 (1952)
18. Wei, B., Zhu, Y.: Hamiltonian k-factors in graphs. J. Graph Theory 25, 217–227 (1997)
19. Yu, J., Liu, G., Cao, B.: Connected factors of graphs. OR Transactions 9(1) (2005)

Yao Graphs Span Theta Graphs[*]

Mirela Damian and Kristin Raudonis

Department of Computer Science
Villanova University, Villanova, USA
{mirela.damian,kristin.raudonis}@villanova.edu

Abstract. The Yao and Theta graphs are defined for a given point set and a fixed integer $k > 0$. The space around each point is divided into k cones of equal angle, and each point is connected to a nearest neighbor in each cone. The difference between Yao and Theta graphs is in the way the nearest neighbor is defined: Yao graphs minimize the Euclidean distance between a point and its neighbor, and Theta graphs minimize the Euclidean distance between a point and the orthogonal projection of its neighbor on the bisector of the hosting cone. We prove that, corresponding to each edge of the Theta graph Θ_6, there is a path in the Yao graph Y_6 whose length is at most 8.82 times the edge length. Combined with the result of Bonichon, Gavoille, Hanusse and Ilcinkas, who prove an upper bound of 2 on the stretch factor of Θ_6, we obtain an upper bound of 17.7 on the stretch factor of Y_6.

Keywords: Yao graph; Theta graph; spanner.

1 Introduction

Let $\mathcal{P}$ be a set of points in the plane. The Yao and Theta graphs for $\mathcal{P}$ are both geometric graphs with vertex set $\mathcal{P}$ and edges defined by an integer parameter $k > 0$ as follows. Divide the space around each point $a \in \mathcal{P}$ into cones of (equal) angle $2\pi/k$, using k rays rooted at a. Then the Yao and Theta graphs connect a to a nearest neighbor in each of its cones, using directed edges rooted at a. This yields an out-degree of at most k. The difference between the Yao and Theta graphs lies in the way the nearest neighbor is defined. In the case of Yao graphs, the nearest neighbor of a in a cone C is a point $b \neq a$ that lies in C and minimizes the Euclidean distance $|ab|$ between a and b. In the case of Theta graphs, the nearest neighbor of a is a point $b \neq a$ that lies in C and minimizes the Euclidean distance between a and the orthogonal projection of b onto the bisector of C. Henceforth, we will refer to the Yao graph as Y_k and the Theta graph as Θ_k.

A t-spanner of a graph G is a spanning subgraph H such that, for each edge $ab \in G$, the length of the path in H between a and b is at most $t|ab|$. If G is the complete Euclidean graph of a plane point set $\mathcal{P}$, then H is also referred to as a t-spanner of $\mathcal{P}$. The value t is called *stretch factor* of the spanner.

---

[*] Supported by NSF grant CCF-0728909.

W. Wu and O. Daescu (Eds.): COCOA 2010, Part II, LNCS 6509, pp. 181–194, 2010.
© Springer-Verlag Berlin Heidelberg 2010

It is a long standing open problem to determine whether the Yao graph Y_6 is a spanner or not. In this paper we settle this problem by showing that Y_6 is a 17.7-spanner. Our result relies on a recent result by Bonichon et al. [1], who prove an upper bound of 2 on the stretch factor of the Theta graph Θ_6. We show that, corresponding to each edge $ab \in \Theta_6$, there is a path between a and b in Y_6 no longer than $t|ab|$, for $t = 8.82$. Combined with the 2-spanner result for the Θ_6 graph [1], this yields an upper bound of 17.7 for the stretch factor of the Yao graph. As far as we know, this is the first result showing that Y_6 is a spanner.

1.1   Existing Results

In this section we summarize existing results on spanning properties of Y_k only. For a comprehensive discussion of spanners in general, we refer the reader to the books by Peleg [7] and Narasimhan and Smid [6].

For a concise presentation, set $\theta = 2\pi/k$. Bose et al. [3] show that, for $k \geq 9$, Y_k is a spanner with stretch factor $\frac{1}{\cos\theta - \sin\theta}$. In [2], Bose et al. improve the stretch factor to

$$\frac{1 + \sqrt{2 - 2\cos\theta}}{2\cos\theta - 1}$$

and show that, in fact, Y_k is a spanner for any $k \geq 7$. In the same paper they show that Y_4 is a spanner with stretch factor $8(29 + 23\sqrt{2})$. Molla [5] shows that Y_2 and Y_3 are not spanners, and that Y_4 is a spanner with stretch factor $4(2 + \sqrt{2})$, for the special case when the input nodes are in convex position (see also [4]). In this paper, we settle that Y_6 is a spanner. The question whether Y_5 is a spanner or not remains open.

1.2   Notation and Definitions

Fix $k = 6$. Let $r_1, r_2, \ldots, r_6$ be the rays separating the cones at each point. We assume without loss of generality that r_1 is horizontal and points to the right. For each node $a \in \mathcal{P}$, we use $C_i(a)$ to denote the half-open cone delimited by r_i and r_{i+1}, including r_i but excluding r_{i+1} (note that r_j refers to $r_{(j \mod 7)+1}$, for any j). See Fig. 1(a). For a fixed point x, let $T_i(a, x)$ denote the *equilateral* triangle delimited by the two bounding rays for $C_i(a)$, and the line through x that cuts $C_i(a)$ at a $60°$ angle (and is, therefore, parallel to r_{i+2}). This definition is illustrated in Fig. 1(b).

For any real $\xi \geq 0$, let $D(a, \xi)$ denote the closed disk with center a and radius ξ. Sometimes, we need to refer to a disk piece that lies inside a Yao cone, case in which we attach the cone index to the disk notation: $D_i(a, \xi)$ refers to the piece of $D(a, \xi)$ that lies inside $C_i(a)$ (see Fig. 1c).

Let $P_i(a, x)$ denote the parallelogram with diagonal ax and edges parallel to r_i and r_{i+1} (see Fig. 2a). We will use the notation $\mathtt{yao}_i(a, x)$ to refer to the directed path in the Y_6 graph that starts at a and follows the Y_6 edges that lie in cones C_i, until the path hits x or exits the parallelogram $P_i(a, x)$ (refer to Fig. 2b). Similarly, we use $\mathtt{theta}_i(a, x)$ to refer to the directed path in the Θ_6 graph that starts at a and follows the Θ_6 edges that lie in cones C_i, until the path hits x or exits the parallelogram $P_i(a, x)$.

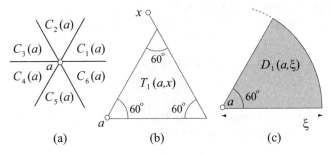

Fig. 1. Definitions (a) Cones (b) Equilateral triangle $T_1(a, x)$ (c) Disk sector $D_1(a, \xi)$

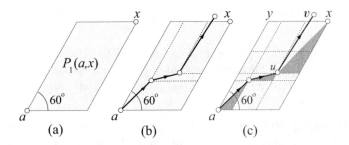

Fig. 2. (a) Parallelogram $P_1(a, x)$ (b) Path $\mathtt{yao}_1(a, x)$ (c) Lem. 2: an upper bound on $|\mathtt{yao}_1(a, x)|$

Most often we ignore the direction of edges (the spanning paths we identify in Y_6 are undirected). For an edge $ab \in Y_6$, we refer to the directed version $(\overrightarrow{ab}$ or $\overrightarrow{ba})$ of ab only if important in the context.

We say that two edges *intersect* each other if they share a common point. If the common point is not an endpoint, the edges *cross* each other. For any pair of points a and b, we use the notation $\pi(a, b)$ to refer to a path in Y_6 between a and b, and $|\pi(a, b)|$ to denote the length of this path. To refer to a path in Θ_6, we use the subscript Θ: $\pi_\Theta(a, b)$ refers to a path in Θ_6 between a and b. Throughout the paper, we use $\oplus$ to denote the concatenation operator.

2   Basic Lemmas

In this section we provide a few isolated lemmas that will be used in the main proof. Readers can skip ahead to section 3 describing the main result if they wish to, and refer back to these lemmas when called upon from the main proof. We begin with a statement of a result established in [1].

Theorem 1. *For any pair of points $a, b \in \mathcal{P}$, there is a path in Θ_6 whose total length is bounded above by $2|ab|$. [1]*

Proposition 1. *The sum of the lengths of crossing diagonals of a convex quadrilateral abcd is strictly greater than the sum of the lengths of either pair of opposite sides:*

$$|ac| + |bd| > |ab| + |cd|$$
$$|ac| + |bd| > |bc| + |da|$$

This can be derived by applying the triangle inequality on pairs of opposite triangles formed by diagonals ac and bd.

Lemma 1. *Two edges* $\vec{ab}, \vec{xy} \in Y_6$, *with* $b \in C_i(a)$ *and* $y \in C_i(x)$, *for some* $1 \leq i \leq 6$, *cannot cross each other.*

This follows immediately from Prop. 1, and the fact that b is closest to a and y is closest to x.

Lemma 2. *The length of* $\mathtt{yao}_i(a, x)$ *does not exceed half the perimeter of* $P_i(a, x)$. *Furthermore, if* $v \neq a$ *is the other endpoint of* $\mathtt{yao}_i(a, x)$, *then* xv *is strictly shorter than the side of* $P_i(a, x)$ *that crosses* $\mathtt{yao}_i(a, x)$. *Both arguments hold for* $\mathtt{theta}_i(a, x)$ *as well.*

Proof. Refer to Fig. 2(c). Let uv be the last edge of $\mathtt{yao}_i(a, x)$ that exits $P_i(a, x)$. Then $|uv| \leq |ux|$, since $\vec{uv} \in Y_6$. This implies that the length of the path $\pi = \mathtt{yao}_i(a, u) \oplus ux$ is an upper bound on the length of $\mathtt{yao}_i(a, x)$. For each edge $e \in \pi$, we derive an upper bound on the length of e by using the triangle inequality on the triangle delimited by e, and edges parallel to the rays of C_i (these are the triangles shaded in Fig. 2c). Summing up these inequalities yields the first claim of the lemma. Let xy be the side of $P_i(a, x)$ crossed by $\mathtt{yao}_i(a, x)$. The fact that $|uv| \leq |ux|$, along with Prop. 1 applied on quadrilateral $uxvy$, yields $|vy| < |xy|$. This means that $v \in D_i(y, |xy|)$, and since vx is a chord in this $60°$ sector, we get $|xv| < |xy|$. This settles the second claim of the lemma.

Lemma 3. *Let* $\triangle abc \subset \triangle ab_1c_1 \subset C_i(a)$ *be equilateral triangles with side lengths* $|ab|$ *and* $|ab| + \delta_{ab}$, *and* b_1c_1 *tangent to* $D_i(a, |ab|)$ *(see Fig. 3a). Then*

$$\delta_{ab} = |ab|(\frac{2}{\sqrt{3}} - 1) \tag{1}$$

This follows from the trigonometric formula $\cos(30°) = |ab|/(|ab| + \delta_{ab})$ applied on the triangle $\triangle aa_1b_1$ from Fig. 3(a).

Lemma 4. *Let* $c \in P$, $\triangle abc = T_5(c, a)$ *empty of points in* P, *and* $u \in P$ *inside* $D_1(a, |ab|)$. *Then there is a path* $\pi_\Theta(u, c)$ *in* Θ_6 *of length*

$$\pi_\Theta(u, c) \leq |ab| + 2\delta_{ab}.$$

Furthermore, each edge of $\pi_\Theta(u, c)$ *is shorter than ab.*

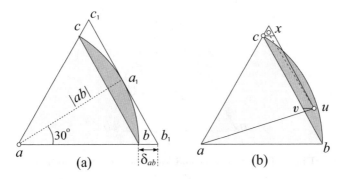

Fig. 3. Basic lemmas (a) Computation of δ_{ab} (b) Lem. 4

Proof. Refer to Fig. 3(b). Let v be the right corner of $T_5(c, u)$. By definition, $\text{theta}_3(u, c)$ cannot exit $\triangle abc$ (otherwise, the projection of c on the bisector of $C_3(u)$ would be closer to u than the projection of x). Also, since $\triangle abc$ is empty of points in $\mathcal{P}$, $\text{theta}_3(u, c)$ cannot enter $\triangle abc$. This enables us to use Lem. 2 to show that $|\text{theta}_3(u, c)| \leq |uv| + |vc|$. Note that $|uv| < |vb|$. (This follows from $|au| \leq |ab|$, since $u \in D_1(a, |ab|)$, and Prop. 1 applied on quadrilateral $avub$.) Then $|\text{theta}_3(u, c)| \leq |vb| + |vc| = |ab|$, and each edge on this path is smaller than $|ab|$.

Let $x \neq u$ be the other endpoint of $\text{theta}_3(u, c)$. Recall that $\text{theta}_3(u, c)$ does not intersect $\triangle abc$. This implies that $\text{theta}_3(u, c)$ exits $P_3(u, c)$ through its top edge, meaning that x lies inside $T_1(c, u)$ (tiny top triangle in Fig. 3b). Then $|cx| \leq |uv| \leq \delta_{ab}$. By Thm. 1, there is a path $\pi_\Theta(x, c)$ in Θ_6 no longer than $2\delta_{ab} < |ab|$, so each edge on this path is also smaller than ab. We concatenate the two paths together to obtain $\pi_\Theta(u, c) = \text{theta}_3(u, c) \oplus \pi_\Theta(x, c)$, no longer than $|ab| + 2\delta_{ab}$.

Lemma 5. *Let $\overrightarrow{ae} \in \Theta_6$ be an edge in the upper half of $C_1(a)$ (see Fig. 4). Let $\triangle abc = T_1(a, e)$, with ab horizontal. Fix $x \in D_1(a, |ae|)$ and assume that $\overrightarrow{xy} \in C_3(x)$ is an edge in Y_6 that crosses ae. Then y lies inside $D_3(b, |ab|)$ and below e, and the following inequalities hold:*

$$(i) \; |ce| < |ab|(1 - \frac{\sqrt{3}}{2})$$

$$(ii) \; |ae| > |ab|\frac{\sqrt{7 - 2\sqrt{3}}}{2}$$

In other words, ae cannot lie too close to the bisector of $C_1(a)$.

Proof. Since $\overrightarrow{ae} \in \Theta_6$, $\triangle abc$ is empty, therefore x lies right of bc. This implies that $e \in C_3(x)$, and since $\overrightarrow{xy} \in Y_6$ is also in $C_3(x)$, by definition $|xy| \leq |xe|$. This along with Prop. 1 applied on quadrilateral $byex$ implies

$$|by| < |be| \tag{2}$$

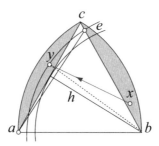

Fig. 4. Lem. 5: Upper and lower bounds on ce and ae

It follows that y lies inside $D_3(b, |be|) \subset D_3(b, |ab|)$. If y were to lie above e, then $\angle xey$ would be obtuse, meaning that $|xy| > |xe|$, a contradiction. This settles the first claim of the lemma.

To derive the bounds on ce and ae, observe that by must be at least as long as the height $h = |ab|\sqrt{3}/2$ of $\triangle abc$, since y lies outside this triangle. This along with inequality (2) above implies $|be| > h$, which is equivalent to $|ce| < |ab| - h = |ab|(1 - \sqrt{3}/2)$. This establishes inequality (i). We use this inequality, along with the Law of Cosines applied on $\triangle ace$ ($|ae|^2 = |ac|^2 + |ce|^2 - |ac||ce|$), and the fact that ae gets shorter as ce gets longer, in deriving inequality (ii).

3 Y_6 Paths Span Θ_6 Edges

We now turn to proving the main result of the paper.

Theorem 2. *For any edge* $ae \in \Theta_6$*, there is a path in* Y_6 *between* a *and* e *no longer than* $t \cdot |ae|$*, for* $t = 8.82$.

Proof. Fix $ae \in \Theta_6$. Throughout this proof, we assume without loss of generality the setting from Fig. 5: $\vec{ae} \in C_1(a)$, above the bisector of $C_1(a)$; $\triangle abc = T_1(a, e)$, with ab horizontal; b_1c_1 is parallel to bc and tangent to $D_1(a, |ab|)$; and b_2c_2 is the vertical reflection of b_1c_1, tangent to $D_3(b, |ab|)$. The quantity $\delta_{ab} = |bb_1| = |ab_2|$, introduced in Lem. 3, will be useful in computing the value of t.

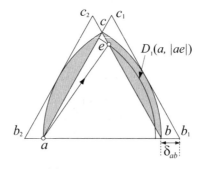

Fig. 5. (a) General setting for Thm. 2

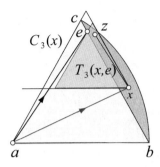

Fig. 6. Base case for Thm. 2: $|xz| < |ae|$

The proof is by induction on the length of edges in Θ_6. Consider first the case where ae is a shortest edge in Θ_6 (base case). We show that $\overrightarrow{ae} \in Y_6$ as well. Assume the contrary, and let $\overrightarrow{ax} \in C_1(a)$ be in Y_6, with $x \neq e$ (see Fig. 6). By definition, $|ax| \leq |ae|$. The following three arguments imply that there is an edge in Θ_6 shorter than ae: (i) $|ae| > |be|$, which follows from the fact that $\angle abe = 60° > \angle bae$, and the Law of Sines applied on $\triangle abe$ (ii) be is longer than one side of $T_3(x, e)$, since $x \in D_1(a, |ae|) \subset D_1(a, |ab|)$ lies to the right of be, and (iii) $C_3(x)$ contains an edge $\overrightarrow{xz} \in \Theta_6$ no longer than one side of $T_3(x, e)$ (by definition). It follows that $|xz| < |ae|$, contradicting the fact that ae is a shortest edge in Θ_6.

Assume now that $ae \in \Theta_6$ is not a shortest edge, and that the theorem holds for all edges in Θ_6 shorter than ae. We determine a path in Y_6 with endpoints a and e that is no longer than $t|ae|$. As in the base case, let $\overrightarrow{ax} \in C_1(a)$ be in Y_6. The case when x and e coincide is trivial, so assume $x \neq e$.

The simplest situation occurs when the path $\mathtt{yao_3}(x, e)$ does not cross ae. We will encounter this scenario in various disguises throughout the proof, so we pause to prove two small results related to this case (refer to Fig. 7).

Lemma 6. *Let $p, x \in \mathcal{P}$ be such that $p \in C_3(x)$ and $T_5(p, x)$ is empty of points in $\mathcal{P}$. Let q be the right corner of $T_5(p, x)$. Assume that Thm. 2 holds for all edges shorter than $2|qx|$. If $\mathtt{yao_3}(x, p)$ does not cross pq, then Y_6 contains a path $\pi(x, p)$ of length*

$$|\pi(x, p)| \leq |pq| + |qx| + 2t|qx| \tag{3}$$

Proof. This situation is illustrated in Fig. 7(a). By Lem. 2, $\mathtt{yao_3}(x, p)$ is no longer than half the perimeter of $P_3(x, p)$, which equals $|pq| + |qx|$. The lemma states that $\mathtt{yao_3}(x, p)$ does not cross pq, which means that $\mathtt{yao_3}(x, p)$ must exit $P_3(x, p)$ through its top edge. Let $z \neq x$ be the other endpoint of $\mathtt{yao_3}(x, p)$. By Lem. 2, $|pz| < |qx|$. By Thm. 1, Θ_6 contains a path p_Θ from z to p no longer than $2|pz| < 2|qx|$. This means that each edge on p_Θ is no longer than $2|qx|$, therefore we can use the inductive hypothesis to claim the existence of a path

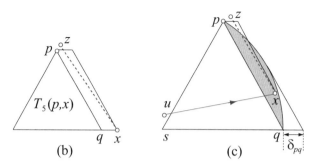

Fig. 7. Assisting lemmas. (a) Lem. 6 (c) Lem. 7.

$\pi(z, p)$ in Y_6 no longer than $2t|qx|$. We concatenate these two paths together to determine a path

$$\pi(x, p) = \mathtt{yao}_3(x, p) \oplus \pi(z, p),$$

whose length is bounded above by the quantity in (3).

Lemma 7. *Let $p \in \mathcal{P}$ and let q be an arbitrary point (not necessarily in $\mathcal{P}$) on the right ray of $C_5(p)$. Let $\overrightarrow{ux} \in C_1(u)$ cross the slanted edges of $T_5(p, q)$. Assume that Thm. 2 holds for all edges shorter than pq. If $\mathtt{yao}_3(x, p)$ does not cross pq, then Y_6 contains a path $\pi(u, p)$ of length*

$$|\pi(u, p)| \le |ux| + |pq| + \delta_{pq} + 2t\delta_{pq} \qquad (4)$$

Proof. This situation is illustrated in Fig. 7(b). Let $s \ne p, q$ be the left corner of $T_5(p, q)$. Note that both points p and x lie in $C_1(u)$. Since $\overrightarrow{ux} \in Y_6$, by definition $|ux| \le |up|$. This along with Prop. 1 applied on quadrilateral $supx$ implies $|sx| < |sp|$. This means that x lies inside $D_1(s, |pq|)$, therefore the horizontal side of $P_3(x, p)$ is no longer than δ_{pq}. This along with Lem. 6 implies that Y_6 contains a path $\pi(x, p)$ no longer than $|pq| + \delta_{pq} + 2t\delta_{pq}$, which concatenated with ux yields the upper bound claimed by the lemma.

Back to the main theorem: the situation in which $\overrightarrow{ax} \in Y_6$ and $\mathtt{yao}_3(x, e)$ does not cross $\triangle abc$ is handled immediately by Lem. 7, which proves the existence of path $\pi(a, e)$ in Y_6 of length

$$|\pi(a, e)| \le 2|ab| + \delta_{ab} + 2t\delta_{ab}$$

(This inequality uses the fact that $|ax| \le |ab|$.) Using inequality (1) and the fact that ae is at least as long as the height of $\triangle abc$, it can be verified that $|\pi(a, e)| < t|ae|$ for any $t \ge 3.42$.

The situation in which $\mathtt{yao}_3(x, e)$ crosses ae is more involved and requires a careful case analysis. We proceed with a discussion of the worst case scenario that yields the stretch factor value $t = 8.82$. For the other cases, a fairly loose analysis yields a bound still below $t = 8.82$.

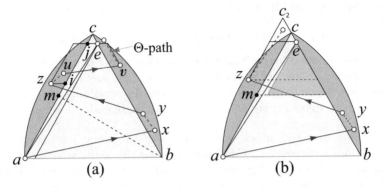

Fig. 8. (a) Worst case scenario. (b) Case 1: $\mathsf{yao}_1(z, e)$ does not cross ae.

Worst Case Scenario. This occurs under the following conditions (see Fig. 8a).

(a) $D_5(c, |ab|)$ contains some points from $\mathcal{P}$.
(b) The edge $\vec{yz} \in \mathsf{yao}_3(x, e)$ that crosses $\triangle abc$ first, does not cross the bisector of $\angle abc$.
(c) $\mathsf{yao}_1(z, e)$ crosses ae (contingent upon (b)).

Under these conditions, we determine a path from a to e no longer than $t|ae|$. By Lem. 5, $z \in D_3(b, |ab|)$, and ce and ae have length restrictions given by (i) and (ii) from Lem. 5. Let $uv \in \mathsf{yao}_1(z, e)$ be the first edge that crosses $\triangle abc$. Let m be the midpoint of ac. Since z (and therefore u) is above m, uv crosses $T_1(m, c)$. By Lem. 5, $v \in D_1(m, |mc|)$. Also note that v lies below e (otherwise, $\angle uev$ would be obtuse, meaning that $|uv| > |ue|$, contradicting $\vec{uv} \in Y_6$.) These two observations, together with Lem. 4, imply that there is a path $\pi_\Theta(v, e)$ no longer than $|mc| + 2\delta_{mc}$, and each edge on this path is smaller than mc. This enables us to apply the inductive hypothesis to prove the existence of a path $\pi(v, e)$ of length

$$|\pi(v, e)| \leq t(|mc| + 2\delta_{mc}) = t(|ab|/2 + \delta_{ab}) \qquad (5)$$

This latter inequality follows from the fact that $2|mc| = |ab|$. Consider now the Yao path

$$\pi(a, v) = ax \oplus \mathsf{yao}_3(x, z) \oplus \mathsf{yao}_1(z, v)$$

By Lem. 2, the length of $\mathsf{yao}_3(x, z) \subset \mathsf{yao}_3(x, e)$ is bounded above by half the perimeter of $P_3(x, e)$, which is no greater than $|ab|$ (as shown in the proof of Lem. 4). Similarly, since $\mathsf{yao}_1(z, v) \subseteq \mathsf{yao}_1(z, e)$, by Lem. 2 we have that $|\mathsf{yao}_1(z, v)|$ does not exceed half the perimeter of $P_1(z, e)$, which we claim to be bounded above by $(|ab| + \delta_{ab})/2$. (To see this, let i and j be the lower and upper right corners of $P_1(z, a)$. Then $|im| \geq |iz| \sin 30° = |iz|/2$, since z is in the upper half of $C_3(b)$, by assumption (b). Also $|jc| = |je|$. Half the perimeter of $P_1(z, e)$ is smaller than $|iz| + |ij| + |je| = |iz| + |ic| = |iz| + |mc| - |im| \leq |mc| + |iz|/2 \leq |ab|/2 + \delta_{ab}/2$.) Then

$$|\pi(a, v)| \leq |ae| + 3|ab|/2 + \delta_{ab}/2 \qquad (6)$$

Summing up (5) and (6), we get that the path $\pi(a, e) = \pi(a, v) \oplus \pi(v, e)$ has length

$$|\pi(a, e)| \leq |ae| + 3|ab|/2 + \delta_{ab}/2 + t(|ab|/2 + \delta_{ab}) \tag{7}$$

Using the lower bound for ae from Lem. 5 and the upper bound for δ_{ab} from (1) in the inequality above, we obtain $|\pi(a, e)| \leq t|ae|$ for any $t \geq 8.82$. This is the only case that yields the stretch factor 8.82. For the remaining cases we adopt a relaxed analysis that only seeks to stay within this bound.

Case 1. This case eliminates condition (c) of the worst case scenario (i.e., $\mathsf{yao}_1(z, e)$ does *not* cross ae). Conditions (a) and (b) of the worst case scenario are assumed to hold (see Fig. 8b). Note that z must lie below e. Otherwise, $\angle yez$ would be obtuse, meaning $|yz| > |ye|$, contradicting the assumption that $\overrightarrow{yz} \in Y_6$. Since $\mathsf{yao}_1(z, e)$ does not cross $T_5(e, m)$ (since it does not cross ae), we can use Lemma 6 to claim the existence of a path $\pi(z, e)$ in Y_6 no longer than $|mc| + |c_2e| + 2t|c_2e|$. Then $p(a, e) = ax \oplus \mathsf{yao}_3(x, z) \oplus \pi(z, e)$ is no longer than

$$|p(a, e)| \leq |ae| + 3|ab|/2 + |ce| + \delta_{ab} + 2t(|ce| + \delta_{ab})$$

Using the bounds for ce, ae and δ_{ab} from Lem. 5 and (1) in the inequality above yields $|\pi(a, e)| \leq t|ae|$ for any $t \geq 7.6$.

Case 2. This case eliminates condition (b) of the worst case scenario (i.e., z lies in the lower half of $C_3(b)$). See Fig. 9. Under condition (a) of the worst case scenario, there is $\overrightarrow{eu} \in Y_5(e)$, with $|eu| \leq |ae|$. We claim that $\mathsf{yao}_3(u, a)$ is restricted to the lower half of $C_3(b)$. This is because (i) by Prop. 1 $\mathsf{yao}_3(u, a)$ cannot cross $\overrightarrow{yz}$, and (ii) if there were an edge $\overrightarrow{vw} \in \mathsf{yao}_3(u, a)$ that did not cross yz, with $v \in P_3(u, a)$ and w in the upper half of $C_3(b)$, then $\angle vzw$ would be obtuse, meaning that $|vw| > |vz|$, contradicting $\overrightarrow{vw} \in Y_6$.

If $\mathsf{yao}_3(u, a)$ does not cross $\triangle abc$, the analysis is similar to the one used in Case 1 (imagine left of Fig. 9 rotated clockwise by 90°, and compare with Fig. 8b); the extra path $\mathsf{yao}_3(x, z)$ (with $|\mathsf{yao}_3(x, z)| < |ab|$, as shown before) used in computing the upper bound in Case 1 offsets the longer side of $P_3(u, a)$, which is also upper bounded by $|ab|$.

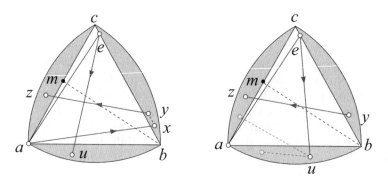

Fig. 9. Case 2: yz lies in the lower half of $C_3(b)$

The case when $\mathrm{yao}_3(u, a)$ crosses $\triangle abc$ is very similar to the worst case scenario, with eu here playing the role of ax in the worst case; an upper bound of $|ae|$ applies to each. So this case is settled.

Case 3. This case eliminates condition (a) of the worst case scenario. So the assumption here is that the disk sector $D_5(c, |ab|)$ contains no points from $\mathcal{P}$. The importance of this assumption will become clear shortly. We adopt a slightly different approach for this case, and reassign $x \in \mathcal{P}$ to be a highest point in the parallelogram $P_5(c_2, a)$ (see Fig. 10a). Ties are broken in favor of a rightmost point. We claim that $\mathrm{yao}_5(x, a)$ does not cross $\triangle abc$. Indeed, if there were an edge $\overrightarrow{zu} \in \mathrm{yao}_5(x, a)$ crossing $\triangle abc$, then by Lem. 5 u would end up in $D_5(c, |ab|)$, contradicting the fact that $D_5(c, |ab|)$ is empty. This enables us to use Lem. 6 to claim the existence of a path $\pi(x, a) \in Y_6$ between x and a of length

$$|\pi(x, a)| \leq |ia| + \delta_{ab} + 2t\delta_{ab} \tag{8}$$

Here i is used to denote the upper right corner of $P_5(x, a)$, as marked in Fig. 10(b). Next we seek a path in Y_6 between x and e. We discuss two cases, depending on the relative position of x and e.

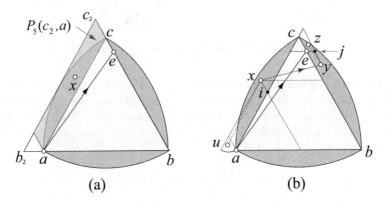

Fig. 10. (a) $x \in P_5(c_2, a)$ (b,c) Case 3a, xy crosses $\triangle abc$

Case 3a. x lies below e. Then there is an edge $\overrightarrow{xy} \in Y_6$ that lies in $C_1(x)$, since $e \in C_1(x)$. Since x is a highest point in $P_5(c_2, a)$, xy must exit $P_5(c_2, a)$. We first discuss the situation where xy exits $P_5(c_2, a)$ through its side along ac, as depicted in Fig. 10(b). In this case, xy also crosses $\triangle abc$, which is empty of points in $\mathcal{P}$. Now note that $\mathrm{yao}_3(y, e)$ cannot cross $\triangle abc$ again, since by Lem. 5 it would end up in a point in $C_3(b, |ab|)$ higher than x, a contradiction. This enables us to use Lem. 6 to claim the existence of a path $\pi(y, e) \in Y_6$ of length

$$|\pi(y, e)| \leq |jy| + \delta_{ab} + 2t\delta_{ab} \tag{9}$$

Here j is used to denote the upper right corner of $P_3(y, e)$. We concatenate these paths together to determine a path $\pi(a, e) = \pi(x, a) \oplus xy \oplus \pi(y, e)$ of length

$$|\pi(a, e)| \leq 2|ab| + 2\delta_{ab} + 4t\delta_{ab} \tag{10}$$

The bound above combines the bounds from (8) and (9). The term $2|ab|$ above accounts for the inequalities $|ia| + |jy| < |ab|$ and $|xy| < |ab|$. It can be verified that $2|ab| + 2\delta_{ab} + 4t\delta_{ab} < t|ae|$ for any $t \geq 7.2$.

We now turn to the case where xy exits $P_5(c_2, a)$ through its side c_2c, as illustrated in Fig. 11(a). In this case, xy crosses $T_5(c_2, e)$. To apply the induction hypothesis, we need to move closer to e (within a distance of δ_{ab}), so we consider $\mathsf{yao}_5(y, e)$.

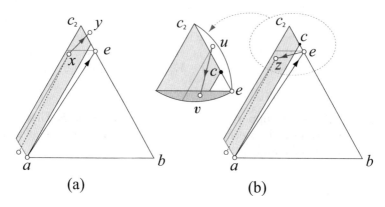

(a)   (b)

Fig. 11. Case 3a, xy does not cross $\triangle abc$

Assume first that $\mathsf{yao}_5(y, e)$ does not cross c_2e (i.e., it exits $P_5(y, e)$ through its lower horizontal side). This situation is similar to the one described in Lem. 7, with $T_3(e, c_2)$ playing the role of $T_5(p, q)$. We apply Lem. 7 to show that Y_6 contains a path $\pi(x, e)$ of length $|\pi(x, e)| \leq |xy| + |c_2e| + \delta_{c_2e} + 2t\delta_{c_2e}$, which is bounded above by

$$|\pi(x, e)| \leq |ab| + |c_2e| + \delta_{ab} + 2t\delta_{ab}$$

This along with (8) shows that $\pi(a, e) = \pi(x, a) \oplus \pi(x, e)$ is bounded above by

$$|\pi(a, e)| \leq 2|ab| + 2\delta_{ab} + 4t\delta_{ab}$$

(Here we used the fact that $|ia| + |c_2e| < |ab|$.) This bound is identical to the one in (10).

Assume now that there is $uv \in \mathsf{yao}_5(y, e)$ that crosses c_2e (see Fig. 11b). By Lem. 5, $v \in D_3(c_2, |c_2e|)$ and therefore $|ev| < |c_2e|$. This in turn implies that Y_6 contains an edge $\vec{ez} \in C_4(e)$, with $|ez| \leq |ev| < |c_2e|$. Using the inequalities from Lem. 3 and 5, it can be verified that $2|c_2e| \leq 2(|ce| + \delta_{ab}) < |ab|$. This enables us to use Lemma 6 to determine a path $\pi(e, a) = ez \oplus \pi(z, a)$ of length $|\pi(e, a)| < 2|ab| + \delta_{ab} + 2t\delta_{ab}$, with an upper bound lower than the one in (10). (Here we used the fact that $|ez| < |c_2e| < |ab|$.)

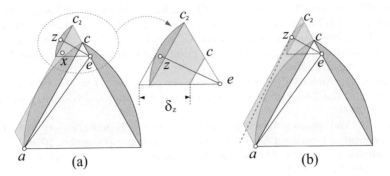

Fig. 12. Case 3b: x above e

Case 3b. x lies above e. This situation is depicted in Fig. 12a. Since $x \in C_3(e)$, there is an edge edge $\overrightarrow{ez} \in Y_6$ in $C_3(e)$, with $|ez| \leq |ex| < |ec_2|$, so $z \in D_3(e, |ec_2|)$ It follows that the shorter side of $P_5(z, a)$ is no longer than $\delta_z = \delta_{ab} + \delta_{ec_2}$. Substituting $|ec_2| = |ec| + \delta_{ab}$, the upper bound for ec from Lem. 5, and the inequality $\delta_{ec_2} = |ec_2|(2/\sqrt{3} - 1)$ corresponding to (1), we obtain

$$\delta_z < 4\delta_{ab}/3$$

Recall that we are in the situation where $D_3(c, |ab|)$ is empty and $\mathsf{yao}_3(z, a)$ cannot cross ac. By Lem. 6, there is a path $\pi(z, a)$ in Y_6 with a loose upper bound of $|\pi(z, a)| \leq |ab| + \delta_z + 2t\delta_z$. Then the path $\pi(a, e) = ez \oplus \pi(z, a)$ is no longer than $2|ab| + \delta_z + 2t\delta_z$, which is smaller than $t|ae|$ for any $t > 3.8$.

Having exhausted all cases, we conclude that the claim of the theorem holds.

The results of theorems 1 and 2 combined yield our main result:

Theorem 3. Y_6 *is a* 17.64-*spanner.*

4   Conclusions

In this paper we establish that the Yao graph Y_6 is a spanner. It is known that Y_2 and Y_3 are not spanners, Y_4 is a spanner, and Y_k is a spanner for $k \geq 7$. We conjecture that Y_5 is also a spanner. Settling the truth value of this conjecture remains an interesting open problem, whose resolution would complete our understanding of Yao spanners.

Acknowledgement. We thank Michiel Smid for pointing us to the result of Bonichon et al. [1].

References

1. Bonichon, N., Gavoille, C., Hanusse, N., Ilcinkas, D.: Connections between Theta-graphs, Delaunay triangulations, and orthogonal surfaces. In: 36th International Workshop on Graph-Theoretic Concepts in Computer Science (2010) (to appear); Also as Technical Report hal-00454565_v1, HAL (February 2010)

2. Bose, P., Damian, M., Douïeb, K., O'Rourke, J., Seamone, B., Smid, M.H.M., Wuhrer, S.: Pi/2-angle yao graphs are spanners. CoRR, abs/1001.2913 (2010)
3. Bose, P., Maheshwari, A., Narasimhan, G., Smid, M., Zeh, N.: Approximating geometric bottleneck shortest paths. Computational Geometry: Theory and Applications 29, 233–249 (2004)
4. Damian, M., Molla, N., Pinciu, V.: Spanner properties of $\pi/2$-angle yao graphs. In: Proc. of the 25th European Workshop on Computational Geometry, pp. 21–24 (March 2009)
5. Molla, N.: Yao spanners for wireless ad hoc networks. Technical report, M.S. Thesis, Department of Computer Science, Villanova University (December 2009)
6. Narasimhan, G., Smid, M.: Geometric Spanner Networks. Cambridge University Press, New York (2007)
7. Peleg, D.: Distributed computing: a locality-sensitive approach. Society for Industrial and Applied Mathematics, Philadelphia (2000)

A Simpler Algorithm for the All Pairs Shortest Path Problem with $O(n^2 \log n)$ Expected Time

Tadao Takaoka and Mashitoh Hashim

Department of Computer Science
University of Canterbury
Christchurch, New Zealand

Abstract. The best known expected time for the all pairs shortest path problem on a directed graph with non-negative edge costs is $O(n^2 \log n)$ by Moffat and Takaoka. Let the solution set be the set of vertices to which the given algorithm has established shortest paths. The Moffat-Takaoka algorithm maintains complexities before and after the critical point in balance, which is the moment when the size of the solution set is $n - n/\log n$. In this paper, we remove the concept of critical point and the data structure, called a batch list, whereby we make the algorithm simpler and seamless, resulting in a simpler analysis and speed-up.

1 Introduction

The research in the all pairs shortest path (APSP) algorithms has several categories, depending on the type of graphs. Let m and n be the number of edges and vertices of the given directed graph. One is for sparse graphs, where we use two parameters m and n. The best known result in this area, based on the traditional comparison-based model, is $O(mn + n^2 \log \log n)$ by Seth Pettie [10]. If we go to the area of a dense graph, we use only n for the complexity parameter. Within this category, we have two problems. One is for the worst case analysis. The best known result are slightly sub-cubic by Chan [3], which is $O(n^3 (\log \log n)^3 / \log^2 n)$, and $O(n^3 / \log^2 n)$ by Blelloch, et. al. [2]. The other area is for the average case analysis, which is the main theme of this paper.

Spira [11] in 1973 was the pioneer who brought us a surprising result of $O(n^2 \log^2 n)$ expected time, a significant improvement from the classical $O(n^3)$. Basically he expands the solution set one by one, similar to Dijkstra's algorithm. He maintains the sorted list of edges from each vertex, or equivalently the end points of those edges, and a priority queue for the solution set. Each vertex in the solution set has its own candidate, a member of the sorted list, for possible inclusion in the solution set. The key for vertex v is the established distance from source to v plus the edge cost from v to its candidate. To check the edge list one by one he maintains a pointer on the list. When a vertex is taken from the queue as minimum, and if the candidate is outside the solution set, it is successfully included into the solution set. Otherwise the target vertex of the next edge in the list is chosen as the next candidate, and the vertex is put back to

W. Wu and O. Daescu (Eds.): COCOA 2010, Part II, LNCS 6509, pp. 195–206, 2010.
© Springer-Verlag Berlin Heidelberg 2010

the queue with the new key value, regardless whether it is in the solution set or not. Choice of the next edge for the new candidate is not the best choice. Some effort to find a better candidate may be preferred. We call this effort to find a better candidate that is more likely outside the solution set than just the next candidate the scanning effort. The scanning effort is small at an early stage of the expansion process, whereas towards the end it becomes a great effort as it is hard to find a candidate outside the solution set. Thus there is a stage when we need to give up the candidate outside the solution set with probability one. We call this stage the critical point.

Since Spira published it, there have been many results with the idea of limited scan of edge lists. Early results achieved $O(n^2 \log n \log \log n)$ [12] and $O(n^2 \log n \log^* n)$ [2], where $\log^* n$ is the minimum number of logarithms on n that brings the value below 1. These algorithms are one-phase algorithms, so to speak, as there is no critical point to separate the computation.

The best known result is the complexity of $O(n^2 \log n)$ by Moffat and Takaoka (MT) [9]. This algorithm changes the scanning strategy at the critical point, meaning it can be regarded as a two-phase algorithm. Under some reasonable assumption, it is proven in [8] that this is a lower bound for a comparison-based computational model. An over-sight in the analysis in the original paper by Moffat and Takaoka was fixed by Mehlhorn and Priebe [8]. The analysis for the limited scan of edge lists was tricky after the critical point. Specifically the over-run of scanning that does not contribute to find the vertices to be included into the solution set is hard to analyze.

We remove the concept of critical point in this paper, so that the analysis of edge list scanning becomes simpler. As there is no critical point, we do not switch at the critical point, enabling us to use the same strategy for the priority queue operations throughout the computation. Our algorithm is seamless, so to speak. We also remove the data structure L, called the batch list, which was needed in the MT algorithm for maintaining completely clean candidates before the critical point. For the priority queue, we use the classical binary heap. As Goldberg and Tarjan [7] point out, the binary heap is the best choice from a practical point of view for Dijkstra's algorithm, since *decrease-key* in a Fibonacci heap, which takes $O(1)$ amortized time, is not performed frequently on average. In our framework of algorithms, too, we show that the binary heap works well.

The amount of scanning is determined by the bound on pointer movements in [12] and [2], which we call bound-oriented scanning, whereas in [9] and [8] scanning is done until a specified destination is found, which we call destination-oriented scanning. Our new algorithm in this paper goes back to the category of the former two, that is, a one-phase algorithm with bound-oriented scanning. Spira is a special case in this category with the bound equal to one. The contribution of this paper is to show that a small modification of Spira's algorithm can achieve the optimal complexity of $O(n^2 \log n)$ with a simpler analysis.

In Section 2, we describe the basic definition of the APSP problem, and the probabilistic assumption used in this paper. Also Spira's algorithm is described as the starting point of this area of research.

In Section 3, we describe the Moffat-Takaoka algorithm as the target algorithm to be improved upon in this paper. How the concept of critical point works is explained. Also why we need the data structure called the batch list is explained.

In Section 4, the new algorithm is described. It is explained how we can get rid of the critical point, and instead use the concept of clean-up booster to get a cleaner candidate for expanding the solution set. Also it is explained why we can go without the batch list.

Section 5 and Section 6 are devoted to the correctness and analysis of the new algorithm, and Section 7 is for concluding remarks and future research.

2 Spira's Algorithm

Let $G = (V, E)$ be a directed graph with non-negative edge costs with no self-loop. Here V and E are the sets of vertices and edges such that $|V| = n$ and $|E| = m$. The cost of edge (u, v) is given by $c(u, v)$. The cost of a path is the sum of the costs of edges that form the path. The shortest path from u to v is the path with the minimum cost. The single source shortest path (SSSP) problem is to compute shortest paths from a specified vertex, called the source, to all other vertices. The all pairs shortest path (APSP) problem is to compute shortest paths for all pairs of vertices. In the area of average case for the APSP problem, we normally run an SSSP algorithm n times by changing the source, meaning that we try to speed up the SSSP computation after preprocessing is done.

The edges from vertex v are sorted in non-decreasing order of edge costs. We call this pre-sort or pre-processing. A pointer is maintained for the sorted list. We actually maintain the sorted edge list from each vertex v by putting the end points of the sorted edges from v. The edge pointed to by the pointer is called the current edge. The endpoint of the current edge is denoted by $ce(v)$. Thus we can say $(v, ce(v))$ is the current edge. We move pointer by one in *update* to get the next edge.

Spira's algorithm maintains the solution set, denoted by S, which is the set of vertices to which shortest paths have been established by the algorithm, in a priority queue Q. The key for u in the queue, $key(u)$, is given by $key(u) = d[u] + c(u, ce(u))$, where $d[u]$ is the known shortest distance from the source to u.

The queue is initialized with one element of s, the source. Let $key(s) = c(s, t)$, where (s, t) is the minimum cost edge from s. Obviously t is included in the solution set as the second member. In general, suppose u is the minimum of the queue. If $v = ce(u)$ is not in S, it can be included in S with $d[v] = key(u)$, and then included in Q with $key(v) = d[v] + c(v, w)$, where (v, w) is the shortest edge from v.

Regardless of whether v is in S or not, the pointer on the edge list from u is advanced by one because edge (u, v) is no longer useful, meaning this edge is not going to be examined for other shortest paths.

The priority queue Q needs to support *find-min, increase-key* and *insert* operations efficiently, which we express by the repertory (*find-min, increase-key, insert*). For Spira's algorithm, we use an ordinary binary heap, which supports all of those operations in $O(\log n)$ time. All pointers for edge lists are initialized to 0. The function "next of $ce(v)$" is to advance the pointer, $point[v]$, by one and take the $point[v]$-th member in the list. As described below, we can assume an artificial edge with cost of infinity at the end of each edge list for a stopper. The algorithm follows.

Algorithm 1

1. $S = \{s\}; j = 1; ce(s) = $ next of $ce(s)$;
2. $t = ce(s)$;
3. $Q = \{s\}$ with $key(s) = c(s, t)$; /* heap initialization */
4. **while** $|S| < n$ **do begin**
5. $u=$find-min(Q);
6. $v = ce(u)$;
7.   **if** $v \notin S$ **then begin**
8. $S = S \cup \{v\}; j = j + 1$;
9. $update(v)$;
10   **end if**;
11   $update(u)$;
12 **end while**
13 **procedure** $update(v)$ **begin**
14   $ce(v) =$next of $ce(v); w = ce(v)$;
15   $d[w] = min\{d[w], d[v] + c(v, w)\}$;
16   $key(v) = d[v] + c(v, w)$;
17   **if** $v \in Q$ **then** increase-key(v);
18   **else** insert(v);
19 **procedure end**

The probabilistic assumption used in this paper is the end-point independence model [2]. In this model, when we scan the edge list, any vertex appears independently with the probability of $1/n$. When there are less than n edges, we assume edges with costs of infinity attached at the end of the list randomly and independently.

Let T_1, ..., T_{n-1} be the times for expanding the solution set by one at each stage of the size. Then, ignoring some overhead time between expansion processes, we have for the expected value $E[T]$ of the total time T

$$E[T] = E[T_1 + ... + T_{n-1}] = E[T_1] + ... + E[T_{n-1}].$$

From the theorem of total expectation, we have $E[E[X|Y]] = E[X]$ where $X|Y$ is the conditional random variable of X conditioned by Y. The first E in the left hand side goes over the sample space of Y and the second over that of X. In our analysis, X can represent a particular T_i and Y the rest. We use the same idea in later sections for analysis.

Now we analyze T_j where the solution set is expanded from size j to $j+1$. The probability that v is outside S at line 7 is $(n-j)/n$. The number of executions of $find\text{-}min$ at line 5 is given by the reciprocal of this probability, that is, $n/(n-j)$, which corresponds to the above $E[E[X|Y]]$. Each time $find\text{-}min$ is executed, we spend $O(\log n)$ time in $find\text{-}min$ and $O(\log n)$ time in $update$ in line 11. Thus from the above total expectation, the expected time for lines 5 and 11 is

$$n \log n \Sigma_{j=1}^{n-1} \frac{1}{(n-j)} = O(n^2 \log^2 n)$$

The $update$ at line 9 is executed exactly $n-1$ times. Thus a separate analysis can give us $O(n \log n)$ time, which is absorbed into the above main complexities.

3   Moffat-Takaoka Algorithm

Let us say the candidate $ce(u)$ of u is clean if it is outside S, and non-clean, otherwise. When we take the next edge from the edge list in $update$ in Spira's algorithm, the new candidate may be non-clean. It may be expensive to scan the edge list until we find a clean candidate as in [4]. However a careful design of scanning strategy may bring down the complexity. We simplify the Moffat-Takaoka (MT) algorithm a little in this section, and show more simplification in the next section.

We define the critical point to be the moment when the size of the solution set is equal to $n - n/\log n$. We round up any fraction in this paper. This algorithm does unlimited search for clean candidates before the critical point, and limited search after the critical point. To identify the critical point, we maintain array $T[v]$, which gives the order in which v is included in S, and is called the time stamp of v. Like Spira's algorithm, members of S are organized in a binary heap. The times for heap operations are measured by the number of (key) comparisons. As in Algorithm 1, all pointers for edge lists are initialized to 0.

We maintain the list $L[v]$, called the batch list, for each vertex v whose members are vertices u such that $ce(u) = v$. The key for vertex u in the priority queue, $key(u)$, is given by $key(u) = d[u] + c(u, v)$. Whether v is found to be a member of S at line 7 or not, those members in $L[v]$ need to be updated at line 13 to have more promising candidates. Also v itself needs to be treated to have a reasonable candidate at line 9 when v is included in S. How much scanning needs to be done for a good candidate is the major problem hereafter.

Algorithm 2

1. **for** $v \in V$ **do** $T[v] = \infty$;
2. $S = \{s\}; j = 1; T[s] = 1; ce(s) = $ next of $ce(s)$;
3. $t = ce(s)$;
4. $Q = \{s\}$ with $key(s) = c(s, t)$; /* heap initialization */
5. **while** $|S| < n$ **do**
6. u=find-min$(Q); v = ce(u)$;
7.   **if** $v \notin S$ **then begin**
8. $S = S \cup \{v\}; j = j + 1; T[v] = j$;
9. $update(v)$;
10.   **end;**

```
11   for u ∈ L[v] do
12       delete u from L[v];
13       update(u);
14   end for
15 end while;
16 procedure update(v) begin
17   w = ce(v);
18   if j ≤ n − n/ log n then limit = ∞ else limit = n − n/ log n;
19   while w ∈ S and T[w] ≤ limit do begin
20       ce(v)= next of ce(v); w = ce(v);
21   end while;
22   append v to L[w];
23   d[w] = min{d[w], d[v] + c(v, w)};
24   key(v) = d[v] + c(v, w);
25   if v ∈ Q then increase-key(v)
26                else insert(v);
27 end procedure
```

We sometimes omit "expected" from "expected time". The MT algorithm has two phases; Phase 1 before the critical point (CP) and Phase 2 after CP [1]. The computing time consists of two major components. One is the number of key comparisons in the heap operations and the other is the time for the scanning effort on the edge lists. The times before CP and after CP are both $O(n \log n)$ and balanced in both comparisons and scanning. If *limit* is set to infinity for all the computation, that is we do unlimited search for clean candidates, the resulting algorithm is called Dantzig's algorithm [4].

Let $U = V − S$ when $|S| = n − n/ \log n$, that is, $|U| = n/ \log n$. Before CP all candidates are clean, meaning the if statement at line 7 is entered with probability 1 and $O(n \log n)$ heap operations are done in total. Scanning effort to go outside S is $O(\log n)$ before CP, resulting in $O(n \log n)$ time. We can say the phase before CP is similar to Dantzig.

Labeling vertices as members in S is modeled as the coupon collector's problem [5]. To collect n different coupons, we need $O(n \log n)$ coupons. After CP, all candidates are limited to U, meaning the process is modeled as collecting $n/ \log n$ coupons. Thus we need $O((n/ \log n) \log(n/ \log n)) = O(n)$ trials, meaning we need $O(n \log n)$ comparisons.

The analysis of *increase-key* in *update* before CP involves some probabilistic analysis on members in the batch list. In [9], vertices u in the batch list $L[v]$ are processed for *increase-key* in a bottom-up fashion, and the time for this is shown to be $O(pn + \log n)$ where p is the probability that a vertex in S falls on any u in the list at line 11. This complexity remains $O(\log n)$ until the critical point is reached, but exceeds our target complexity after it. To avoid this analysis, [8]

---

[1] We notice that the condition for while in line 19 can be simplified to $T[w] \leq n − n/ \log n$, and line 18 can be removed. The above version clarifies the meaning of CP better.

uses a Fibonacci heap with (*delete-min, decrease-key, insert*) for maintaining candidates of vertices in a queue. This simplifies the analysis for the *update* for $L[v]$, but after CP, the heap must be re-initialized to include S and operations must be switched to (*delete-min, increase-key, insert*).

The scanning effort is not easy to analyze after CP, as the last movement of the pointer at each vertex, which we call an over-run, does not always lead to successful inclusion of the candidate vertex. In [9] the probabilistic dependence before and after CP regarding the amount of over-run was overlooked, and in [8] an analysis on this part is given, where the over-run associated with each vertex is regarded as a random variable conditioned by the behavior of Spira's algorithm.

It is this analysis of "over-run" that motivates the new algorithm in the next section for a simpler analysis.

4 New Algorithm

This algorithm does limited search for clean candidates in the edge list. The bound is dynamically changing and given by $n/(n-j+N)$, where $N = n/\log n$ and j is the size of the solution set. The fact that $n/(n-j+N) \leq \log n$ for all j is important, as the over-run can be bounded by $O(\log n)$ deterministically, and need not be analyzed as a random variable. Another simplification is that we get rid of the batch list L, as there is no CP and thus we scrap the policy of complete clean-ness of candidates before CP. Candidates are organized in a binary heap as in Algorithm 2. We call the while loop starting from line 4 the main iteration. Note that Algorithm 1 and Algorithm 3 are almost identical. Only difference is the clean-up booster at lines 15-17 in the latter.

Algorithm 3

```
1 S = {s}; j = 1;
2. t = ce(s);
3. Q = {s} with key(t) = c(s,t); /* heap initialization */
4. while |S| < n do
5.     u=find-min(Q);
6.     v = ce(u);
7.     if v ∉ S then begin
8.         S = S ∪ {v}; j = j + 1;
9.         update(v);
10    end;
11    update(u);
12 end while;
13 procedure update(v) begin
14    i = 0; w = ce(v); /* i is a counter */
15    while w ∈ S and i ≤ n/(n − j + N) do begin
16       ce(v)= next of ce(v); w = ce(v); i = i + 1;
```

17   **end while;**
18    $d[w] = min\{d[w], d[v] + c(v, w)\}$;
19    $key(v) = d[v] + c(v, w)$;
20   **if** $v \in Q$ **then** increase-key(v)
21   **else** insert(v);
22 **end procedure**

5   Correctness

The correctness of a generic algorithm with limited scan including our Algorithms 1-3 comes from the following two lemmas borrowed from [2]. Spira is a special case of limited search. Proof can be given by induction following the execution of the algorithm.

Lemma 1. *Suppose vertex* $v \in S$ *is such that* $ce(v)$ *is not in* S *and*

$$d[v] + c(v, ce(v)) = min\{d[u] + c(u, ce(u))|u \in S\}.$$

Then the final distance from the source to $ce(v)$ *is given by* $d[v] + c(v, ce(v))$. *Also* $d[u]$ *for* u *in* S *are all correct shortest distances from the source.*

Proof. If there is a shorter distance to $ce(v)$, it must come from some u in S with $d[u] + c(path(u, v))$, where $c(path(u, v))$ is the cost of some path, $path(u, v)$, from u to v and the first edge on the path goes out of S. From Lemma 2, the end points of edges from u shorter than $(u, ce(u))$ are all in S, and thus this first edge must be longer than or equal to $(u, ce(u))$. Then this distance must be greater than or equal to $d[v] + c(v, ce(v))$ defined above, a contradiction. Thus the shortest distance to $ce(v)$ is correctly computed and S is a correct solution set after inclusion of $ce(v)$.

Lemma 2. *For any* $v \in S$, *vertices in the edge list of* v *from position 1 to* $point[v] - 1$ *are all in* S.

Proof. From the nature of the algorithm, the pointer movement stops whenever the algorithm finds a candidate outside S. It may stop without finding a candidate outside S.

Theorem 1. *Any algorithm that is a variation of Spira's algorithm with limited scan is correct.*

6   Analysis

Lemma 3. *If* $(m - 1)p \le 1$ *and* $m \ge 1$, *then* $(1/2)mp \le 1 - (1 - p)^m$. *That is, if an event occurs with probability* p, *the probability that the event occurs within* m *trials is at least* $(1/2)mp$ *if the above condition is satisfied.*

Proof is by induction on m. Basis $m = 1$ is clear. Assume the lemma is true for $m - 1$.

$(1/2)(m - 1)p \le 1 - (1 - p)^{m-1}$
$(1/2)(m - 1)p - 1 \le -(1 - p)^{m-1}$
$((1/2)(m - 1)p - 1)(1 - p) \le -(1 - p)^m$

$$((1/2)(m-1)p - 1)(1-p) + 1 \leq 1 - (1-p)^m$$
$$(1/2)mp + (1/2)p(1 - (m-1)p) \leq 1 - (1-p)^m$$
$$(1/2)mp \leq 1 - (1-p)^m$$

Lemma 4. *Let $clean(ce(v))$ be the probability that $ce(v)$ is clean. Probability $p_j = clean(ce(v))$ for any $v \in S$ at the end of the main iteration when $|S| = j$ is at least $(1/2)\frac{n-j}{n-j+1+N}$*

Proof. We prove by induction based on the execution of the main iteration with $j = |S|$. At the beginning when $j = 1$, Lemma holds. When we execute the main iteration, S goes from $|S| = j - 1$ to $|S| = j$, or remains the same. For u and $v \in S$, we have two cases:

(1) *update* is performed on v and/or u at line 9 and/or 11.
(2) v is not touched by *update*.

(1) We consider the first case. Let us consider executing line 9. Let us call lines 15-17 in *update(v)* the clean-up booster, as this part of while-loop increases the probability that $ce(v)$ is outside S. Once the pointer for the edge list of v is increased by one, the probability that $ce(v)$ is clean becomes $(n - j)/n$. We boost this probability by limited scanning of m tests. We have $clean(ce(v)) = 1 - (1 - (n-j)/n)^{n/(n-j+N)}$. Letting $p = (n-j)/n$ and $m = n/(n - j + N)$ in the above lemma we have $clean(ce(v)) \geq (1/2)\frac{n-j}{n-j+N}$. The condition is satisfied since $pm \leq 1$. Note that $(1/2)\frac{n-j}{n-j+N} \geq (1/2)\frac{n-j}{n-j+1+N}$.
  Now consider executing line 11. The booster is performed with j that is one less than that for line 9 or the same value depending on whether S is expanded. In either case, we can show $clean(ce(u))$ satisfies the lemma for u.
  (2) Suppose S expands and the lemma holds for any $v \in S$ when $|S| = i = j - 1$, that is, the probability that $ce(v)$ is clean is at least $(1/2)\frac{n-i}{n-i+N}$. As v is not touched by *update*, this probability of $ce(v)$ being clean is at least $(1/2)\frac{n-i}{n-i+N}\frac{n-i-1}{n-i}$, since $(n - i - 1)/(n - i)$ is the conditional probability that $ce(v)$ is clean at $|S| = j = i + 1$ on condition that it is clean at $|S| = i$. Thus

$$clean(ce(v)) \geq (1/2)\frac{n-i-1}{n-i+N} \geq (1/2)\frac{n-j}{n-j+1+N}$$

We can show when S does not expand, the probability $clean(ce(u))$ satisfies the lemma.

Lemma 5. *Find-min at line 5 is executed $O(n)$ times on average.*

Proof. Let p_j be the probability that $v = ce(u)$ is clean at line 7 when $|S| = j$. From the previous lemma, we have $p_j \geq (1/2)\frac{n-j}{n-j+1+N}$. Since the expected number of trials for $ce(u)$ being clean is $1/p_j$, we have the expected number of *find-min* executions as

$$\Sigma_{j=1}^{n-1} 1/p_j \leq \Sigma_{j=1}^{n-1} 2\frac{n-j+N+1}{n-j} = \Sigma_{j=1}^{n-1} 2(1 + \frac{N+1}{n-j}) = O(n)$$

As each *find-min* requires $O(1)$ time, the expected time for total *find-min* is $O(n)$.

Now let us analyze *update* in two components. One is the time for heap operations, the other being the scanning efforts.

Lemma 6. *The expected time for comparisons in update is $O(n \log n)$ in total.*

Proof. Increase-key or *insert* is performed at the end of each *update*, spending $O(\log n)$ time. The *update* at line 9 is done $n - 1$ times, meaning $O(n \log n)$ time for this part. The number of the second *update* executions is $O(n)$ on average. Thus the expected total time for comparisons in *update* is $O(n \log n)$.

Next we analyze the time for scanning effort.

Lemma 7. *The scanning effort is $O(n \log n)$ on average.*

From the above analysis, the number of updates executed at line 11 when $|S| = j$ is bounded by $1/p_j = 2(n - j + 1 + N)/(n - j)$ on average. Each update moves its pointer at most by $n/(n - j + N)$. Thus the total expected number of pointer movements is bounded by

$$\sum_{j=1}^{n-1} 2 \frac{n-j+1+N}{n-j} \frac{n}{n-j+N} = \sum_{j=1}^{n-1} \frac{2n}{n-j}\left(1 + \frac{1}{n-j+N}\right) = O(n \log n)$$

Line 9 is executed $n - 1$ times, each moving the pointer by at most $\log n$.

Since the cost for comparisons and scanning are both $O(n \log n)$, we have the following theorem.

Theorem 2. *The expected running time of Algorithm 3 is $O(n \log n)$ after the presort is done. The time for the APSP problem based on Algorithm 3 is $O(n^2 \log n)$ including the time for presort.*

The main analysis of expected time ends here. The following is an alternative analysis on the scanning for a generic algorithm with limited scan, including Algorithm 3. This is similar to the analysis in [8], which analyzes the amount of over-runs as random variables conditioned by the scanning of Spira's algorithm. In our case, the over-runs are deterministic quantities bounded by $O(\log n)$ each.

Lemma 8. *Let the pointer, point[u], for ce(u) in update(u) in the new algorithm come to p. If this movement of pointer is not the last, that is, point[v] moves at least once more, the pointer for ce(u) in Spira's algorithm comes to p also.*

Proof. Suppose $v_1, v_2, ..., v_n$ are the vertices chosen by Spira's algorithm for shortest paths from the source v_1 in this order. Let $DIS(v_i)$ be the distances $\{d[v_i] + c(v_i, w_{ij}) | j = 1, ..., point[v_i]\}$, where w_{ij} are the end points of scanned edges from v_i. That is, $DIS(v_i)$ is the set of all distances from v_i checked by Spira's algorithm. Let DIS be defined by $DIS = DIS(v_1) \cup ... \cup DIS(v_n)$. DIS includes at least all shortest distances from the source. Let $SORTED\_DIS$ be the sorted set of DIS. We observe that the whole set of $SORTED\_DIS$ is checked by Spira's algorithm, that is, they are returned by *find-min* from the priority queue in the sorted order, or as $d[v_i] + c(v_i, w_{ij})$ for some v_i and w_{ij} by the last scanned edge from v_i in *update*. We observe that as the shortest distance from the source to some vertex there is a longer or equal distance, say, x, in DIS than the distance $d[u] + c(u, ce(u))$ by the new algorithm since p is not the last position, that is, distance x to v at line 6 will be found after the pointer comes to p. This distance x is obtained by Spira's algorithm as well. Then the pointer for $ce(u)$ is reached by Spira since the end points of edge list from first to the current are all in the solution set and Spira pushes the pointer over those endpoints at least until it gets to x.

Theorem 3. *The scanning effort in Algorithm 3 is $O(n \log n)$*

Proof. The scanning of each edge list has the same tracking record as that in Spira's algorithm apart from the last scanning. The total scanning effort in Spira is $O(n \log n)$. The last scanning from each vertex is bounded by $O(\log n)$. The vertex obtained by the last scanning may not be used for shortest distances. The total time for the last scanning efforts for all edge lists is thus bounded by $O(n \log n)$. Therefore the total scanning efforts cost us $O(n \log n)$.

7   Concluding Remarks

We showed the expected times for *find-min, increase-key, insert* and scanning effort are all $O(n \log n)$, when an SSSP problem is solved. The presort takes $O(m \log n)$ time. Thus the expected time of our algorithm for the APSP problem is $O(n^2 \log n)$. Computer experiment is easy since Algorithm 3 is obtained from Algorithm 2 by deleting a few lines and changing the scanning condition, giving us a good test bed for fair comparison. We did experiment with a run of 100 times on a random complete graph with $n = 100$ under Linux gcc on the machine INTEL QUAD 2.66 GHz with 2048 MB cache. Our experiments show that cpu times are 0.73 sec and 0.47 sec for Algorithms 2 and 3, about 36% reduction in computing time for the latter algorithm. We conjecture this reduction is mainly achieved by removing the batch list L at a slight increase of the times of the main iterations. This number is around $1.8n$, suggesting the factor $1/2$ in Lemma 3 may be an underestimation.

We note that we make a balance between (*find-min, increase-key, insert*), which are measured by the number of comparisons, and scanning effort, which is measured by pointer movements. In the approach by the critical point, we maintained balance on the numbers of comparisons before and after CP, and scanning efforts before and after CP, and also between comparisons and scanning for the total computation. In our new algorithm, we balance between the number of key comparisons and the scanning effort through the whole computation. In other words, we balance two different complexities of different nature, whereas in the CP approach we keep balance on more complexities. In fact, $\log n$ in $n \log n$ in comparison analysis comes from data structure, whereas $n \log n$ in scanning comes from the coupon collector's problem.

This leads to the observation that the scanning bound of $n/(n - j + N)$ in *update* can be parameterized as $kn/(n - j + N)$, and we can do some best tuning by changing the parameter k, depending on the characteristic of the computer used with specific speeds for comparisons and pointer movements. We see that the greater k, the more scanning and the less comparisons. Our experiments show that the cpu time becomes lowest when k is between 1 and 2, and a sharp rise occurs for small k (closer to Spira) and large k (closer to Dantzig). This suggests that on our machine there is not much difference in time between comparisons and scanning.

References

1. Blelloch, G.E., Vassilevska, V., Williams, R.: A New Combinatorial Approach for Sparse Graph Problems. In: Aceto, L., Damgård, I., Goldberg, L.A., Halldórsson, M.M., Ingólfsdóttir, A., Walukiewicz, I. (eds.) ICALP 2008, Part I. LNCS, vol. 5125, pp. 108–120. Springer, Heidelberg (2008)
2. Bloniarz, P.: A shortest path algorithm with expected time $O(n^2 \log n \log^* n)$. SIAM Journal on Computing 12, 588–600 (1983)
3. Chan, T.: More algorithms for all pairs shortest paths. In: STOC 2007, pp. 590–598 (2007)
4. Dantzig, G.: On the shortest route in a network. Management Science 6, 269–271 (1960)
5. Feller, W.H.: An Introduction to Probability and its Applications, 3rd edn., vol. 1. John-Wiley, New York (1968)
6. Fredman, M., Tarjan, R.: Fibonacci heaps and their uses in improved network optimization problems. JACM 34, 596–615 (1987)
7. Goldberg, A.V., Tarjan, R.E.: Expected Performance of Dijkstra's Shortest Path Algorithm, Technical Report 96-062, NEC Research Institute, Inc. (June 1996)
8. Mehlhorn, K., Priebe, V.: On the All-Pairs Shortest Path Algorithm of Moffat and Takaoka. Random Structures and Algorithms 10, 205–220 (1997)
9. Moffat, A., Takaoka, T.: An all pairs shortest path algorithm with expected running time $O(n^2 \log n)$. SIAM Journal on Computing 16, 1023–1031 (1987)
10. Pettie, S.: A new approach to all pairs shortest paths on real weighted graphs. Theoretical Computer Science 312, 47–74 (2004)
11. Spira, P.: A new algorithm for finding all shortest paths in a graph of positive arcs in average time $O(n^2 \log^2 n)$. SIAM Journal on Computing 2, 28–32 (1973)
12. Takaoka, T., Moffat, A.: An $O(n^2 \log n \log \log n)$ expected time algorithm for the all pairs shortest path problem. In: Dembinski, P. (ed.) MFCS 1980. LNCS, vol. 88, pp. 643–655. Springer, Heidelberg (1980)

New Min-Max Theorems for Weakly Chordal and Dually Chordal Graphs

Arthur H. Busch[1,*], Feodor F. Dragan[2], and R. Sritharan[3,**]

[1] Department of Mathematics, The University of Dayton, Dayton, OH 45469
art.busch@notes.udayton.edu
[2] Department of Computer Science, Kent State University, Kent, OH 44242
dragan@cs.kent.edu
[3] Department of Computer Science, The University of Dayton, Dayton, OH 45469
srithara@notes.udayton.edu

Abstract. A distance-k matching in a graph G is matching M in which the distance between any two edges of M is at least k. A distance-2 matching is more commonly referred to as an induced matching. In this paper, we show that when G is weakly chordal, the size of the largest induced matching in G is equal to the minimum number of co-chordal subgraphs of G needed to cover the edges of G, and that the co-chordal subgraphs of a minimum cover can be found in polynomial time. Using similar techniques, we show that the distance-k matching problem for $k > 1$ is tractable for weakly chordal graphs when k is even, and is NP-hard when k is odd. For dually chordal graphs, we use properties of hypergraphs to show that the distance-k matching problem is solvable in polynomial time whenever k is odd, and NP-hard when k is even. Motivated by our use of hypergraphs, we define a class of hypergraphs which lies strictly in between the well studied classes of acyclic hypergraphs and normal hypergraphs.

1   Background and Motivation

In this paper, all graphs are undirected, simple, and finite. That is, a graph $G = (V, E)$ where V is a finite set whose elements are called vertices together with a set E of unordered pairs of vertices. We say $H = (V', E')$ is a subgraph of $G = (V, E)$ if $V' \subseteq V$ and $E' \subseteq E$, and we say that H is an induced subgraph if $E' = \{uv \in E \mid \{u, v\} \subseteq V'\}$. We use P_k to denote an induced path on k vertices and C_k is an induced cycle on k vertices. A graph is chordal if it does not contain any C_k, $k \geq 4$. A graph is co-chordal if its complement is chordal. A graph G is weakly chordal if neither G nor $\overline{G}$ contains any C_k, $k \geq 5$. For background on these and other graph classes referenced below, we refer the interested reader to [6].

An *induced matching* in a graph is a matching that is also an induced subgraph, i.e., no two edges of the matching are joined by an edge in the graph.

---

* Acknowledges support from the University of Dayton Research Institute.
** Acknowledges support from The National Security Agency, USA.

W. Wu and O. Daescu (Eds.): COCOA 2010, Part II, LNCS 6509, pp. 207–218, 2010.
© Springer-Verlag Berlin Heidelberg 2010

The size of an induced matching is the number of edges in it. Let $im(G)$ denote the size of a largest induced matching in G. Given G and positive integer k, the problem of deciding whether $im(G) \geq k$ is NP-complete [8] even when G is bipartite.

For vertices x and y of G, let $dist_G(x, y)$ be the number of edges on a shortest path between x and y in G. For edges e_i and e_j of G let $dist_G(e_i, e_j) = \min\{dist_G(x, y) \mid x \in e_i \text{ and } y \in e_j\}$. For $M \subseteq E(G)$, M is a *distance-k matching* for a positive integer $k \geq 1$ if for every $e_i, e_j \in M$ with $i \neq j$, $dist_G(e_i, e_j) \geq k$. For $k = 1$, this gives the usual notion of matching in graphs. For $k = 2$, this gives the notion of induced matching. The *distance-k matching problem* is to find, for a given graph G and an integer $k \geq 1$, a distance-k matching with the largest possible number of edges.

A bipartite graph $G = (X, Y, E)$ is a *chain graph* if it does not have a $2K_2$ as an induced subgraph. Bipartite graph $G' = (X', Y', E')$ is a *chain subgraph* of bipartite graph $G = (X, Y, E)$, if G' is a subgraph of G and G' contains no $2K_2$. For a bipartite graph $G = (X, Y, E)$, let $ch(G)$ denote the fewest number of chain subgraphs of G the union of whose edge-sets is E. A set of $ch(G)$ chain subgraphs of bipartite graph $G = (V, E)$ whose edge-sets cover E is a *minimum chain subgraph cover* for G. Yannakakis showed [23] that when $k \geq 3$, deciding whether $ch(G) \leq k$ for a given bipartite graph G is NP-complete. An efficient algorithm to determine whether $ch(G) \leq 2$ for a given bipartite graph G is known [16].

It is clear that for any bipartite graph G, $im(G) \leq ch(G)$. Families of bipartite graphs where equality holds have been considered in literature. For example, it was shown in [24] that when G is a convex bipartite graph, $im(G) = ch(G)$. A bipartite graph is chordal bipartite if it does not contain any induced cycles on 6 or more vertices. It is known that every convex bipartite graph is also chordal bipartite.

The following more general result was recently shown:

Proposition 1. [1] For a chordal bipartite graph G, $im(G) = ch(G)$.

Let us move away from the setting of bipartite graphs and consider graphs in general. We say H is a *co-chordal subgraph* of G if H is a subgraph of G and also H is co-chordal. Let $coc(G)$ be the minimum number of co-chordal subgraphs of G needed to cover all the edges of G. As a chain subgraph of a bipartite graph G is a co-chordal subgraph of G and vice versa, the parameter $coc(G)$ when restricted to a bipartite graph G is essentially the same as $ch(G)$.

Again, it is clear from the definitions that for any graph G, $im(G) \leq coc(G)$. In Section 2, we show that when G is weakly chordal, $im(G) = coc(G)$ and that the co-chordal subgraphs of a minimum cover can be found in polynomial time. As every chordal bipartite graph is weakly chordal and as a chain subgraph of a bipartite graph is a co-chordal subgraph and vice versa, our result generalizes Proposition 1. In Section 3, we use similar techniques to show that the distance-k matching problem for $k > 1$ is tractable for weakly chordal graphs when k is even, and NP-hard when k is odd. Next, in Section 4 we use techniques from the study of hypergraphs to show that the opposite holds for the class of dually chordal graphs; the distance-k matching problem can be solved in polynomial time for dually chordal graphs if a k is odd, and is NP-hard for all even k. Motivated by our results and by the use of hypergraphs in Section 4, we define a class of hypergraphs in Section 5 which lies strictly in between the well studied classes of acyclic hypergraphs and normal hypergraphs.

2 A Min-Max Theorem for Weakly Chordal Graphs

For a graph G, let G^* denote the square of the line graph of G. More explicitly, vertices of G^* are edges of G. Edges e_i, e_j of G are nonadjacent in G^* if and only if they form a $2K_2$ in G.

It is clear from the construction of G^* that the set of edges of a co-chordal subgraph of G maps to a clique of G^*. Further, $\mathrm{im}(G) = \alpha(G^*)$, where $\alpha(G^*)$ is the size of a largest independent set in G^*.

The following is known:

Proposition 2. [9] If G is weakly chordal, then G^* is weakly chordal.

Also, it is well known that every weakly chordal graph is perfect [13]. Therefore, when G is weakly chordal, $\mathrm{im}(G) = \alpha(G^*) = \theta(G^*)$, where $\theta(G^*)$ is the minimum clique cover number of G^*. Thus, when G is weakly chordal $\theta(G^*) \leq \mathrm{coc}(G)$.

We will show that when G is weakly chordal, $\mathrm{coc}(G) \leq \theta(G^*)$ also holds and therefore we have the following:

Proposition 3. If G is weakly chordal, then $\mathrm{coc}(G) = \mathrm{im}(G)$.

The proof of Proposition 3 utilizes the following edge elimination scheme for weakly chordal graphs. Edge xy is a co-pair of graph G, if vertices x and y are not the endpoints of any P_k, $k \geq 4$, in $\overline{G}$.

Proposition 4. [19] Suppose e is a co-pair of graph G. Then, G is weakly chordal if and only if $G - e$ is weakly chordal.

The following is implied by Corollary 2 in [12]:

Proposition 5. [12] Suppose G is a weakly chordal graph that contains a $2K_2$. Then, G contains co-pairs e and f such that e and f form a $2K_2$ in G.

Lemma 6. If e is a co-pair of a weakly chordal graph G, then $G^* - e = (G-e)^*$.

Proof. Deleting an edge xy from G will never destroy a $2K_2$, unless it is one of the edges of the $2K_2$. If deleting xy creates a new $2K_2$ then xy must be the middle edge of a P_4 in G, or equivalently, x and y are the end vertices of a P_4 in $\overline{G}$. Thus, when e is a co-pair, two edges form a $2K_2$ in $G - e$ if and only if they form a $2K_2$ in G that does not include the edge e. Since the vertices of $(G^* - e)$ and $(G - e)^*$ both consist of the edges of $G - e$, this guarantees that the edge sets of $(G^* - e)$ and $(G - e)^*$, are identical as well. Hence the graphs are identical. □

In order to establish that when G is weakly chordal, $\mathrm{coc}(G) \leq \theta(G^*)$, first observe that every member of a clique cover of G^* can be assumed to be a maximal clique of G^*. We have the following:

Theorem 7. *Let G be weakly chordal. Then, every maximal clique of G^* is the edge-set of a maximal co-chordal subgraph of G.*

Proof. Proof is by induction on the number of edges in the graph. Clearly, the statement is true when G has no edges.

Assume the statement is true for all weakly chordal graphs with up to $k - 1$ edges, and let G be a weakly chordal graph with k edges. If G contains no $2K_2$, then G is co-chordal and G^* is a clique and the theorem holds.

Now, suppose G contains a $2K_2$. Then, from Proposition 5, G contains a $2K_2$ e_1, e_2 each of which is a co-pair of G.

Let M be a maximal clique of G^*. As no maximal clique of G^* contains both e_1 and e_2, we can choose $i \in \{1, 2\}$ such that $e_i \notin M$. As a result, M is a maximal clique of $G^* - e_i$ which equals $(G - e_i)^*$ by Lemma 6. Also, by Proposition 4, $G - e_i$ is weakly chordal. It then follows by the induction hypothesis that M is the edge set of a maximal co-chordal subgraph of $G - e_i$.

Clearly, this subgraph remains co-chordal in G, so it remains to show that this subgraph is, in fact, maximal. If this is not the case, then there exists a co-chordal subgraph M' of G such that $M \subset M'$. As every co-chordal subgraph of G maps to a clique of G^*, it follows that M and M' are cliques of G^* such that $M \subset M'$; this contradicts M being a maximal clique of G^*.   □

Thus, $\theta(G^*) = \mathrm{coc}(G)$, establishing Proposition 3. As an efficient algorithm exists [14] to compute a minimum clique cover of a weakly chordal graph, we have the following:

Corollary 8. *When G is weakly chordal, $\mathrm{coc}(G)$ and a minimum cover of G by co-chordal subgraphs of G can be found in polynomial time.*

We recently learned of a surprising application of this result: the parameters $\mathrm{coc}(G)$ and $\mathrm{im}(G)$ yield upper and lower bounds, respectively, on the Castelnuovo-Mumford regularity of the edge ideal of G [22]. Thus, when G is weakly chordal, this parameter can be computed efficiently.

Another application of Corollary 8 utilizes the complement of G. As the complement of a weakly chordal graph remains weakly chordal, after taking the complement of each graph in a cover by co-chordal subgraphs, we have a set of chordal graphs whose edge-intersection is the edge-set of a weakly chordal graph. The study of a variety of similar parameters, known as the *intersection dimension* of a graph G with respect to a graph class $\mathcal{A}$, was introduced in [15]. The problem when $\mathcal{A}$ is the set of chordal graphs was termed the *chordality* of G in [17]. We use $\dim_{\mathrm{CH}}(G)$ to denote the chordality or chordal dimension of a graph G. In this context, we have another corollary of Theorem 7.

Corollary 9. *When G is weakly chordal, $\dim_{\mathrm{CH}}(G) = \mathrm{im}(\overline{G})$ and a minimum set of chordal graphs whose edge-intersection give the edge-set of G can be found in polynomial time.*

A chordal graph that does not contain a $2K_2$ is a split graph, and it has been shown in [8] that a split graph cover of a chordal graph can be computed in polynomial time.

Proposition 10. [8] Let G be a chordal graph. Then, a minimum cover of edges of G by split subgraphs of G can be found in polynomial time.

The proof of Proposition 10 in [8] utilizes the clique tree of a chordal graph G and the Helly property. An alternate proof can be given by showing that the edges referred to in Proposition 5 can be chosen so that each edge is incident with a simplicial vertex of G. Since no such edge is the only chord of a cycle, this guarantees that $G - e$ will be chordal whenever G is chordal. As every chordal graph is also weakly chordal, Lemma 6 and a slightly modified version of Theorem 7 then imply Proposition 10.

3 Distance-k Matchings in Weakly Chordal Graphs

In this section, we observe that the correspondence between maximal cliques of G^* and maximal co-chordal subgraphs of G can be adapted to find maximum a distance-k matching in a weakly chordal graph G for any positive even integer k. We then show that finding a largest distance-k matching when k is odd and $k \geq 3$ is NP-hard.

We begin by noting the fundamental connection between distance-k matchings in a graph G and independent sets in the kth power of the line graph of G, which we denote $L^k(G)$.

Proposition 11. [7] For $k \geq 1$ and graph G, the edge set M is a distance-k matching in G if and only if M is an independent vertex set in $L^k(G)$.

As a result, identifying a largest distance-k matching in a graph G is no more difficult than constructing the kth power of the line graph of G and finding a maximum independent set in $L^k(G)$. Clearly, for any edge e, the set of edges within distance k of e can be computed in linear time, and as a result a polynomial time algorithm exists for the distance-k matching problem whenever an efficient algorithm exists for finding a largest independent set in $L^k(G)$. Proposition 2 guarantees that such an efficient algorithm exists for induced matchings, as efficient algorithms to compute the largest independent sets of weakly chordal graphs are well known. For $k > 1$, the existence of a polynomial algorithm for computing distance-$2k$ matchings in weakly chordal graphs is guaranteed by combining Proposition 2 with the following result.

Proposition 12. [5] Let G be a graph and $k \geq 1$ be a fixed integer. If G^2 is weakly chordal, then so is G^{2k}.

For distance-k matchings when k is odd, we note that the case $k = 3$ was recently shown to be NP-complete for the class of chordal graphs, which is properly contained in the class of weakly chordal graphs.

Proposition 13. [7] The largest distance-3 matching problem is NP-hard for chordal graphs.

We will extend this result to distance-$(2k+1)$ matchings for every positive integer k. This extension is done by showing that the distance-$(2k+1)$ matching problem can be transformed into the distance-$(2k + 3)$ matching problem in polynomial time, for any positive integer k.

Theorem 14. *For any positive integer k, there exists a polynomial time transformation from the distance-$(2k + 1)$ matching problem to the distance-$(2k + 3)$ matching problem.*

Proof. Let $G = (V, E)$ be a graph, and let k be a positive integer. We will define the graph $G^+ = (V^+, E^+)$ from G as follows. For each edge $e = uv$ of G, we introduce two new vertices x_e and y_e and add edges to make the subgraph induced by $\{u, v, x_e, y_e\}$ a clique. Formally, $V^+ = V \cup \{x_e, y_e \mid e \in E\}$, and

$$E^+ = E \cup \left(\bigcup_{e=uv \in E} \{x_e y_e, \ x_e u, \ x_e v, \ y_e u, \ y_e v\} \right).$$

We will show that a distance-$(2k + 1)$ matching of size p exists in G if and only if a distance-$(2k + 3)$ matching of size p exists in G^+. Since G^+ can clearly be constructed from G in linear time, this will establish the theorem.

First, we note that every matching M in G corresponds to a matching $M^+ = \{x_e y_e \mid e \in M\}$ in G^+, and that the distance between two edges $x_e y_e$ and $x_f y_f$ from M^+ is clearly $\text{dist}_G(e, f) + 2$. Thus, if M is a distance-$(2k + 1)$ matching of size p in G, M^+ is a distance-$(2k+3)$ matching of size p in G^+. Next, note that if M is a distance-$(2k + 3)$ matching of G^+, then for each $e = uv$, at most one edge of M is incident with any vertex of $\{u, v, x_e, y_e\}$ as these vertices induce a clique in G^+. We construct a set M^- by removing from M any edge incident with a vertex x_e or y_e of $V^+ \setminus V$ and replacing it with the associated edge e. Since no other edge of M is incident with either vertex of e, we conclude that $|M^-| = |M|$ and that M^- is a matching in G. The pairwise distance between any two edges in M^- is at least $(2k+3) - 2 = 2k+1$. Thus, if a distance-$(2k+3)$ matching of size p exists in G^+, then a distance-$(2k+1)$ matching of size p exists in G. $\qquad\square$

Combining Proposition 13 and Theorem 14 we conclude the following:

Corollary 15. *For $k > 0$, the distance-$(2k + 1)$ matching problem is NP-hard for chordal graphs.*

Proof. We use the well known result of Dirac [10] that a graph is chordal if and only if it has a simplicial elimination ordering. Since each vertex of $V^+ \setminus V$ is simplicial in G^+, we can easily extend any simplicial elimination ordering of G to a simplicial elimination ordering of G^+. Therefore G^+ is chordal whenever G is chordal. The corollary now follows immediately from Proposition 13 and repeated application of Theorem 14. $\qquad\square$

We summarize the consequences of Propositions 2, 11, 12 and Corollary 15 below.

Proposition 16. *Suppose G is a weakly chordal graph and k is a positive integer. Then, when k is even, a largest distance-k matching in G can be found in polynomial time. When k is odd, finding a largest distance-k matching in G is NP-hard for any $k > 1$.*

4 A Min-Max Theorem for Distance-(2k+1) Matching in Dually Chordal Graphs

As we saw in the previous section, for a weakly chordal graph G and any integer $k \geq 1$, a largest distance-$(2k)$ matching in G can be found in polynomial time, whereas computing a largest distance-$(2k + 1)$ matching in G is an NP-hard problem. In this section, we show that for the class of dually chordal graphs the opposite holds.

Dually chordal graphs were introduced in [11] as a generalization of strongly chordal graphs (which are the hereditary dually chordal graphs) where the Steiner tree problem and many domination-like problems still (as in strongly chordal graphs) have efficient solutions. It also was shown recently in [7] that the distance-k matching problem is solvable in polynomial time for strongly chordal graphs for any integer k. Here, we extend this result to all doubly chordal graphs by showing that the distance-k matching problem is solvable in polynomial time for any odd k for all dually chordal graphs.

To define dually chordal graphs and doubly chordal graphs, we need some notions from the theory of hypergraphs [2]. Let $\mathcal{E}$ be a hypergraph with underlying vertex set V, i.e. $\mathcal{E}$ is a collection of subsets of V (called *hyperedges*) The *dual hypergraph* $\mathcal{E}^*$ has $\mathcal{E}$ as its vertex set and for every $v \in V$ a hyperedge $\{e \in \mathcal{E} : v \in e\}$. The *line graph* $L(\mathcal{E}) = (\mathcal{E}, E)$ of $\mathcal{E}$ is the intersection graph of $\mathcal{E}$, i.e. $ee' \in E$ if and only if $e \cap e' \neq \emptyset$. A *Helly hypergraph* is one whose edges satisfy the Helly property, that is, any subfamily $\mathcal{E}' \subseteq \mathcal{E}$ of pairwise intersecting edges has a nonempty intersection. A hypergraph $\mathcal{E}$ is a *hypertree* if there is a tree T with vertex set V such that every edge $e \in \mathcal{E}$ induces a subtree in T. Equivalently, $\mathcal{E}$ is a hypertree if and only if the line graph $L(\mathcal{E})$ is chordal and $\mathcal{E}$ is a Helly hypergraph. A hypergraph $\mathcal{E}$ is a *dual hypertree* (also known as α–*acyclic hypergraph*) if there is a tree T with vertex set $\mathcal{E}$ such that, for every vertex $v \in V$, $T_v = \{e \in \mathcal{E} : v \in e\}$ induces a subtree of T. Observe that $\mathcal{E}$ is a hypertree if and only if $\mathcal{E}^*$ is a dual hypertree.

For a graph $G = (V, E)$ by $\mathcal{C}(G) = \{C : C$ is a maximal clique in $G\}$ we denote the *clique hypergraph*. Let also $\mathcal{D}(G) = \{D_r(v) : v \in V, r$ a non-negative integer$\}$ be the *disk hypergraph* of G. Recall that $D_r(v) = \{u \in V : \text{dist}_G(u, v) \leq r\}$ is a disk of radius r centered at vertex v. A graph G is called *dually chordal* if the clique hypergraph $\mathcal{C}(G)$ is a hypertree [4, 11, 20]. In [4, 11] it is shown that dually chordal graphs are exactly the graphs G whose disk hypergraphs $\mathcal{D}(G)$ are hypertrees (see [4, 11] for other characterizations, in particular in terms of certain elimination schemes, and [3] for their algorithmic use). From the definition of hypertrees we deduce that for dually chordal graphs the line graphs of the clique and disk hypergraphs are chordal. Conversely, if G is a chordal graph then $\mathcal{C}(G)$ is a dual hypertree, and, therefore, the line graph $L(\mathcal{C}(G))$ is a dually chordal graph, justifying the term "dually chordal graphs". Hence, the dually chordal graphs are exactly the intersection graphs of maximal cliques of chordal graphs (see [20, 4]). Finally note that graphs being both chordal and dually chordal were dubbed *doubly chordal* and investigated in [4, 11, 18]. The class of dually chordal graphs contains such known graph families as interval graphs, ptolemaic graphs, directed path graphs, strongly chordal graphs, doubly chordal graphs, and others.

Let $G = (V, E)$ be an arbitrary graph and r be a non-negative integer. For an edge $uv \in E$, let $D_r(uv) := \{w \in V : \text{dist}_G(u, w) \leq r \text{ or } \text{dist}_G(v, w) \leq r\} = D_r(v) \cup D_r(u)$ be a disk of radius r centered at edge uv. For the edge-set E of graph $G = (V, E)$, we define a hypergraph $\mathcal{D}_{E,r}(G)$ as follows:

$$\mathcal{D}_{E,r}(G) = \{D_r(uv) : uv \in E\}.$$

Along with the distance-k matching problem we will consider also a problem which generalizes the (minimum) vertex cover problem. A *distance-k vertex cover* for a non-negative integer $k \geq 0$ in an undirected graph $G = (V, E)$ is a set of vertices $S \subseteq V$ such that for every edge $uv \in E$ there exists a vertex $x \in S$ with $D_k(x) \cap \{u, v\} \neq \emptyset$. For $k = 0$, this gives the usual notion of vertex cover in graphs. The *distance-k vertex cover problem* is to find, for a given graph G and an integer $k \geq 0$, a distance-k vertex cover with the fewest number of vertices.

Using the hypergraph $\mathcal{D}_{E,k}(G)$ of G (where $k \geq 0$ is an integer), the distance-$(2k + 1)$ matching problem and the distance-k vertex cover problem on G can be formulated as the transversal and matching problems, respectively, on the hypergraph $\mathcal{D}_{E,k}(G)$. Recall that a *transversal* of a hypergraph $\mathcal{E}$ is a subset of vertices which meets all edges of $\mathcal{E}$. A *matching* of $\mathcal{E}$ is a subset of pairwise disjoint edges of $\mathcal{E}$. For a hypergraph $\mathcal{E}$, the *transversal problem* is to find a transversal with minimum size $\tau(\mathcal{E})$ and the *matching problem* is to find a matching with maximum size $\nu(\mathcal{E})$.

Denote by $\text{dm}_k(G)$ the size of a largest distance-k matching in G, and by $\text{dvc}_k(G)$ the size of a smallest distance-k vertex cover in G. From the definitions we obtain

Lemma 17. *Let $G = (V, E)$ be an arbitrary graph and $k \geq 0$ be a non-negative integer. S is a distance-k vertex cover of G if and only if S is a transversal of $\mathcal{D}_{E,k}(G)$. M is a distance-$(2k + 1)$ matching of G if and only if $\{D_k(uv) : uv \in M\}$ is a matching of $\mathcal{D}_{E,k}(G)$. Thus, $\tau(\mathcal{D}_{E,k}(G)) = \text{dvc}_k(G)$ and $\nu(\mathcal{D}_{E,k}(G)) = \text{dm}_{2k+1}(G)$ hold for every graph G and every non-negative integer k.*

The parameters $\text{dm}_{2k+1}(G)$ and $\text{dvc}_k(G)$ are always related by a min–max duality inequality $\text{dvc}_k(G) \geq \text{dm}_{2k+1}(G)$. The next result shows that for dually chordal graphs and $k \geq 1$ equality holds.

Theorem 18. *Let $G = (V, E)$ be a dually chordal graph and $k \geq 1$ be an integer. Then, the hypergraph $\mathcal{D}_{E,k}(G)$ is a hypertree and, as a consequence, $\mathrm{dvc}_k(G) = \mathrm{dm}_{2k+1}(G)$ holds.*

Proof. Since G is a dually chordal graph, the disk hypergraph $\mathcal{D}(G)$ of G is a hypertree. That is, there is a tree T with vertex set V such that every disk $D_k(v)$, $v \in V$, induces a subtree in T. Consider an arbitrary edge uv of G. Since both disks $D_k(v)$ and $D_k(u)$ induce subtrees in T and $k \geq 1$ (i.e., $v \in D_k(u)$ as well as $u \in D_k(v)$), vertices of $D_k(v) \cup D_k(u)$ induce a subtree in T (the union of subtrees induced by $D_k(v)$ and $D_k(u)$). Hence, disk $D_k(uv)$ of G centered at any edge $uv \in E$ induces a subtree in T, implying that the hypergraph $\mathcal{D}_{E,k}(G)$ of G is a hypertree.

It is well known [2] that the equality $\tau(\mathcal{E}) = \nu(\mathcal{E})$ holds for every hypertree. Hence, by Lemma 17, we obtain the required equality.   □

For any hypertree $\mathcal{E}$ with underlying set V, a transversal with minimum size $\tau(\mathcal{E})$ and a matching with maximum size $\nu(\mathcal{E})$ can be found in time $O(|V| + \sum_{e \in \mathcal{E}} |e|)$ [21]. Thus, we conclude.

Corollary 19. *Let G be a dually chordal graph and $k \geq 1$ be an integer. Then, a smallest distance-k vertex cover of G and a largest distance-$(2k+1)$ matching of G can be found in $O(|V||E|)$ time.*

We notice now that any graph G can be transformed into a dually chordal graph G' by adding to G two new, adjacent to each other, vertices x and y and making one of them, say vertex x, adjacent to all vertices of G (and leaving y as a pendant vertex in G'). Using this transformation, it is easy to show that the classical vertex cover problem (i.e., the distance-k vertex cover problem with $k = 0$) is NP-hard for dually chordal graphs. It is enough to notice that $\mathrm{dvc}_0(G) = \mathrm{dvc}_0(G') - 1$. Adding a new vertex z to G' and making it adjacent to only y transforms dually chordal graph G' into a dually chordal graph G''. It is easy to see also that $\mathrm{dm}_2(G) = \mathrm{dm}_2(G'') - 1$, implying that the induced matching problem (i.e., the distance-2 matching problem) is NP-hard on dually chordal graphs.

Proposition 20. *The vertex cover problem (i.e., the distance-k vertex cover problem with $k = 0$) and the induced matching problem (i.e., the distance-2 matching problem) both are NP-hard on dually chordal graphs.*

We can prove also a more general result.

Theorem 21. *For every integer $k \geq 1$, the distance-$(2k)$ matching problem is NP-hard on dually chordal graphs.*

Proof. By Proposition 20, we can assume that $k \geq 2$. Let $G = (V, E)$ be a dually chordal graph. We construct a new graph G^+ by attaching to each vertex $v \in V$ an induced path $P_v = (v = v_0, v_1, \ldots, v_k)$ of length k. Note that each path P_v shares only vertex v with G and $P_v \cap P_u = \emptyset$ if $v \neq u$. Denote by e_v the last edge $v_{k-1}v_k$ of P_v. Since adding a pendant vertex to a dually chordal graph results in a dually chordal graph (see [4, 3]), the graph G^+ is dually chordal. It is enough to show now that $\alpha(G) = \mathrm{dm}_{2k}(G^+)$, where $\alpha(G)$ is the size of a largest independent set of G. It is known that the maximum independent set problem is NP-hard on dually chordal graphs [3].

Let $S \subseteq V$ be a largest independent set of G, i.e., $|S| = \alpha(G)$. Consider two arbitrary vertices x and y from S. We have $\mathrm{dist}_G(x,y) \geq 2$. But then, by construction of G^+, $\mathrm{dist}_{G^+}(e_x, e_y) \geq k - 1 + 2 + k - 1 = 2k$. Hence, $\alpha(G) \leq \mathrm{dm}_{2k}(G^+)$. Let now M be a set of edges of G^+ forming a largest distance-$(2k)$ matching of G^+. We can assume that $M \subseteq \{e_v : v \in V\}$ since for any edge $e \in M$ there must exist a vertex $v \in V$ such that $P_v \cap e \neq \emptyset$ and therefore e can be replaced in M by e_v. Now, if e_x and e_y $(x \neq y)$ are in M, then $\mathrm{dist}_{G^+}(e_x, e_y) \geq 2k$ implies $\mathrm{dist}_G(x,y) \geq 2$. That is $\alpha(G) \geq \mathrm{dm}_{2k}(G^+)$.   □

Figure 1 shows the containment relationships between the classes of Weakly Chordal, Chordal, Dually Chordal, Strongly Chordal and Interval Graphs. Since doubly chordal graphs are both chordal and dually chordal, and all chordal graphs are weakly chordal, we conclude this section with the following corollary of Proposition 16 and Corollary 19.

Corollary 22. *A maximum distance-k matching of a doubly chordal graph G can be found in polynomial time for every integer $k \geq 1$.*

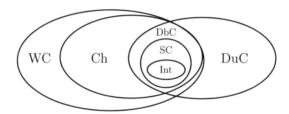

Fig. 1. The containment relationships between the classes of Weakly Chordal (WC), Chordal (Ch), Dually Chordal (DuC), Doubly Chordal (DbC), Strongly Chordal (SC) and Interval (Int) Graphs

5 A Class of Hypergraphs

As we have seen in the previous section, the hypertrees were very useful in obtaining the duality result between distance-k vertex cover and distance-$(2k+1)$ matching on dually chordal graphs. Recall that the dual hypertrees (called also α–acyclic hypergraphs) are exactly the clique hypergraphs of the chordal graphs (and the hypertrees are exactly the clique hypergraphs of dually chordal graphs). The duality result obtained for weakly chordal graphs in Section 2 can also be interpreted in terms of transversal and matching of a special class of hypergraphs.

We will need few more definitions from hypergraph theory. Let $\mathcal{E}$ be a hypergraph with underlying vertex set X. The *2–section graph* $2SEC(\mathcal{E})$ of the hypergraph $\mathcal{E}$ has vertex set X and two distinct vertices are adjacent if and only if they are contained in a common edge of $\mathcal{E}$. A hypergraph $\mathcal{E}$ is *conformal* if every clique C in $2SEC(\mathcal{E})$ is contained in an edge $e \in \mathcal{E}$.

Three well–known properties of hypergraphs will be helpful (for these and other properties cf. [2]).

(i) For a hypergraph $\mathcal{E}$, the graphs $L(\mathcal{E})$ and $2SEC(\mathcal{E}^*)$ are isomorphic;
(ii) A hypergraph $\mathcal{E}$ is conformal if and only if its dual hypergraph $\mathcal{E}^*$ has the Helly property;
(iii) Taking the dual of a hypergraph twice is isomorphic to the hypergraph itself.

Using these properties, α–acyclic hypergraphs can be equivalently defined as conformal hypergraphs with chordal 2–section graphs.

We now define a hypergraph $\mathcal{E}$ (with underlying vertex set X) to be *weakly acyclic* if $2SEC(\mathcal{E})$ is weakly chordal and $\mathcal{E}$ is conformal. In other words, weakly acyclic hypergraphs are exactly the clique hypergraphs of weakly chordal graphs. Note that since weakly chordal graphs are perfect, $\alpha(2SEC(\mathcal{E})) = \theta(2SEC(\mathcal{E}))$ and, as a consequence, $\alpha(L(\mathcal{E}^*)) = \theta(L(\mathcal{E}^*))$. The latter implies that equality $\nu(\mathcal{E}^*) = \tau(\mathcal{E}^*)$ holds for every weakly acyclic hypergraph $\mathcal{E}$. Indeed, $\alpha(L(\mathcal{E}^*)) = \nu(\mathcal{E}^*)$ holds for every hypergraph $\mathcal{E}^*$, and $\theta(L(\mathcal{E}^*)) = \tau(\mathcal{E}^*)$ holds for every Helly hypergraph $\mathcal{E}^*$ (equivalently, for every conformal hypergraph $\mathcal{E}$) [2]. For hypergraph $\mathcal{E}$, $\tau(\mathcal{E}^*)$ is equal to the minimum number of hyperedges needed to cover entire underlying vertex set X of $\mathcal{E}$, $\nu(\mathcal{E}^*)$ is equal to the maximum number of vertices $S \subseteq X$ such that no two of them can be covered by an hyperedge $e \in \mathcal{E}$.

Let now $G = (V, E)$ be a weakly chordal graph. Define a hypergraph $\mathcal{E}$ with underlying vertex set E as follows (note that the vertices of $\mathcal{E}$ are exactly the edges of G). Let $\mathcal{E}$ be the set of all maximal co-chordal subgraphs of G. Firstly, $2SEC(\mathcal{E}) = G^*$. Therefore, by Proposition 2, $2SEC(\mathcal{E})$ is weakly chordal. Further, from Theorem 7, $\mathcal{E}$ is conformal. Therefore, $\mathcal{E}$ is a weakly acyclic hypergraph. Consequently, since $\mathrm{coc}(G) = \tau(\mathcal{E}^*)$, $\mathrm{im}(G) = \nu(\mathcal{E}^*)$ and $\nu(\mathcal{E}^*) = \tau(\mathcal{E}^*)$, we obtain $\mathrm{coc}(G) = \mathrm{im}(G)$.

References

[1] Abuieda, A., Busch, A., Sritharan, R.: A min-max property of chordal bipartite graphs with applications. Graphs Combin. 26(3), 301–313 (2010)
[2] Berge, C.: Hypergraphs. North-Holland Mathematical Library, vol. 45. North-Holland Publishing Co., Amsterdam (1989); Combinatorics of finite sets, Translated from the French
[3] Brandstädt, A., Chepoi, V.D., Dragan, F.F.: The algorithmic use of hypertree structure and maximum neighbourhood orderings. Discrete Appl. Math. 82(1-3), 43–77 (1998)
[4] Brandstädt, A., Dragan, F., Chepoi, V., Voloshin, V.: Dually chordal graphs. SIAM J. Discrete Math. 11(3), 437–455 (1998) (electronic)

[5] Brandstädt, A., Dragan, F.F., Xiang, Y., Yan, C.: Generalized powers of graphs and their algorithmic use. In: Arge, L., Freivalds, R. (eds.) SWAT 2006. LNCS, vol. 4059, pp. 423–434. Springer, Heidelberg (2006)

[6] Brandstädt, A., Van Bang, L., Spinrad, J.P.: Graph classes: a survey. SIAM Monographs on Discrete Mathematics and Applications. Society for Industrial and Applied Mathematics (SIAM), Philadelphia (1999)

[7] Brandstädt, A., Mosca, R.: On distance-3 matchings and induced matchings. In: Lipshteyn, M., Levit, V.E., McConnell, R.M. (eds.) Graph Theory. LNCS, vol. 5420, pp. 116–126. Springer, Heidelberg (2009)

[8] Cameron, K.: Induced matchings. Discrete Appl. Math. 24(1-3), 97–102 (1989); First Montreal Conference on Combinatorics and Computer Science (1987)

[9] Cameron, K., Sritharan, R., Tang, Y.: Finding a maximum induced matching in weakly chordal graphs. Discrete Math. 266(1-3), 133–142 (2003); The 18th British Combinatorial Conference, Brighton (2001)

[10] Dirac, G.A.: On rigid circuit graphs. Abh. Math. Sem. Univ. Hamburg 25, 71–76 (1961)

[11] Dragan, F.F., Prisakar', K.F., Chepoǐ, V.D.: The location problem on graphs and the Helly problem. Diskret. Mat. 4(4), 67–73 (1992)

[12] Eschen, E., Sritharan, R.: A characterization of some graph classes with no long holes. J. Combin. Theory Ser. B 65(1), 156–162 (1995)

[13] Hayward, R., Hoàng, C., Maffray, F.: Optimizing weakly triangulated graphs. Graphs Combin. 5(4), 339–349 (1989)

[14] Hayward, R.B., Spinrad, J., Sritharan, R.: Weakly chordal graph algorithms via handles. In: Proceedings of the Eleventh Annual ACM-SIAM Symposium on Discrete Algorithms, San Francisco, CA, pp. 42–49. ACM, New York (2000)

[15] Kratochvíl, J., Tuza, Z.: Intersection dimensions of graph classes. Graphs Combin. 10(2), 159–168 (1994)

[16] Ma, T.H., Spinrad, J.P.: On the 2-chain subgraph cover and related problems. J. Algorithms 17(2), 251–268 (1994)

[17] McKee, T.A., Scheinerman, E.R.: On the chordality of a graph. J. Graph Theory 17(2), 221–232 (1993)

[18] Moscarini, M.: Doubly chordal graphs, Steiner trees, and connected domination. Networks 23(1), 59–69 (1993)

[19] Spinrad, J., Sritharan, R.: Algorithms for weakly triangulated graphs. Discrete Appl. Math. 59(2), 181–191 (1995)

[20] Szwarcfiter, J.L., Bornstein, C.F.: Clique graphs of chordal and path graphs. SIAM J. Discrete Math. 7(2), 331–336 (1994)

[21] Tarjan, R.E., Yannakakis, M.: Simple linear-time algorithms to test chordality of graphs, test acyclicity of hypergraphs, and selectively reduce acyclic hypergraphs. SIAM J. Comput. 13(3), 566–579 (1984)

[22] Woodroofe, R.: Personal Communication

[23] Yannakakis, M.: The complexity of the partial order dimension problem. SIAM J. Algebraic Discrete Methods 3(3), 351–358 (1982)

[24] Yu, C.-W., Chen, G.-H., Ma, T.-H.: On the complexity of the k-chain subgraph cover problem. Theoret. Comput. Sci. 205(1-2), 85–98 (1998)

A Simpler and More Efficient Algorithm for the Next-to-Shortest Path Problem

Bang Ye Wu

National Chung Cheng University, ChiaYi, Taiwan 621, R.O.C.
bangye@cs.ccu.edu.tw

Abstract. Given an undirected graph $G = (V, E)$ with positive edge weights and two vertices s and t, the next-to-shortest path problem is to find an st-path which length is minimum among all st-paths of lengths strictly larger than the shortest path length. In this paper we give an $O(|V| \log |V| + |E|)$ time algorithm for this problem, which improves the previous result of $O(|V|^2)$ time for sparse graphs.

Keywords: algorithm, shortest path, time complexity, next-to-shortest path.

1 Introduction

Let $G = (V, E, w)$ be an undirected graph, in which w is a positive edge weight. For $s, t \in V$, an st-path is a simple path from s to t, in which "simple" means there is no repeated vertex in the path. In this paper, a path always means a simple path. The length of a path is the total weight of all edges in the path. An st-path is a shortest st-path if its length is minimum among all possible st-paths. In a positive weight graph, a shortest path is always simple. The shortest path length from s to t is denoted by $d(s, t)$ which is the length of their shortest path. A *next-to-shortest* st-path is an st-path which length is minimum among those the path lengths *strictly larger* than $d(s, t)$. And the next-to-shortest path problem is to find a next-to-shortest st-path for given G, s and t.

   While the shortest path problem has been widely studied and efficient algorithms have been proposed, the next-to-shortest path problem attracts researchers just in the last decade. The problem was first studied by Lalgudi and Papaefthymiou in the directed version with no restriction to positive edge weight [6]. They showed that the problem is intractable for path and can be efficiently solved for walk (allowing repeated vertices). Algorithms for the problem on special graphs were also studied [2,8]. The first polynomial algorithm for undirected positive version, i.e., the next-to-shortest path defined in this paper, was developed by Krasikov and Noble, and their algorithm takes $O(n^3m)$ time [5], in which n and m are the number of vertices and edges, respectively. The time complexity was then reduced to $O(n^3)$ by Li et al. [7]. Recently, Kao et al. further improved the time complexity to $O(n^2)$ [9]. In this paper, we give an algorithm with time complexity $O(n \log n + m)$, which is the same as the currently best

W. Wu and O. Daescu (Eds.): COCOA 2010, Part II, LNCS 6509, pp. 219–227, 2010.

algorithm for shortest path problem (when there is no other assumption to the input graph). Since by a simple reduction we can show that the next-to-shortest path problem is at least as hard as the shortest path problem, the time complexity of our algorithm is near optimal. Furthermore, our algorithm is simpler than the one in [9].

Let D be the union of all shortest st-paths. For convenience let D^+ be the digraph obtained from D by orientating all edge toward t. Apparently s and t are in $V(D^+)$ and, for any $x, y \in V(D^+)$, any (directed) xy-path in D^+ is a shortest xy-path (undirected) in G. Since the optimal path either contains an edge in $E - E(D)$ or not, we divide the problem into two subproblems, and the better of the solutions of the two subproblems is the optimal path. The first subproblem looks for a shortest path using at least one edge not in $E(D)$, which is named as the case of *outward path* in [9]; and the second subproblem looks for a shortest path consisting of only edges in $E(D)$ and with length larger than $d(s, t)$, which is named as the case of *backward path* in the previous paper. Since any st-path in D^+ has length $d(s, t)$, the optimal path of the second subproblem uses at least one edge with reverse direction. By the optimality the following two observations can be easily shown.

Claim. The optimal path of the first subproblem contains exactly one subpath in G^-, in which $G^- = (V, E - E(D), w)$.

Claim. The optimal path of the second subproblem contains exactly one subpath in reverse direction.

In the second subproblem we treat an undirected path as a directed one for the convenience of explanation. The reason for the two observations is the same: If there are two non-consecutive such subpaths, we can replace one of them with a subpath in D^+ to obtain a better one. The two observations were also shown in the previous paper and they are the basis of the algorithms in the previous and this papers. Following the previous names, we name the optimal paths of the first and the second subproblems as "optimal outward path" (optimal path with a outward subpath) and "optimal backward path" (optimal path with a backward subpath), respectively. Due to [9], the first subproblem can be solved in $O(m + n \log n)$ time. But, for the second subproblem, they only gave an $O(n^2)$ time algorithm. The contribution of this paper is an $O(m + n \log n)$ time algorithm for the second subproblem, which also reduces the total time complexity of the whole algorithm.

In the remaining paragraphs, we consider the shortest path containing a backward subpath. In this case, only vertices and edges in D^+ need considering, and any vertex is assumed in D^+. Since the numbers of vertices and edges of D^+ are bounded by n and m respectively, we shall represent the time complexities as functions of n and m. We shall first derive some properties and the objective function in Section 2, and then show the algorithm in Section 3. The correctness and time complexity are given in Sections 4, and concluding remarks are given in Section 5.

2   The Objective Function and the Constraints

Let $d_H(x, y)$ denote the shortest path length from x to y in a graph H. For convenience, $d(x, y) = d_G(x, y)$. Let P_2^* denote an optimal backward path. By the claim in the introduction, P_2^* has the form $Q_1 \circ Q_2^{-1} \circ Q_3$, in which "$\circ$" means concatenation, Q_i are paths in D^+ and Q_2^{-1} means the reverse path of Q_2. Since the optimal path is required to be simple, the three subpaths must be simple and internally disjoint, in which two paths are internally disjoint if they have no common vertex except for their endpoints. Therefore our goal is to find $x, y \in V(D)$ minimizing

$$d(s, x) + d(x, y) + d(y, t) \tag{1}$$

subject to that there are an sx-path, a yx-path and a yt-path in D^+ which are mutually internally disjoint. If x and y satisfy the constraint, we say (x, y) is valid. Since D^+ is a *directed acyclic graph* (DAG), we can use the terms such as parent, child, ancestor and descendant as in a rooted tree. Also, for convenience, we abuse notation of the distances in D and in D^+, as long as an endpoint is an ancestor of the other endpoint, i.e., $d_{D^+}(x, y) = d(x, y)$ if x is an ancestor of y. Since all paths in D^+ are shortest, we have $d(y, t) = d(y, x) + d(x, t)$ and $d(s, x) + d(x, t) = d(s, t)$, and the objective function can be simplified to $d(s, t) + 2d(x, y)$ and also equivalent to $d(x, y)$ since $d(s, t)$ is independent on x and y.

To efficiently determine the constraint, a good idea of *immediate dominator* is used in [9]. We follow their idea. A vertex $p \in V(D^+)$ is an s-dominator of another vertex x iff all paths from s to x contain p. An s-dominator p is an s-immediate-dominator of x, denoted by $I_s(x)$, if it is the one closest to x, i.e., any other s-dominator of x is an s-dominator of p. We remind that, in D^+, s is the unique vertex of in-degree 0 and t is the unique vertex of out-degree 0. Also, for any vertex $x \in V(D^+)$, s is an ancestor of x and t is a descendant of x. Apparently any vertex has a unique s-immediate-dominator and all s-dominators, as well as the s-immediate-dominator, are ancestors of the vertex. Similarly we define the t-dominator, i.e., c is a t-dominator of x iff any shortest xt-path contains c, and $I_t(x)$ is the closest t-dominator of x. Finding the immediate dominator is one of the most fundamental problems in the area of global flow analysis and program optimization. The first algorithm for the problem was proposed in 1969, and then had been improved several times. A linear time algorithm for finding the immediate dominator for each vertex was given in [1]. For details of the problem, see [1].

We define a binary relation on $V(D^+)$: $x \prec y$ iff x is an ancestor of y. In our definition, a vertex is not an ancestor of itself. Also define $x \preceq y$ iff x is an ancestor of y or $x = y$. We derive some properties used in this paper.

Lemma 1. *For any vertices x and y in D^+ and $y \prec x$, either $I_s(x) \preceq y$ or $y \prec I_s(x)$. Similarly either $I_t(y) \preceq x$ or $x \prec I_t(y)$.*

Proof. We show the first statement and the second statement is similar. If neither of the two conditions holds, there is an sx-path passing through y and avoiding $I_s(x)$, which contradicts the definition of dominator. □

Let $d_s(v) = d(s,v)$ for any $v \in V(D^+)$. The following corollary comes from Lemma 1.

Corollary 1. *If $y \prec x$ and $d_s(I_s(x)) < d_s(y)$, then $I_s(x) \prec y$. Similarly, if $y \prec x$ and $d_s(I_t(y)) > d_s(x)$, then $x \prec I_t(y)$.*

For any vertex x, let $C(x) = \{v | I_s(x) \prec v \prec x\}$. The vertices in $C(x)$ form a closed region in the sense that no path can enter this region without passing through $I_s(x)$, and it is easy to see that, for any vertex $x \neq s$, $C(x) = \emptyset$ iff x has only one parent. The most important thing is that, as shown later, the vertices which are valid for x must be in $C(x)$.

Lemma 2. *For any $y \in C(x)$, $I_s(x) \preceq I_s(y)$.*

Proof. Since $I_s(x) \prec y$, we have $I_s(x) \preceq I_s(y)$ or $I_s(y) \preceq I_s(x)$ by Lemma 1. By the definition of $I_s(x)$, the in-neighbors of $C(x)$ are contained in $C(x) \cup \{I_s(x)\}$. Therefore it is impossible that $I_s(y) \prec I_s(x)$. □

Lemma 3. *If $y \in C(x)$, there are two internally disjoint paths from $I_s(x)$, and y respectively, to x.*

Proof. Let $p = I_s(x)$. By the definition of immediate dominator, no vertex in $C(x)$ is a px-cut and therefore there are two internally disjoint px-paths, said P_1 and P_2. If y is on one of them, we have done. Otherwise, let P_3 be any yx-path and q be the first vertex on P_3 and also in $V(P_1) \cup V(P_2)$. W.l.o.g. let $q \in V(P_1)$. Then, the subpath from y to q of P_3 concatenating the subpath from q to x of P_1 is a yx-path disjoint to P_2. □

Next we derive the objective function and its constraints.

Lemma 4. *If the pair (x,y) is valid, $y \in C(x)$ and $x \prec I_t(y)$.*

Proof. By definition, $y \prec x$. By Lemma 1, either $I_s(x) \prec y$ or $y \preceq I_s(x)$. Apparently y cannot be an s-dominator of x. If $y \preceq I_s(x)$, by the definition of immediate dominator, any sx-path and yx-path cannot be disjoint. Therefore we have $I_s(x) \prec y$. The statement $x \prec I_t(y)$ can be shown similarly. □

Define
$$g(x,y) = \begin{cases} d(y,x) & \text{if } y \in C(x) \text{ and } x \prec I_t(y) \\ \infty & \text{otherwise} \end{cases} \tag{2}$$

and let $g^*(x) = \min_y g(x,y)$.

Lemma 5. *If $g^*(x) \neq \infty$ and $y^* = \arg\min_y g(x,y)$, then (x,y^*) is valid.*

Proof. By Lemma 3, since $y^* \in C(x)$, there are two disjoint paths from $I_s(x)$, and y^* respectively, to x. Therefore we have a simple path, said Q_1, from s, passing through $I_s(x)$ to x, and then from x to y^* by backward edges. Since $x \prec I_t(y^*)$, there must be a path, said Q_2, from y^* to t and avoiding x. We shall show that Q_1 and Q_2 are disjoint, which completes the proof. Suppose to the contrary that $p \neq y^*$ is the last common vertex of Q_1 and Q_2, i.e., any other common vertex precedes p in Q_2. Since $p \in V(Q_2)$, we have $y^* \prec p$, and $p \prec x$ because $p \in V(Q_1)$. So, we have $p \in C(x)$. By the definition of Q_2 and $p \prec x \prec I_t(y^*)$, there are two disjoint paths from p to $I_t(y^*)$, and thus $x \prec I_t(p)$. Therefore $g(x, p) \neq \infty$ and $d(p, x) < d(y^*, x)$, a contradiction to the optimality of y^*. That is, Q_1 and Q_2 must be disjoint. □

By Lemmas 4 and 5, we can rewrite our objective function and constraints.

$$\text{OPT}_2 = \min_x \min_{y \in C(x)} \{d_s(x) - d_s(y) | x \prec I_t(y)\} \qquad (3)$$

Or, by Corollary 1, it can be also written as

$$\text{OPT}_2 = \min_x \min_{y \in C(x)} \{d_s(x) - d_s(y) | d_s(x) < d_s(I_t(y))\} \qquad (4)$$

The convenience of the latter form is that we can easily determine the ancestor relation by simply comparing their d_s values. The above formula provides us a way to find the optimal backward path: for each vertex x, checking each vertex $y \in C(x)$. But the naive method takes at least $\Theta(n^2)$ time in worst case since $|C(x)|$ may be up to $\Omega(n)$.

3   The Efficient Algorithm

We say "the pair (x, y) is feasible" or "y is feasible for x" if $g(x, y) \neq \infty$. We also say "x is feasible" if there exists y which is feasible for x. Note that a feasible (x, y) may be not valid. However, our algorithm find the (x, y) minimizing function g, and by Lemma 5 it must be valid.

Let $\mathcal{A}(x)$ be the set of parents of x. Our algorithm basically finds $g^*(x)$ for each x according to the following formula:

$$g^*(x) = \min_{p \in \mathcal{A}(x)} \min_{y \preceq p} g(x, y), \qquad (5)$$

and $\text{OPT}_2 = \min_x \{g^*(x)\}$. We denote by F the set of all the feasible vertices, i.e.,

$$F = \{x | g^*(x) \neq \infty\}$$

To make the algorithm efficient, we derive some properties to avoid some non-necessary search.

Lemma 6. *If $y \prec x$ and $y \in F$, then $\min_{u \prec y}\{g(x, u)\} > g^*(y)$.*

Proof. If $y \preceq I_s(x)$, $\min_{u \prec y}\{g(x, u)\} = \infty$ and the result holds since $y \in F$. We only need to consider the remaining case that $I_s(x) \prec y$. Let $u \prec y$ and $g(x, u) \neq \infty$. By definition, $u \in C(x)$ and $x \prec I_t(u)$. Since $y \in C(x)$, by Lemma 2, $I_s(x) \preceq I_s(y)$. If $I_s(y) \prec u$, u is also feasible for y and $d(u, y) < d(u, x) = d(u, y) + d(y, x)$. Otherwise $d(u, x) > d(I_s(y), y)$. Since $y \in F$, by definition $g^*(y) < d(I_s(y), y)$. □

Lemma 7. *If* $y \prec x$ *and* $y \notin F$, $\min_{u \prec y}\{g(x, u)\} = \min_{u \preceq I_s(y)}\{g(x, u)\}$.

Proof. Since $y \notin F$, for any vertex $v \in C(y)$, i.e., $I_s(y) \prec v \prec y$, we have $I_t(v) \preceq y$. Since $y \prec x$, $I_t(v) \prec x$ and v cannot be feasible for x. Therefore if $u \prec y$ and $g(x, u) \neq \infty$, we have $u \preceq I_s(y)$. □

Lemma 8. *Let* v_j *and* v_i *be two descendants of* y *and* $d_s(v_j) \leq d_s(v_i)$. *If* $g^*(v_j) \neq \infty$, $\min_{u \prec y}\{g(v_i, u)\} \geq g^*(v_j)$.

Proof. If $I_s(v_j) \prec u \prec y$ and $g(v_i, u) \neq \infty$, then $d_s(I_t(u)) > d_s(v_i) \geq d_s(v_j)$. Since u is also an ancestor of v_j, $v_j \prec I_t(u)$ by Corollary 1. Therefore $g(v_i, u) = d_s(v_i) - d_s(u) \geq d_s(v_j) - d_s(u) \geq g^*(v_j)$. For otherwise there is no such u, and we have $\min_{x \prec y}\{g(v_i, x)\} \geq d_s(v_i) - d_s(I_s(v_j)) \geq d_s(v_j) - d_s(I_s(v_j)) > g^*(v_j)$. □

Corollary 2. *Let* v_j *and* v_i *be two vertices and* $d_s(v_j) \leq d_s(v_i)$. *If* $y \in C(v_j) \cap C(v_i)$ *and* v_j *is not an ancestor of* v_i, *then* $g^*(v_j) \neq \infty$ *and* $\min_{u \prec y}\{g(v_i, u)\} \geq g^*(v_j)$.

Proof. Since $y \in C(v_j) \cap C(v_i)$, y is a common ancestor of v_j and v_i. Let y' be a lowest common ancestor of them. Since v_j is not an ancestor of v_i, we have $y' \neq v_j$ and there is a path from y' to v_i avoiding v_j. By definition $v_j \prec I_t(y')$ and $g^*(v_j) \neq \infty$. The inequality follows directly from Lemma 8. □

Our algorithm for the optimal backward path is as follows.

Algorithm Bk_N2SP
Input: The digraph D^+.
Output: The length of the optimal backward path.
1: find $d_s(v)$ for each v and label the vertices such that
     $d_s(v_i) \leq d_s(v_{i+1}) \ \forall i$;
2: find $I_s(v)$ and $I_t(v)$ for each v;
3: $Best \leftarrow \infty$;
     $color(v) \leftarrow white, \ \forall v \in V(D^+)$;
     $color(s) \leftarrow black$;
4: for $i \leftarrow 2$ to $n - 1$ do
5:      for each parent p of v_i do
6:          $y \leftarrow p$;
7:          while $I_s(v_i) \prec y$ and $g(v_i, y) = \infty$ and $color(y) = white$ do
8:              $color(y) \leftarrow black$; $y \leftarrow I_s(y)$;
9:          if $g(v_i, y) \neq \infty$ then

10: $Best \leftarrow \min\{Best, d_s(v_i) - d_s(y)\};$
              $color(v_i) \leftarrow black;\ color(y) \leftarrow black;$
11:   end for next parent;
12: end for next i;
13: output $d(s, t) + 2 \times Best$

In the algorithm, each vertex v is associated with a color, which is white initially and may be set to black as the algorithm runs. The algorithm begins with a preprocessing stage at Steps 1–3. We first find and sort the distances from s to all the other vertices. Let $V(D^+) = \{v_1, v_2 \ldots\}$, in which $d_s(v_i) \leq d_s(v_{i+1})$ for any i. Note that it is a topological order since D^+ is the union of all shortest st-paths and all edges have positive weights. Then we find the s- and t-immediate dominators for each vertex. All vertices are assigned white color except that s is colored black. The variable $Best$ is used to keep the objective value of the best solution found so far. In the main loop from Steps 4 to 12, we deal with all the vertices one by one. In the i-th iteration, we try to find any feasible y for v_i from each parent of v_i (Steps 5–11).

4   Correctness and Time Complexity

We shall show the correctness of the algorithm by examining the feasibility and the optimality.

Feasibility. The algorithm finds solutions only at Step 10. By Lemma 5, the final solution is feasible as long as its minimality can be ensured.

Optimality. Apparently neither s nor t can be a feasible vertex. By (5) what we need to show is that the solutions we skipped are really not better. By (3), we should check all $y \in C(v_i)$ for each v_i. There are two kinds of solutions skipped by the algorithm.

- **Type 1:** The first kind of possible solutions ignored by the algorithm is at Step 8, where we look for feasible solution $g(v_i, y)$ for any $y \preceq p$ but do not try all such y. Instead, we jump to $I_s(y)$ after checking y and skip the vertices in $C(y)$. If y is feasible, by Lemma 6, we do not need to check $g(v_i, u)$ for any ancestor u of y; and if y is not feasible, by Lemma 7, ignoring (v_i, u) for any $u \in C(y)$ does not affect the optimality.
- **Type 2:** The second kind of skipped solutions is due to the conditions of the while-loop at Step 7. The while-loop stops when $y \preceq I_s(v_i)$ or $g(v_i, y) \neq \infty$ or $color(y) = black$. Except for the first condition, the loop may terminate before reaching $I_s(v_i)$. For the second condition, if $g(v_i, y) \neq \infty$, the optimality is ensured by Lemma 6; and we divide the third condition $color(y) = black$ into two sub-cases according to when it turns black.

  • If y is colored black in this iteration, y must have been checked at this iteration from another parent of v_i. For the same reason of Type 1, it is also not necessary to check any $u \prec y$.

- Otherwise y is colored black before the i-th iteration, and this implies that $y \in C(v_j)$ for some $j < i$. If v_j is not an ancestor of v_i, we have that v_j must be feasible and $\min_{u \prec y}\{g(v_i, u)\} \geq g^*(v_j)$ by Corollary 2. The remaining case is that $v_j \prec v_i$ and $y \in C(v_j)$. Let $(y_1 = p, y_2, \ldots, y_k = y)$ be the sequence of vertices checked at the while-loop. Since $y_{q+1} = I_s(y_q)$ for $1 \leq q \leq k - 1$ and $I_s(v_j) \prec y_k \prec v_j$, we have that v_j does not appear in the sequence. Hence, there exists y_q which is in $C(v_j)$ and has a path to v_i avoiding v_j. Therefore v_j is feasible, and then similar to Corollary 2, it is not necessary to check any ancestor of y for v_i.

By the above explanation, we conclude the correctness of the algorithm.

Lemma 9. *Algorithm Bk_N2SP computes the optimal backward path correctly.*

Time complexity. We shall show that the time complexity of our algorithm is $O(n \log n + m)$. The algorithm assumes that D^+ is the input. It does not matter since D^+ can be constructed in $O(n \log n + m)$ time by the Dijkstra's algorithm using a Fibonacci Heap [4]. Step 1 takes $O(n \log n)$ time for sorting and Step 2 takes $O(m)$ time [1]. Step 3 takes $O(n)$ time. The inner loop (Steps 6–10) is entered $O(m)$ times since the total number of parents of all vertices is bounded by the number of edges. Since y is always an ancestor of v_i, the conditions "$I_s(v_i) \prec y$" and "$g(v_i, y) = \infty$", as well as to compute $g(v_i, y)$, can all be done in constant time by checking the d_s values as in (4). The remaining question is how many times Step 8 is executed. By the condition of the while loop, only white vertices will be colored black at Step 8, and therefore it is executed at most n times in total.

Lemma 10. *The time complexity of the algorithm Bk_N2SP is $O(m + n \log n)$.*

Summarizing this section and together with the result of the outward path in [9], we obtain the following theorem.

Theorem 1. *The next-to-shortest path problem on undirected graphs with positive edge weights can be solved in $O(n \log n + m)$ time.*

5   Concluding Remarks

It is easy to show that the next-to-shortest path problem is at least as hard as the shortest path problem. Given an instance of the shortest path problem, we add a dummy edge between s and t with sufficient small weight. Then if there is an algorithm for the next-to-shortest path problem, we can solve the shortest path problem with the same time complexity since the above reduction is linear time. Interesting future works include the directed version and the undirected case with zero or negative edge weights.

Acknowledgment

The author would like to thank Y.-L. Wang and the anonymous referees for their helpful comments. This work was supported in part by NSC 97-2221-E-194-064-MY3 and NSC 98-2221-E-194-027-MY3 from the National Science Council, Taiwan.

References

1. Alstrup, S., Harel, D., Lauridsen, P.W., Thorup, M.: Dominators in linear time. SIAM J. Comput. 28(6), 2117–2132 (1999)
2. Barman, S.C., Mondal, S., Pal, M.: An efficient algorithm to find next-to-shortest path on trapezoid graphs. Adv. Appl. Math. Anal. 2, 97–107 (2007)
3. Cormen, T.H., Leiserson, C.E., Rivest, R.L., Stein, C.: Introduction to Algorithms. MIT Press and McGraw-Hill (2001)
4. Fredman, M.L., Tarjan, R.E.: Fibonacci heaps and their uses in improved network optimization algorithms. J. ACM 34, 209–221 (1987)
5. Krasiko, I., Noble, S.D.: Finding next-to-shortest paths in a graph. Inf. Process. Lett. 92, 117–119 (2004)
6. Lalgudi, K.N., Papaefthymiou, M.C.: Computing strictly-second shortest paths. Inf. Process. Lett. 63, 177–181 (1997)
7. Li, S., Sun, G., Chen, G.: Improved algorithm for finding next-to-shortest paths. Inf. Process. Lett. 99, 192–194 (2006)
8. Mondal, S., Pal, M.: A sequential algorithm to solve next-to-shortest path problem on circular-arc graphs. J. Phys. Sci. 10, 201–217 (2006)
9. Kao, K.-H., Chang, J.-M., Wang, Y.-L., Juan, J.S.-T.: A quadratic algorithm for finding next-to-shortest paths in graphs. Algorithmica (2010) (in press), doi:10.1007/s00453-010-9402-4

Fast Edge-Searching and Related Problems

Boting Yang

Department of Computer Science,
University of Regina
boting@cs.uregina.ca

Abstract. Given a graph $G = (V, E)$ in which a fugitive hides on vertices or along edges, graph searching problems are usually to find the minimum number of searchers required to capture the fugitive. In this paper, we consider the problem of finding the minimum number of steps to capture the fugitive. We introduce the fast edge-searching problem in the edge search model, which is the problem of finding the minimum number of steps (called the fast edge-search time) to capture the fugitive. We establish relations between the fast edge-search time and the fast search number. While the family of graphs whose fast search number is at most k is not minor-closed for any positive integer $k \geq 2$, we show that the family of graphs whose fast edge-search time is at most k is minor-closed. We establish relations between the fast (edge-)searching and the node searching. These relations allow us to transform the problem of computing node search numbers to the problem of computing fast edge-search time or fast search numbers. Using these relations, we prove that the problem of deciding, given a graph G and an integer k, whether the fast (edge-)search number of G is less than or equal to k is NP-complete; and it remains NP-complete for Eulerian graphs. We also prove that the problem of determining whether the fast (edge-)search number of G is a half of the number of odd vertices in G is NP-complete; and it remains NP-complete for planar graphs with maximum degree 4. We present a linear time approximation algorithm for the fast edge-search time that always delivers solutions of at most $(1 + \frac{|V|-1}{|E|+1})$ times the optimal value. This algorithm also gives us a tight upper bound on the fast search number of the graph. We also show a lower bound on the fast search number using the minimum degree and the number of odd vertices.

1 Introduction

Given a graph in which a fugitive hides on vertices or along edges, graph searching problems are usually to find the minimum number of searchers required to capture the fugitive. The edge searching problem and the node searching problem are two major graph searching problems. The edge searching problem was introduced by Megiddo et al. [10]. They showed that determining the edge search number of a graph is NP-complete. They also gave a linear time algorithm to compute the edge search number of a tree. The node searching problem was

W. Wu and O. Daescu (Eds.): COCOA 2010, Part II, LNCS 6509, pp. 228–242, 2010.

introduced by Kirousis and Papadimitriou [8]. They showed that the node search number is equal to the pathwidth plus one and that the edge search number and node search number differ by at most one.

Let $G = (V, E)$ be a graph with vertex set V and edge set E. In the edge search model, initially, G contains no searchers but G contains one fugitive who hides on vertices or along edges. The fugitive is invisible to searchers, and he can move at a great speed at any time from one vertex to another vertex along a searcher-free path between the two vertices. There are three types of actions for searchers in each step, i.e., *placing* a searcher on a vertex, *removing* a searcher from a vertex, and *sliding* a searcher along an edge from one endpoint to the other. An edge is cleared only by a sliding action. An edge the fugitive could be on is said to be *contaminated*, and an edge the fugitive cannot be on is said to be *cleared*. A contaminated edge uv can be cleared in one of two ways by one sliding action: (1) sliding a searcher from u to v along uv while at least one searcher is located on u, and (2) sliding a searcher from u to v along uv while all edges incident on u except uv are already cleared. An *edge search strategy* in a k-step search is a sequence of k actions such that the final action leaves all edges of G cleared. The graph G is *cleared* if all of its edges are cleared. The minimum number of searchers needed to clear G in the edge search model is the *edge search number* of G, denoted by $es(G)$. In this paper, we introduce a new searching problem in the edge search model, called *fast edge-searching*, which is the problem of finding the minimum number of steps (or equivalently, actions) to capture the fugitive in the edge search model. In the fast edge-searching problem, the minimum number of steps required to clear G is the *fast edge-search time* of G, denoted by $fet(G)$, and the minimum number of searchers required so that G can be cleared in $fet(G)$ steps is the *fast edge-search number* of G, denoted by $fen(G)$. A fast edge-search strategy that uses $fsn(G)$ searchers to clear G is called an *optimal fast edge-search strategy*.

The motivation to consider the fast edge-searching problem is that, in some real-life scenarios, the cost of a searcher may be relatively low in comparison to the cost of allowing a fugitive to be free for a long period of time. For example, if a dangerous fugitive hiding along streets in an area, policemen always want to capture the fugitive as soon as possible.

The fast edge-searching problem has a strong connection with the fast searching problem, which was first introduced by Dyer et al. [6]. The fast search model has the same setting as the edge search model except that every edge is traversed exactly once by a searcher and searchers cannot be removed. The minimum number of searchers required to clear G in the fast search model is the *fast search number* of G, denoted by $fsn(G)$. A *fast search strategy* in a k-step fast search is a sequence of k actions such that the final action leaves all edges of G cleared. Notice that this definition is slightly different from the one used in [6] [1]. A fast search strategy that uses $fsn(G)$ searchers to clear G is called an *optimal fast search strategy*.

---

[1] In [6], a fast search strategy for graph G is a sequence of $|E(G)|$ sliding actions that clear G.

Note that the goal of the fast edge-searching problem is to find the minimum number of steps to capture the fugitive in the edge search model, while the goal of the fast searching problem is to find the minimum number of searchers to capture the fugitive in the fast search model.

The fast searching problem has a close relation with the graph brushing problem [9] and the balanced vertex-ordering problem [3]. For any graph, the brush number is equal to the total imbalance of an optimal vertex-ordering. For some graphs, such as trees, the fast search number is equal to the brush number. But for some other graphs, the gap between the fast search number and the brush number can be arbitrarily large. For example, for a complete graph K_n with n ($n \geq 4$) vertices, the fast search number is n, and the brush number is $n^2/4$ if n is even, and $(n^2 - 1)/4$ otherwise.

Bonato et al. [4] introduced the capture time on cop-win graphs in the Cops and Robber game. While the capture time of a cop-win graph on n vertices is bounded above by $n - 3$, half the number of vertices is sufficient for a large class of graphs including chordal graphs.

Throughout this paper, all graphs and multigraphs have no loops. We use $G = (V, E)$ to denote a graph with vertex set V and edge set E, and we also use $V(G)$ and $E(G)$ to denote the vertex set and edge set of G respectively. We use uv to denote an edge with endpoints u and v.

For a graph $G = (V, E)$, the *degree* of a vertex $v \in V$, denoted by $\deg_G(v)$, is the number of edges incident on v. A vertex is *odd* when its degree is odd. Similarly, a vertex is *even* when its degree is even. Let $V_{\text{odd}}(G)$ be the set of all odd vertices in G, and $V_{\text{even}}(G) = V \setminus V_{\text{odd}}(G)$. For a vertex $v \in V$, the set $\{u : uv \in E\}$ is the *neighborhood* of v, denoted as $N_G(v)$. In the case with no ambiguity, we use $\deg(v)$ and $N(v)$ without subscripts. Let $\delta(G) = \min\{|N(v)| : v \in V(G)\}$. For a subset $V' \subseteq V$, $G[V']$ denotes the subgraph induced by V'.

A *component* of a graph G is a maximal connected subgraph of G. A *cut-edge* or *cut-vertex* of a graph is an edge or vertex whose deletion increases the number of components. A *block* of a graph G is a maximal connected subgraph of G that has no cut-vertex. If G itself is connected and has no cut-vertex, then G is a block. It is easy to see that an edge of G is a block if and only if it is a cut-edge. If a block has at least 3 vertices, then it is 2-connected. Thus, the blocks of a graph are its isolated vertices, its cut-edges, and its maximal 2-connected subgraphs. The *block graph* of G is a graph T in which each vertex represents a block of G and two vertices are connected by an edge of T if the two corresponding blocks share a vertex of G. The block graph must be a forest. A block of G that corresponds to a vertex of degree one in T is called a *leaf block*. Note that every leaf block has exactly one cut-vertex.

2   Fast Edge-Searching vs. Fast Searching

In this section, we consider the relationship between the fast edge-searching in the edge search model and the fast searching in the fast search model.

Theorem 1. *For any graph $G = (V, E)$, $\mathrm{fet}(G) = \mathrm{fsn}(G) + |E|$.*

Proof. Note that a fast search strategy can be considered as an edge search strategy. Since a fast search strategy consists of $\mathrm{fsn}(G)$ placing actions and $|E|$ sliding actions, we have $\mathrm{fet}(G) \leq \mathrm{fsn}(G) + |E|$. Recall that the fast edge-search time is the minimum number of actions needed to clear G in the edge search model. If an edge search strategy of G containing removing actions, we can delete all removing actions and the remaining actions still form a valid edge search strategy with fewer actions. If an edge search strategy of G containing sliding actions that slide a searcher from u to v along a cleared path between them, we can replace these actions by placing a searcher on vertex v. The resulted strategy is an edge search strategy that may contain fewer actions. So, in an optimal fast edge-search strategy, removing actions are not contained, and traversing along a cleared path is also not necessary. Thus, we can always convert a fast edge-search strategy to a fast search strategy. Hence, $\mathrm{fsn}(G) \leq \mathrm{fet}(G) - |E|$.

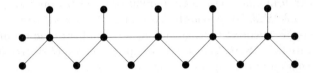

Fig. 1. A graph H with $3n + 6$ vertices and $4n + 5$ edges, where $n = 4$

Corollary 1. *For any graph G, $\mathrm{fen}(G) \leq \mathrm{fsn}(G)$.*

The difference between $\mathrm{fen}(G)$ and $\mathrm{fsn}(G)$ can be large. As illustrated in Figure 1, let H be a graph with $3n + 6$ ($n > 1$) vertices and $4n + 5$ edges. Note that the fast search number of a graph is at least half of the number of odd vertices in the graph [6]. Since H has $2n+6$ odd vertices, we know that $\mathrm{fsn}(H) \geq n+3$. In fact, we can clear H using $n+3$ searchers by a fast search strategy. But we can clear H using 2 searchers by a fast edge-search strategy. Thus the ratio $\mathrm{fsn}(H)/\mathrm{fen}(H) = (n + 3)/2$ can be arbitrarily large. We have the following relation between the fast edge-search number and the fast search number.

Theorem 2. *For any graph G, let $\widehat{G}$ be a graph obtained from G by replacing each edge of G by a path of length 2. Then, $\mathrm{fsn}(G) = \mathrm{fen}(\widehat{G})$.*

Proof. Because every optimal fast search strategy of G can be converted to a fast edge-search strategy of $\widehat{G}$, we have $\mathrm{fen}(\widehat{G}) \leq \mathrm{fsn}(G)$. We now show that $\mathrm{fsn}(G) \leq \mathrm{fen}(\widehat{G})$. Let S be an optimal fast edge-search strategy of $\widehat{G}$. For any edge $uv \in E(G)$ and its corresponding path $uu'v$ in $\widehat{G}$, the following two cases cannot happen in S: (1) if a searcher slides from u to v along $uu'v$ and then back from v to u along $vu'u$, then these 4 sliding actions can be replaced by 3 actions, that is, placing a searcher on v and sliding the searcher from v to u along $vu'u$; and (2) if a searcher slides from u to u' and another searcher slides from v to u' and then one searcher slides from u' to v and the other slides from

u' to v, then these 4 sliding actions can be replaced by 3 actions as in case (1). Because the above two cases cannot happen in S, we can easily convert S to a fast edge-search strategy of G. Thus $\mathrm{fsn}(G) \leq \mathrm{fen}(\widehat{G})$.

Corollary 2. *Let G be a graph such that for every vertex v with $\deg(v) \neq 2$, all neighbors of v have degree 2. Then, $\mathrm{fsn}(G) = \mathrm{fen}(G)$.*

From [6], we know that the family of graphs $\{G : \mathrm{fsn}(G) \leq k\}$ is not minor-closed for any positive integer $k \geq 2$. But for the fast edge-searching, we can show that the family of graphs $\{G : \mathrm{fet}(G) \leq k\}$ is minor-closed.

Theorem 3. *Given a graph G, if H is a minor of G, then $\mathrm{fet}(H) \leq \mathrm{fet}(G)$.*

3   Node Searching vs. Fast (Edge-)Searching

In this section, we establish relations between the node search number and the fast (edge-)search number. Using these relations, we can prove that both fast edge-search problem and fast search problem are NP-hard. In the node search model [8], there are only two types of actions for searchers: placing and removing. An edge is cleared if both endpoints are occupied by searchers. We use $place_X(u)$ to denote the action of placing a searcher on vertex u in the strategy X, and use $remove_X(u)$ to denote the action of removing a searcher from vertex u in the strategy X. For a graph G, the minimum number of searchers needed to clear G in the node search model is the *node search number* of G, denoted by $\mathrm{ns}(G)$. In the fast search model, we use $place_Y(u)$ to denote the action of placing a searcher on vertex u in the strategy Y, and use $slide_Y(u, v)$ to denote the action of sliding a searcher from u to v along edge uv in the strategy Y. In the case with no ambiguity, we use $place(u)$, $remove(u)$ and $slide(u, v)$ without subscripts.

For a path P of length at least 1, we know that $\mathrm{ns}(P) = 2$ and $\mathrm{fsn}(P) = \mathrm{fen}(P) = 1$. For any graph G, it is easy to see that $\mathrm{ns}(G) \leq \mathrm{fsn}(G) + 1$ and $\mathrm{ns}(G) \leq \mathrm{fen}(G) + 1$. The gap between the node search number and the fast (edge-)search number can be arbitrarily large for some graphs. For example, for a complete bipartite graph $K_{1,n}$, we have $\mathrm{fsn}(K_{1,n}) = \mathrm{fen}(K_{1,n}) = \lceil \frac{n}{2} \rceil$ whereas $\mathrm{ns}(K_{1,n}) = 2$.

From [8], we know that node search strategies can be standardized as follows.

Lemma 1. [8] *For any graph G, there always exists a monotonic node search strategy satisfying the following conditions:*

(i) *it clears G using $\mathrm{ns}(G)$ searchers;*

(ii) *every vertex is visited exactly once by one searcher;*

(iii) *every searcher is removed immediately after all the edges incident on it have been cleared (ties are broken arbitrarily); and*

(iv) *a searcher is removed from a vertex only when all the edges incident on it are cleared.*

An optimal node search strategy satisfying the properties in Lemma 1 is called a *standard node search strategy*. For a graph with n vertices, any standard node search strategy is monotonic and has $2n$ actions. It is easy to see the first action is placing and the last action is removing.

For a graph G, let G' be a graph obtained from G by adding a vertex a and connecting it to each vertex of G. Let A'_G be a multigraph obtained from G' by replacing each edge with 4 parallel edges. Let A_G be a graph obtained from A'_G by replacing each edge of A'_G with a path of length 2. In graphs G', A'_G and A_G, the vertex a is called *apex*. It is easy to see that $\mathrm{fsn}(A'_G) = \mathrm{fsn}(A_G)$.

Lemma 2. *For a complete graph K_n with $n \geq 2$, $\mathrm{fsn}(A_{K_n}) = \mathrm{fen}(A_{K_n}) = n+2$.*

We have the following relation between the node search number of G and the fast search number of A_G.

Lemma 3. *For a graph G and its corresponding graph A_G described above, $\mathrm{fsn}(A_G) \leq \mathrm{ns}(G) + 2$.*

Proof. Since $\mathrm{fsn}(A'_G) = \mathrm{fsn}(A_G)$, we will show that $\mathrm{fsn}(A'_G) \leq \mathrm{ns}(G) + 2$. Let $\mathrm{ns}(G) = k$ and $X = (X_1, \ldots, X_{2n})$ be a standard node search strategy, where n is the number of vertices in G. Each X_i is one of the two actions: placing and removing. There is no searcher on G before X_1 and X_1 is a placing-action. Let $E_i(X), 1 \leq i \leq 2n$, be the set of cleared edges just after X_i and $E_0(X)$ be the set of cleared edges just before X_1. We will show that $\mathrm{fsn}(A'_G) \leq k + 2$ by constructing a fast search strategy Y that uses $k + 2$ searchers to clear A'_G. For each action X_i, $1 \leq i \leq 2n$, we use a sequence of actions, denoted as $y(X_i)$, to simulate the action X_i. So Y is the concatenation of all $y(X_i)$ and can be expressed as $(y_0, y(X_1), \ldots, y(X_{2n}))$, where y_0 is a sequence of $k+2$ actions that place $k + 2$ searchers on the apex a and each $y(X_i)$, $1 \leq i \leq 2n$, is a sequence of sliding actions. Let $E_i(Y)$ be the set of all cleared edges by strategy Y just after $y(X_i)$ and $E_0(Y)$ be the set of cleared edges just after y_0. Note that $E_i(Y)$ is not a multiset, that is, a multiple edge pq appears in $E_i(Y)$ only when all parallel edges between p and q are cleared. Let E_a be a set of all edges incident on a.

We now construct Y from X inductively such that $E_i(X) = E_i(Y) \setminus E_a$ for each i satisfying $1 \leq i \leq 2n$. It is easy to see that $E_0(X) = E_0(Y) = \emptyset$. Initially, if the action X_1 is $place_X(u)$, then let $y(X_1) = (slide_Y(a, u))$. Thus, $E_1(X) = E_1(Y) \setminus E_a = \emptyset$.

Suppose that $E_{j-1}(X) = E_{j-1}(Y) \setminus E_a$ and the set of vertices in G occupied by searchers just after X_{j-1} is equal to the set of vertices in $A'_G - a$ occupied by searchers just after the last action of $y(X_{j-1})$. We now consider $E_j(X)$ and $E_j(Y) \setminus E_a$. There are two cases regarding the action of X_j.

CASE 1. $X_j = place_X(v)$. If $E_j(X) \setminus E_{j-1}(X) = \emptyset$, then no edge is cleared by X_j, and no recontamination happens. Thus we set $y(X_j) = (slide_Y(a, v))$. It is easy to see that $E_j(X) = E_{j-1}(X) = E_{j-1}(Y) \setminus E_a = E_j(Y) \setminus E_a$. If $E_j(X) \setminus E_{j-1}(X) \neq \emptyset$, the graph G_j formed by the edges of $E_j(X) \setminus E_{j-1}(X)$ is a star with the center v. It is easy to see that each vertex of $G_j - v$ is occupied by a searcher just before X_j. Let $V(G_j - v) = \{u_1, u_2, \ldots, u_m\}$. We can construct $y(X_j) = ((slide_Y(a, v))^2, (slide_Y(v, u_1), slide_Y(u_1, v))^2, \ldots, (slide_Y(v, u_m), slide_Y(u_m, v))^2, (slide_Y(v, a))^2)$, where $(slide_Y(a, v))^2$ means that the action $slide_Y(a, v)$ contiguously appears two times to clear two parallel edges between a and v, and $(slide_Y(v, u_1), slide_Y(u_1, v))^2$ means that a pair of actions $(slide_Y(v, u_1), slide_Y(u_1, v))$ contiguously appears two times to clear four parallel edges between v and u_1. Since X is a standard node search strategy, the apex a is occupied by at least three searchers just before two of them moves from a to v in the first two actions of $y(X_i)$. Thus $E_j(X) \setminus E_{j-1}(X) = (E_j(Y) \setminus E_a) \setminus (E_{j-1}(Y) \setminus E_a)$. It follows from the inductive hypothesis that $E_j(X) = E_j(Y) \setminus E_a$ and the set of vertices in G occupied by searchers just after X_j is equal to the set of vertices in $A'_G - a$ occupied by searchers just after the last action of $y(X_j)$.

CASE 2. $X_j = remove_X(v)$. We set $y(X_j) = (slide_Y(a, v), (slide_Y(v, a))^2)$. Since no edge can be cleared by a removing-action in X, we have $E_j(X) \setminus E_{j-1}(X) = (E_j(Y) \setminus E_a) \setminus (E_{j-1}(Y) \setminus E_a) = \emptyset$. Since X is a standard node search strategy, each edge incident on v is cleared just before X_{j-1}. Thus, just before the first action of $y(X_i)$, each edge in $A'_G - a$ incident on v is cleared and the apex a is occupied by at least two searchers and one of them moves from a to v in the first action of $y(X_i)$. From the inductive hypothesis, we know that $E_j(X) = E_j(Y) \setminus E_a$ and the set of vertices in G occupied by searchers just after X_j is equal to the set of vertices in $A'_G - a$ occupied by searchers just after the last action of $y(X_j)$.

Lemma 4. *For a graph G, let G' be a graph obtained from G by adding a vertex a and connecting it to each vertex of G. Then $ns(G') = ns(G) + 1$.*

Proof. For G', we first place one searcher on a, and then use an optimal node search strategy of G to clear G'. Thus, $ns(G') \leq ns(G) + 1$.

We now show that $ns(G) \leq ns(G') - 1$. Let S be a standard node search strategy of G'. Since a is adjacent to all other vertices in G', it follows from Lemma 1(iv) that no searcher is removed from any vertex before a searcher is placed on a and no searcher is placed on any vertex after the searcher on a is removed. Suppose that there is a moment t at which $G' - a$ contains $ns(G')$ searchers. Note that the last action before t must be placing a searcher on a vertex in $G' - a$. Since no searcher is placed on any vertex after the searcher on a is removed, it follows from Lemma 1(ii) that a has not been occupied before t. Thus, all edges incident on a are dirty. Note that the first action after t must be removing a searcher from a vertex in $G' - a$. From Lemma 1(iv), all the edges incident on this vertex are cleared. This is a contradiction. Thus, at any moment when G' contains $ns(G')$ searchers, there is a searcher on a. Let S' be a strategy obtained from S by deleting the actions $place(a)$ and $remove(a)$. Then S' is a monotonic node search strategy that can clear the graph $G' - a$ (i.e., G) using $ns(G') - 1$ searchers. Hence, $ns(G) \leq ns(G') - 1$. Therefore, $ns(G') = ns(G) + 1$.

Lemma 5. *For a graph G and its corresponding graph A_G, $ns(G) \leq fsn(A_G) - 2$.*

Proof. It follows from Lemma 4 that $ns(G) = ns(G') - 1$. Since $ns(G') = ns(A'_G)$ and $fsn(A'_G) = fsn(A_G)$, we only need to show that $ns(A'_G) \leq fsn(A'_G) - 1$.

Let $S = (S_0, s_1, \ldots, s_m)$ be an optimal fast search strategy of A'_G that clears A'_G using k searchers, where S_0 is a sequence of k placing actions. We can construct a monotonic node search strategy T by modifying S in the following way. For each action s_i ($i \geq 1$) that slides a searcher from u to v, if u is occupied by only one searcher just before sliding, then we delete this action; otherwise, we replace s_i by the actions $remove(u)$ and $place(v)$.

For any multiple edge between two vertices u and v, when a searcher slides the second time from one endpoint to the other by S, both u and v must be occupied by searchers. Thus, all four parallel edges are cleared by T. Hence, T is a monotonic node search strategy that clears A'_G using k searchers.

We now show that we can modify T to obtain a monotonic node search strategy that clears A'_G using $k - 1$ searchers. Note that some actions in T may be redundant, that is, placing a searcher on an occupied vertex. For any multiple edge uv, when a searcher slides the first time from one endpoint to the other by S, all four parallel edges are cleared by the actions $remove(u)$ and $place(v)$ in T. Thus, any moment when S requires a searcher to slide along the second parallel edge between u and v, T does not need such a searcher. Therefore, we can delete redundant actions from T to obtain a monotonic node search strategy that clears A'_G using $k - 1$ searchers.

From Lemmas 3, 5 and Corollary 2, we have the main result of this section.

Theorem 4. *For a graph G and its corresponding graph A_G, $\mathrm{ns}(G) = \mathrm{fsn}(A_G) - 2 = \mathrm{fen}(A_G) - 2$.*

For a graph G, we use $\mathrm{pw}(G)$ to denote the pathwidth of G. Since $\mathrm{pw}(G) = \mathrm{ns}(G) - 1$, we have $\mathrm{pw}(G) = \mathrm{fsn}(A_G) - 3$.

Given a graph G and an integer k, the fast search (edge-search) problem is to determine whether G can be cleared by k searchers in the fast (edge) search model. Then we have the following result[2].

Corollary 3. *The fast search problem and the fast edge-search problem are NP-complete. They remain NP-complete for Eulerian graphs.*

From Theorem 1, we can also show that, given a graph G and an integer k, it is NP-complete to determine whether $\mathrm{fet}(G) \leq k$. It remains NP-complete for Eulerian graphs.

4   An Approximation Algorithm

Since the family of graphs $\{G : \mathrm{fsn}(G) \leq k\}$ is not minor-closed for any positive integer $k \geq 2$ [6], we cannot obtain an upper bound on the fast search number using the fast search number of complete graphs. In this section, we present a linear time algorithm that can compute a fast search strategy for any connected graph $G = (V, E)$, which is also a fast edge-search strategy because any fast search strategy is also a fast edge-search strategy. We can use this algorithm

---

[2] Dereniowski et al. [5] independently proved the fast search problem is NP-complete by a "weak search" approach that is different from our method.

to show that the number of vertices in a graph is an upper bound on the fast search number of the graph. Since the fast search number of a complete graph K_n ($n \geq 4$) is n, we know this upper bound is tight. Using this algorithm, we can also compute a fast edge-search strategy of G whose length (i.e., the number of actions) is at most $(1 + \frac{|V|-1}{|E|+1})$ times the fast edge-search time of G.

If G is not connected, the fast search number of G is the sum of fast search numbers of all components. So we only consider connected graphs. The input of the algorithm is a connected graph G with at least 4 vertices. The output of the algorithm is a fast (edge-)search strategy $\langle V_p, A_s \rangle$, where V_p is a multiset of vertices on which we place searchers and A_s is a sequence of arcs corresponding to sliding actions, that is, an arc (u, v) corresponds to sliding along the arc from tail u to head v. Given two vertices u and v in G, the distance between them, denoted $\text{dist}_G(u, v)$, is the number of edges on the shortest path between them.

Algorithm FASTSEARCH(G)
Input: A connected graph $G = (V, E)$ with at least 4 vertices.
Output: A fast (edge-)search strategy $\langle V_p, A_s \rangle$ of G.

1. Compute a block graph T of G.
2. Arbitrarily pick a leaf t of T. Let B be a block of G corresponding to t and a be a vertex of B which is not a cut-vertex of G. Call FASTSEARCHBLOCK (B, a).
3. Update T by deleting the leaf t, and update G by deleting all vertices of B except the vertex that is a cut-vertex of G and is incident with a dirty edge of G. If G contains no edges, then stop and output the multiset of vertices V_p on which searchers are placed and output the sequence of arcs A_s in the order when searchers slide along them from tail to head; otherwise, go to step 2.

Algorithm FASTSEARCHBLOCK(B, a)

1. $B' \leftarrow B - a$, $H \leftarrow B'$, and $\mathcal{P} \leftarrow \emptyset$.
2. If $V_{\text{odd}}(B') = \emptyset$, then place searchers on a if necessary so that a is occupied by at least $\deg_B(a)$ searchers. Slide searchers from a to every vertex in $N_B(a)$ to clear a. If $V(B')$ contains only one vertex, then return to FASTSEARCH; otherwise, place a searcher on each unoccupied vertex in $V(B')$. If there is a vertex occupied by at least two searchers, then slide one of them along all edges of B'; otherwise, place a searcher on an arbitrary vertex of B' and slide it along all edges of B'. Return to FASTSEARCH.
3. Arbitrarily pick a vertex $u \in V_{\text{odd}}(H)$ and find a vertex $v \in V_{\text{odd}}(H)$ such that $\text{dist}_H(u, v) = \min\{\text{dist}_H(u, w) : w \in V_{\text{odd}}(H) \text{ and } w \neq u\}$. Let P_{uv} be the shortest path between u and v. Update $\mathcal{P} \leftarrow \mathcal{P} \cup \{P_{uv}\}$ and $H \leftarrow H - E(P_{uv})$. If $V_{\text{odd}}(H) \neq \emptyset$, repeat Step 3.
4. If H has only one component, then place searchers on a if necessary so that a is occupied by at least $|V(H) \cap N_B(a)|$ searchers. Slide $|V(H) \cap N_B(a)|$ searchers from a to every vertex in $V(H) \cap N_B(a)$. Place a searcher on each

unoccupied vertex of H. If a vertex of H is occupied by at least two searchers, then slide one of them along all edges of H; otherwise, place a searcher on a vertex of H and slide it along all edges of H to clear H. For each path in $\mathcal{P}$, we slide the searcher from one end of the path to the other. Return to FastSearch.

5. Let h be the number of components in H. Construct a graph H' such that each vertex v of $V(H')$ represents a component H_v of H and two vertices u and v are connected by an edge of H' if there is a path in $\mathcal{P}$ which contains a vertex of the component H_u corresponding to u and contains a vertex of the component H_v corresponding to v, and the subpath between u and v does not contain any vertex of other components (different from H_u and H_v). Assign a direction to each path in $\mathcal{P}$ such that each path in $\mathcal{P}$ becomes a directed path and H' becomes an acyclic graph. Let $H_1, H_2, \ldots, H_h$ be a sequence of all components in H such that the corresponding sequence of all vertices of H' forms an acyclic ordering. Set $i \leftarrow 1$.

6. If $V(H_i) \cap N_B(a) \neq \emptyset$, then go to Step 9.

7. If H_i contains a single vertex, then slide all searchers on this vertex along untraversed edges to the other endpoints complying with the direction of edges. $i \leftarrow i + 1$ and go to Step 6.

8. Place a searcher on each unoccupied vertex of H_i. If a vertex of H_i is occupied by at least two searchers, then slide one of them along all edges of H_i; otherwise, place a searcher on a vertex of H_i and slide it along all edges of H_i to clear H_i. Go to Step 11.

9. If H_i contains more than one vertex, then go to Step 10. Let x be the unique vertex in H_i. Place searchers on a if it is occupied by less than two searchers so that a is occupied by two searchers. Slide a searcher from a to x. Slide all searchers on x along untraversed edges to the other endpoints complying with the direction of edges. If $i = h$, then return to FastSearch; otherwise, $i \leftarrow i + 1$ and go to Step 6.

10. If $i < h$ and a is occupied by less than $|V(H_i) \cap N_B(a)| + 1$ searchers, then place searchers on a so that a is occupied by $|V(H_i) \cap N_B(a)| + 1$ searchers. If $i = h$ and a is occupied by less than $|V(H_i) \cap N_B(a)|$ searchers, then place searchers on a so that a is occupied by $|V(H_i) \cap N_B(a)|$ searchers. Slide $|V(H_i) \cap N_B(a)|$ searchers from a to every vertex in $V(H_i) \cap N_B(a)$. Place a searcher on each unoccupied vertex of H_i. If a vertex of H_i is occupied by at least two searchers, then slide one of them along all edges of H_i; otherwise, place a searcher on a vertex of H_i and slide it along all edges of H_i.

11. For each pair of vertices $u, v \in V(H_i)$ satisfying that the shortest path P_{uv} between them is a subpath of a path in $\mathcal{P}$, we slide the searcher from one end of P_{uv} to the other complying with the direction of edges. If $i = h$, then return to FastSearch; otherwise, $i \leftarrow i + 1$ and go to Step 6.

Theorem 5. *For any connected graph $G = (V, E)$, Algorithm* FastSearch(G) *outputs a fast search strategy that clears G using at most $|V|$ searchers in the fast search model.*

Proof. In FastSearch(G), we first decompose G into blocks. Then we choose a leaf block B, clear B and leave one searcher on the vertex of B which is a cut-vertex of G. Since the block graph of G is a tree, we can repeat this process until G is cleared. If each leaf block B can be cleared using at most $|V(B)|$ searchers, then G can be cleared using $|V|$ searchers.

We now consider how to clear a leaf block B using at most $|V(B)|$ searchers such that the cut-vertex of G in B is occupied by at least one searcher when B is cleared. Note that B has only one cut-vertex of the current G since B is a leaf block in G. If B is an edge uv, where u is a leaf of the current G then uv can be cleared by sliding a searcher from u to v. Thus, B (i.e., uv) can be cleared using one searcher such that the cut-vertex v of G in B is occupied by one searcher when B is cleared.

Suppose that B contains at least three vertices. Pick a vertex a of B which is not a cut-vertex of the current G. Let $B' = B - a$. Then B' is connected since B is a block. We have two cases on $V_{\mathrm{odd}}(B')$.

Case 1. $V_{\mathrm{odd}}(B') = \emptyset$. Then B' is an Eulerian graph. Clear a by sliding searchers from a to every vertex in $N_B(a)$. Place a searcher on each unoccupied vertex in $V(B')$. If there is a vertex occupied by at least two searchers, then slide one of them along all edges of B', and thus the total number of searchers used to clear B is at most $|V(B)| - 1$; otherwise place an additional searcher on an arbitrary vertex of B' and slide it along all edges of B'. Thus, the total number of searchers used to clear B is at most $|V(B)|$.

Case 2. $V_{\mathrm{odd}}(B') \neq \emptyset$. Note that every graph has even number of odd vertices. Let u and v be two vertices in $V_{\mathrm{odd}}(B')$ and P_{uv} be the shortest path between them. Since v is the closest vertex to u in $V_{\mathrm{odd}}(B')$, we know that $V(P_{uv}) \cap V_{\mathrm{odd}}(B') = \{u, v\}$. Let B'' be the graph obtained from B' by deleting all edge of P_{uv}. Note that both u and v have even degree in B''. Thus, $|V_{\mathrm{odd}}(B'')| = |V_{\mathrm{odd}}(B')| - 2$. We can repeat the above process until we obtain an even graph H and the set of all deleted shortest paths $\mathcal{P}$. If H has only one component, similar to Case 1, we can clear H using at most $|V(H)|$ searchers. Since all end vertices of paths in $\mathcal{P}$ are different, we can clear each path of $\mathcal{P}$ by sliding a searcher from one endpoint to the other.

Suppose that H contains at least two components, i.e., $h \geq 2$. Since B' is connected, each component H_i, $1 \leq i \leq h$, in H must contain at least one vertex of a path in $\mathcal{P}$. We clear each $H_1, H_2, \ldots, H_h$ in the acyclic ordering of H'. If H_i contains a single vertex v, then v cannot be a leaf of B because B is a block containing at least 3 vertices. Note that v becomes a single vertex in H_i because we delete all edges of $\mathcal{P}$ from H. Thus at least one path in $\mathcal{P}$ contains v as an interior vertex, and furthermore, v cannot be the end vertex of a path in $\mathcal{P}$ because no path in $\mathcal{P}$ contains an odd vertex of H as an interior vertex. If $v \in N_B(a)$, then place searchers on a if necessary so that a is occupied by two searchers, and slide one searcher from a to v. Because the number of in-edges of v is equal to the number of out-edges of v and we clear $H_1, \ldots, H_h$ in the acyclic ordering of H', we can slide searchers from v along all untraversed edges to the other endpoints complying with the edge directions. Suppose that H_i

contains at least two vertices. If $V(H_i) \cap N_B(a) \neq \emptyset$, then place searchers on a if necessary so that we can slide $|V(H_i) \cap N_B(a)|$ searchers from a to every vertex in $V(H_i) \cap N_B(a)$. We have two subcases.

CASE 2.1. $i < h$. In this case, we place a searcher on each unoccupied vertex of H_i, and place another searcher on a vertex of H_i and slide it along all edges of H_i to clear H_i. Since $i < h$, we have enough searchers to clear H_i and leave at most $|V(H_i)|$ searchers on vertices $V(H_i)$ when B is cleared.

CASE 2.2. $i = h$. Since $h > 1$, there is a vertex u of H_h that is an end vertex of a path in $\mathcal{P}$ and is occupied before we place searchers on H_h. We place a searcher on each unoccupied vertex of H_h, and place another searcher on the vertex u and slide it along all edges of H_h to clear H_h. Thus, we can use $|V(H_h)| + 1$ searchers to clear H_h and at least one searcher comes from another component.

For each pair of vertices $u, v \in V(H_i)$ satisfying that the shortest path P_{uv} between them is a subpath of a path in $\mathcal{P}$, we slide the searcher from one end of P_{uv} to the other complying with the edge directions. We clear $B'[V(H_i)]$ that is a subgraph of B' induced from $V(H_i)$ using at most $|V(H_i)|$ searchers.

From cases 2.1 and 2.2, B can be cleared using at most $|V(B)|$ searchers. Therefore, it follows from cases 1 and 2 that $\mathrm{fsn}(G) \leq |V|$.

Theorem 6. *Algorithm* FASTSEARCH*(G) can be implemented with linear time.*

For any connected graph G, since each placing-action places a new searcher in the fast search model, $|V_p|$ is the number of searchers required by FASTSEARCH(G). Then G can be cleared in $|V_p| + |A_s|$ steps by FASTSEARCH(G). Since any fast search strategy is also an edge search strategy, G can be cleared in at most $|V_p| + |A_s|$ steps in the edge search model. Thus, FASTSEARCH(G) is also an approximation algorithm for the fast edge-search time with the following approximation ratio.

Theorem 7. *For any connected graph G with n vertices and m edges,*

$$\frac{|V_p| + |A_s|}{\mathrm{fet}(G)} \leq (1 + \frac{n-1}{m+1}).$$

For odd graphs, the approximation ratio for the fast search number is 2, and for fast edge-search time is $1 + \frac{n}{n+2m}$.

Corollary 4. *For any connected odd graph G with n vertices and m edges,*

$$\frac{|V_p|}{\mathrm{fsn}(G)} \leq 2 \quad and \quad \frac{|V_p| + |A_s|}{\mathrm{fet}(G)} \leq (1 + \frac{n}{n + 2m}).$$

5   A Lower Bound

In this section, we give a new lower bound that is related to both the number of odd vertices and the minimum degree.

Theorem 8. *For a connected graph G with $\delta(G) \geq 3$,*

$$\text{fsn}(G) \geq \max\{\delta(G) + 1, \lceil \frac{\delta(G) + |V_{\text{odd}}(G)| - 1}{2} \rceil\}.$$

Proof. Note that $\text{fsn}(G) \geq \text{es}(G)$ for any graph G. From Theorem 2.4 in [2], we know that $\text{es}(G) \geq \delta(G) + 1$ for any connected graph G with $\delta(G) \geq 3$. If $|V_{\text{odd}}(G)| \leq \delta(G) + 3$, then $\text{fsn}(G) \geq \delta(G) + 1 \geq \lceil \frac{\delta(G) + |V_{\text{odd}}(G)| - 1}{2} \rceil$, which completes the proof.

Suppose that $|V_{\text{odd}}(G)| > \delta(G) + 3$. Let S be an optimal fast search strategy of G such that searchers are placed on vertices only when it is necessary. Let v be the first vertex cleared by S. When v is cleared, each vertex in $N(v)$ must contain at least one searcher. Let $V' \subseteq V(G)$ be the set of occupied vertices just after v is cleared and k be the total number of searchers on V'. If v is occupied by searchers after it is cleared, then these searchers will stay on v until the end of the search. For each vertex $u \in V' \setminus \{v\}$, if $\deg(u)$ is even, then each searcher on u maybe move to an odd vertex in the rest of the searching process; if $\deg(u)$ is odd, either a searcher was placed on u, or a searcher slid to u and this searcher will stay on u until the end of the search. Thus, just after v is cleared, we need at least $\frac{1}{2}\max\{(|V_{\text{odd}}(G) \setminus \{v\}| - k), 0\}$ additional searchers to clear G. Notices that $N(v) \subseteq V'$ and $|V'| \leq k$. Therefore, $\text{fsn}(G) \geq k + \frac{1}{2}\max\{(|V_{\text{odd}}(G) \setminus \{v\}| - k), 0\}$ $\geq \frac{1}{2}\max\{(|V_{\text{odd}}(G) \setminus \{v\}| + k), 0\} \geq \lceil \frac{\delta(G) + |V_{\text{odd}}(G)| - 1}{2} \rceil$

From Theorem 8, we can improve the approximation ratio of FASTSEARCH(G).

Theorem 9. *If G is connected graph with n vertices and m edges, and $\delta(G) \geq 3$, then*

$$\frac{|V_p| + |A_s|}{\text{fet}(G)} \leq (1 + \frac{n - \delta(G) - 1}{m + \delta(G) + 1}).$$

6 Planar Graphs

Let $F = \{f_1, f_2, \ldots, f_k\}$ be a family of plane curves satisfying the following conditions: (1) each f_i is the graph of a continues function of time with domain $[s_i, t_i]$, $-\infty < s_i < t_i < +\infty$, (2) any pair of curves do not share an endpoint, and (3) each pair of curves have a finite number of intersection points. From condition (2) we know that at each intersection point, at most one curve starts from or ends on this intersection point. From condition (3) we know that no pair of curves overlap over any period of time.

A graph of F, denoted by $G_F = (V_F, E_F)$, is the graph formed from F such that V_F is the set of all endpoints and intersection points of curves in F and $E_F = \{f : f$ is a subcurve of a curve in F whose endpoints belong to V_F and no interior point of f belongs to $V_F\}$. Note that the definition of the edge set E_F can be easily converted to the traditional definition, that is, a set of pairs of vertices.

Theorem 10. *Let $F = \{f_1, f_2, \ldots, f_k\}$ be a set of plane curves satisfying the above three conditions, and let $G_F = (V_F, E_F)$ be the graph of F. Then $\mathrm{fsn}(G_F) = k$ and $\mathrm{fet}(G_F) = k + |E_F|$.*

From [11], we know that the fast search number of cubic graphs can be found in $O(n^2)$ time. Similar to [7], we now show that the fast search problem is NP-complete for planar graphs with maximum degree 4. We first show a property of variable gadgets as follows.

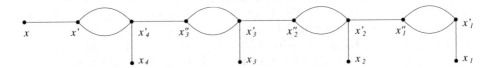

Fig. 2. A variable gadget G_x^k with $k = 4$

Lemma 6. *Let G_x^k be a multigraph as illustrated in Figure 2. For any optimal fast search strategy of G_x^k, if a searcher slides from x to its neighbor x', then for each leaf x_i $(1 \le i \le k)$ there is a searcher sliding to x_i from its neighbor x_i'; and if a searcher slides to x from its neighbor x', then for each leaf x_i $(1 \le i \le k)$ there is a searcher sliding from x_i to its neighbor x_i'.*

We can use the graph G_x^k in Lemma 6 as a variable gadget to show the NP-completeness for planar graphs with maximum degree 4. The reduction is the same as the one used in Theorem 1 of [7].

Theorem 11. *Given a planar graph G with maximum degree 4, the problem of determining whether $\mathrm{fsn}(G) = |V_{\mathrm{odd}}(G)|/2$ is NP-complete.*

From corollary 2, we can show that, given a planar graph G with maximum degree 4, it is NP-complete to determine whether $\mathrm{fen}(G) = |V_{\mathrm{odd}}(G)|/2$.

Corollary 5. *Given a planar graph G with maximum degree 4, the problem of determining whether $\mathrm{fet}(G) = \frac{1}{2}|V_{\mathrm{odd}}(G)| + |E(G)|$ is NP-complete.*

7   Conclusions

Many graph searching problems have been introduced. Almost all of these problems only consider the minimum number of searchers required to capture the fugitive. In this paper, we consider the minimum number of steps to capture the fugitive. We introduce the fast edge-searching problem in the edge search model. We establish relations between the fast edge-search time and the fast search number. We also establish relations between the fast (edge-)searching and the node searching. By these relations, the problem of computing the fast search number, edge search number, node search number, or pathwidth of a graph is equivalent to that of computing the fast edge-search time of a related graph. This makes the fast edge-searching is more versatile than others. We can use the fast edge-searching to investigate either how to draw a graph "evenly" (an extended version of the balanced vertex-ordering), or how to decompose a graph into a "path" (i.e., pathwidth, which is related to many graph parameters). We show that the family of graphs whose fast edge-search time is at most k is minor-closed. This makes arguments for upper bounds and lower bounds of the fast edge-search time less complicated, comparing with the fast search number. We prove NP-completeness results for computing the fast (edge-)search number, and the fast edge-search time, respectively. We also prove that the problem of determining whether $\text{fsn}(G) = |V_{\text{odd}}(G)|/2$ or $\text{fet}(G) = \frac{1}{2}|V_{\text{odd}}(G)| + |E(G)|$ is NP-complete; and it remains NP-complete for planar graphs with maximum degree 4. For connected graphs with $\delta(G) \geq 3$, we present a linear time approximation algorithm for the fast edge-search time that can give solutions of at most $(1 + \frac{|V| - \delta(G) - 1}{|E| + \delta(G) + 1})$ times the optimal value. This algorithm also gives us a tight upper bound on the fast search number of graphs.

References

1. Alon, N., Pralat, P., Wormald, R.: Cleaning regular graphs with brushes. SIAM Journal on Discrete Mathematics 23, 233–250 (2008)
2. Alspach, B., Dyer, D., Hanson, D., Yang, B.: Lower bounds on edge searching. In: Chen, B., Paterson, M., Zhang, G. (eds.) ESCAPE 2007. LNCS, vol. 4614, pp. 516–527. Springer, Heidelberg (2007)
3. Biedl, T., Chan, T., Ganjali, Y., Hajiaghayi, M.T., Wood, D.: Balanced vertex-ordering of graphs. Discrete Applied Mathematics 148, 27–48 (2005)
4. Bonato, A., Golovach, P., Hahn, G., Kratochvíl, J.: The capture time of a graph. Discrete Mathematics 309, 5588–5595 (2009)
5. Dereniowski, D., Diner, Ö., Dyer, D.: Three-fast-searchable graphs (manuscript)
6. Dyer, D., Yang, B., Yaşar, Ö.: On the fast searching problem. In: Fleischer, R., Xu, J. (eds.) AAIM 2008. LNCS, vol. 5034, pp. 143–154. Springer, Heidelberg (2008)
7. Kára, J., Kratochvíl, J., Wood, D.: On the complexity of the balanced vertex ordering problem. Discrete Mathematics and Theoretical Computer Science 9, 193–202 (2007)
8. Kirousis, L., Papadimitriou, C.: Searching and pebbling. Theoretical Computer Science 47, 205–218 (1986)
9. Messinger, M.E., Nowakowski, R.J., Pralat, P.: Cleaning a network with brushes. Theoretical Computer Science 399, 191–205 (2008)
10. Megiddo, N., Hakimi, S., Garey, M., Johnson, D., Papadimitriou, C.: The complexity of searching a graph. Journal of ACM 35, 18–44 (1988)
11. Stanley, D., Yang, B.: Lower bounds on fast searching and their applications. In: Dong, Y., Du, D.-Z., Ibarra, O. (eds.) ISAAC 2009. LNCS, vol. 5878, pp. 964–973. Springer, Heidelberg (2009)

Diameter-Constrained Steiner Tree

Wei Ding[1], Guohui Lin[2], and Guoliang Xue[3]

[1] Zhejiang Water Conservancy and Hydropower College
Hangzhou, Zhejiang 310000, China
dingweicumt@163.com
[2] Department of Computing Science, University of Alberta
Edmonton, Alberta T6G 2E8, Canada
ghlin@cs.ualberta.ca
[3] Department of Computer Science and Engineering, Arizona State University
Tempe, AZ 85287-8809, USA
xue@asu.edu

Abstract. Given an edge-weighted undirected graph $G = (V, E, c, w)$, where each edge $e \in E$ has a cost $c(e)$ and a weight $w(e)$, a set $S \subseteq V$ of terminals and a positive constant $\mathbf{D}_0$, we seek a minimum cost Steiner tree where all terminals appear as leaves and its diameter is bounded by $\mathbf{D}_0$. Note that the diameter of a tree represents the maximum weight of path connecting two different leaves in the tree. Such problem is called the minimum cost diameter-constrained Steiner tree problem. This problem is NP-hard even when the topology of Steiner tree is fixed. In present paper we focus on this restricted version and present a fully polynomial time approximation scheme (FPTAS) for computing a minimum cost diameter-constrained Steiner tree under a fixed topology.

Keywords: Diameter-constrained Steiner tree, fully polynomial time approximation scheme, fixed topology.

1 Introduction

The *Steiner minimum tree* (SMT) problem in graphs asks for a minimum length connected subgraph of the given graph that spans all given *terminals*. This problem has been widely applied in many fields [4,9], such as communication networks, computational biology. It is NP-hard in the strong sense [6]. However many approximation algorithms with a constant performance ratio have been proposed, see [4,12,19].

With the development of applications of SMT, its many variants have arisen rapidly and been studied extensively. In this paper, we are interested in a variant of SMT which requires all terminal to be its leaves, called *terminal Steiner tree* (TeST), which is proposed by Lin and Xue in [11], also see [3,5].

Let $G = (V, E, c, w)$ be an undirected edge-weighted graph, where each edge $e \in E$ is associated with a cost $c(e)$ and another weight $w(e)$. Let $T = (U, F, c, w)$ be a TeST, where $U \subseteq V$ and $F \subseteq E$. The *cost* of T is the sum of all costs over its all edges, denoted by $c(T)$, i.e. $c(T) = \sum_{e \in F} c(e)$. For any two vertices $u, v \in U, u \neq v$, let $\pi[u, v]$ denote the unique $u-v$ path of T. Obviously $\pi[u, v] = \pi[v, u]$.

W. Wu and O. Daescu (Eds.): COCOA 2010, Part II, LNCS 6509, pp. 243–253, 2010.

We use $w(\pi[u,v])$ to denote the weight of $\pi[u,v]$ and similarly have $w(\pi[u,v]) = \sum_{e \in \pi[u,v]} w(e)$. Specially we call $\pi[u,v]$ a *leaf-path* of T if $u,v \in S, u \neq v$. The maximum weight of leaf-path of T is called the *diameter* of T, denoted by $w(T)$, that is,

$$w(T) = \max \left\{ \sum_{e \in \pi[u,v]} w(e) : u,v \in S, u \neq v \right\} . \tag{1}$$

An extension of TeST, called *diameter-constrained Steiner tree* (DCST), is a TeST subject to the constraint that $w(T) \leq \mathbf{D}_0$. The problem of computing a minimum cost diameter-constrained Steiner tree is NP-hard (since the terminal Steiner tree problem, which is NP-hard [11], can be viewed as its a special case of $\mathbf{D}_0 = \infty$). Diameter-constrained minimum spanning tree and Steiner tree have been studied [1,7,13]. Note that the diameter of tree in their studies is the maximum hop count between two different vertices instead of the maximum weight connecting them.

Now we concern with a variant of TeST, called the *realization of objective tree* (ROT), which is a tree in G with a same topology as a sample TeST, spanning all terminals, see Fig. 1 and Sect. 3 for more details. Let $\mathcal{T} = (\mathcal{V}, \mathcal{E})$ be a sample TeST. Its a realization is called $\mathcal{T}$–ROT. This variant has many applications in telecommunication, distributed computing, etc. There have been few related studies. Wang and Jia [16] studied the minimum cost delay-constrained Steiner tree with a fixed topology and proposed a pseudo-polynomial-time algorithm. Xue and Xiao [17] studied the minimum cost delay-constrained multicast tree under a Steiner topology and presented a fully polynomial time approximation scheme (FPTAS).

In present paper we study the following problem: Given an undirected edge-weighted complete graph $G = (V, E, c, w)$, a subset $S \subseteq V$ of terminals, a non-negative constant $\mathbf{D}_0$, a sample TeST $\mathcal{T} = (\mathcal{V}, \mathcal{E})$, we seek for a minimum cost $\mathcal{T}$–ROT as $T = (U, F, c, w)$ in G subject to the constraint of $w(T) \leq \mathbf{D}_0$. This problem is called the *minimum cost diameter-constrained realization of objective tree problem* (MCDCRP) (formally defined in Sect. 2), which can be applied in the design of quality of service (QoS) telecommunication and distributed computing. It is fortunate that MCDCRP is easier than the optimization problem of DCST. We will present an FPTAS for MCDCRP in the following.

The rest of this paper is organized as follows. In Sect. 2, we formally define MCDCRP as well as its decision version and demonstrate its intractability. In Sect. 3, we complete some fundamental preliminaries. Next, we present a pseudo-polynomial-time algorithm for MDCCRP in Sect. 4, then present an FPTAS for MCDCRP in Sect. 5. Finally, we conclude this paper with some future research topics in Sect. 6.

2   The Problems and Intractability

In this section, we formally define the *minimum cost diameter-constrained real-ization of objective tree problem* as well as its decision version, and prove it to be

NP-hard. Besides, we define the *minimum diameter cost-constrained realization of objective tree problem.*

For ease of presentation, we employ an INPUT as follows: an undirected edge-weighted complete graph $G = (V, E, c, w)$ where each edge $e \in E$ is associated with an integer cost $c(e) \geq 0$ and a real-number weight $w(e) \geq 0$, a subset $S \subseteq V$ of terminals, a sample TeST $\mathcal{T} = (\mathcal{V}, \mathcal{E})$.

Problem 1. Given an INPUT and a real number $\mathbf{D}_0 \geq 0$, the *minimum cost diameter-constrained realization of objective tree problem* (MCDCRP) asks for a minimum cost $\mathcal{T}$–ROT as $T = (U, F, c, w)$ in G subject to the constraint that $w(T) \leq \mathbf{D}_0$.

The decision version of MCDCRP is formally defined in the following.

Problem 2. Given an INPUT, a real number $\mathbf{D}_0 \geq 0$, and a constant $\mathcal{B} \geq 0$, is there a $\mathcal{T}$–ROT as $T = (U, F, c, w)$ in G subject to the constraint that $w(T) \leq \mathbf{D}_0$ such that $c(T) \leq \mathcal{B}$?

Theorem 1. *Problem 2 is NP-hard.*

Proof. Let us consider a special case of Problem 2 in which there are only two terminals and $\mathcal{T}$ has $|V| - 2$ nonleaves and $\mathbf{D}_0 = \sum_{e \in E} w(e)$. There is a trivial polynomial-time reduction from the Hamilton Path problem, which is known to be NP-hard [6], to such a special case. □

Problem 3. Given an INPUT and an integer $\zeta \geq 0$, the *minimum diameter cost-constrained realization of objective tree problem* (MDCCRP) asks for a minimum diameter $\mathcal{T}$–ROT as $T = (U, F, c, w)$ in G subject to the constraint that $c(T) \leq \zeta$.

3 Preliminaries

Given an INPUT, then $V \setminus S$ is a set of *nonterminals* in G and $\mathcal{V} \setminus S$ is a set of *nonleaves* of $\mathcal{T}$. Note that the objctive tree discussed in the rest of this paper is a rooted version since an unrooted tree can be transformed into a rooted tree by assigning any nonleaf as its root (let ν be the root vertex of $\mathcal{T}$). This rooted tree is either a *binary tree* or a *general tree*. In this section, we will discuss these two cases respectively. Besides, we use $\mathcal{T}(\alpha)$ to denote the *subtree* of $\mathcal{T}$ rooted at α for any vertex $\alpha \in \mathcal{V}$. Accordingly, the realization of subtree $\mathcal{T}(\alpha)$ is called $\mathcal{T}(\alpha)$–ROT.

3.1 Realizing an Objective Tree

The essence of realizing $\mathcal{T}$ in G is to construct a tree in G with a topology of $\mathcal{T}$, spanning all terminals. The critical task is to allocate all vertices of $\mathcal{T}$ at some locations of G. After two adjacent vertices $\alpha, \beta \in \mathcal{V}$ are respectively allocated at two different vertices $u, v \in U$, the edge $\{\alpha, \beta\} \in \mathcal{E}$ accordingly corresponds to $\{u, v\} \in E$ since G is a complete graph. This is essentially the edge between u and v in $\mathcal{T}$–ROT. As a consequence, all allocations of vertex of $\mathcal{T}$ result in a realization of $\mathcal{T}$.

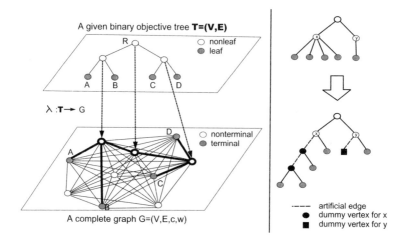

Fig. 1. Illustration of realizing a binary objective tree

Since all leaves (terminals) of T are fixed at known locations in G, we only need to allocate all nonleaves of T at some locations in G. We observe that every nonleaf of T is required to be allocated at a location of nonterminal in G. These form an *allocating function*, denoted by $\lambda : \mathcal{V} \setminus S \to V \setminus S$ (see the left-hand graph on Fig. 1), where $\lambda(\alpha) = v$ means that $\alpha \in \mathcal{V} \setminus S$ is allocated at $v \in V \setminus S$.

We call T–ROT *degenerate* if different nonleaves of T are allocated at a same location of G and *undegenerate* otherwise. Furthermore, we call a degenerate T–ROT *strongly degenerate* if two adjacent vertices of T are allocated at a same location of G and *weakly degenerate* otherwise. The strongly degenerate T–ROT is easy to be avoided by setting $w(v, v) = \infty$ additionally for every vertex v of G. Essence of Theorem 1 states that the undegenerate MCDCRP is NP-hard (its hardness may be no less than that of Hamilton Path Problem). Considering that the weakly degenerate T–ROT is not only more representative in practical settings but also much easier to be solved than the undegenerate one, we will concentrate on the weakly degenerate MCDCRP in the rest of this paper.

3.2   Binary Tree

When T is a binary tree (see Fig. 1), α has its *left child* denoted by α_l and *right child* denoted by α_r for each $\alpha \in \mathcal{V} \setminus S$. Thus, $T(\alpha)$ consists of its *left branch* denoted by $T_l(\alpha)$ and *right branch* denoted by $T_r(\alpha)$ (i.e. $T(\alpha) = T_l(\alpha) \cup T_r(\alpha)$), $T_l(\alpha)$ consists of $T(\alpha_l)$ and the edge $\{\alpha_l, \alpha\}$ (i.e. $T_l(\alpha) = T(\alpha_l) \cup \{\alpha_l, \alpha\}$). Similarly, $T_r(\alpha) = T(\alpha_r) \cup \{\alpha_r, \alpha\}$. Consequently, $T(\alpha)$ can be partitioned recursively by Lemma 1.

Lemma 1. *For every* $\alpha \in \mathcal{V} \setminus S$, $T(\alpha)$ *can be partitioned into*

$$T(\alpha) = \left(T(\alpha_l) \cup \{\alpha_l, \alpha\} \right) \bigcup \left(T(\alpha_r) \cup \{\alpha_r, \alpha\} \right) . \tag{2}$$

3.3   General Tree

When $\mathcal{T}$ is a general tree, we assume that α has $\kappa(\alpha)$ children for any $\alpha \in V \setminus S$. We can employ the method in [15] to transform $\mathcal{T}$ into a binary tree $\mathcal{T}^{\mathrm{B}}$, where, for processing α, we add $\kappa(\alpha) - 2$ dummy vertices and $\kappa(\alpha) - 2$ artificial edges if $\kappa(\alpha) \geq 3$ and add one dummy vertex and one artificial edge if $\kappa(\alpha) = 1$ (see the right-hand graph on Fig. 1).

We must allocate α as well as all dummy vertices for α at a same location of nonterminal in G, and set both cost and weight on every artificial edge of $\mathcal{T}^{\mathrm{B}}$ to zero in constructing a realization of $\mathcal{T}^{\mathrm{B}}$. As a consequence, to compute a $\mathcal{T}$–ROT is equivalent to compute a $\mathcal{T}^{\mathrm{B}}$–ROT. Without specified otherwise, $\mathcal{T}$ is always a binary tree in the rest of this paper.

4   A Pseudo-Polynomial-Time Algorithm for MDCCRP

In this section, we present an efficient algorithm for MDCCRP, which can compute the minimum diameter of $\mathcal{T}$–ROT with a cost of no more than ζ in a pseudo polynomial time, for given $\zeta \geq 0$. On basis of this algorithm, we describe another algorithm as Algorithm 1, which can decide whether $c(\mathcal{T}) > \zeta$ or $c(\mathcal{T}) \leq \zeta$ in a pseudo polynomial time for MCDCRP. Note that $c(\mathcal{T})$ denote the minimum cost of $\mathcal{T}$–ROT with a diameter of no more than $\mathbf{D}_0$.

Given an objective tree $\mathcal{T}$, we realize it by using the bottom-up dynamic programming. For any $\alpha \in V$, when $\lambda(\alpha) = v$, we use $D[\alpha, v, \mathcal{C}]$ to denote the minimum diameter of all $\mathcal{T}(\alpha)$–ROT's with a cost of no more than $\mathcal{C}$ and $R[\alpha, v, \mathcal{C}]$ to denote the minimum radius of all $\mathcal{T}(\alpha)$–ROT's with a cost of no more than $\mathcal{C}$. Note that the *radius* of $T = (U, F, c, w)$ is the maximum weight of $\pi[t, r], t \in S$ in T provided that $\lambda(\nu) = r$.

When $\alpha \in V$ is a leaf of $\mathcal{T}$, considering that $\mathcal{T}(\alpha)$ has single vertex, for all $\mathcal{C} \in \{0, 1, \ldots, \zeta\}$, we initialize $D[t, t, \mathcal{C}] = R[t, t, \mathcal{C}] = 0$ if $\alpha = t \in S$ and set $D[\alpha, v, \mathcal{C}] = R[\alpha, v, \mathcal{C}] = 0$ for all $v \in V \setminus S$ if α is a dummy vertex.

When $\alpha \in V$ is a nonleaf of $\mathcal{T}$, it is required that $\lambda(\alpha) = v \in V \setminus S$. Recall that $\lambda(\beta) = \lambda(\alpha)$ if α is a dummy vertex for β. In the following, we present two recurrence equations for computing $D[\alpha, v, \mathcal{C}]$ as in Eq. (3) and $R[\alpha, v, \mathcal{C}]$ as in Eq. (4) for all $\mathcal{C} \in \{0, 1, \ldots, \zeta\}$. In addition, we set $D[\alpha, v, \mathcal{C}] = R[\alpha, v, \mathcal{C}] = \infty$ when $\mathcal{C} < 0$. Given an edge $\{u, v\} \in E$, we use $c(u, v)$ and $w(u, v)$ to denote the cost and weight on $\{u, v\}$ respectively.

$$D[\,\alpha, v, \mathcal{C}] = \min_{\substack{\mathcal{C}_l + \mathcal{C}_r \leq \mathcal{C}, \\ \mathcal{C}_l \geq, \mathcal{C}_r \geq 0}} \min_{\substack{v_l \in V \setminus S, \\ v_r \in V \setminus S}} \max \Big\{$$

$$D[\alpha_l, v_l, \mathcal{C}_l - c(v_l, v)],$$

$$D[\alpha_r, v_r, \mathcal{C}_r - c(v_r, v)], \tag{3}$$

$$R[\alpha_l, v_l, \mathcal{C}_l - c(v_l, v)] + R[\alpha_r, v_r, \mathcal{C}_r - c(v_r, v)] + w(v_l, v) + w(v_r, v) \Big\} \,,$$

and

$$R[\alpha, v, \mathcal{C}] = \min_{\substack{\mathcal{C}_l + \mathcal{C}_r \leq \mathcal{C}, \\ \mathcal{C}_l \geq, \mathcal{C}_r \geq 0}} \max \Bigg\{$$

$$\min_{v_l \in V \setminus S} \{R[\alpha_l, v_l, \mathcal{C}_l - c(v_l, v)] + w(v_l, v)\}, \tag{4}$$

$$\min_{v_r \in V \setminus S} \{R[\alpha_r, v_r, \mathcal{C}_r - c(v_r, v)] + w(v_r, v)\}\Bigg\} .$$

Let $D^{\mathrm{E}}[\alpha, v, \mathcal{C}]$ denote the minimum diameter of all $\mathcal{T}(\alpha)$–ROT's with cost $\mathcal{C}$, and $R^{\mathrm{E}}[\alpha, v, \mathcal{C}]$ denote the minimum radius of all $\mathcal{T}(\alpha)$–ROT's with cost $\mathcal{C}$. We conclude that $D[\alpha, v, \mathcal{C}] = \min\{D[\alpha, v, \mathcal{C} - 1], D^{\mathrm{E}}[\alpha, v, \mathcal{C}]\}$ from Eq. (3) and $R[\alpha, v, \mathcal{C}] = \min\{R[\alpha, v, \mathcal{C} - 1], R^{\mathrm{E}}[\alpha, v, \mathcal{C}]\}$ from Eq. (4). This leads to a faster computation of $D[\alpha, v, \mathcal{C}]$ and $R[\alpha, v, \mathcal{C}]$.

When $\alpha \in \mathcal{V}$ is the root ν of $\mathcal{T}$, we compute $\min_{v \in V \setminus S} D[\nu, v, \mathcal{C}]$ for all $\mathcal{C} \in \{0, 1, \ldots, \zeta\}$. The value of $\min_{v \in V \setminus S} D[\nu, v, \zeta]$ is the minimum diameter of $\mathcal{T}$–ROT with cost bounded by ζ. Above discussions form an efficient algorithm for computing a minimum diameter $\mathcal{T}$–ROT with cost bounded by ζ for MDCCRP, whose time complexity is *pseudopolynomial* according to the analysis of Step 2 in the proof of Theorem 2. In all values of $\min_{v \in V \setminus S} D[\nu, v, \mathcal{C}], \mathcal{C} \in \{0, 1, \ldots, \zeta\}$, either none of them is no more than $\mathbf{D}_0$ or some of them are no more than $\mathbf{D}_0$. Let $c(\mathcal{T}, \zeta) = \min\{\mathcal{C} \in \{0, 1, \ldots, \zeta\} : \min_{v \in V \setminus S} D[\nu, v, \mathcal{C}] \leq \mathbf{D}_0\}$. We set $c(\mathcal{T}, \zeta) = \infty$ and output NO if the former occurs, and record $c(\mathcal{T}, \zeta)$ and output YES if the latter occurs. This leads us to Algorithm 1, the step 2 of which uses the dynamic programming method in [2], for deciding whether $c(\mathcal{T}) > \zeta$ or $c(\mathcal{T}) \leq \zeta$ for MCDCRP. The time complexity of Algorithm 1 is also pseudopolynomial, see Theorem 2.

Let $H(\mathcal{T})$ denote the height of $\mathcal{T}$ (the bottom of $\mathcal{T}$ is labeled as 1-st level), h denote the variable of current height, and $\mathcal{V}_h$ denote the subset of vertices of $\mathcal{T}$ on the h-th level. Let $|S| = k, |V| = n$, and then $|\mathcal{V} \setminus S| = k - 1, |V \setminus S| = n - k$.

Algorithm 1. Pseudo-polynomial-time algorithm for deciding whether $c(\mathcal{T}) > \zeta$ or $c(\mathcal{T}) \leq \zeta$ for MCDCRP.

Input: An INPUT, a positive real number $\mathbf{D}_0$, and a positive integer ζ.
Output: Either YES (meaning $c(\mathcal{T}) \leq \zeta$) or NO (meaning $c(\mathcal{T}) > \zeta$) and $c(\mathcal{T}, \zeta)$.

Step 1 for $\{$all $\alpha \in \mathcal{V}, v \in V, \mathcal{C} \in \{0, 1, \ldots, \zeta\}\}$ do
        Initialize $D[\alpha, v, \mathcal{C}] := 0, R[\alpha, v, \mathcal{C}] := 0$ as discussed above.
    endfor
Step 2 for h from 1 up to $H(\mathcal{T})$ do
        for every $\alpha \in \mathcal{V}_h$ do
            if $\alpha \in S$ then break;
            else for $\{$every $v \in V \setminus S$; $\mathcal{C}$ from 0 up to $\zeta\}$ do
                    Compute $D[\alpha, v, \mathcal{C}], R[\alpha, v, \mathcal{C}]$ by Eq. (3) and Eq. (4);
        endfor
Step 3 $c(\mathcal{T}, \zeta) := \min\{\mathcal{C} \in \{0, 1, \ldots, \zeta\} : \min_{v \in V \setminus S} D[\nu, v, \mathcal{C}] \leq \mathbf{D}_0\}$;
    if $c(\mathcal{T}, \zeta) = \infty$ then output NO; else output YES;
    When the answer is YES, a minimum cost $\mathcal{T}$–ROT with diameter bounded by $\mathbf{D}_0$ can be traced out top-down from ν of $\mathcal{T}$.

Theorem 2. *Given an MCDCRP where G has n vertices and k terminals, the time complexity of Algorithm 1 is $O(kn^3\zeta^2)$. Furthermore, we infer that $c(\mathcal{T}) \leq \zeta$ if the output is* YES *and $c(\mathcal{T}) > \zeta$ if the output is* NO.

Proof. Step_1 of Algorithm 1 spends $O(nk\zeta)$ time to make all initializations. Step_2 of Algorithm 1 computes all $D[\alpha, v, \mathcal{C}]$ bottom-up in $\mathcal{T}$, whose time complexity is $\sum_{\alpha \in S} O(1) + \sum_{\alpha \in V\setminus S} \sum_{v \in V\setminus S} \sum_{\mathcal{C}=0}^{\zeta}(O(\mathcal{C}n^2) + O(\mathcal{C}n)) \leq O(k) + O(kn) \cdot (O(n^2)\sum_{\mathcal{C}=0}^{\zeta} O(\mathcal{C})) = O(kn^3\zeta^2)$. Step_3 of Algorithm 1 only requires $O(k)$ time if we save some book-keepings information during the computation in Step_2. Therefore the time complexity of Algorithm 1 is $O(kn^3\zeta^2)$.

Recall the definition of $c(\mathcal{T}, \zeta)$, we infer that $c(\mathcal{T}) = c(\mathcal{T}, \zeta) \leq \zeta$ when $c(\mathcal{T}, \zeta) < \infty$ and $c(\mathcal{T}) > \zeta$ when $c(\mathcal{T}, \zeta) = \infty$. Consequently, the output YES implies that $c(\mathcal{T}) \leq \zeta$ and the output NO implies that $c(\mathcal{T}) > \zeta$. □

5 An FPTAS for MCDCRP

In this section we present an FPTAS for computing a $(1+\epsilon)$-approximation of MCDCRP, on basis of Algorithm 1, using standard technique of scaling and rounding [8,10,14,17,18]. To prepare for the FPTAS, we need several auxiliary algorithms which are used as subroutines in the FPTAS. We will present these in the following subsections.

5.1 Polynomial Time Approximate Testing

Given a real number $\mathbf{C} > 0$, deciding whether $c(\mathcal{T}) > \mathbf{C}$ or $c(\mathcal{T}) < \mathbf{C}$ is NP-hard. Using the standard technique of scaling and rounding [8,10,14,17,18], we can decide, in a fully polynomial time, whether $c(\mathcal{T}) > \mathbf{C}$ or $c(\mathcal{T}) < (1+\epsilon) \times \mathbf{C}$, for any given constant $\epsilon > 0$. This technique plays an important role in our FPTAS. We describe this approximate testing in Algorithm 2 as TEST.

Let c_θ be the *scaled edge cost function* such that $c_\theta(e) = \lfloor c(e) \times \theta \rfloor$ for every $e \in E$. Then we construct an auxiliary graph $G_\theta = (V, E, c_\theta, w)$ which is the same as $G = (V, E, c, w)$ except that the cost $c(e)$ is changed to $c_\theta(e)$ for every $e \in E$. Correspondingly, we have $c_\theta(\mathcal{T}, \zeta)$ and $c_\theta(\mathcal{T})$. Similarly, let $c'(e) = c(e) \times \theta$ and construct $G' = (V, E, c', w)$. Let T be the minimum cost $\mathcal{T}$–ROT with diameter bounded by $\mathbf{D_0}$ in G, and T' and T_θ be that in G' and G_θ respectively. Note that $c'(T') = c'(T), c'(T') \leq c'(T_\theta)$ and $c_\theta(T_\theta) \leq c_\theta(T)$.

Algorithm 2. TEST$(\mathbf{C}, \epsilon)$.

Input: An INPUT, two positive constants $\mathbf{D_0}$ and $\mathbf{C}$, and a positive real number $\epsilon \in (0, 4k-4]$.
Output: Either YES (meaning $c(\mathcal{T}) < (1+\epsilon) \times \mathbf{C}$) or NO (meaning $c(\mathcal{T}) > \mathbf{C}$).

Step_1 Set $\theta := \frac{4k-4}{\mathbf{C}\times\epsilon}$; $c_\theta(e) := \lfloor c(e) \times \theta \rfloor$ for each $e \in E$; Set $\zeta := \lfloor \mathbf{C} \times \theta \rfloor$;
Step_2 Replace G by G_θ into INPUT; Apply Algorithm 1;
        if $c_\theta(\mathcal{T}, \zeta) \leq \zeta$ then output YES; else output NO;

Theorem 3. *Given an INPUT, two positive constants* $\mathbf{C}$ *and* ϵ, *if* $\mathsf{TEST}(\mathbf{C}, \epsilon) = \mathsf{YES}$ *then* $c(\mathcal{T}) < (1+\epsilon) \times \mathbf{C}$ *and if* $\mathsf{TEST}(\mathbf{C}, \epsilon) = \mathsf{NO}$ *then* $c(\mathcal{T}) > \mathbf{C}$. *In addition, the worst-case time complexity of* $\mathsf{TEST}(\mathbf{C}, \epsilon)$ *is* $O(\frac{k^3 n^3}{\epsilon^2})$.

Proof. Firstly, we assume $\mathsf{TEST}(\mathbf{C}, \epsilon) = \mathsf{NO}$, which means $c_\theta(\mathcal{T}) > \zeta$. Since $\zeta = \lfloor \mathbf{C} \times \theta \rfloor$ and $c_\theta(\mathcal{T})$ is an integer, we have $c_\theta(\mathcal{T}_\theta) = c_\theta(\mathcal{T}) \geq \lfloor \mathbf{C} \times \theta \rfloor + 1 > \mathbf{C} \times \theta$. For every $e \in E$, since $c_\theta(e) = \lfloor c(e) \times \theta \rfloor \leq c(e) \times \theta$, we have $c(e) \geq \frac{c_\theta(e)}{\theta}$. Since $c_\theta(\mathcal{T}) \geq c_\theta(\mathcal{T}_\theta)$, we have $c(\mathcal{T}) = c(\mathcal{T}) \geq \sum_{e \in T} \frac{c_\theta(e)}{\theta} = \frac{c_\theta(\mathcal{T})}{\theta} \geq \frac{c_\theta(\mathcal{T}_\theta)}{\theta} > \frac{\mathbf{C} \times \theta}{\theta} = \mathbf{C}$. Therefore $\mathsf{TEST}(\mathbf{C}, \epsilon) = \mathsf{NO}$ implies that $c(\mathcal{T}) > \mathbf{C}$.

Next, we assume $\mathsf{TEST}(\mathbf{C}, \epsilon) = \mathsf{YES}$, which means $c_\theta(\mathcal{T}) \leq \zeta$. Since $\zeta = \lfloor \mathbf{C} \times \theta \rfloor$, we have $c_\theta(\mathcal{T}_\theta) = c_\theta(\mathcal{T}) \leq \mathbf{C} \times \theta$. For every $e \in E$, since $c_\theta(e) = \lfloor c(e) \times \theta \rfloor > c(e) \times \theta - 1$, we have $c(e) < \frac{c_\theta(e)}{\theta} + \frac{1}{\theta}$. Since $c_\theta(e) \leq c'(e) < c_\theta(e) + 1$ for every $e \in E$ and $\mathcal{T}$ has exactly $2k - 2$ edges, we have $c_\theta(T) \leq c'(T)$ and $c'(T_\theta) < c_\theta(T_\theta) + (2k - 2)$. Recall that $c'(T) = c'(T')$ and $c'(T') \leq c'(T_\theta)$, then it follows that $c_\theta(T) \leq c'(T) = c'(T') \leq c'(T_\theta) < c_\theta(T_\theta) + (2k - 2)$. Consequently, we have $c(\mathcal{T}) = c(T) < \sum_{e \in T}(\frac{c_\theta(e)}{\theta} + \frac{1}{\theta}) = \frac{c_\theta(T)}{\theta} + \frac{2k-2}{\theta} < \frac{c_\theta(T_\theta)}{\theta} + 2 \times \frac{2k-2}{\theta} \leq \frac{\mathbf{C} \times \theta}{\theta} + \frac{4k-4}{\theta}$. Recall that $\theta = \frac{4k-4}{\mathbf{C} \times \epsilon}$, therefore $\mathsf{TEST}(\mathbf{C}, \epsilon) = \mathsf{YES}$ implies that $c(\mathcal{T}) < (1 + \epsilon) \times \mathbf{C}$.

The time complexity of Algorithm 2 can be obtained by substituting $\zeta = \lfloor \frac{4k-4}{\epsilon} \rfloor$ into the time complexity of Algorithm 1.   □

5.2   Fully Polynomial Time Approximation Scheme

We conclude Theorem 4 from Theorem 3 using the techniques in [8,17].

Theorem 4. *Given an INPUT, a positive constant* ϵ, *a known lower bound* LB *and an upper bound* UB *on* $c(\mathcal{T})$ *such that* $\mathsf{LB} \leq c(\mathcal{T}) \leq \mathsf{UB}$, $\mathsf{TEST}(\mathbf{C}, \epsilon)$ *can compute a* $\mathcal{T}$–*ROT as* T^{A} *such that* $c(T^{\mathrm{A}}) < (1+\epsilon) \times c(\mathcal{T})$ *in* $O(\frac{k^3 n^3 \times \mathsf{UB}^2}{\epsilon^2 \times \mathsf{LB}^2})$ *time, provided that we get a scaled cost function* $c_\theta(e) = \lfloor c(e) \times \theta \rfloor$ *using* $\theta = \frac{2k-2}{\mathsf{LB} \times \epsilon}$ *and set* $\zeta = \lfloor \mathsf{UB} \times \theta \rfloor$.

Proof. Recall that $c_\theta(e) = \lfloor c(e) \times \theta \rfloor \leq c(e) \times \theta$ for every $e \in E$. Then it follows that $c_\theta(T) \leq c(T) \times \theta$. Since $c(T) = c(\mathcal{T}) \leq \mathsf{UB}$ and $c_\theta(T)$ is an integer, we have $c_\theta(T) \leq \lfloor \mathsf{UB} \times \theta \rfloor$. Therefore, we can apply Algorithm 1 (in G_θ and with $\zeta = \lfloor \theta \times \mathsf{UB} \rfloor$) to compute a minimum cost $\mathcal{T}$–ROT as T^{A}, i.e. $c_\theta(T^{\mathrm{A}}) = c_\theta(T_\theta)$. Recall that $c(e) < \frac{c_\theta(e)}{\theta} + \frac{1}{\theta}$ for every $e \in E$. Since T has exactly $2k - 2$ edges and $c_\theta(T_\theta) \leq c_\theta(T)$, we have $c(T^{\mathrm{A}}) = c(T_\theta) < \sum_{e \in T_\theta}(\frac{c_\theta(e)}{\theta} + \frac{1}{\theta}) = \frac{c_\theta(T_\theta)}{\theta} + \frac{2k-2}{\theta} \leq \frac{c_\theta(T)}{\theta} + \frac{2k-2}{\theta} \leq \frac{c(T) \times \theta}{\theta} + \frac{2k-2}{\theta}$. Since $\theta = \frac{2k-2}{\mathsf{LB} \times \epsilon}$ and $c(T) = c(\mathcal{T}) \geq \mathsf{LB}$, we have $c(T^{\mathrm{A}}) < (1 + \epsilon) \times c(\mathcal{T})$.

The time complexity can be obtained by substituting $\zeta = \lfloor \frac{(2k-2) \times \mathsf{UB}}{\epsilon \times \mathsf{LB}} \rfloor$ into the time complexity of Algorithm 1.   □

It is easy to see from Theorem 4 that the time complexity is related to the ratio $\frac{\mathsf{UB}}{\mathsf{LB}}$. We can reduce this time complexity by reducing the ratio. This idea can be implemented by first initializing LB and UB as easily computable values and then using bisection to reduce the ratio.

An initial value of LB can be computed as follows. Compute the minimum cost of $\mathcal{T}$–ROT ignoring the bound on diameter, which requires $O(kn^2)$ time. We can take this minimum as the initial value of LB. If $\mathbf{C} <$ LB (sufficient but unnecessary condition), there is no $\mathcal{T}$–ROT with diameter bounded by $\mathbf{D_0}$. An initial value of UB can be computed as follows. Find the maximum edge cost over all edges of G, which requires $O(\log n)$ time. We can take $(2k-2) \times \max_{e \in E} c(e)$ as the initial value of UB.

Let $\mathbf{B}$ be a chosen real number of greater than $1 + \epsilon$. We apply the bisection method to drive $\frac{\mathsf{UB}}{\mathsf{LB}}$ down to some number below $\mathbf{B}$. Suppose that our lower bound LB and upper bound UB satisfy that $\frac{\mathsf{UB}}{\mathsf{LB}} > \mathbf{B} > 1 + \epsilon$. Let $\mathbf{C} = \sqrt{\frac{\mathsf{LB} \times \mathsf{UB}}{1 + \epsilon}}$. If $\mathsf{TEST}(\mathbf{C}, \epsilon) = \mathsf{NO}$ then $\mathbf{C}$ is a new lower bound and UB is also an upper bound for $c(\mathcal{T})$. If $\mathsf{TEST}(\mathbf{C}, \epsilon) = \mathsf{YES}$ then $(1 + \epsilon) \times \mathbf{C}$ is a new upper bound and LB is also a lower bound for $c(\mathcal{T})$. Therefore, the ratio of the new upper bound over the new lower bound will be no more than $\sqrt{\frac{\mathsf{UB}}{\mathsf{LB}}} \times (1 + \epsilon)$. Above process is called an *iteration*. Such an iteration can be accomplished in a fully polynomial time (according to Theorem 3). Moreover, $\frac{\mathsf{UB}}{\mathsf{LB}}$ will be reduced to a number below $\mathbf{B}$ in $\log \mathcal{S}$ iterations ($\mathcal{S}$ is the input size of the given instance), see [17]. Above analysis leads to our FPTAS, described as Algorithm 3. The time complexity of Algorithm 3, shown in Theoerm 5, follows from Theorem 3 and 4.

Algorithm 3. FPTAS for computing a $(1 + \epsilon)$-approximation of MCDCRP.

Input: An INPUT, two positive constants $\mathbf{D_0}$ and ϵ.
Output: A $\mathcal{T}$–ROT as T^{A} with diameter bounded by $\mathbf{D_0}$ such that $c(T^{\mathrm{A}}) < (1 + \epsilon) \times c(\mathcal{T})$.

Step_1 Set $\mathbf{B} := 2 \times (1 + \epsilon)$; Set both LB and UB to their initial values;
Step_2 if UB $\leq \mathbf{B} \times$ LB then
        goto Step_3;
    else
        Let $\mathbf{C} := \sqrt{\frac{\mathsf{LB} \times \mathsf{UB}}{1 + \epsilon}}$;
        if $\mathsf{TEST}(\mathbf{C}, \epsilon) = \mathsf{NO}$, set $\mathsf{LB} = \mathbf{C}$;
        if $\mathsf{TEST}(\mathbf{C}, \epsilon) = \mathsf{YES}$, set $\mathsf{UB} = \mathbf{C} \times (1 + \epsilon)$;
        goto Step_2;
    endif
Step_3 Set $\theta := \frac{2k-2}{\mathsf{LB} \times \epsilon}$; $c_\theta(e) := \lfloor c(e) \times \theta \rfloor$ for each $e \in E$; Set $\zeta := \lfloor \mathsf{UB} \times \theta \rfloor$;
        Replace G by G_θ into INPUT; Apply Algorithm 1;

Theorem 5. *Given an INPUT and two positive constants $\mathbf{D_0}$ and ϵ, if there is a diameter-constrained $\mathcal{T}$–ROT, Algorithm 3 will find a diameter-constrained $\mathcal{T}$–ROT as T^{A} such that $c(T^{\mathrm{A}}) < (1 + \epsilon) \times c(\mathcal{T})$. Furthermore, the time complexity of Algorithm 3 is $O(\frac{k^3 n^3}{\epsilon^2} \times \log \mathcal{S})$, where $\mathcal{S}$ is the input size of the given instance.*

6   Conclusions

In this paper, we have presented an FPTAS for MCDCRP in a complete graph. In practice, there are lots of topologies available of objective tree. How to determine a nearly best topology of objective tree remains as a future research topic. It is also interesting to study MCDCRP under a more general topology of objective tree in uncomplete connected graphs.

References

1. Deo, N., Abdalla, A.: Computing a Diameter-Constrained Minimum Spanning Tree in Parallel. In: Bongiovanni, G., Petreschi, R., Gambosi, G. (eds.) CIAC 2000. LNCS, vol. 1767, pp. 17–31. Springer, Heidelberg (2000)
2. Ding, W., Xue, G.: A Linear Time Algorithm for Computing a Most Reliable Source on a Tree Network with Faulty Nodes. Theor. Comput. Sci. (2009), doi:10.1016/j.tcs.2009.08.003
3. Drake, D.E., Hougrady, S.: On Approximation Algorithms for the Terminal Steiner Tree Problem. Information Processing Letters 89, 15–18 (2004)
4. Du, D.Z., Smith, J.M., Rubinstein, J.H.: Advances in Steiner Trees. Kluwer Academic Publishers, Dordrecht (2000)
5. Fuchs, B.: A Note on the Terminal Steiner tree Problem. Information Processing Letters 87, 219–220 (2003)
6. Garey, M.R., Johnson, D.S.: Computers and Intractability: A Guide to the Theory of NP-Completeness. Freeman, San Francisco (1979)
7. Gouveia, L., Magnanti, T.L.: Network Flow Models for Designing Diameter-Constrained Minimum Spanning and Steiner Trees. In: Operations Research Center Working Papers. Operations Research Center, Massachusetts Institute of Technology (2001)
8. Hassin, R.: Approximation Schemes for the Restricted Shortest Path Problem. Mathematics of Operations Research 17, 36–42 (1992)
9. Hwang, F.K., Richards, D.S., Winter, P.: The Steiner Tree Problem. Annals of Discrete Mathematics 53 (1992)
10. Ibarra, O., Kim, C.: Fast Approximation Algorithms for the Knapsack and Sum of Subset Problems. Journal of the ACM 22(4), 463–468 (1975)
11. Lin, G.H., Xue, G.: On the Terminal Steiner Problem. Information Processing Letters 84, 103–107 (2002)
12. Robins, G., Zelikovsky, A.: Improved Steiner Tree Approximation in Graphs. In: Proceedings of the 11th Annual ACM-SIAM Symposium on Discrete Algorithm (SODA 2000), pp. 770–779 (2000)
13. Dos Santos, A.C., Lucena, A., Ribeiro, C.C.: Solving Diameter Constrained Minimum Spanning Tree Problems in Dense Graphs. In: Ribeiro, C.C., Martins, S.L. (eds.) WEA 2004. LNCS, vol. 3059, pp. 458–467. Springer, Heidelberg (2004)
14. Sahni, S.: General Techniques for Combinatorial Approximations. Operations Research 35, 70–79 (1977)
15. Tamir, A.: An $O(pn^2)$ Algorithm for the p-Median and Related Problems on Tree Graphs. Operations Research Letters 19, 59–64 (1996)

16. Wang, L.S., Jia, X.H.: Note Fixed Topology Steiner Trees and Spanning Forests. Theoretical Computer Science 215(1-2), 359–370 (1999)
17. Xue, G., Xiao, W.: A Polynomial Time Approximation Scheme for Minimum Cost Delay-Constrained Multicast Tree under a Steiner Topology. Algorithmica 41(1), 53–72 (2004)
18. Xue, G., Zhang, W., Tang, J., Thulasiraman, K.: Polynomial Time Approximation Algorithms for Multi-Constrained QoS Routing. IEEE/ACM Transactions on Networking 16, 656–669 (2008)
19. Zelikovsky, A.: An $\frac{11}{6}$-Approximation Algorithm for the Network Steiner Problem. Algorithmica 9(5), 463–470 (1993)

Minimizing the Maximum Duty for Connectivity in Multi-Interface Networks

Gianlorenzo D'Angelo[1], Gabriele Di Stefano[1], and Alfredo Navarra[2]

[1] Dipartimento di Ingegneria Elettrica e dell'Informazione,
Università degli Studi dell'Aquila, Italy
gianlorenzo.dangelo@univaq.it, gabriele.distefano@univaq.it
[2] Dipartimento di Matematica e Informatica,
Università degli Studi di Perugia, Italy
navarra@dmi.unipg.it

Abstract. In modern networks, devices are equipped with multiple wired or wireless interfaces. By switching among interfaces or by combining the available interfaces, each device might establish several connections. A connection is established when the devices at its endpoints share at least one active interface. Each interface is assumed to require an activation cost. In this paper, we consider the problem of guarantee the connectivity of a network $G = (V, E)$ while keeping as low as possible the maximum cost set of active interfaces at the single nodes. Nodes V represent the devices, edges E represent the connections that can be established. We study the problem of minimizing the maximum cost set of active interfaces among the nodes of the network in order to ensure connectivity. We prove that the problem is NP-hard for any fixed $\Delta \geq 3$ and $k \geq 10$, with Δ being the maximum degree, and k being the number of different interfaces among the network. We also show that the problem cannot be approximated within $O(\log |V|)$. We then provide approximation and exact algorithms for the general problem and for special cases, respectively.

1 Introduction

Wireless networks certainly provide intriguing problems for the scientific community due to the wide range of real-world applications. A very important issue recently addressed is constituted by the heterogeneity of the devices. Different computational power, energy consumption, radio interfaces, supported communication protocols, and other peculiarities can characterize the involved devices. In this paper, we are mainly interested in devices equipped with multiple interfaces. An example of a network instance is shown in Figure 1, where mobile phones, smart-phones and laptops can communicate by means of different interfaces and protocols such as IRdA, Bluetooth, WiFi, GSM, Edge, UMTS and Satellite. All the possible connections can be covered by means of at least one interface. Note that, some devices are not directly connected even though they share some interfaces. This can be due to many factors like for instance obstacles or distances.

W. Wu and O. Daescu (Eds.): COCOA 2010, Part II, LNCS 6509, pp. 254–267, 2010.

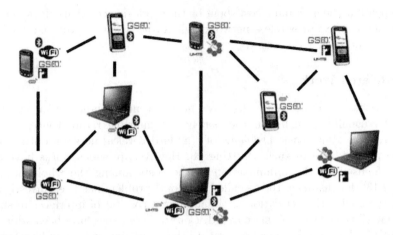

Fig. 1. The composed network according to available interfaces and proximities

A connection between two or more devices might be accomplished by means of different communication networks according to provided requirements. The selection of the most suitable interface for a specific connection might depend on various factors. Such factors include: its availability in specific devices, the cost (in terms of energy consumption) of maintaining an active interface, the available neighbors, and so forth. While managing such connections, a lot of effort must be devoted to energy consumption issues. Devices are, in fact, usually battery powered and the network survivability might depend on their persistence in the network.

We study communication problems in wireless networks supporting multiple interfaces. In the considered model, the input network is described by a graph $G = (V, E)$, where V represents the set of wireless devices and E is the set of possible connections according to proximity of devices and the available interfaces that they may share. Each $v \in V$ is associated with a set of available interfaces $W(v)$. The set of all the possible interfaces available in the network is then determined by $\bigcup_{v \in V} W(v)$; we denote the cardinality of this set by k. We say that a connection is satisfied (or covered) when the endpoints of the corresponding edge share at least one active interface. If an interface x is activated at some node u, then u consumes some energy $c(x)$ for maintaining x as active. In this setting, we study the problem of establishing a connected spanning subgraph of G by minimizing the maximum cost required at the single nodes. In other words, we look for the set of active interfaces among V, in such a way that for each pair of nodes $(u, v) \in V$ there exists a path of covered edges leading from u to v such that the maximum cost required for a single node is minimized. This implies that the cost provided by all the interfaces activated in the whole network to accomplish the connectivity requirement might not be the global minimum. Indeed, the chosen requirement is in favor of a uniform energy consumption among the devices, as it tries to maintain as low as possible the maximum cost spent by the single devices. This plays a central role in the context of wireless networks where the whole network survivability might depend on few devices.

1.1 Related Work

Multi-interface wireless networks have been recently studied in a variety of contexts, usually focusing on the benefits of multiple radio devices of each node [7,9,10]. Many basic problems of standard wireless network optimization can be reconsidered in such a setting [4]. However, previous works have been always focused on the minimization of the costs among the whole network. In [6,12,13], for instance, the so called *Coverage* problem has been investigated, where the goal is the activation of the minimum cost set of interfaces in such a way that all the edges of G are covered. *Connectivity* issues have been addressed in [3,14,15]. The goal becomes to activate the minimum cost set of interfaces in G in order to guarantee a path of communication between every pair of nodes. In [5,15], the attention has been devoted to the so called *Cheapest path* problem. This corresponds to the well-known shortest path problem but in the context of multi-interface networks. To the best of our knowledge, problems requiring to minimize the maximum cost at single nodes have not been treated before.

1.2 Our Results

In this paper, we study the problem of establishing a connected spanning subgraph of G by minimizing the maximum cost required at the single nodes. We call this problem the Minimum Maximum Cost Connectivity problem in Multi-Interface Networks (*MMCC* for short). The chosen requirement is a first step toward distributed environments where the objective function refers to local properties rather than global costs.

We consider two variants of the above problem: the parameter k is either considered as part of the input (this is called the *unbounded case*), or k is a fixed constant (the *bounded case*). The case where the cost function is constant for each interface is called the *unit cost* case.

First, we prove that the problem is NP-hard, even for the unit cost case and even when the number of interfaces k and the maximum node degree Δ are fixed. In particular, we prove that the problem remains NP-hard for any fixed $\Delta \geq 3$ and $k \geq 10$. Then, we present efficient algorithms that optimally solve the problem in some relevant special cases. In detail, we focus on instances where the input graph is a tree, by giving a linear time algorithm for fixed k or fixed Δ. By using this algorithm we can derive polynomial time algorithms for $\Delta \leq 2$ and for polynomially recognizable Hamiltonian graphs. Furthermore, we give a polynomial time algorithm for $k \leq 2$.

Concerning approximation results for *MMCC*, we show that the problem is not approximable within an $\eta \ln(\Delta)$ factor for a certain constant η, unless $P = NP$. This result holds even in the unit cost case and when the input graph is a tree but only when k or Δ are unbounded. Hence, we give some simple approximation algorithms which guarantee a factor of approximation of $\frac{c_{max}}{c_{min}} k$ or $\frac{c_{max}}{c_{min}} \Delta$, where c_{min} and c_{max} are the minimum and the maximum cost associated with an interface, respectively. In the unit cost case, we improve this result providing $\frac{k}{2}$- and $\frac{\Delta}{2}$-approximation algorithms. When $k = O(1)$ or $\Delta = O(1)$ these algorithms achieve a $O(1)$-approximation factor. As the inapproximability result holds when the input graphs are restricted to trees, we provide a $(\ln(\Delta) + 1 + \min\{\ln(\Delta) + 1, c_{max}\})$-approximation algorithm for this special case which guarantees a $(\ln(\Delta) + 2)$-approximation factor for the unit cost case. Note that, the obtained approximation factor for non-unit cost trees is optimal within a factor of 2, while in the unit cost case we only have an additive factor of 2.

In summary, *MMCC* is *NP*-hard for any fixed $\Delta \geq 3$, while it is polynomially solvable for $\Delta \leq 2$. Moreover, it is *NP*-hard for any fixed $k \geq 10$ while it is polynomially solvable for $k \leq 2$. For fixed k, $3 \leq k \leq 9$, the complexity of *MMCC* remains open.

Concerning approximability results, *MMCC* is not approximable within $\eta \ln(\Delta)$ when k and Δ are both unbounded and even in the unit cost case. When one among k or Δ is bounded, the problem can be approximated with a $O(1)$-approximation factor. When the input graph is a tree, the problem is still not approximable within $\eta \ln(\Delta)$, even in the unit cost case. However, for trees, we provide a $(\ln(\Delta) + 1 + \min\{\ln(\Delta) + 1, c_{max}\})$-approximation algorithm which guarantees a $(\ln(\Delta) + 2)$-approximation factor for the unit cost case.

1.3   Structure of the Paper

In the next section, we formally define the problem of establishing a connected spanning subgraph by minimizing the maximum cost required at the single nodes problem. In Section 3, we study the complexity of *MMCC* by analyzing the cases where the problem is *NP*-hard and the cases where it is polynomially solvable. In Section 4, we study the approximability of *MMCC* by giving a lower bound to the best approximation factor achievable and by giving polynomial time algorithms which match this bound in some cases. In Section 5 we outline some conclusion and possible future research.

2   Definitions and Notation

For a graph G, we denote by V its node set, by E its edge set. We denote the sizes of V and E by n and m, respectively. For any $v \in V$, let $N(v)$ be the set of its neighbors, and $deg(v) = |N(v)|$ be its degree in G. The maximum degree of G is denoted by $\Delta = \max_{v \in V} deg(v)$. Unless otherwise stated, the graph $G = (V, E)$ representing the network is always assumed to be simple (i.e., without multiple edges and loops), undirected and connected.

A global assignment of the interfaces to the nodes in V is given in terms of an appropriate interface assignment function W, according to the following definition.

Definition 1. *A function* $W\colon V \to 2^{\{1,2,\ldots,k\}}$ *is said to* cover *graph* G *if for each* $\{u,v\} \in E$ *we have* $W(u) \cap W(v) \neq \emptyset$.

The cost of activating an interface i is given by the cost function $c\colon \{1,2,\ldots,k\} \to \mathbb{R}_+$ and it is denoted as $c(i)$. It follows that each node holding an interface i pays the same cost $c(i)$ by activating i. The considered *MMCC* optimization problem is formulated as follows.

MMCC: Minimum Maximum Cost Connectivity in Multi-Interface Networks

---

Input:	A graph $G = (V,E)$, an allocation of available interfaces $W\colon V \to 2^{\{1,2,\ldots,k\}}$ covering graph G, an interface cost function $c\colon \{1,2,\ldots,k\} \to \mathbb{R}_+$.
Solution:	An allocation of active interfaces $W_A\colon V \to 2^{\{1,2,\ldots,k\}}$ covering a connected subgraph $G' = (V,E')$ of G such that $W_A(v) \subseteq W(v)$ for all $v \in V$, and $E' \subseteq E$.
Goal:	Minimize the maximum cost of the active interfaces among all the nodes, i.e. $\min_{W_A} \max_{v \in V} \sum_{i \in W_A(v)} c(i)$.

We recall that two variants of the above problem are considered: when the parameter k is part of the input (i.e., the unbounded case), and when k is a fixed constant (i.e., the bounded case). In both cases we assume $k \geq 2$, since the case $k = 1$ admits the obvious solution provided by activate the only interface at all the nodes.

3   Complexity

In this section we study the complexity of *MMCC*. First, we prove that the problem is *NP*-hard, even for the unit cost case and even when the number of interfaces k and the maximum node degree Δ are fixed. In particular, we prove that the problem remains *NP*-hard for any fixed $\Delta \geq 3$ and $k \geq 10$. Then, we present efficient algorithms that optimally solve the problem in some relevant special cases. In detail, we focus on instances where the input graph is a tree, by giving a polynomial time algorithm for fixed k or fixed Δ. By using this algorithm we can derive polynomial time algorithms for $\Delta \leq 2$ and for polynomially recognizable Hamiltonian graphs. Furthermore, we give a polynomial time algorithm for $k \leq 2$.

In summary, the problem is *NP*-hard for any fixed $\Delta \geq 3$, while it is polynomially solvable for $\Delta \leq 2$. Moreover, *MMCC* is *NP*-hard for any fixed $k \geq 10$ and it is polynomially solvable for $k \leq 2$. For fixed k, $3 \leq k \leq 9$, the complexity of *MMCC* remains open.

Theorem 1. *MMCC is NP-hard even when restricted to the bounded unit cost case, for any fixed $\Delta \geq 3$ and $k \geq 10$.*

Proof. We prove that the underlying decisional problem, denoted by $MMCC_D$, is in general *NP*-complete. We need to add one bound $B \in \mathbb{R}$ such that the problem will be to ask whether there exists an activation function which induces a maximum cost of the active interfaces per node of at most B.

The problem is in NP as, given an allocation function of active interfaces for an instance of $MMCC_D$, to check whether it covers a connected spanning subgraph of G with a maximum cost of active interfaces per node of at most B is linear in the size of the instance.

The proof then proceeds by a polynomial reduction from the well-known *Hamiltonian Path* problem. The problem is known to be NP-complete [11] and it can be stated as follows:

HP: Hamiltonian Path

Input: Graph $H = (V_H, E_H)$
Question: Does H contain a Hamiltonian path?

Given an instance of HP, we can build an instance of $MMCC_D$ in polynomial time as follows. Let $B = 2$. The graph G of $MMCC_D$ is the input graph H of HP. Regarding the interfaces, we associate a distinct interface to each edge in G. That is, the set of interfaces $W(v)$ of each node v in G is given by the interfaces associated to the edges incident on v. Then, if $e = \{u, v\}$ is an edge of G, we define $W(e) = W(u) \cap W(v)$. By construction, for each edge e in G we have:

1. $|W(e)| = 1$;
2. $W(e) \neq W(e')$, for each pair of edges e, e' such that $e \neq e'$.

Let us assume that H admits an Hamiltonian path P. Then the connected subgraph of G to be covered is $G' \equiv P$ and, for each node $v \in G'$, the set of active interface $W_A(v)$ is given by the interfaces associated to the edges in G' incident on v. As G' is a path, there are at most two active interfaces for each node, and then, being $B = 2$, $MMCC_D$ has a positive answer.

On the contrary, let us assume that $MMCC_D$ has a positive answer. Let G' be the covered spanning subgraph and let $W_A(v)$, $v \in V$, be such that $|W_A(v)| \leq B = 2$. As a consequence of the above properties 1) and 2), the degree of v in G' is less than or equal to 2. Since G' is connected, then G' must be either a path or a cycle. This implies that H admits a Hamiltonian path.

Now, let us show that the problem is NP-complete even if k and Δ are bounded. We assume here that each vertex of H has degree 3, in fact HP remains NP-complete even with this restriction [11].

Clearly, the above proof remains valid if we assign the interfaces in such a way that for each node v, each interface in $W(v)$ covers only one edge incident on v. Formally, we have to find an assignment of the interfaces to the nodes such that

1. $|W(e)| = 1$, for each e in G;
2. $W(u) \cap W(v) \cap W(w) = \emptyset$, for each pair of distinct edges $\{u, v\}$, $\{v, w\}$ in G.

To this end, we consider the strong edge-coloring of graph G, that is, an edge-coloring in which every color class is an induced matching. In other words, any two vertices belonging to distinct edges with the same color are not adjacent. It is known [2] that cubic graphs admit a strong edge-coloring by means of 10 colors.

Now, let us associate to each edge e in G the interface corresponding to the color assigned to e in the above coloring. Then $W(v)$, for each $v \in G$, is given by the interfaces associated to the edges incident on v, as above. It remains to show that this interface assignment fulfills property number 2), as by construction $|W(e)| = 1$, for each e in G.

We write e_1^x, e_2^x, and e_3^x to denote the three edges of each node $x \in V$. Then $W(x) = W(e_1^x) \cup W(e_2^x) \cup W(e_3^x)$. By contradiction, let us assume that there are two edges $\{u,v\}$, $\{v,w\}$ in G such that $W(u) \cap W(v) \cap W(w)$ is not empty. This means that there are three edges e_j^u, e_k^v and e_l^w, for suitable values $j, k,$ and l, such that $W(e_j^u) = W(e_k^v) = W(e_l^w)$. As at least two edges among e_j^u, e_k^v and e_l^w are distinct, then the associated interfaces are different, a contradiction.   □

In the case of bounded number of interfaces or bounded degree, we are able to show a dynamic programming algorithm which optimally solves $MMCC$ when the input graph is a tree.

Let us consider a node $v \in V$, we introduce the following notation: T^v is the rooted undirected tree obtained from G by using v as a root; for each $u \in V$, $T^v(u)$ is the subtree of T^v rooted in u. Given a set of interfaces S, the cost of activating all the interfaces in S is denoted by $c(S) = \sum_{i \in S} c(i)$.

Note that, for any optimal solution $W_A \colon V \to 2^{\{1,2,\dots,k\}}$, then $|W_A(u)| \leq \deg(u)$, for each $u \in V$. Therefore, we define for each node $u \in V$, the set $\mathcal{W}(u)$ of subsets of $W(u)$ covering $N(u)$ whose size is at most $\deg(u)$, formally:

$$\mathcal{W}(u) = \{S \subseteq W(u) \mid \forall z \in N(u), W(z) \cap S \neq \emptyset \text{ and } |S| \leq \deg(u)\}.$$

Given a rooted tree T^v and a node $u \in T^v$, for each set of interfaces $S \in \mathcal{W}(u)$, we introduce a data structure $C^v[u, S]$ which stores the minimal cost that has to be paid to cover the subtree $T^v(u)$ if we choose S to cover $N(u)$. Intuitively, $C^v[u, S]$ is given by the maximum among $c(S)$ and, for each $z \in N(u)$, the minimal cost that has to be paid to cover $T^v(z)$ by activating in z a set of interfaces $S_z \in \mathcal{W}(z)$ which shares at least one interface with S (i.e. $S_z \cap S \neq \emptyset$). Formally, $C^v[u, S]$ is defined as,

$$C^v[u, S] = \max_{z \in N(u) \cap T^v(u)} \left\{ c(S), \min_{S_z \in \mathcal{W}(z), S_z \cap S \neq \emptyset} \{C^v[z, S_z]\} \right\}. \tag{1}$$

Note that, in the above expression, if u is a leaf of T_v, then $N(u) \cap T^v(u) = \emptyset$. It follows that, in this case, $C^v[u, S] = c(S)$.

The following lemma allows us to compute an optimal solution by recursively compute $C^v[u, S]$, starting from the leaves of T^v and going upwards until we reach v.

Lemma 1. *If the input graph is a tree, then the optimal value of MMCC is given by* $\text{OPT} = \min_{S \in \mathcal{W}(v)} C^v[v, S]$.

Proof. Given a node u, and a set of interfaces $S \in \mathcal{W}(u)$, let us define as $\text{OPT}^v(u, S)$ the optimal value of the subproblem consisting of tree $T^v(u)$ assuming that u activates the set of interfaces in S. The optimal value of the original instance is then given by $\text{OPT} = \min_{S \in W(v)} \text{OPT}^v(v, S)$. Hence, it is sufficient to show that $\text{OPT}^v(u, S) = C^v[u, S]$, for each $u \in V$ and $S \in W(u)$. The proof is by induction on the height h of the tree $T^v(u)$.

If $h = 1$, then the subproblem considered is made only of node u and its neighbors $N(u)$. Hence $\text{OPT}^v(u, S) = c(S)$. By Equation 1, $C^v[u, S] = \max_{z \in N(u) \cap T^v(u)} \{c(S), \min_{S_z \in W(z), S_z \cap S \neq \emptyset} \{C^v[z, S_z]\}\}$. Since nodes z are leaves of T^v, then $S_z \subseteq S$, for each $S_z \in W(z)$. Hence, $C^v[z, S_z] \leq c(S)$ and then $C^v[u, S] = c(S)$.

If $h > 1$, by inductive hypothesis, let us assume that $\text{OPT}^v(z, S_z) = C^v[z, S_z]$, for each $z \in N(u) \cap T^v(u)$ and $S_z \in W(z)$. By cut-and-paste arguments we can show that:

$$\text{OPT}^v(u, S) = \max_{z \in N(u) \cap T^v(u)} \left\{ c(S), \min_{S_z \in W(z), S_z \cap S \neq \emptyset} \{\text{OPT}^v[z, S_z]\} \right\}. \quad (2)$$

In fact, if we consider an assignment of interfaces $\hat{W} \colon V \to 2^{\{1, \dots, k\}}$ such that $\hat{W}(u) = S$ and $\sum_{x \in T^v(u)} c(\hat{W}(x)) = \text{OPT}^v(u, S)$ and we suppose that, in $\hat{W}$, z is the node of $N(u) \cap T^v(u)$ that maximizes Equation 2 and that $\hat{W}$ does not induce a cost $\text{OPT}^v(z, S_z)$ over $T^v(z)$, where $S_z = \hat{W}(z)$, that is $\sum_{x \in T^v(z)} c(\hat{W}(x)) > \text{OPT}^v(z, S)$. Then, we can cut out from $\hat{W}$ the part defined for $T^v(z)$ and paste in an optimal assignment for $T^v(z)$ of cost $\text{OPT}^v(z, S_z)$, hence obtaining a value that is smaller than $\text{OPT}^v[u, S]$, a contradiction.

By inductive hypothesis and equation 1,

$$\text{OPT}^v(u, S) = \max_{z \in N(u) \cap T^v(u)} \left\{ c(S), \min_{S_z \in W(z), S_z \cap S \neq \emptyset} \{C^v[z, S_z]\} \right\} = C^v[u, S].$$

$\square$

Theorem 2. *If the input graph is a tree and $k = O(1)$ or $\Delta = O(1)$, MMCC can be optimally solved in $O(n)$ or $O(k^{2\Delta} n)$ time, respectively.*

Proof. By Lemma 1, it is sufficient to compute $C^v[u, S]$, for each $u \in V$ and $S \in W(u)$. Hence a straightforward algorithm is given by the definition of $C^v[u, S]$, that is $C^v[u, S]$ is computed recursively, starting from the leaves of T^v and going upwards to v. Moreover, while computing $C^v[u, S]$, we need to store, for each $z \in N(u)$, the sets $S_z \in W(z)$ which give the minimum value in Equation 1.

Note that, for each $u \in V$, $|W(u)| \leq k$, and hence $|\mathcal{W}(u)| \leq 2^{|W(u)|} \leq 2^k$. Therefore, if $k = O(1)$, then $|\mathcal{W}(u)| = O(1)$. It follows that, for each node $u \in V$ and $S \in \mathcal{W}(u)$, computing $C^v[u, S]$ by using Equation 1 requires $O(|\mathcal{W}(z)| \deg(u)) = O(\deg(u))$. Since $|\mathcal{W}(u)| = O(1)$, computing $C^v[u, \cdot]$ requires $O(\deg(u))$ for each node $u \in V$. Therefore, the overall computational time is $\sum_{u \in V} O(\deg(u)) = O(\sum_{u \in V} \deg(u)) = O(n)$.

Moreover, for each $u \in V$, $\deg(u) \leq \Delta$. Hence $|\mathcal{W}(u)| \leq |W(u)|^{\deg(u)} \leq k^\Delta$, implying that, if $\Delta = O(1)$, then $|\mathcal{W}(u)| = O(k^\Delta)$. It follows that, for each node $u \in V$ and $S \in \mathcal{W}(u)$, computing $C^v[u, S]$ by using Equation 1 requires $O(|\mathcal{W}(z)| \deg(u)) = O(k^\Delta \deg(u))$. Since $|\mathcal{W}(u)| = O(k^\Delta)$, computing $C^v[u, \cdot]$

requires $O(k^{2\Delta} deg(u))$ for each node $u \in V$. Therefore, the overall computational time is $\sum_{u \in V} O(k^{2\Delta} deg(u)) = O(\sum_{u \in V} k^{2\Delta} deg(u)) = O(k^{2\Delta} n)$.   □

By using the algorithm given in Theorem 2, we can derive two simple polynomial time algorithms for fixed $\Delta \leq 2$ that is, for paths and cycles.

Corollary 1. *If the input graph is a path, MMCC can be optimally solved in $O(k^4 n)$ time.*

Proof. It is enough to note that a path is a tree with $\Delta = 2$.   □

Corollary 2. *If the input graph is a cycle, MMCC can be optimally solved in $O(k^4 n^2)$ time.*

Proof. It suffices to choose the cheapest solutions among the ones obtained by solving the n possible paths obtained from the input cycle by excluding one different edge at time.   □

The next theorems allow us to optimally solve the problem in two special cases: when the input graph is a polynomially recognizable Hamiltonian graph in the unit cost case; and when $k \leq 2$.

Theorem 3. *In the unit cost case, if the input graph is a polynomially recognizable Hamiltonian graph, MMCC can be optimally solved in $O(k^4 n)$ time.*

Proof. If all the nodes hold one common interface, then an optimal solution is given by activating such interface and the minimum cost is 1. Otherwise, an optimal solution costs at least 2. Hence, by solving the problem restricted to an Hamiltonian path of the input graph we obtain solution of cost 2 which is optimal.   □

Theorem 4. *MMCC is polynomially solvable when $k \leq 2$.*

Proof. If an optimal solution uses only one interface, then it can be easily found by checking whether all the nodes hold the same interface. Otherwise any optimal solution needs to activate both the available interfaces at some node, regardless the cost of the interfaces.   □

4   Approximation

In this section, we study the approximation bounds for *MMCC*. In particular, we show that the problem is not approximable within an $\eta \ln(\Delta)$ factor for a certain constant η, unless $P = NP$. This result holds even for the unit cost case and when the input graph is a tree but only when k or Δ are unbounded. Hence, we give some simple approximation algorithms which guarantee a factor of approximation of $\frac{c_{max}}{c_{min}} k$ or $\frac{c_{max}}{c_{min}} \Delta$, where c_{min} and c_{max} are the minimum and the maximum cost associated with an interface, respectively. For the unit cost case we improve this result providing $\frac{k}{2}$- and $\frac{\Delta}{2}$-approximation algorithms. When

$k = O(1)$ or $\Delta = O(1)$ these algorithms achieve an $O(1)$-approximation factor. As the inapproximability result holds when the input graphs is restricted to trees, we provide a $(\ln(\Delta) + 1 + \min\{\ln(\Delta) + 1, c_{max}\})$-approximation algorithm for this special case. Note that, this bound is $\ln(\Delta) + 2$ for the unit cost case.

Summarizing, when k and Δ are both unbounded $MMCC$ is not approximable within $\eta \ln(\Delta)$, even in the unit cost case. However, when k or Δ are bounded, the problem can be approximated with an $O(1)$-approximation factor. Even if we restrict the input graph to a tree with unitary costs, the problem is not approximable within $\eta \ln(\Delta)$. For trees, we provide a $(\ln(\Delta) + 1 + \min\{\ln(\Delta) + 1, c_{max}\})$-approximation algorithm which is optimal within a factor of 2. For the unit cost case the approximation factor decreases to $(\ln(\Delta) + 2)$ which is optimal within an additive factor of 2.

Theorem 5. *MMCC in the unit cost unbounded case cannot be approximated within an $\eta \ln(\Delta)$ factor for a certain constant η, unless $P = NP$.*

Proof. The proof provides a polynomial time algorithm that transforms any instance I_1 of *Set Cover* (SC) into an instance I_2 of $MMCC$ with unit costs such that the optimum value SOL^*_{SC} on I_1 for the problem SC is equal to the optimum value SOL^*_{MMCC} on I_2 for the problem $MMCC$.

SC: Set Cover

Input:	A set U with n elements and a collection $S = \{S_1, S_2, \ldots, S_q\}$ of subsets of U.
Solution:	A cover for U, i.e. a subset $S' \subseteq S$ such that every element of U belongs to at least one member of S'.
Goal:	Minimize $|S'|$.

The graph G is a star of $n + 1$ nodes, that is, one for each element of U and a node connected to all the other ones constituting the center. There are $k = q$ interfaces of unitary cost, one for each subset in S. Each node corresponding to an element belonging to a subset S_i holds interface i.

Let $SOL_{SC}(I_1, \sigma_1)$ be the cost of a solution σ_1 for the instance I_1 of SC, and let $SOL^*_{SC}(I_1)$ be the optimal cost for instance I_1. Moreover, let $SOL_{MMCC}(I_2, \sigma_2)$ be the cost of a solution σ_2 for the instance I_2 of $MMCC$, and let $SOL^*_{MMCC}(I_2)$ be the optimal cost for $MMCC$ on instance I_2.

Let us assume that we have an optimal solution $\{S_{i_1}, S_{i_2}, \ldots, S_{i_m}\}$ for SC. Then, by activating all the available interfaces $i_1, i_2, \ldots, i_m$ in G, and in particular at the central node, we obtain a feasible solution σ for I_2 such that $SOL(I_2, \sigma) = SOL^*_{SC}(I_1)$, hence: $SOL^*_{MMCC}(I_2) \leq SOL^*_{SC}(I_1)$.

Now we show that it is possible to transform in polynomial time any solution σ_2 for the instance I_2 of $MMCC$ into a solution σ_1 for the instance I_1 of SC such that $SOL_{SC}(I_1, \sigma_1) = SOL_{MMCC}(I_2, \sigma_2)$. As the connectivity requirement imposes the covering of all the edges of G, a solution σ_2 consists in activating the minimal set of interfaces $\{i_1 \ldots i_m\}$ at the central node in such a way that each leaf shares at least one active interface with the center. We obtain

---

Algorithm 1
1.   **if** $\exists\, i \in \bigcap_{v \in V} W(v)$ **then**
2. $W_A(v) = i$ for each $v \in V$
3.   **else**
4. $W_A(v) \equiv W(v)$, for each $v \in V$

Fig. 2.

a covering of all the elements of U by means of subsets $S_{i_1}, S_{i_2}, \ldots, S_{i_m}$ corresponding to the subsets of nodes holding interfaces $\{i_1 \ldots i_m\}$. As a consequence, $SOL_{SC}(I_1, \sigma_1) = SOL_{MMCC}(I_2, \sigma_2)$.

If there exists an α factor approximation algorithm A for $MMCC$, we would obtain an α factor approximation algorithm for SC. In fact, given an instance I_1 of SC we could find a solution σ_1 by using the above transformation in an instance I_2 of $MMCC$ and applying A to find an α-approximate solution σ_2. Hence $SOL_{SC}(I_1, \sigma_1) = SOL_{MMCC}(I_2, \sigma_2) \leq \alpha SOL^*_{MMCC}(I_2) \leq \alpha SOL^*_{SC}(I_1)$. In [1], the authors show that no approximation algorithm for SC exists with an approximation factor less than $\eta \ln |U|$, for a certain constant η. Then there is no algorithm for $MMCC$ with an approximation factor less than $\eta \ln |U| = \eta \ln(n-1) = \eta \ln(\Delta)$. □

As simple approximation algorithms, we can easily guarantee a factor of approximation of $\frac{c_{max}}{c_{min}} k$ or $\frac{c_{max}}{c_{min}} \Delta$ by simply activating all the interfaces available at all nodes or by activating the cheapest interface for each edge, respectively. For the unit cost case, something a bit better can be ensured by the next two theorems.

Theorem 6. *In the unit cost case, MMCC is $\frac{k}{2}$-approximable in $O(n)$ time.*

Proof. Algorithm 1 provides a $\frac{k}{2}$-approximation for $MMCC$ in the unit cost case. In fact, if the optimum is given by activating just one interface among all the nodes of the network, then the algorithm provides the optimal solution by code lines 1–2. If the optimum must activate at least two interfaces at some node, then Algorithm 1 activates all the interfaces at all the nodes. Hence, the maximum number of interfaces that a single node can activate is k. It follows that the cost of Algorithm 1 is at most k while the cost of an optimal solution is at least 2. □

Corollary 3. *In the unit cost case, in $O(nk)$ time it is possible to check whether there exists an optimal solution of unitary cost.*

Theorem 7. *In the unit cost case MMCC is $\frac{\Delta}{2}$-approximable in $O(n+m)$ time.*

Proof. As for Theorem 6, if an optimal solution costs 1, then Algorithm 2 finds it at lines 1–2. If the optimum costs at least two, then Algorithm 2 activates for each node, at most one interface for each neighbor. Hence, the maximum number of interfaces that a single node can activate is Δ. □

Algorithm 2
1.   **if** $\exists\, i \in \bigcap_{v \in V} W(v)$ **then**
2. $W_A(v) = i$ for each $v \in V$
3.   **else**
4.   **for each** $e = \{u, v\} \in E$
5.   choose an interface $i \in W(v) \cap W(u)$
6. $W_A(v) := W_A(v) \cup \{i\}$
7. $W_A(u) := W_A(u) \cup \{i\}$
8. $E := E \setminus \{e\}$

Fig. 3.

Algorithm 3
1.   **for each** $v \in V$ starting from the leaves up to the root
2.   Apply the best approximation algorithm available for solving
         the weighted SC problem arising from the star composed of v and its children,
         and activate at v the corresponding set S' of interfaces
3.   Activate at the children of v the minimal cost interface available among S'

Fig. 4.

For trees, Algorithm 3 provides a $(\ln(\Delta) + 1 + \min\{\ln(\Delta) + 1, c_{max}\})$-approximation, with c_{max} being the maximum cost among the available interfaces. The algorithm makes use of a similar transformation to the one applied in Theorem 5 from $MMCC$ in star networks to the weighted version of SC. The difference just resides in the fact that each subset, corresponding to one interface, is associated with a weight, that is the activation cost of the corresponding interface. The goal is then to minimize the sum of the weights of the subsets chosen to accomplish the set cover.

Theorem 8. *When the input graph is a tree, there exists a polynomial algorithm for MMCC which provides a $(\ln(\Delta) + 1 + \min\{\ln(\Delta) + 1, c_{max}\})$-approximation, with $c_{max} = \max_{i \in \{1,\ldots k\}} c(i)$.*

Proof. As shown in Theorem 5, when the input graph is a star of n' nodes and maximum degree $\Delta' = n' - 1$, $MMCC$ corresponds to an instance of SC having a universe U of Δ elements. As now we are considering the non-unit cost case, we associate as weight to each subset the cost of the corresponding interface. From [8], there exists a $(\ln|U| + 1)$-approximation algorithm for the weighted SC that can be applied by Algorithm 3 at code line 2. Hence, each node of the tree will activate a set of interfaces of cost at most $\ln(\Delta) + 1$ times the optimum plus one interface for the connection to its parent node at code line 3. The cost of the interface induced to connect a node to its parent might be c_{max} but it cannot be bigger than the whole cost of the solution evaluated by Algorithm 3 at code line 2 for the parent node, hence obtaining $\min\{\ln(\Delta) + 1, c_{max}\}$.   □

For the unit cost case, the following corollary holds.

Corollary 4. *In the unit cost case, when the input graph is a tree, Algorithm 3 provides a* $(\ln(\Delta) + 2)$*-approximation.*

5 Conclusion

We have considered the Connectivity problem in Multi-Interface Networks. The new objective function with respect to previous works in this area considers the minimization of the maximum cost required by the single nodes of the network in order to accomplish the connectivity task. We focused on problem hardness and approximation factors in general and more specific settings. The obtained results have shown that the problem is *NP*-hard to be optimally or approximately solved. Polynomial algorithms for special cases have been provided. However, the lack of a general approximation algorithm, apart from trivial ones, encourages further investigations.

References

1. Alon, N., Moshkovitz, D., Safra, S.: Algorithmic construction of sets for k-restrictions. ACM Transactions on Algorithms 2(2), 153–177 (2006)
2. Andersen, L.: The strong chromatic index of a cubic graph is at most 10. Discrete Mathematics 108(1-3), 231–252 (1992)
3. Athanassopoulos, S., Caragiannis, I., Kaklamanis, C., Papaioannou, E.: Energy-efficient communication in multi-interface wireless networks. In: Královič, R., Niwiński, D. (eds.) MFCS 2009. LNCS, vol. 5734, pp. 102–111. Springer, Heidelberg (2009)
4. Bahl, P., Adya, A., Padhye, J., Walman, A.: Reconsidering wireless systems with multiple radios. SIGCOMM Comput. Commun. Rev. 34(5), 39–46 (2004)
5. Barsi, F., Navarra, A., Pinotti, M.: Cheapest paths in multi-interface networks. In: Garg, V., Wattenhofer, R., Kothapalli, K. (eds.) ICDCN 2009. LNCS, vol. 5408, pp. 37–42. Springer, Heidelberg (2008)
6. Caporuscio, M., Charlet, D., Issarny, V., Navarra, A.: Energetic Performance of Service-oriented Multi-radio Networks: Issues and Perspectives. In: 6th Int. Workshop on Software and Performance (WOSP), pp. 42–45. ACM Press, New York (2007)
7. Cavalcanti, D., Gossain, H., Agrawal, D.: Connectivity in multi-radio, multi-channel heterogeneous ad hoc networks. In: IEEE 16th Int. Symp. on Personal, Indoor and Mobile Radio Communications (PIMRC), pp. 1322–1326. IEEE, Los Alamitos (2005)
8. Chvatal, V.: A greedy heuristic for the set covering problem. Mathematics of Operations Research 4, 233–235 (1979)
9. Draves, R., Padhye, J., Zill, B.: Routing in multi-radio, multi-hop wireless mesh networks. In: 10th Annual International Conference on Mobile Computing and Networking (MobiCom), pp. 114–128. ACM, New York (2004)
10. Faragó, A., Basagni, S.: The effect of multi-radio nodes on network connectivity—a graph theoretic analysis. In: IEEE Int. Workshop on Wireless Distributed Networks (WDM). IEEE, Los Alamitos (2008)

11. Garey, M.R., Johnson, D.S.: Computers and Intractability, A Guide to the Theory of NP-Completeness. W.H. Freeman and Company, New York (1979)
12. Klasing, R., Kosowski, A., Navarra, A.: Cost minimisation in multi-interface networks. In: Chahed, T., Tuffin, B. (eds.) NET-COOP 2007. LNCS, vol. 4465, pp. 276–285. Springer, Heidelberg (2007)
13. Klasing, R., Kosowski, A., Navarra, A.: Cost Minimization in Wireless Networks with a Bounded and Unbounded Number of Interfaces. Networks 53(3), 266–275 (2009)
14. Kosowski, A., Navarra, A.: Cost minimisation in unbounded multi-interface networks. In: Wyrzykowski, R., Dongarra, J., Karczewski, K., Wasniewski, J. (eds.) PPAM 2007. LNCS, vol. 4967, pp. 1039–1047. Springer, Heidelberg (2007)
15. Kosowski, A., Navarra, A., Pinotti, M.: Exploiting Multi-Interface Networks: Connectivity and Cheapest Paths. Wireless Networks 16(4), 1063–1073 (2010)

A Divide-and-Conquer Algorithm for Computing a Most Reliable Source on an Unreliable Ring-Embedded Tree

Wei Ding[1] and Guoliang Xue[2]

[1] Zhejiang Water Conservancy and Hydropower College,
Hangzhou, Zhejiang, China
dingweicumt@163.com

[2] Department of Computer Science and Engineering, Arizona State University
Tempe, AZ 85287-8809, USA
xue@asu.edu

Abstract. Given an unreliable communication network, we seek a most reliable source (MRS) of the network, which maximizes the expected number of nodes that are reachable from it. The problem of computing an MRS in general graphs is #P-hard. However, this problem in tree networks has been solved in a linear time. A tree network has a weakness of low capability of failure tolerance. Embedding rings into it by adding some additional certain edges to it can enhance its failure tolerance, resulting in another class of sparse networks, called the ring-tree networks. This class of network also has an underlying tree-like topology, leading to its advantage of being easily administrated. This paper concerns with an important case whose underlying topology is a strip graph, called λ–rings network, and focuses on an unreliable λ–rings network where each link has an independent operational probability while all nodes are immune to failures. We apply the Divide-and-Conquer approach to design a fast algorithm for computing its an MRS, and employ a binary division tree (BDT) to analyze its time complexity to be $O(\|\lambda\|_2^2 + \lceil \log|\lambda| \rceil \cdot \|\lambda\|_1)$.

Keywords: Most reliable source, ring-tree, underlying topology, Divide-and-Conquer algorithm.

1 Introduction

A computer network or communication network is often represented as an undirected graph $G = (V, E)$, where n nodes in V represent processing or switching elements and m edges in E represent communication links [3]. For any given pair of nodes u and v, the communication between u and v is achieved by a u–v path. Failures may occur to links or nodes [4–7, 11, 17]. As networks grow in size, they become increasingly vulnerable to failures of some links and/or nodes. In the past decade, a large number of network reliability problems have been extensively studied, see [1, 2, 8, 12, 13]. Many of them focused on the computation of a most reliable source, defined in the following.

W. Wu and O. Daescu (Eds.): COCOA 2010, Part II, LNCS 6509, pp. 268–280, 2010.

Let u and v be two nodes in an unreliable communication network. We use $\Pr(u,v)$ to denote the probability that a message can be transmitted correctly from u to v. The expected number of nodes reachable from u is called the *reachability* of u, which is denoted by $Z(u)$. So we have

$$Z[u] = \sum_{v \in V} \Pr(u,v) \ . \tag{1}$$

A node with the maximum reachability is called a *most reliable source* (MRS) of the network. Consequently, to compute an MRS of a given network is essentially to determine a node u^* in the network such that

$$Z[u^*] = \max_{u \in V} \sum_{v \in V} \Pr(u,v) \ . \tag{2}$$

In an unreliable network, an MRS is a good candidate as the source for data broadcast, as the expected number of nodes reachable from an MRS is maximum. The problem of computing an MRS of an unreliable network is one of the network reliability problems, which has attracted attention of many researchers. Several papers studied the case in which each link has an independent operational probability while all nodes are immune to failures. It is #P-hard in general graphs [4, 14]. However, for tree networks, Melachrinoudis and Helander [11] presented a quadratic time algorithm. Xue [17] developed an improved linear time algorithm. Colbourn and Xue [5] presented a linear time algorithm for computing an MRS on series-parallel graphs. Recently, Ding and Xue [7] studied another case where each node has an independent probability of being faulty while all links are immune to failures, and proposed a linear time algorithm.

As tree network is one of the most important sparse networks, a large number of problems on them have been studied. Its characteristic of being easily administrated benefits from its sparse and recursive structure. However, it has a weakness of low connectivity (there is a single path between its each node pair), which leads to its low capability of failure tolerance in an unreliable network setting. A general (dense) network has a higher this capability, but it is hard to administrate it, e.g. it is #P-hard [4, 14] to find an MRS of a general network. Therefore, we seek for a compromise between the reliability of a network and the ease of network administration. Embedding rings into a tree by adding some additional edges can help to achieve this aim, e.g. adding five dashed edges $\{a,e\}, \{f,g\}, \{h,i\}, \{k,l\}, \{m,n\}$ in Fig. 1–(b) results in five embedded rings $abcdea, bfgb, chijc, jklj, dmnd$. The resulting network is called a *ring-tree* network, formally defined in Sect. 2. It is clear that a ring-tree not only has a higher capability of failure tolerance than the original tree, e.g. there is a single f–v path fbc in the original tree while there are four f–v paths $fbc, fgbc, fbaedc, fgbaedc$ in the ring-tree, but also holds an underlying tree topology, which leads to its advantage of being easily administrated. Note that we can derive different ring-trees from any given tree by adding different edges to it.

In this paper, we study a special class of ring-trees, called λ–*rings* network, see Fig. 1–(a), whose underlying topology is a strip graph. Given an unreliable λ–rings network where each link has an independent operational probability while

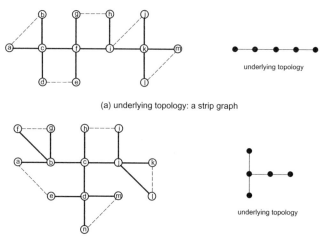

(a) underlying topology: a strip graph

(b) underlying topology: a tree graph

Fig. 1. Bold solid edges of two left-hand graphs form two sample tree networks. Adding several dashed edges to them yields two ring-tree networks, whose underlying topologies are obtained by shrinking every ring into a vertex and using an edge to represent the adjacency relationship of two rings, shown as right-hand graphs: (a) underlying topology is a strip graph; (b) underlying topology is a tree graph. It is evident that (a) is a special case of (b).

all nodes are immune to failures, we present a Divide-and-Conquer algorithm for computing its an MRS, with a time complexity of $O(\|\lambda\|_2^2 + \lceil \log|\lambda| \rceil \cdot \|\lambda\|_1)$.

The rest of this paper is organized as follows. In Sect. 2, we formally define the unreliable λ–rings network. In Sect. 3, we accomplish preliminaries, including some definitions and basic computations of probability. In Sect. 4, we present a Divide-and-Conquer algorithm for determining an MRS of a given unreliable λ–rings network, including a procedure for computing all reachabilities of node on a ring, Divide subroutine, and Merge subroutine. We conclude the paper with some research directions in Sect. 5.

2   The Unreliable λ-Rings Network

Definition 1. *Let $T = (V_T, E_T)$ be an undirected tree graph. A ring-tree graph $N = (V, E)$ with an underlying topology T is constructed in the following way: (i) each node $v_j \in V_T$ is expanded into an undirected ring $N_j = (V_j, E_j)$; (ii) each edge $e \in E_T$ is removed. (see Fig. 1–(b))*

We concern with a special case of ring-tree graph, whose underlying topology is a strip graph $S = (V_S, E_S)$ with m nodes, called m–*rings* graph, see Fig. 1–(a). An m–rings graph is called a λ-*rings* graph where $\lambda = (\lambda_1, \ldots, \lambda_m)$ if its every $N_j, j \in \{1, \ldots, m\}$ contains λ_j nodes. Also, let $\mathcal{N}(j_1, j_2) = (\mathcal{V}(j_1, j_2), \mathcal{E}(j_1, j_2))$ where $j_1 \leq j_2$ denote a subgraph of N containing a cluster of rings with the consecutive indices $j_1, \ldots, j_2$. Obviously $\mathcal{N}(j_1, j_1) = N_{j_1}$ when $j_2 = j_1$ and $\mathcal{N}(1, m) = N$ when $j_1 = 1, j_2 = m$. Hence both N_j and N can be regarded as a special case of $\mathcal{N}(j_1, j_2)$. Definition 1 states that every node of the underlying topology S of an m–rings graph represents a ring of the m–rings, every edge of S represents an adjacency relationship of two rings of the m–rings. We refer readers to [3, 15] for other graph theoretic notations not defined here. Also we will use vertex and node interchangeably, as well as edge and link.

The definition of an m–rings shows that it contains m rings and any two consecutive rings have a single common node, called a *joint* of m–rings. It is evident that an m–rings has $m - 1$ joints. Let H denote the set, composed of all joints, i.e. $H = \{h_1, \ldots, h_{m-1}\}$. Every internal ring N_j where $j \in \{2, \ldots, m-1\}$ has two joints, the *left joint* denoted by h_j^l and *right joint* denoted by h_j^r. The left end ring N_1 has only right joint h_1^r and the right end ring N_m has only left joint h_m^l. Clearly, $h_j = h_j^r = h_{j+1}^l, j \in \{1, \ldots, m-1\}$ (see Fig. 1–(a)).

Now, we consider a weighted version of λ–rings. Given $j \in \{1, \ldots, m\}$, every $e = \{x, y\} \in E_j$ is associated with a weight $p(e)$ representing the *edge operational probability* of e and associated with two arcs of (x, y) from x to y and (y, x) from y to x. All arcs form $A_j = \{(x, y), (y, x) : \{x, y\} \in E_j\}$. Let $p(x, y)$ denote the *arc operational probability* of (x, y). The relationship of $p(x, y) = p(y, x) = p(e)$ holds. All arc operational probabilities form $P_j = \{p(x, y) : (x, y) \in A_j\}$. As a consequence, we construct a bi-directed weighted ring $C_j = (V_j, A_j, P_j)$, which is symmetric in terms of arc operational probability. Accordingly, we obtain a bi-directed symmetric-weighted λ–rings $C = (V, A, P)$ and a cluster of bi-directed symmetric-weighted rings $\mathcal{C}(j_1, j_2) = (\mathcal{V}(j_1, j_2), \mathcal{A}(j_1, j_2), \mathcal{P}(j_1, j_2))$.

Definition 2. *Given $j \in \{1, \ldots, m\}$ and any pair of nodes $u, v \in V_j, u \neq v$, let $Q_j^+(u, v), Q_j^-(u, v)$ denote the probability that a message is correctly transmitted from u to v along C_j in the clockwise direction and in the anticlockwise direction respectively, and $Q_j(u, v)$ denote the probability that a message is correctly transmitted from u to v along C_j.*

Definition 3. *Given a node $u \in V$, let $\mathbb{R}_j[u]$ denote the expected number of nodes in C_j other than u which are reached from u, $\mathbb{E}[u]$ denote the expected number of nodes in C other than u which are reached from u, and $\mathbb{E}[u; \mathcal{C}(j_1, j_2)]$ denote the expected number of nodes in $\mathcal{C}(j_1, j_2)$ other than u which are reached from u.*

3   Preliminaries

We formulate $\mathbb{E}[u]$ by its definition as follows

$$\mathbb{E}[u] = \sum_{v \in V \setminus \{u\}} \Pr(u, v), \quad \forall u \in V . \tag{3}$$

According to Eq. (1) and (3), we conclude the relationship of $Z[u] = 1 + \mathbb{E}[u]$. Then an MRS of C is just a node which maximizes $\mathbb{E}(u)$. So our critical task is to compute all $\mathbb{E}[u]$ for all $u \in V$ and determine the maximum.

We formulate $\mathbb{R}_j[u]$ by its definition as follows

$$
\mathbb{R}_j[u] = \begin{cases} \displaystyle\sum_{v \in V_j} \Pr(u,v) & \text{if } u \notin V_j \\[4mm] \displaystyle\sum_{v \in V_j \setminus \{u\}} Q_j(u,v) & \text{if } u \in V_j \end{cases} , \quad \forall j \in \{1, \ldots, m\} . \tag{4}
$$

Clearly, the expected number of nodes in C_j which are reached from u is $1 + \mathbb{R}_j[u]$ if $u \in V_j$ and just $\mathbb{R}_j[u]$ if $u \notin V_j$.

We formulate $\mathbb{E}[u; \mathcal{C}(j_1, j_2)]$ by its definition as follows

$$
\mathbb{E}[u; \mathcal{C}(j_1, j_2)] = \begin{cases} \displaystyle\sum_{v \in \mathcal{V}(j_1,j_2)} \Pr(u,v) & \text{if } u \notin \mathcal{V}(j_1, j_2) \\[4mm] \displaystyle\sum_{v \in \mathcal{V}(j_1,j_2) \setminus \{u\}} \Pr(u,v) & \text{if } u \in \mathcal{V}(j_1, j_2) \end{cases} , \quad 1 \le j_1 \le j_2 \le m .
$$

$$\tag{5}$$

Clearly, the expected number of nodes in $\mathcal{C}(j_1, j_2)$ which are reached from u is $1 + \mathbb{E}[u; \mathcal{C}(j_1, j_2)]$ if $u \in V_j$ and just $\mathbb{E}[u; \mathcal{C}(j_1, j_2)]$ if $u \notin V_j$. In particular, $\mathbb{E}[u; \mathcal{C}(j_1, j_2)] = \mathbb{R}_{j_1}[u]$ when $j_1 = j_2$, and $\mathbb{E}[u; \mathcal{C}(1, m)] = \mathbb{E}[u]$ when $j_1 = 1, j_2 = m$. Hence, Eq. (4) and (3) can be both taken as a special case of Eq. (5).

The following Lemma 1 shows the formulas to compute basic probabilities, which form the basis of our algorithm.

Lemma 1. *Given any* $j_1, j_2 \in \{1, \ldots, m\}, j_1 < j_2$ *and* $j_1 \le k \le j_2$,

(i) given $u, v \in V_k, k \in \{j_1, \ldots, j_2\}$, *we have*

$$
Q_k(u,v) = Q_k^+(u,v) + Q_k^-(u,v) - Q_k^+(u,v) \cdot Q_k^-(u,v) . \tag{6}
$$

(ii) given $u \in V_j, j \in \{j_1, \ldots, k\}$ *and* $v \in V_d, d \in \{k+1, \ldots, j_2\}$, *we have*

$$
\Pr(u,v) = Q_j(u, h_j^r) \cdot \Pr(h_{j+1}^l, h_k^r) \cdot \Pr(h_{k+1}^l, v) . \tag{7}
$$

(iii) given $u \in V_j, j \in \{k+1, \ldots, j_2\}$ *and* $v \in V_d, d \in \{j_1, \ldots, k\}$, *we have*

$$
\Pr(u,v) = Q_j(u, h_j^l) \cdot \Pr(h_{j-1}^r, h_{k+1}^l) \cdot \Pr(h_k^r, v) . \tag{8}
$$

Proof. Given an m–rings graph where every edge has an independent operational probability and all nodes are immune to failures, for any pair of nodes u and v of $V_j, j \in \{1, \ldots, m\}$, we use $\pi_j^C[u, v]$ to denote the u–v path on C_j in the clockwise direction and $\pi_j^A[u, v]$ denote that in the anticlockwise direction. Then we have:

(i) For any pair of nodes u and v of $V_k, k \in \{j_1, \ldots, j_2\}$, $\pi_j^C[u, v]$ and $\pi_j^A[u, v]$ have no edge in common. So the event that a message is correctly transmitted from u to v along $\pi_j^C[u, v]$ is independent with that along $\pi_j^A[u, v]$. Based on the related property of probability, we immediately obtain Eq. (6).

(ii) For any pair of nodes u_1 and v_1 of $C_{j_1}, j_1 \in \{1, \ldots, m\}$ and any pair of nodes u_2 and v_2 of $C_{j_2}, j_2 \in \{1, \ldots, m\}, j_2 > j_1$, all of $\pi_{j_1}^C[u_1, v_1]$, $\pi_{j_1}^A[u_1, v_1]$,

$\pi_{j_2}^C[u_2, v_2]$, $\pi_{j_2}^A[u_2, v_2]$ have no edge in common. So the event that a message is correctly transmitted from u_1 to v_1 along C_{j_1} is independent with that from u_2 to v_2 along C_{j_2}. Furthermore, given any $j_1 \leq k \leq j_2$ and $u \in V_j, j \in \{j_1, \ldots, k\}$ and $v \in V_d, d \in \{k+1, \ldots, j_2\}$, the following three events are independent, i.e. the event that a message is correctly transmitted from u to h_j^r, from h_{j+1}^l to h_k^r, from h_{k+1}^l to v. This completes the proof of Eq. (7).

(iii) Similar to above (ii), Eq. (8) follows.   □

The following Lemma 2 shows an approach to compute all $\mathbb{E}[u; \mathcal{C}(j_1, j_2)]$ for all $u \in \mathcal{V}(j_1, j_2)$ where $j_1, j_2 \in \{1, \ldots, m\}, j_1 < j_2$.

Lemma 2. *Given any $j_1, j_2 \in \{1, \ldots, m\}, j_1 < j_2$ and $j_1 \leq k \leq j_2$,*
(i) for all $u \in V_j, j \in \{j_1, \ldots, k\}$, we have

$$\mathbb{E}[u; \mathcal{C}(j_1, j_2)] = \mathbb{E}[u; \mathcal{C}(j_1, k)]$$
$$+ Q_j(u, h_j^r) \cdot \Pr(h_{j+1}^l, h_k^r) \cdot \mathbb{E}[h_{k+1}^l; \mathcal{C}(k+1, j_2)] \quad . \tag{9}$$

(ii) for all $u \in V_j, j \in \{k+1, \ldots, j_2\}$, we have

$$\mathbb{E}[u; \mathcal{C}(j_1, j_2)] = \mathbb{E}[u; \mathcal{C}(k+1, j_2)]$$
$$+ Q_j(u, h_j^l) \cdot \Pr(h_{j-1}^r, h_{k+1}^l) \cdot \mathbb{E}[h_k^r; \mathcal{C}(j_1, k)] \quad . \tag{10}$$

Proof. (i) Based on Eq. (5) and (7), we conclude that

$$\mathbb{E}[u; \mathcal{C}(j_1, j_2)] \overset{\text{Eq.(5)}}{=} \sum_{v \in \mathcal{V}(j_1, j_2) \setminus \{u\}} \Pr(u, v)$$

$$= \sum_{v \in \mathcal{V}(j_1, k) \setminus \{u\}} \Pr(u, v) + \sum_{v \in \mathcal{V}(k+1, j_2) \setminus \{h_{k+1}^l\}} \Pr(u, v)$$

$$\overset{\text{Eq.(7)}}{=} \mathbb{E}[u; \mathcal{C}(j_1, k)] + \sum_{v \in \mathcal{V}(k+1, j_2) \setminus \{h_{k+1}^l\}} Q_j(u, h_j^r) \cdot \Pr(h_{j+1}^l, h_k^r) \cdot \Pr(h_{k+1}^l, v)$$

$$= \mathbb{E}[u; \mathcal{C}(j_1, k)] + Q_j(u, h_j^r) \cdot \Pr(h_{j+1}^l, h_k^r) \cdot \mathbb{E}[h_{k+1}^l; \mathcal{C}(k+1, j_2)] \quad .$$

(ii) Similarly, Eq. (10) follows from Eq. (5) and (8).   □

4   A Divide-and-Conquer Algorithm

The Divide-and-Conquer method is one of the most important techniques of algorithm design [9, 10, 16]. In this section, we set a bi-directed symmetric-weighted λ–rings $C = (V, A, P), \lambda = \{\lambda_1, \ldots, \lambda_m\}$ and its *index sequence* $I = \{1, \ldots, m\}$ as an example. We will present a Divide-and-Conquer algorithm for computing an MRS of C, comprising two main steps—Divide subroutine and Merge subroutine. To determine all $\mathbb{E}[u; \mathcal{C}(j_1, j_2)]$ for all $u \in \mathcal{V}(j_1, j_2)$ is called to achieve the *solution* of $\mathcal{C}(j_1, j_2)$ or to *solve* $\mathcal{C}(j_1, j_2)$. Divide subroutine bisects a large-size cluster of rings into two small-size clusters of rings recursively until every cluster contains one ring. Then we compute the solution of every $C_j, j \in I$ respectively. Merge subroutine combines the solutions of two small-size clusters of rings into the solution of the merged large-size cluster of rings recursively until the solution of C is achieved. The implementations of both Divide subroutine and Merge subroutine are based on I in our algorithm.

For ease of presentation, we use $J = \{j_1, \ldots, j_2\}$ to denote a subsequence of I ($J \subseteq I$, corresponding to $\mathcal{C}(j_1, j_2)$), Size(J) to denote the size of J (the number of rings in $\mathcal{C}(j_1, j_2)$), and Sol(J) to denote the solution of $\mathcal{C}(j_1, j_2)$.

4.1   An MRS on a Ring

For every $C_j = (V_j, A_j, P_j), j \in I$, Eq. (4) shows that we are required to compute all $Q_j(u, v)$ for all $v \in V_j \setminus \{u\}$ so as to compute each $\mathbb{R}_j[u], u \in V_j$. This directly leads to the following Procedure MRSR (see [6] for more details). Note that the input J of MRSR is a single-element subsequence $\{j\}, j \in I$ which corresponds to a ring C_j.

MRSR(J):
BEGIN
    for each $u \in V_j$ do
        Compute all $Q_j^+(u, v)$ consecutively in the clockwise direction;
        Compute all $Q_j^-(u, v)$ consecutively in an anticlockwise direction;
        Compute all $Q_j(u, v)$ using Eq. (6), record $Q_j(u, h_j^l)$ and $Q_j(u, h_j^r)$;
        Compute $\mathbb{R}_j[u]$ using Eq. (4);
    end for
    Return Sol(J) and $Q_j(u, h_j^l), Q_j(u, h_j^r)$ for all $u \in V_j$;
END

Theorem 1. *Procedure* MRSR *requires* $O(\lambda_j^2)$ *time.*

Proof. For each $u \in V_j$, Procedure MRSR spends $O(\lambda_j)$ time to compute all $Q_j^+(u, v)$ and $O(\lambda_j)$ time to compute all $Q_j^-(u, v)$ for all $v \in V_j \setminus \{u\}$, $O(1)$ time to compute $Q_j(u, v)$, and $O(\lambda_j)$ time to compute $\mathbb{R}_j[u]$. So the total time of Procedure MRSR is $O(\lambda_j^2)$.   □

4.2   Divide Subroutine

A longer sequence can be divided into a group of shorter subsequences by many division schemes. In the following, we present a *bisection scheme* as Eq. (11) for bisecting a longer sequence into two shorter subsequences recursively, which finally produces a group of single-element subsequences. Such bisection scheme forms a resultant rooted tree, called the *binary division tree* (BDT), see Fig. 2.

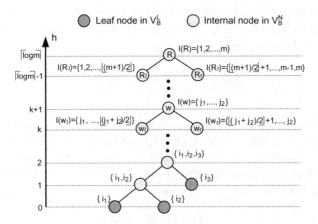

Fig. 2. BDT $= (V_B, E_B)$, where each node $w \in V_B$ is associated with a subsequence $I(w) \subseteq I$, illustrates the whole process of Procedure DIVD

Without loss of generality, we bisect $J = \{j_1, \ldots, j_2\}$ as an example

$$\{j_1, \ldots, j_2\} = \{j_1, \ldots, \lfloor \frac{j_1 + j_2}{2} \rfloor\} \oplus \{\lfloor \frac{j_1 + j_2}{2} \rfloor + 1, \ldots, j_2\} , \qquad (11)$$

where let $J_l = \{j_1, \ldots, \lfloor \frac{j_1+j_2}{2} \rfloor\}$ and $J_r = \{\lfloor \frac{j_1+j_2}{2} \rfloor + 1, \ldots, j_2\}$.

DIVD(J):
BEGIN
    if Size(J) = 1 then
       break;
    else
       Divide J into $J := J_l \oplus J_r$ using Eq. (11);
       DIVD(J_l) and DIVD(J_r) respectively;
    end if
END

Procedure DIVD presents a recursive bisection scheme for dividing I into m single-element subsequences, resulting in a BDT $= (V_B, E_B)$ rooted at R (see Fig. 2). Initially, we set J to I. It is easy to see that Procedure DIVD performs $m - 1$ divide operations in total to achieve m single-element subsequences each of which corresponds to a leaf of BDT, and the height of BDT is $\lceil \log m \rceil$ provided that the bottom of BDT is labeled as its 0–level. Each internal node of BDT corresponds to a non-single-element subsequence of I. All leaves of BDT form a set denoted by V_B^L and all internal nodes of BDT form a set denoted by V_B^N. Obviously, $V_B = V_B^L \cup V_B^N$ and $|V_B^L| = m, |V_B^N| = m - 1$.

Let h be the variable of height and $\mathcal{H}(h)$ be the set of all nodes on the h–level of BDT. Considering $V_B = \bigcup_{h=0}^{\lceil \log m \rceil} \mathcal{H}(h)$ and all internal nodes of BDT lie on its h–level, $h \geq 1$, we obtain

$$V_B^N = \bigcup_{h=1}^{\lceil \log m \rceil} \{V_B^N \cap \mathcal{H}(h)\} \ . \tag{12}$$

Let $I(w)$ be the subsequence of I, associated with a node $w \in V_B$. It is clear that every $I(w), w \in V_B^L$ is a single-element subsequence, every $I(w), w \in V_B^N$ is a non-single-element subsequence, and $I(w) = I(w_l) \oplus I(w_r)$ for every $w \in V_B^N$. Particularly, $I(\mathrm{R}) = I$. Combing the fact that all leaves of BDT lie on its 0–level or 1–level, we conclude from $I = \bigoplus_{w \in V_B^L} I(w)$ that

$$\bigoplus_{w \in \{V_B^N \cap \mathcal{H}(h)\}} I(w) : \begin{cases} \subseteq I & \text{if} \quad h = 1 \\ = I & \text{if} \quad h \in \{2, \ldots, \lceil \log m \rceil\} \end{cases} . \tag{13}$$

4.3   Merge Subroutine

In this subsection, we present an efficient algorithm for combining two shorter subsequences J_l and J_r into a longer sequence J. Fig. 3 helps to illustrate the Merge subroutine. We see that it is required to achieve $\mathrm{Sol}(J)$ and some related probabilities. Theorem 2 shows formulas for computing all $\mathbb{E}[u; \mathcal{C}(j_1, j_2)]$ for all $u \in \mathcal{V}(j_1, j_2)$ when combining $\mathrm{Sol}(J_l)$ with $\mathrm{Sol}(J_r)$ into $\mathrm{Sol}(J)$.

Theorem 2. *Given a sequence* $J = \{j_1, \ldots, j_2\}$ *and the bisection scheme as Eq. (11), for all* $u \in V_j, j \in J$,
(i) if $j \in J_l$, *we have*

$$\mathbb{E}[u; \mathcal{C}(j_1, j_2)] = \mathbb{E}\left[u; \mathcal{C}\left(j_1, \left\lfloor \tfrac{j_1+j_2}{2} \right\rfloor\right)\right]$$

$$+ Q_j(u, h_j^r) \cdot \Pr(h_{j+1}^l, h_{\lfloor \frac{j_1+j_2}{2} \rfloor}^r) \cdot \mathbb{E}\left[h_{\lfloor \frac{j_1+j_2}{2} \rfloor+1}^l; \mathcal{C}\left(\left\lfloor \tfrac{j_1+j_2}{2} \right\rfloor + 1, j_2\right)\right] \ . \tag{14}$$

Fig. 3. Set a combination of two 2–rings as an example to illustrate the whole process of Procedure MERG for all nodes $u \in \mathcal{V}(j_1, g)$, where let $g = \lfloor \frac{j_1+j_2}{2} \rfloor$

(ii) if $j \in J_r$, we have

$$\mathbb{E}[u; \mathcal{C}(j_1, j_2)] = \mathbb{E}\left[u; \mathcal{C}\left(\lfloor \tfrac{j_1+j_2}{2} \rfloor + 1, j_2\right)\right]$$
$$+ Q_j(u, h_j^l) \cdot \Pr(h_{j-1}^r, h_{\lfloor \frac{j_1+j_2}{2} \rfloor + 1}^l) \cdot \mathbb{E}\left[h_{\lfloor \frac{j_1+j_2}{2} \rfloor}^r; \mathcal{C}\left(j_1, \lfloor \tfrac{j_1+j_2}{2} \rfloor\right)\right] . \tag{15}$$

Proof. This theorem follows from Lemma 2 where let $k = \lfloor \tfrac{j_1+j_2}{2} \rfloor$. □

Both Eq. (14) and (15) show that we are also required to achieve some related probabilities for computing $\mathbb{E}[u; \mathcal{C}(j_1, j_2)]$. Actually, $Q_j(u, h_j^l), Q_j(u, h_j^r)$ can be achieved by Procedure MRSR, and $\Pr(h_{j+1}^l, h_{\lfloor \frac{j_1+j_2}{2} \rfloor}^r), \Pr(h_{j-1}^r, h_{\lfloor \frac{j_1+j_2}{2} \rfloor + 1}^l)$ have been achieved previously by the following Eq. (16) and (17). It follows directly from Eq. (7) where let $u = h_{j+1}^l, v = h_{j_2}^r, k = \lfloor \tfrac{j_1+j_2}{2} \rfloor$ that, for all $j \in J_l$,

$$\Pr(h_{j+1}^l, h_{j_2}^r) = \Pr\left(h_{j+1}^l, h_{\lfloor \frac{j_1+j_2}{2} \rfloor}^r\right) \cdot \Pr\left(h_{\lfloor \frac{j_1+j_2}{2} \rfloor + 1}^l, h_{j_2}^r\right) . \tag{16}$$

Also, it follows directly from Eq. (8) where let $u = h_{j-1}^r, v = h_{j_1}^l, k = \lfloor \tfrac{j_1+j_2}{2} \rfloor$ that, for all $j \in J_r$,

$$\Pr(h_{j-1}^r, h_{j_1}^l) = \Pr\left(h_{j-1}^r, h_{\lfloor \frac{j_1+j_2}{2} \rfloor + 1}^l\right) \cdot \Pr\left(h_{\lfloor \frac{j_1+j_2}{2} \rfloor}^r, h_{j_1}^l\right) . \tag{17}$$

Above discussion leads to Procedure MERG in the following, whose time complexity is presented in Theorem 3.

MERG(J_l, J_r):
BEGIN
    for each $j \in J_l$ do
        for each $u \in V_j$ do
            Use Eq. (14) to compute $\mathbb{E}[u; \mathcal{C}(j_1, j_2)]$;
        end for
        Compute $\Pr(h_j^r, h_{j_2}^r)$ by Eq. (16);
    end for
    for each $j \in J_r$ do
        for each $u \in V_j$ do
            Use Eq. (15) to compute $\mathbb{E}[u; \mathcal{C}(j_1, j_2)]$;
        end for
        Compute $\Pr(h_j^l, h_{j_1}^l)$ by Eq. (17);
    end for
    Return Sol(J), all $\Pr(h_j^r, h_{j_2}^r), j \in J_l$ and $\Pr(h_j^l, h_{j_1}^l), j \in J_r$;
END

Theorem 3. *Procedure MERG requires $O(\sum_{j=j_1}^{j_2} \lambda_j)$ time.*

Proof. Procedure MERG needs to compute all $\mathbb{E}[u; \mathcal{C}(j_1, j_2)]$ for all $u \in \mathcal{V}(j_1, j_2)$ and spend $O(1)$ time for each $\mathbb{E}[u; \mathcal{C}(j_1, j_2)]$, as well as no more than $O(m)$ time to compute all related probabilities. Therefore, it requires $O(\sum_{j=j_1}^{j_2} \lambda_j)$ time for combining J_l with J_r into J. □

4.4   A Divide-and-Conquer Algorithm

Based on discussion and procedures in above subsections, we propose a Divide-and-Conquer algorithm for computing all reachabilities of a node of C, described as Algorithm MRS_Rings. Then we can easily achieve an MRS of C.

MRS_Rings(J):
BEGIN
    if Size(J) = 1 then
        MRSR(J) to achieve Sol(J);
    else
        Divide J into $J := J_l \oplus J_r$ using Eq. (11);
        MRS_Rings(J_l) and MRS_Rings(J_r) respectively;
        MERG(J_l, J_r) to achieve Sol(J);
    end if
END

The implementation of Algorithm MRS_Rings can be taken as based on BDT in the following way: each $I(w), w \in V_B^N$ performs a divide operation and each $I(w), w \in V_B^L$ performs Procedure MRSR once during a top-down process among BDT, as well as each $I(w), w \in V_B \setminus \{R\}$ performs a merge operation during a bottom-up process among BDT. Initially, we input $J := I$.

Given a vector $\alpha = (\alpha_1, \alpha_2, \dots, \alpha_m)$, we introduce 1–norm $\|\cdot\|_1$, 2–norm $\|\cdot\|_2$ and length $|\cdot|$ of α, formulated by Eq. (18) as follows

$$\|\alpha\|_1 = \sum_{j=1}^{m} \alpha_j, \qquad \|\alpha\|_2 = \sqrt{\sum_{j=1}^{m} \alpha_j^2}, \qquad |\alpha| = m \ . \tag{18}$$

Theorem 4. *Given a λ–rings graph $C = (V, A, P), \lambda = (\lambda_1, \dots, \lambda_m)$, Algorithm MRS_Rings can compute an MRS of C correctly in $O(\|\lambda\|_2^2 + \lceil \log|\lambda| \rceil \cdot \|\lambda\|_1)$ time.*

Proof. The correctness of Algorithm MRS_Rings follows from the discussion before this theorem in this section. Below is shown the discussion of time complexity of Algorithm MRS_Rings.

Let T[J] be the time occupied for achieving Sol(J), D[J] be the time occupied for dividing J into J_l and J_r, and M[J] be the time occupied for combining J_l with J_r into J. Also, let T[C_j] be the time occupied by Procedure MRSR for achieving Sol($\{j\}$). So we have

$$\mathrm{T}[J] = \begin{cases} \mathrm{D}[J] + \mathrm{T}[J_l] + \mathrm{T}[J_r] + \mathrm{M}[J] & \text{if } J \in \{I(w): \ w \in V_B^N\} \\ \mathrm{T}[C_j], j \in J & \text{if } J \in \{I(w): \ w \in V_B^L\} \end{cases} \ . \tag{19}$$

Eq. (19) is a recursive equation. Initially, we input $J := I$. Fig. 2 shows that it is required to perform $m - 1$ divide operations until we obtain m single-element subsequences, and correspondingly $m-1$ merge operations until we obtain Sol(I).

For each single-element subsequence $J' \in \{I(w) : w \in V_B^L\}$ (corresponding to C_j when $J' = \{j\}$), Procedure MRSR spends $O(\lambda_j^2)$ time to achieve $\mathrm{Sol}(J')$ according to Theorem 1, i.e. $\mathrm{T}[J'] = \mathrm{T}[C_j] = O(\lambda_j^2)$. For each non-single-element subsequence $J'' \in \{I(w) : w \in V_B^N\}$, it spends $O(1)$ time to perform a divide operation, i.e. $\mathrm{D}[J''] = O(1)$. For merge operations, we conclude from Eq. (13) that

$$\sum_{w \in \{V_B^N \cap \mathcal{H}(h)\}} \mathrm{M}[I(w)] : \begin{cases} \leq \mathrm{M}[I(\mathrm{R})] & \text{if} \quad h = 1 \\ \\ = \mathrm{M}[I(\mathrm{R})] & \text{if} \quad h \in \{2, \ldots, \lceil \log m \rceil\} \end{cases} \tag{20}$$

and from Theorem 3 that the time required for combining $I(\mathrm{R}_l)$ with $I(\mathrm{R}_r)$ into $I(\mathrm{R})$ is $\mathrm{M}[I] = \mathrm{M}[I(\mathrm{R})] = O(\sum_{j=1}^{m} \lambda_j)$. Therefore, it follows after $m - 1$ iterations of Eq. (19) that

$$
\begin{aligned}
\mathrm{T}[I(\mathrm{R})] \quad &= \quad \sum_{w \in V_B^L} \mathrm{T}[I(w)] + \sum_{w \in V_B^N} \mathrm{D}[I(w)] + \sum_{w \in V_B^N} \mathrm{M}[I(w)] \\
&\overset{\mathrm{Eq.(12)}}{=} \sum_{j \in I} O(\lambda_j^2) + \sum_{w \in V_B^N} O(1) + \sum_{h=1}^{\lceil \log m \rceil} \sum_{w \in \{V_B^N \cap \mathcal{H}(h)\}} \mathrm{M}[I(w)] \\
&\overset{\mathrm{Eq.(20)}}{\leq} O\left(\sum_{j=1}^{m} \lambda_j^2 \right) + O(m) + \sum_{h=1}^{\lceil \log m \rceil} \mathrm{M}[I(\mathrm{R})] \\
&\overset{\mathrm{Eq.(18)}}{=} O(\|\lambda\|_2^2 + \lceil \log|\lambda| \rceil \cdot \|\lambda\|_1) \ .
\end{aligned}
$$

□

5   Conclusions

In this paper, we concentrate on a class of networks containing embedded rings with an underlying topology of a strip graph, and present a Divide-and-Conquer algorithm for determining an MRS on such unreliable networks. Furthermore we can develop a parallel or distributed algorithm based on the basic results presented in this paper. It is also of interest to study networks containing embedded rings with other classes of underlying topologies, and compute an MRS on these unreliable networks.

References

1. Ball, M.O., Lin, F.L.: A Reliability Model Applied to Emergency Service Vehicle Location. Oper. Res. 41(1), 18–36 (1993)
2. Ball, M.O., Provan, J.S., Shier, D.R.: Reliability Covering Problems. Networks 21(3), 345–357 (1991)
3. Bondy, J.A., Murty, U.S.R.: Graph Theory with Application. Macmillan, London (1976)
4. Colbourn, C.J.: The Combinatorics of Network Reliability. Oxford University Press, New York (1987)

5. Colbourn, C.J., Xue, G.: A Linear Time Algorithms for Computing the Most Reliable Source on a Series-Parallel Graph with Unreliable Edges. Theor. Comput. Sci. 209, 331–345 (1998)
6. Ding, W.: Computing the Most Reliable Source on Stochastic Ring Networks. In: WRI World Congress on Software Engineering 2009, Xiamen, China, May 19-21, vol. 1, pp. 345–347 (2009)
7. Ding, W., Xue, G.: A Linear Time Algorithm for Computing a Most Reliable Source on a Tree Network with Faulty Nodes. Theor. Comput. Sci. (2009), doi:10.1016/j.tcs.2009.08.003
8. Eiselt, H.A., Gendreau, M., Laporte, G.: Location of Facilities on a Network Subject to a Single-Edge Failure. Networks 22(3), 231–246 (1992)
9. Even, G., Naor, J.S., Rao, S., Schieber, B.: Divide-and-conquer approximation algorithms via spreading metrics. Journal of the ACM 47(4), 585–616 (2000)
10. Hoare, C.A.R.: Quicksort. The Computer Journal 5(1), 10–16 (1962)
11. Melachrinoudis, E., Helander, M.E.: A Single Facility Location Problem on a Tree with Unreliable Edges. Networks 27(3), 219–237 (1996)
12. Mirchandani, P.B., Odoni, A.R.: Locations of Medians on Stochastic Networks. Transport. Sci. 13, 85–97 (1979)
13. Nel, L.D., Colbourn, C.J.: Locating a Broadcast Facility in an Unreliable Network. INFOR. 28, 363–379 (1990)
14. Shier, D.R.: Network Reliability and Algebraic Structure. Oxford University Press, New York (1991)
15. West, D.B.: Introduction to Graph Theorey. Prentice Hall, Englewood Cliffs (2001)
16. Wu, I.-C., Kung, H.T.: Communication Complexity for Parallel Divide-and-Conquer. In: Proceedings of the 32nd Annual Symposium on Foundations of Computer Science (FOCS 1991), San Juan, Puerto Rico, pp. 151–162 (October 1991)
17. Xue, G.: Linear Time Algorithms for Computing the Most Reliable Source on an Unreliable Tree Network. Networks 30(1), 37–45 (1997)

Constrained Low-Interference Relay Node Deployment for Underwater Acoustic Wireless Sensor Networks

Deying Li, Zheng Li, Wenkai Ma, and Wenping Chen

Key Laboratory of Data Engineering and Knowledge Engineering, MOE,
School of Information, Renmin University of China, China

Abstract. An Underwater Acoustic Wireless Sensor Network (UA-WSN) consists of many resource-constrained Underwater Sensor Nodes (USNs), which are deployed to perform collaborative monitoring tasks over a given region. One way to preserve network connectivity while guaranteing other network QoS is to deploy some Relay Nodes (RNs) in the networks, in which RNs' function is more powerful than USNs and their cost is more expensive. This paper addresses Constrained Low-interference Relay Node Deployment (C-LRND) problem for 3-D UA-WSNs in which the RNs are placed at a subset of candidate locations to ensure connectivity between the USNs, under both the number of RNs deployed and the value of total incremental interference constraints. We first prove that it is NP-hard, then present a general approximation algorithm framework and get two polynomial time $O(1)$-approximation algorithms.

Keywords: Underwater acoustic wireless sensor network, Relay node deployment, Connectivity, Low-interference, Approximation Algorithm.

1 Introduction and Motivations

Underwater Acoustic Wireless Sensor Networks (UA-WSNs) have attracted a great deal of research attentions due to their wide-range applications including oceanographic data collection, pollution monitoring, offshore exploration, disaster prevention, assisted navigation and tactical surveillance. UA-WSNs consist of Underwater Sensor Nodes (USNs) that are deployed to perform collaborative monitoring tasks over a given region [1–3].

The network topology is in general a crucial factor in determining the energy consumption, the capacity and the communication delay of a network. Hence, the network topology should be carefully engineered, and post-deployment topology optimization should be performed, when possible [4, 5]. One approach to ensure connectivity and improve network performance for UA-WSNs is to deploy a small number of costly, but more powerful Relay Nodes (RNs) whose main task is to communicate with other USNs or RNs [6].

However, deploying extra RNs to assist the communication between partitioned USNs will result in the increment of inherent interference in the UA-WSNs. The impacts of interference in multihop wireless networks (e.g. UA-WSNs) have

W. Wu and O. Daescu (Eds.): COCOA 2010, Part II, LNCS 6509, pp. 281–291, 2010.
© Springer-Verlag Berlin Heidelberg 2010

been observed and studied both theoretically and empirically in the literature [7]. The interference in underwater acoustic communication may lead to some negative influences such as energy waste, impaired underwater channel, limited available bandwidth, high bit error rates and so on. These negative factors have to reduce the system performance.

Most of the existed works on the RNs deployment problems are the *unconstrained* version, i.e., the RNs can be placed anywhere. In practice, however, there are some physical constraints on the placement of the RNs [8]. For example, there may be some forbidden regions where relay nodes cannot be placed. For solving this challenging problem, we study the *constrained* RNs deployment problem where the RNs can only be placed at a subset of candidate locations.

This paper mainly addresses the Constrained Low-interference Relay Nodes Deployment (C-LRND) problem for the UA-WSNs, which meets connectivity requirement under both the number of RNs deployed and the value of total incremental interference constraints. The deployment strategie studied in this paper is trying to guarantee both less number of deployed RNs and lower total incremental interference simultaneously. We first discuss its computational complexity, then present a general approximation algorithm framework and get two polynomial time $O(1)$-approximation algorithms.

The rest of this paper is organized as follows. In Section 2 we present related works. Section 3 describes the network model and basic notations. In section 4, we investigate the connected C-LRND problem. Section 5 concludes our paper.

2   Related Works

RNs placement problems have been well studied in the 2-D Wireless Sensor Networks (WSNs) [9–11]. Lin *et al.* [10] proved the problem to be NP-hard, and proposed a MST-based 5-approximation algorithm. Chen *et al.* [9] proved that the algorithm in [10] is an 4-approximation algorithm. Cheng *et al.* [11] proposed a faster 3-approximation algorithm and a randomized algorithm with a performance ratio of 2.5. In addition, there are many works on fault-tolerance RNs placement problem[12, 13]. Bredin *et al.* [13] studied the fault-tolerance (k-connected) RNs placement problem, which aims to deploy minimum number of RNs to ensure the resulting network contains k node-disjoint paths between every pair of sensors and RNs. The authors [13] presented polynomial time $O(1)$-approximation algorithms for any fixed k. Kashyap *et al.* [14] presented an 10-approximation algorithm to ensure 2-connectivity. In [6, 15], the authors studied the RNs placement problems to ensure 1-connectivity and 2-connectivity for the case that the RNs have longer transmission range than sensors, and presented approximation algorithms with constant performance ratio.

All of the above works studied the *unconstrained* version RNs placement problem, in the sense that the RNs can be placed in anywhere. However, in reality there are some physical constraints on the placement of the RNs. In ref. [8], the authors formulated the *constrained* RNs placement problems, i.e., placing the minimum number of RNs at a subset of candidate locations to ensure the 1-connectivity and 2-connectivity between the sensor nodes and the base station, respectively.

The majority of the existing works in relay node deployment problem are based on the 2-D network model derived from the terrestrial wireless sensor networks. The work in [1] introduced a type of 3-D UA-WSNs architecture, consisting of USNs and RNs. The role of RNs is to communicate with USNs and other RNs. The works in [16] mainly focused on the surface gateway placement and pointed out the tradeoff between the number of surface gateways and the expected delay and energy consumption. Seah et al.[17] proposed a novel virtual sink architecture for UA-WSNs that aims to achieve robustness and energy efficiency in harsh under water channel conditions.

In this paper, we focus on the Constrained Low-interference Relay Node deployment (C-LRND) strategy for 3-D US-WSNs to meet 1-connectivity between all USNs, under both the number of RNs deployed and total incremental interference constraints. This problem is different from the problems in [8, 16]. The authors in [16] only formulated the problems by Integer Linear Programming model. In [8], the authors studied the constrained relay node placement problem in 2-D wireless sensor networks to meet 1-connectivity and 2-connectivity requirements. However, the 2-D assumption may no longer be valid if RNs are deployed in 3-D underwater environment. Furthermore, the deployment strategy studied in this paper is concerned with not only the number of RNs deployed, but also the total incremental interference, i.e., trying to guarantee both less number of deployed RNs and lower interference simultaneously.

3   Notations and Network Model

In this section, we will formally define the problem and notations that will be used throughout the paper. Let us consider a 3-D UA-WSN consisting of Underwater Sensor Nodes (USNs) and Relay Nodes (RNs). The USNs are pre-deployed in the sensing area and floated at different depths, each of them is equipped with an acoustic communicator which has communication range R_A. On the other hand, RNs only can be deployed at the candidate locations. RNs are also equipped with acoustic communicators with communication ranges R_A. Denote S to the USNs set, and L a set of candidate locations where RNs can be placed. We will use u to denote node u's location, if no confusion arises. The notations can be summarized as follows:

$d_{Euc}(u, v)$ Euclidean distance between two nodes u and v.

$d_T(u)$ degree of node u in T.

$\Delta(T)$ the maximum degree among all nodes in T.

$N(u)$ a set of node u's neighbors.

R_A acoustic communication range.

KN_3 3-dimension kissing number.

Any two nodes u, v (which could be a RN or an USN) can communicate directly with each other if and only if $d_{Euc}(u,v) \leq R_A$. We use an unweighted undirected graph $G(V,E)$ in 3-D space to model the network architecture of a 3-D UA-WSN, where $V(G)=S \cup Y$ and $Y \subseteq L$. The edge set $E(G)$ defined as follows:

- For any two nodes u, $v \in S \cup L$, E contains the undirected edge (u,v) if and only if $d_{Euc}(u,v) \leq R_A$.

Note that $G(S \cup L, E)$ is a 3-D graph corresponding to the 3-D UA-WSNs architecture when every candidate location in L is placed a RN. This topology graph is an Unit Ball Graph (UBG) and can model the 3-D UA-WSN and simplify the problem specifications without losing generality. The graph $G(S \cup L, E)$ defines all possible pair-wise communications between pairs of nodes. For the design and analysis of our algorithms, we will need to define the *relay value* of a node $u \in S \cup L$. Let $G(V,E)$ be a 3-D graph corresponding to the 3-D UA-WSNs architecture when every candidate location in L is placed a RN, where $V = S \cup L$. The *relay value* of a node $u \in V(G)$ can be defined as:

$$R(u) = \begin{cases} 0 & u \in S \\ 1 & u \in L \end{cases}$$

The *relay value* of a subgraph H of G, denoted by $R(H)$, is the number of RNs in H, i.e., $R(H) = \sum_{u \in V(H)} R(u) = |V(H) \cap L|$.

As mentioned earlier, the goal of our RNs deployment strategy is to guarantee both less number of deployed RNs and lower incremental interference simultaneously. When a RN u is placed at a candidate location, we use the number of sensors which can communicate with u in G to define the incremental inherent interference. The *interference value* of a node u, denoted by $I(u)$, i.e.,

$$I(u) = \begin{cases} 0 & u \in S \\ |N(u) \cap S| & u \in L \end{cases}$$

The *interference value* of a subgraph H of G can be denoted as $I(H) = \sum_{u \in V(H)} I(u)$.

A 3-D UBG $G(S \cup L, E)$ corresponding to the 3-D UA-WSN, together with the definitions of node-weight functions $R(u)$ and $I(u)$ collectively induces a node-weight UBG $G(S \cup L, E, R, I)$. For each node $u \in S \cup L$, we assign two node weights $R(u)$ and $I(u)$ to u. Note that the network topology must be a tree which is a subgraph of $G(S \cup L, E, R, I)$ spanning S if we use a minimum number of RNs to connect the partitioned network. Then a RNs deployment strategy for UA-WSNs will correspond to a tree T in $G(S \cup L, E, R, I)$. The deployed number of RNs and total incremental incremental interference value can be denoted as $R(T)$ and $I(T)$ respectively.

The optimization objective of our RNs deployment strategy is to guarantee both less number of deployed RNs and lower incremental interference simultaneously. For a subgraph T of $G(S \cup L, E, R, I)$, we define the *cost* of T

as $C(T) = max\{R(T)/W_1, I(T)/W_2\}$, W_1 and W_2 are given positive constant constraints. The problem studied in this paper can be formally represented as follows:

Definition 1: The Constrained Low-interference Relay Node Deployment (C-LRND) problem: Given an UA-WSN (R_A, S, L), the C-LRND problem is to find a subset Y of L in which each candidate location in Y is placed at a RN such that the network $G(S \cup Y, E)$ is 1-connected and the *cost* $C(G)$ of G is minimized, where $Y = V(G) \cap L$.

Note that the tolerance values W_1 and W_2 are the user-defined system parameters. By carefully setting the values of W_1 and W_2, deployment strategy obtained will not only have lower RNs *relay value*, but also have lower *interference value*. For example, if we set W_1 and W_2 to the solutions of *minimize* $R(T)$ and *minimize* $I(T)$, respectively. With the decrement of the value of $C(T)$, the *relay value* and *interference value* of T will not exceed W_1 and W_2 too much.

4   Algorithms for the Constrained Low-Interference Relay Node Deployment Problem

In this section, we discuss the computational complexity of the C-LRND problem and present a general approximation algorithm framework, based on approximation algorithms for the Steiner Minimum Tree (SMT) problem, for it. Finally, we analyze the quality of the result produced by the algorithm with respect to the optimal solution.

4.1   Computational Complexity and Discussions

Theorem 1: The C-LRND problem is NP-hard.

Proof: It is easy to know the Node-Weighted Steiner Minimum Tree (NW-SMT) problem is a special case of C-LRND problem since the C-LRND problem is for given a node-weight graph $G(S \cup L, E, R, I)$, to find a tree T of $G(S \cup L, E, R, I)$ which spans all nodes in S such that the total cost $C(T)$ is minimized. The NW-SMT problem is proved to be NP-hard [18], then the C-LRND problem is NP-hard. ∎

To our best knowledge, this paper is the first effort to address constrained RNs deployment to ensure small number of deployed RNs and low interference simultaneously. There have been previous studies on RNs deployment for wireless networks [6, 10], most of which focused on maintaining network connectivity with minimum number of RNs. However, if the negative influence of communication interference between nodes can not be neglected, directly applying the existing deployment strategy will only give suboptimal results for prolonging network lifetime and improving QoS. In [3, 4], the authors stated that major challenges in the design of underwater acoustic networks including: the underwater channel is severely impaired; the available bandwidth is severely limited; high bit error rates and so on. Communication interference is responsible for much of these negative influence [3, 4]. Hence, we will try to design the deployment strategy, which has to not only ensure network connectivity, but also achieve low interference between network nodes.

4.2   A General Approximation Algorithm Scheme

In this component, we present a framework of polynomial time approximation algorithm for the C-LRND problem. We prove that the *cost* $C(T)$ computed by our algorithm is no more 12α times the *cost* of optimal solution, where α is the ratio of the approximation algorithm A for the undirected graph Steiner Minimum Tree (SMT) problem. Our approximation algorithm for C-LRND is presented as Algorithm 1.

Algorithm 1. Approximation algorithm for the connected C-LRND Problem

---

Input: An UA-WSNs (R_A, S, L). An approximation algorithm for the Steiner Minimum Tree problem.
Output: An feasible solution Y_A for the C-LRND.

Begin:
1: Construct the initial UA-WSN model graph $G = (V, E, R, I)$, where $V = S \cup L$, and convert it to the single node weighted undirected graph $G^1 = (V, E, w)$, where $w(u) = max\{R(u)/W_1, I(u)/W_2\}$ for $\forall u \in V$.
2: Construct an edge weighted undirected graph $G^2 = (V, E, f)$, while setting the edge-weight $f(e)$ to $\frac{1}{2}(w(u) + w(v))$ for $\forall e \in E$.
3: Apply an approximation algorithm A for the Steiner Minimum Tree problem to compute a low weight Steiner Tree subgraph T of $G^2 = (V, E, f)$, which spans all nodes in S.
4: Output $Y_A = V(T) \cap L$.
End.

The major steps of the algorithm are as follows. First, we construct $G = (V, E, R, I)$, as if we were placing a RN at every candidate location in L. And then we convert $G = (V, E, R, I)$ to the single node weighted undirected graph $G^1 = (V, E, w)$ in which nodes set and edges set are same as $G = (V, E, R, I)$, and $w(u) = max\{R(u)/W_1, I(u)/W_2\}$ for $\forall u \in V$. This is accomplished in Line 1 of Algorithm 1. Next, we transform node weight to edge weight, for each edge $e = (u, v) \in E$ we set $f(e) = \frac{1}{2}(w(u) + w(v))$. Then the single node weighted undirected graph $G^1 = (V, E, w)$ can be convert to edge weighted undirected graph $G^2 = (V, E, f)$. This is accomplished in Line 2 of Algorithm 1. Then we apply algorithm A to compute a low weight tree subgraph T_A of $G^2 = (V, E, f)$, spanning all nodes in S. This is accomplished in Line 3 of Algorithm 1. Finally, in Line 4, we identify the locations to deploy the RNs.

4.3   Theoretical Analysis

In this subsection, we will analyze the performance of Algorithm 1. We assume T_{opt} is an optimal solution for the C-LRNP problem in $G = (V, E, R, I)$; T^1_{opt} is a minimum node-weighted Steiner tree for node set S in $G^1 = (V, E, w)$; T_{opt}^2 is the minimum edge-weighted Steiner tree for node set S in $G^2 = (V, E, f)$; And T is the solution computed by Algorithm 1. We have the following lemmas and theorem.

Lemma 1: $w(T) \leq f(T) \leq Q \cdot f(T_{opt}^2) \leq \frac{\Delta(T_{opt}^1)}{2} \cdot Q \cdot w(T_{opt}^1)$.

Proof: Firstly, let us prove the first inequality: $w(T) \leq f(T)$. Note that tree T which is a Steiner tree in $G^2 = (V, E, f)$ spanning all nodes in set S got by Algorithm 1, also is the Steiner tree for set S in $G = (V, E, R, I)$ and $G^1 = (V, E, w)$. The node weight of each node in S in $G^1 = (V, E, w)$ is equal to 0, and the degree of every node in $T - S$ is at least 2. The definition of edge weight of (u, v) in $G^2 = (V, E, f)$ is: $f((u, v)) = \frac{1}{2}(w(u) + w(v))$. We divide the edge weight of (u, v) into two pieces, add the weight $\frac{w(u)}{2}$ to u, add the weight $\frac{w(v)}{2}$ to v. After all edges are looped over, we have shifted the edge weights of T to the nodes in T. It is clear that the shifted node weight of the node u in T always be greater than $w(u)$ in $G^1 = (V, E, w)$. Hence we have:

$$w(T) \leq f(T) \tag{1}$$

Since T_{opt}^2 is the minimum edge-weighted Steiner tree for node set S in $G^2 = (V, E, f)$, and T is a edge-weighted Steiner tree for S in $G^2 = (V, E, f)$, which is computed by a Q-approximation algorithm for SMT problem. So we can have the second inequality.

$$f(T) \leq Q \cdot f(T_{opt}^2) \tag{2}$$

Note that T_{opt}^1 is the minimum node-weighted Steiner tree for node set S in $G^1 = (V, E, w)$ and T_{opt}^2 is the minimum edge-weighted Steiner tree for node set S in $G^2 = (V, E, f)$, but f is about edge weight, therefore

$$f(T_{opt}^2) \leq f(T_{opt}^1) = \sum_{(u,v) \in E(T_{opt}^1)} \frac{1}{2}(w(u) + w(v))$$

$$= \sum_{u \in V(T_{opt}^1)} \frac{1}{2} d_{T_{opt}^1}(u) \cdot w(u) \tag{3}$$

Let $\Delta(T_{opt}^1)$ denote the maximum degree of T_{opt}^1, from inequality (3), we have

$$f(T_{opt}^2) \leq \frac{\Delta(T_{opt}^1)}{2} \cdot w(T_{opt}^1) \tag{4}$$

From inequalities (1) (4), this lemma thus follows.   ∎

Lemma 2: $w(T_{opt}^1) \leq w(T_{opt}) \leq 2 \cdot C(T_{opt})$.

Proof: Since T_{opt}^1 is the minimum node-weighted Steiner tree for node set S in $G^1(V, E, w)$ and T_{opt} also is a Steiner tree for S, the first inequality follows.

Note that T_{opt} is the optimal solution of the C-LRNP problem in $G = (V, E, R, I)$, we have the following formulas

$$\sum_{u \in T_{opt}} \frac{R(u)}{W_1} = \frac{R(T_{opt})}{W_1} \leq max\{\frac{R(T_{opt})}{W_1}, \frac{I(T_{opt})}{W_2}\} = C(T_{opt}) \qquad (5)$$

$$\sum_{u \in T_{opt}} \frac{I(u)}{W_2} = \frac{I(T_{opt})}{W_2} \leq max\{\frac{R(T_{opt})}{W_1}, \frac{I(T_{opt})}{W_2}\} = C(T_{opt}) \qquad (6)$$

Furthermore, we have

$$\sum_{u \in T_{opt}} (\frac{R(u)}{W_1} + \frac{I(u)}{W_2}) =$$

$$= \{\frac{R(T_{opt})}{W_1} + \frac{I(T_{opt})}{W_2}\} \leq 2 \cdot C(T_{opt}) \qquad (7)$$

Note that $w(u) = max\{R(u)/W_1, I(u)/W_2\}$, then we have:

$$w(u) \leq (\frac{R(u)}{W_1} + \frac{I(u)}{W_2}) \qquad (8)$$

Combining formula (5)-(8), we have

$$w(T_{opt}) = \sum_{u \in T_{opt}} w(u) \leq 2 \cdot C(T_{opt}) \qquad (9)$$

This lemma follows.   ∎

Lemma 3: $C(T) \leq w(T)$.

Proof: Since

$$C(T) = max\{\frac{R(T)}{W_1}, \frac{I(T)}{W_2}\}$$

$$= max\{\sum_{u \in T} \frac{R(u)}{W_1}, \sum_{u \in T} \frac{I(u)}{W_2}\}$$

$$\leq \sum_{u \in T} max\{\frac{R(u)}{W_1}, \frac{I(u)}{W_2}\}$$

$$= \sum_{u \in T} w(u) = w(T). \qquad (10)$$

This lemma thus follows.   ∎

Lemma 4: Let T_{opt}^1 be the optimal solution with the shortest total Euclidean edge length among the minimum node-weighted Steiner trees for node set S in $G^1(V, E, w)$, then $\Delta(T_{opt}^1) \leq 12$.

Proof: We first claim that in T_{opt}^1 any two edges incident to a node form an angle of at least $\frac{\pi}{3}$. Suppose (u, v) and (u, w) are any two edges meeting at u in T, and $\angle vuw < \frac{\pi}{3}$, there must be an angle larger than $\frac{\pi}{3}$ in $\triangle uvw$. Without loss of generality, we assume that $\angle uvw > \frac{\pi}{3}$. Then from the **Sine Theorem**, we have

$$\frac{d_{Euc}(v, w)}{\sin \angle vuw} = \frac{d_{Euc}(u, w)}{\sin \angle uvw}$$

Since $0 < \angle vuw < \frac{\pi}{3} < \angle uvw < \pi - \angle vuw < \pi$, we have $\sin \angle vuw < \sin \angle uvw$ and $d_{Euc}(v, w) < d_{Euc}(u, w)$. Replacing the edge (u, w) with (v, w) in T_{opt}^1 results in T^2, which is also the minimum node-weighted Steiner tree for node set S in $G^1(V, E, w)$, but $d_{Euc}(T^2) < d_{Euc}(T_{opt}^1)$. This contradicts with that T_{opt}^1 is the optimal solution with the shortest total Euclidean edge length among the minimum node-weighted Steiner trees for node set S in $G^1(V, E, w)$. This contradiction proves that in T_{opt}^1 any two edges meeting at a node form an angle of at least $\frac{\pi}{3}$.

Next we claim that for any node u of T_{opt}^1, $d_{T_{opt}^1}(u) \le KN_3$, where $KN_3 = 12$ [19]. If not, there is a node u with $d_{T_{opt}^1}(u) > KN_3$. For $\forall x \in N(u)$, we can pull x to x' along $\overrightarrow{ux}$ such that $d_{Euc}(u, x') = R_A$. $\forall x_1, x_2 \in N(u)$, it is clear that $\angle x_1 u x_2 = \angle x_1' u x_2'$. If we can prove $\angle x_1' u x_2' < \frac{\pi}{3}$, we can get a contradiction from above claim. For every $x \in N(u)$, we draw a ball centered in x' with radius $\frac{R_A}{2}$. Since $d_{T_{opt}^1}(u) > KN_3$, there exist two balls centered in x_1', x_2' intersect [19]. Therefore $d_{Euc}(x_1', x_2') < R_A$. Then $\angle x_1' u x_2' < \angle u x_1' x_2' = \angle u x_2' x_1'$. It implies that $\angle x_1' u x_2' < \frac{\pi}{3}$. Therefore, for any node u of T_{opt}^1, we have $d_{T_{opt}^1}(u) \le KN_3$, that is $\Delta(T_{opt}^1) \le 12$. This lemma thus follows. ∎

Theorem 2: $C(T) \le 12 \cdot Q \cdot C(T_{opt})$.

Proof: Combining **Lemma 1, 2, 3**, we have

$$C(T) \le w(T) \le \Delta(T_{opt}^1) \cdot Q \cdot C(T_{opt}) \tag{11}$$

From **Lemma 4**: $\Delta(T_{opt}^1) \le 12$ and (11), we get the theorem follows. ∎

Corollary 1: The C-LRNP problem has a polynomial time 18.6-approximation algorithm.

Proof: According to the conclusion in [20], there is a polynomial time approximation algorithm for the Steiner Minimum Tree problem whose approximation ratio is at most 1.55. This corollary follows from Theorem 2 with Q=1.55. ∎

Corollary 2: The C-LRNP problem has a polynomial time 24-approximation algorithm with time complexity of $O(|S \cup L|^2) \log |S \cup L|$.

Proof: If we take A in Algorithm 1 as the MST based 2-approximation algorithm for the Steiner Minimum Tree problem [21], the time complexity of Algorithm 1 is $O(|S \cup L|^2) \log |S \cup L|$. The corresponding approximation ratio of Algorithm 1 follows from Theorem 2. ∎

5 Conclusions

In this paper, we studied the C-LRND problem in Underwater Acoustic Wireless Sensor Networks (UA-WSNs). We mainly addressed the connected RNs deployment problem under both the number of RNs and the value of total incremental interference constraints. And presented an approximation algorithm framework for this problem.

Acknowledge

This paper was supported in part by the National Natural Science Foundation of China under Grant 61070191 and Renmim University of China under Grant 10XNJ032.

References

1. Akyildiz, I.F., Pompili, D., Melodia, T.: Underwater Acoustic Sensor Networks: Research Challenges. Elesviers Journal of Ad Hoc Networks 3(3), 257–279 (2005)
2. Akyildiz, I.F., Pompili, D., Melodia, T.: State of the Art in Protocol Research for Underwater Acoustic Sensor Networks. In: Proc. of the ACM WUWNet (2006)
3. Akyildiz, I.F., Pompili, D., Melodia, T.: Challenges for Efficient Communication in Underwater Acoustic Sensor Networks. SIGBED Rev. 2(1), 1–6 (2004)
4. Partan, J., Kurose, J., Levine, B.N.: A Survey of Practical Issues in Underwater Networks. In: Proc. of the ACM WUWNet (2006)
5. Pompili, D., Melodia, T., Akyildiz, I.F.: Deployment Analysis in Underwater Acoustic Wireless Sensor Networks. In: Proc. of the ACM WUWNet (2006)
6. Lloyd, E., Xue, G.: Relay Node Placement in Wireless Sensor Networks. IEEE Trans. on Computers 56, 134–138 (2007)
7. Jain, K., Padhye, J., Padmanabhan, V.N., Qiu, L.: Impact of Interference on Multi-hop Wireless Networks. In: Proc. of the ACM MOBICOM (2003)
8. Misra, S., Hong, S.D., Xue, G., Tang, J.: Constrained Relay Node Placement in Wireless Sensor Networks to Meet Connectivity and Survivability Requirements. In: Proc. of the IEEE INFCOM (2008)
9. Chen, D., Du, D., Hu, X., Lin, G., Wang, L., Xue, G.: Approximations for Steiner Trees with Minimum Number of Steiner Points. Journal of Global Optimization 18, 17–33 (2000)
10. Lin, G., Xue, G.: Steiner Tree Problem with Minimum Number of Steiner Points and Bounded Edge-Length. Information Processing Letters 69, 53–57 (1999)
11. Cheng, X., Du, D., Wang, L., Xu, B.: Relay Sensor Placement in Wireless Sensor Networks. In: ACM/Springer WINET (2008)
12. Han, X., Cao, X., Lloyd, E.L., Shen, C.C.: Fault-tolerant Relay Node Placement in Heterogeneous Wireless Sensor Networks. In: Proc. of the IEEE INFOCOM (2007)
13. Bredin, J.L., Demaine, E.D., Hajiaghayi, M., Rus, D.: Deploying Sensor Networks with Guaranteed Capacity and Fault Tolerance. In: Proc. of the ACM MOBIHOC (2005)
14. Kashyap, A., Khuller, S., Shayman, M.: Relay Node Placement for Higher Order Connectivity in Wireless Sensor Networks. In: Proc. of the IEEE INFOCOM (2006)

15. Zhang, W., Xue, G., Misra, S.: Fault-tolerant Relay Node Placement in Wireless Sensor Networks: Problem and Algorithms. In: Proc. of the IEEE INFOCOM (2007)
16. Ibrahim, S., Cui, J.H., Ammar, R.: Surface-level Gateway Deployment for Underwater Sensor Networks. In: Proc. of the IEEE MILCOM (2007)
17. Seah, W.K.G., Tan, H.X.: Multipath Virtual Sink Architecture for Underwater Sensor Networks. In: Proc. of the OCEANS (2006)
18. Du, D., Hu, X.: Steiner Tree Problems in Computer Communication Networks. World Scientific Publishing Co. Pte. Ltd., Singapore (2008)
19. Conway, J.H., Sloane, N.J.A.: Sphere Packing, Lattices and Groups, 3rd edn. Springer, New York (1999)
20. Robins, G., Zelikovsky, A.: Tighter Bound for Graph Steiner Tree Approximation. SIAM J. on Discrete Mathmatics 19, 122–134 (2005)
21. Kou, L.T., Markowsky, G., Berman, L.: A Fast Algorithm for Steiner Tree. Acta Informatica 15, 141–145 (1981)

Structured Overlay Network for File Distribution

Hongbing Fan[1,*] and Yu-Liang Wu[2,**]

[1] Wilfrid Laurier University, Waterloo, ON Canada N2L 3C5
hfan@wlu.ca
[2] The Chinese University of Hong Kong, Shatin, N.T., Hong Kong
ylw@cse.cuhk.edu.hk

Abstract. The file distribution from a source node to n sink nodes along a structured overlay network can be done in time $\Theta(\log n)$. In this paper, we model the problem of finding an optimal overlay network for file distribution as a combinatorial optimization problem, i.e., finding a weighted spanning tree which connects the source node and sink nodes and has the minimum file distribution time. We use an edge-based file distribution protocol, in which after a node receives a file it then transfers the file to its neighbor nodes one after another in a sequential order. We give the formulation of file distribution time, and use it as the objective function. The corresponding combinatorial optimization problem is NP-hard in general. We present a heuristic algorithm which derives an overlay network with file distribution time $\Theta(\log n)$ and show that the derived overlay network is optimal if the file transfer delays between all pairs of nodes are the same.

Keywords: File distribution, Peer-to-Peer, overlay network, parallel computing.

1 Introduction

Peer-to-peer (P2P) technology has been widely used in content distribution on the Internet. The fundamental problem is to distribute a file from a source peer to a large number of sink peers. Popular applications include BitTorrent [2] for file distribution, Skype for VoIP, PPLive for IPTV. The core feature of P2P systems is the capability to build overlay network to carry out computing tasks over multiple peers. The well-known file distribution applications, such as BitTorrent, use non-structured overlay networks where a peer connects to some seed peers to download different parts of a file and disconnects after the download is complete. A lot of work has been done in recent years on the BitTorrent-type protocols to improve performance in file distribution [3,7,4]. On the other hand, the structured overlay network approach distributes files along a pre-given overlay network. The structured overlay network approach is not as flexible and scalable as the BitTorrent approach, but it can be more efficient than the

---

* Research partially supported by the NSERC, Canada.
** Research partially supported by RGC Earmarked Grant 2150500 Hong Kong.

W. Wu and O. Daescu (Eds.): COCOA 2010, Part II, LNCS 6509, pp. 292–302, 2010.
© Springer-Verlag Berlin Heidelberg 2010

non-structured overlay network approach for file distribution when the set of sink peers is known at the beginning. The superior performance is achieved by better overlay network topology, a simple protocol, a parallel processing scheme and a lower overhead on interconnection reconfiguration. The structured overlay network approach has been used in P2P-based stream video data broadcast [8].

One problem of the structured overlay network approach for file distribution is to find an optimal overlay network topology which minimizes file distribution time according to the given file distribution protocol. Similar problems have been studied in the field of graph theory and network optimization. For an example, the broadcast graph problem studied in [1,6] is to find an minimum broadcast graph on n nodes that allows any node to broadcast to the other $n-1$ nodes in time $lceil \log_2 n \rceil$. The problem was studied under the assumption that a message is transferred along an edge in a unit time and a node can send at most one message at a unit time. In [5], a k-broadcasting graph problem was studied, in which a node can send up to k of its neighbors in each time unit. These work assures that with proper choice of structured overlay network, file distribution can be done in time $O(\log n)$ using a simple file distribution protocol similar to the message distribution.

In this paper, we investigate the optimal overlay network topology problem according to an edge-based file distribution protocol (EFDP). For the convenience of description, we use the terms node and edge instead of peer and connection. The EFDP has the following two properties: First, the file distribution process comprises a plurality of single file transfers, each transfer sends the file from one node to another along an edge of the overlay network. Second, after a node receives a file (i.e., the whole file data arrives and is saved completely at the node), it starts to transfer the file to its first neighbor node (if the node does not receive the file yet nor it is in the process of receiving the file) and when the transfer is done, it transfers the file to the second neighbor node, and continues this process till the last neighbor node receives the file. The file distribution time along an overlay network is the total time span from the first file transfer at the source node to the end of last file transfer at all sink nodes. We are to find an overlay network, along which a minimum file distribution time can be achieved-such an overlay network is said to be optimal.

It is noticed that with the EFDP, after two or more nodes receive a file, the file distribution begins to progress in parallel, i.e. multiple file transfers happen simultaneously at multiple nodes and edges. This parallel feature is the key factor that accelerates file distribution along a structured overlay network with EFDP. Therefore, an optimal overlay network must maximize parallel progressing. It is also noticed that an optimal overlay network can be a rooted spanning tree. This is because the best distribution strategy along an overlay network must use a set of edges to transfer the file. The set of edges must form a connected graph covering all nodes. There is no loop in the graph as otherwise a node would receive the file twice. The root is the node which has the file at the beginning. The direction of an edge can be determined by the direction of file transfer, which must be away from the root. The real file distribution time along a tree is affected by many factors: the file size, the topology of the tree, the ordering of children at each node, the file transfer time along each edge, as well as the file i/o time at each node. The file transfer time along an edge and the i/o time are also dynamic. To simplify the problem, we assume that the file to be distributed has a unit size, and we combine the file i/o times on both sides of an edge and file transfer time along the edge into a non-dynamic time measure, called file transfer delay. Specifically, the file transfer delay along edge (p_i, p_j), denoted by $t(p_i, p_j)$, is the expectation (or estimation) of the amount of time to transfer a unit file from p_i to p_j. We also use the maximum outgoing degree b as a parameter to control the number of file transfers from a node so as to balance the amount of data flowing out a node. b can be infinite when there is no such a balance constraint.

With the above observations and assumptions, the optimal overlay network topology problem associated with EFDP is to find a rooted spanning tree such that each node has at most b children and that it minimizes the file distribution time. In Section 2, we will give a formula to compute the file distribution time, followed by the formal description of the optimal overlay network topology problem for EFDP. The proposed problem is NP-hard in general, so we present a heuristic algorithm together with a theorem. The theorem claims that using the tree derived by the heuristic algorithm, the file distribution can be done in time $\Theta(\log n)$, and that the algorithm derives an optimal tree when the file transfer delays along all edges are the same. The detailed proof of the theorem is given Section 3.

2   Problem Formulation

Let $\{p_0, p_1, \ldots, p_{n-1}\}$ denote the set of all nodes. Assume that node p_0 is the source node with a unit file to be distributed, and nodes $p_1, \ldots, p_{n-1}$ are the sink nodes to receive the file. Let $t(p_i, p_j)$ denote the time to transfer a unit file from node p_i to p_j along edge (p_i, p_j), $i, j = 0, 1, \ldots, n-1$. Let $T = (V, E)$ be a directed tree on $V = \{p_0, p_1, \ldots, p_{n-1}\}$ rooted at p_0. Denote by $D_T^+(p)$ the set of children of p in T, i.e., $D_T^+(p) = \{p' : (p, p') \in E\}$.

2.1   File Distribution Time

By EFDP, after a node p of T receives a file, it then starts to send the file to its first child. After the transmission is done, it transfers the file to the second child, and it continues until all of its children receive the file. Let $t_T(p)$ denote the minimum amount of time required to distribute the file from p to all of its descendants. Next, we present a formula to compute $t_T(p)$.

Suppose that $D_T^+(p) = \{p_j : j = i_1, \ldots, i_k\}$ and that p sends the file to its children in order $p_{\pi(i_1)}, p_{\pi(i_2)}, \ldots, p_{\pi(i_k)}$, where $\pi : \{i_1, \ldots, i_k\} \to \{i_1, \ldots, i_k\}$ is a permutation of $\{i_1, \ldots, i_k\}$. Then child $p_{\pi(i_j)}, 1 \le j \le k$, receives the file at time $\sum_{h=1}^{j} t(p, p_{\pi(i_h)})$ and further more $p_{\pi(i_j)}$ and all of its descendants receive the file at time $\sum_{h=1}^{j} t(p, p_{\pi(i_h)}) + t_T(p_{\pi(i_j)})$. Hence with the ordering of children

$p_{\pi(i_1)}, p_{\pi(i_2)}, \ldots, p_{\pi(i_k)}$, the time it takes for all descendants of p to receive the file can be expressed as

$$t_T(p, \pi) = \max\{\sum_{h=1}^{j} t(p, p_{\pi(i_h)}) + t_T(p_{\pi(i_j)}) : j = 1, \ldots, k\}.$$

Therefore, the minimum time $t_T(p)$ required for all descendants of p to receive the file can be expressed as

$$t_T(p) = \min_{\pi} t_T(p, \pi) = \min_{\pi} \max\{\sum_{h=1}^{j} t(p, p_{\pi(i_h)}) + t_T(p_{\pi(i_j)}) : j = 1, \ldots, k\}, \quad (1)$$

where π is over all permutations of $\{i_1, \ldots, i_k\}$. Hence, the total time to distribute a file from p_0 to $p_1, \ldots, p_{n-1}$ is $t_T(p_0)$, and it can be computed in a bottom-up approach by formula (1). Since at each node p the value $t_T(p)$ is calculated over all permutations of its children, the question is if $t_T(p_0)$ can be computed efficiently (i.e., in polynomial time). The answer to this question is yes.

Theorem 2.1. *The following formula for $t_T(p)$ holds.*

$$t_T(p) = \max\{\sum_{h=1}^{j} t(p, p_{\pi_0(i_h)}) + t_T(p_{\pi_0(i_j)}) : j = 1, \ldots, k\}, \quad (2)$$

where π_0 is a permutation of $\{i_1, \ldots, i_k\}$ with sorted values $t_T(p_{\pi_0(i_1)}) \geq \cdots \geq t_T(p_{\pi_0(i_k)})$. $t_T(p_0)$ can be computed by formula (2) in time $O(n \log n)$.

Proof. First, if the ordering of children is not in decreasing order then distribution time can be reduced. Suppose $t_T(p_{\pi(i_{j_0})}) < t_T(p_{\pi(i_{j_0+1})})$, then swap elements $\pi(i_{j_0})$ and $\pi(i_{j_0+1})$, we have a new permutation permutation π'. Since

$\sum_{h=1}^{j_0-1} t(p, p_{\pi(i_h)}) + t(p, p_{\pi(i_{j_0+1})}) + t_T(p_{\pi(i_{j_0+1})}) \leq \sum_{h=1}^{j_0+1} t(p, p_{\pi(i_h)}) + t_T(p_{\pi(i_{j_0+1})})$
and
$\sum_{h=1}^{j_0+1} t(p, p_{\pi(i_h)}) + t_T(p_{\pi(i_{j_0})}) < \sum_{h=1}^{j_0+1} t(p, p_{\pi(i_h)}) + t_T(p_{\pi(i_{j_0+1})})$, we have
$t_T(p, \pi') = \max\{\sum_{h=1}^{j} t(p, p_{\pi(i_h)}) + t_T(p_{\pi(i_j)}) : j = 1, \ldots, j_0 - 1\} \cup$
$\{\sum_{h=1}^{j_0-1} t(p, p_{\pi(i_h)}) + t(p, p_{\pi(i_{j_0+1})}) + t_T(p_{\pi(i_{j_0+1})}), \sum_{h=1}^{j_0+1} t(p, p_{\pi(i_h)}) + t_T(p_{\pi(i_{j_0})})\}$
$\cup \{\sum_{h=1}^{j} t(p, p_{\pi(i_h)}) + t_T(p_{\pi(i_j)}) : j = j_0 + 2, \ldots, k\}$
$\leq$
$\max\{\sum_{h=1}^{j} t(p, p_{\pi(i_h)}) + t_T(p_{\pi(i_j)}) : j = 1, \ldots, j_0 - 1\} \cup$
$\{\sum_{h=1}^{j_0} t(p, p_{\pi(i_h)}) + t_T(p_{\pi(i_{j_0})}), \sum_{h=1}^{j_0+1} t(p, p_{\pi(i_h)}) + t_T(p_{\pi(i_{j_0+1})})\} \cup$
$\{\sum_{h=1}^{j} t(p, p_{\pi(i_h)}) + t_T(p_{\pi(i_j)}) : j = j_0 + 2, \ldots, k\} = t_T(p, \pi).$

Second, if the ordering of children is in decreasing order then distribution time can not be reduced. We only need to show that if $t_T(p_{\pi(i_{j_0})}) = t_T(p_{\pi(i_{j_0+1})})$ then swapping children $p_{\pi(i_{j_0})}$ and $p_{\pi(i_{j_0+1})}$ will not change the distribution time

at p. Let π' be the permutation after the swapping. Since $\sum_{h=1}^{j_0-1} t(p, p_{\pi(i_h)}) + t(p, p_{\pi(i_{j_0+1})}) + t_T(p_{\pi(i_{j_0+1})}) \leq \sum_{h=1}^{j_0+1} t(p, p_{\pi(i_h)}) + t_T(p_{\pi(i_{j_0})})$
$= \sum_{h=1}^{j_0+1} t(p, p_{\pi(i_h)}) + t_T(p_{\pi(i_{j_0+1})})$, we have $t_T(p, \pi') = t_T(p, \pi)$.

Therefore, $t_T(p, \pi_0) = \min_\pi t_T(p, \pi) = t_T(p)$. when $t_T(p_{\pi_0(i_1)}) \geq \cdots \geq t_T(p_{\pi_0(i_k)})$.

Since sorting and finding the maximum element can be done in polynomial time, $t_T(p_0)$ can be calculated efficiently bottom-up by formula (2). Further more, each node p takes $\sum_p |D_T^+(p)| = O(n)$ time to find the maximum value by formula (2), and it takes $\sum_p |D_T^+(p)| \log |D_T^+(p)| = O(\log n)$ time to insert p into the existing ordered list of its siblings. There are n nodes, therefore the bottom-up evaluation of $t_T(p_0)$ can be done in time $O(n \log n)$. ∎

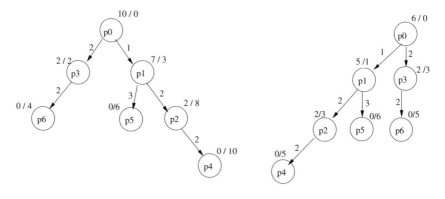

(a) file distribution time evaluation   (b) file distribution time evaualrion
by a given children ordering   by the best children ordering

Fig. 1. File distribution time evaluation along a tree

Fig.1 shows an example of file distribution time evaluation along a tree, in which the number near to an edge (p_i, p_j) is the time delay $t(p_i, p_j)$. At each node, we transfer the file to children in the order of left to right of the tree layout. The number pair x/y at node p means that $x = t_T(p, \pi)$ and y is the time at which p receives the file along the tree. The value of y is obtained by top-down following the child ordering, and the y value of all nodes can be calculated in linear time $O(n)$. Given a tree and children ordering π of each node, we evaluate the $t_T(p, \pi)$ bottom-up for all nodes p of T. Fig.1(a) shows distribution time evaluations by the child ordering from left to right of tree layout. The file distribution time of this ordering is 10. The best ordering can be obtained by the bottom-up method. Starting from leaves, after all children of a node p have been evaluated, we order its children p_{i_j}'s by the decreasing ordering of $t_T(p_{i_j})$ values, then evaluate $t_T(p)$ using this ordering. Fig.1(b) shows the best ordering of children of each node obtained by this method. The file distribution time along the tree is $t_T(p_0) = 6$.

2.2   EFDP Overlay Network Topology Problem

The combinatorial optimization problem of finding an optimal overlay network for file distribution can be described as follows.

Problem 2.2. EFDP Overlay Network Topology Problem
Input: a node set $\{p_0, p_1, \ldots, p_{n-1}\}$ and delays $t(p_i, p_j) > 0, i, j = 0, \ldots, n-1$ and an integer $b \geq 1$ or $b = +\infty$.
Output: a directed tree T on $\{p_0, p_1, \ldots, p_{n-1}\}$ rooted at p_0 such that each node has at most b children and $t_T(p_0)$ is minimum.

We see that problem 2.2 is NP-hard. This is because when $b = 1$, an optimal solution is a shortest Hamiltonian path, while finding a shortest Hamiltonian path in a weighted complete graph is NP-hard. However, for a fixed $b \geq 2$ or $b = +\infty$, it is not known if the problem is NP-hard. We leave it as an open question.

2.3   Heuristic Algorithm

Next we present a heuristic algorithm for the EFDP overlay network topology problem. The idea is to distribute the file to neighbor nodes as early as possible. We will show that the algorithm returns an optimal overlay network when $t(p_i, p_j)$ is a constant for all $i, j = 0, \ldots, n-1$.

EFDP topology algorithm

1. Let $V = \{p_0\}, E = \emptyset, T = (V, E), t(p_0) = 0, s(p_0) = 0$ and $A = \{p_0\}, B = \{p_0, p_1, \ldots, p_{n-1}\} \setminus A$.
2. If $|B| = 0$, output T and stop.
3. If $|A| > 0$, find a $p \in A$ and a $p' \in B$ such that $t(p) + s(p) + t(p, p')$ is minimum among all edges from A to B. Set $V = V \cup \{p\}, E = E \cup \{(p, p')\}, A = A \setminus \{p\}, B = B \setminus \{p'\}, s(p) = s(p) + t(p, p')$ and $t(p') = t(p) + s(p), s(p') = 0$. Go step 2.
4. If $|A| = 0$, reset A to be the set of all nodes of $T = (V, E)$ with outgoing degree less than b. Go to step 2.

The above EFDP topology algorithm progresses by adding new nodes (and edges) one after another. In each adding, it takes at most $O(n^2)$ time to find a pair $p \in A, p' \in B$ such that $t(p) + s(p) + t(p, p')$ is minimum, and constant time to update in step 3. Therefore, the running time of the algorithm is $O(n^3)$. We can modify step 3 of the EFDP topology algorithm to reduce the running time as follows.

Step 3': If $|A| > 0$, randomly choose a $p \in A$ and find a $p' \in B$ such that $t(p, p')$ is minimum among all edges from p to B. Set $V = V \cup \{p\}, E = E \cup \{(p, p')\}, A = A \setminus \{p\}, B = B \setminus \{p'\}$. Go step 2.
Step 3": If $|A| > 0$, randomly choose a $p \in A$ and a $p' \in B$. Set $V = V \cup \{p\}, E = E \cup \{(p, p')\}, A = A \setminus \{p\}, B = B \setminus \{p'\}$. Go step 2.

Clearly, the running time of the modified EFDP topology algorithm is $O(n^2)$ with step 3', and $O(n)$ with step 3". In case when $t(p_i, p_j)$ is equal to the same value for all $i, j = 0, \ldots, n-1$, the modified algorithm with step 3" will be used since it produces a tree with the same distribution time as that produced by the EFDP topology algorithm with step 3.

Theorem 2.3. *When $b \geq 2$, the EFDP topology algorithm returns a tree T with $t_T(p_0) = \Theta(M \log n)$ in time $O(n^3)$ where $M = \max\{t(p_i, p_j) : i, j = 0, \ldots, n - 1\}$. Particularly, when $t(p_i, p_j)$ equals the same value C for all $i, j = 0, \ldots, n-1$, an optimal tree T with $t_T(p_0) = \Theta(C \log n)$ can be derived in time $O(n)$.*

3   Proof of the Theorem

We first show that when the file transfer delays are all of the same value, the EFDP topology algorithm returns an optimal tree. Without loss of generality, assume that $t(p_i, p_j) = 1$ for all $i, j = 0, \ldots, n-1$. We also assume that $b \geq 2$ as when $b = 1$, the algorithm returns a path which is obviously an optimal solution of the case. Let $T_b(n)$ denote the tree of n nodes obtained by the algorithm.

Fig.2 shows the optimal tree topology for $b = 2$ and $n = 1, 2, 4, 7$. From the construction point of view, in the first time period, the file is transferred from p_0 to p_1. In the second time period, both p_0 and p_1 can send the file to new nodes. Intuitively, we can construct an optimal tree iteratively by adding as many new nodes as possible at each iteration and the outgoing degree of a node is at most b. The EFDP topology algorithm follows this idea. Next we show that the tree $T_b(n)$ obtained by the EFDP topology algorithm is an optimal overlay network for EFDP.

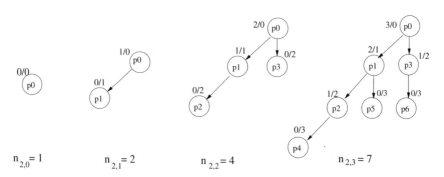

Fig. 2. Optimal overlay network for parallel file distribution

Lemma 3.1. *$T_b(n)$ is an optimal overlay network for EFDP.*

Proof. Assume otherwise, that there is a tree T with outgoing degree at most b and $t_T(p_0) < t_{T_b(n)}(p_0)$ and $|T| = |T_b(n)| = n$. Let $m_T(i)$ and $m_{T_b(n)}(i)$ denote the number of new nodes that have received the file at end of time period i along T and $T_b(n)$ respectively. Let T^i and $T_b^i(n)$ denote the subtree induced by nodes that have received the file at the end of time period i along T and $T_b(n)$ respectively. Then $\sum_{i=0}^{t_T(p_0)} m_T(i) = \sum_{i=0}^{t_{T_b(n)}(p_0)} m_{T_b(n)}(i) = n$. Since $t_T(p_0) < t_{T_b(n)}(p_0)$, there must exist an i' such that $0 < i' \leq t_T(p_0)$ and $m_T(i') > m_{T_b(n)}(i')$. Let i' be the first i' satisfying this condition. That is $0 < i' \leq t_T(p_0)$ and $m_T(i) = m_{T_b(n)}(i), i = 0, \ldots, i' - 1$ and $m_T(i') > m_{T_b(n)}(i')$. By the construction of $T_b(n)$, $T_b^i(n)$ is obtained from $T_b^{i-1}(n)$ by adding new nodes to all nodes of outgoing degree less than b, $i = 1, \ldots, i' - 1$. This $T_b^i(n)$ is unique in terms isomorphism $i = 0, \ldots, i' - 1$. Since $m_T(i) = m_{T_b(n)}(i), i = 0, \ldots, i' - 1$, then T^i is isomorphic to $T_b^i(n)$ for $i = 0, \ldots, i' - 1$. By the construction of $T_b(n)$, $m_{T_b(n)}(i')$ is the maximum number of nodes with outgoing degree less than b in $T_b^{i-1}(n)$. Since $m_T(i') > m_{T_b(n)}(i')$ and that $T^{i'-1}$ is isomorphic to $T_b^{i'-1}(n)$, $T^{i'}$ must have a node with outgoing degree bigger than b, contradicting that T is a tree with outgoing degree of at most b.   ∎

Next we show that the optimal tree $T_b(n)$ has file distribution time $t_{T_b(n)}(p_0) = \Theta(\log n)$. By the EFDP topology algorithm, the tree $T_b(n)$ is constructed iteratively in term of A. Each iteration starts from $|A| > 0$ and ends with $|A| = 0$. In such an iteration (i.e., step 3), some edges are added to the existing tree. Let $T_{b,j}$ denote the tree obtained after the jth iteration.

Lemma 3.2. *If $T_b(n)$ is obtained from $T_{b,i}$ at iteration i, then $t_{T_b(n)}(p_0) = i+1$.*

Proof. From the way that $T_b(n)$ is constructed, we see that at each node p, the $t_{T_b(n)}(p)$ is equal to the time value of its left child plus 1, so it equals the height of the subtree at p. For each iteration, the height of $t_{T_b(n)}(p_0)$ increases by 1, therefore at the end of the ith iteration, the height of the tree is $i + 1$.   ∎

Let $n_{b,j} = |T_{b,j}|$ and $d_{i,j}$ denote the number of nodes with outgoing degree equal to i in $T_{b,j}$ for $i = 0, \ldots, b$. Then the following relations hold.

$$
\begin{aligned}
(d_{0,0}, d_{1,0}, \ldots, d_{b,0}) &= (1, 0, \ldots, 0), \\
n_{b,0} &= \sum_{i=0}^{b} d_{i,0} = 1, \\
d_{0,j+1} &= \sum_{i=0}^{b-1} d_{i,j}, \\
d_{i,j+1} &= d_{i-1,j}, i = 1, \ldots, b - 1, \\
d_{b,j+1} &= d_{b-1,j} + d_{b,j}, \\
n_{b,j+1} &= \sum_{i=0}^{b} d_{i,j+1}.
\end{aligned}
$$

If $0 \leq j \leq b$, then $n_{b,j} = 2^j$, else $j > b$, we have,

$$
\begin{aligned}
d_{b,j} &= d_{b-1,j-1} + d_{b,j-1} \\
&= d_{b-2,j-2} + d_{b-1,j-2} + d_{b,j-2} \\
&= d_{b-3,j-3} + d_{b-2,j-3} + d_{b-1,j-3} + d_{b,j-3} \\
&= \ldots \\
&= d_{0,j-b} + d_{1,j-b} + \ldots + d_{b-1,j-b} + d_{b,j-b} \\
&= n_{b,j-b},
\end{aligned}
$$

$$n_{b,j+1} = \sum_{i=0}^{b} d_{i,j+1} = d_{0,j+1} + \sum_{i=1}^{b-1} d_{i,j+1} + d_{b,j+1}$$
$$= \left(\sum_{i=0}^{b-1} d_{i,j}\right) + \sum_{i=1}^{b-1} d_{i-1,j} + (d_{b-1,j} + d_{b,j})$$
$$= \left(\sum_{i=0}^{b-1} d_{i,j} + d_{b,j}\right) + \left(\sum_{i=0}^{b-2} d_{i,j} + d_{b-1,j} + d_{b,j}\right) - d_{b,j}$$
$$= 2\sum_{i=0}^{b} d_{i,j} - d_{b,j}$$
$$= 2n_{b,j} - d_{b,j}$$
$$= 2n_{b,j} - n_{b,j-b}$$

Therefore, $n_{b,j}$ can be computed efficiently by the following recursive formula.

$$n_{b,0} = 1, n_{b,1} = 2, n_{b,2} = 2^2, \ldots, n_{b,b} = 2^b,$$
$$n_{b,j} = 2n_{b,j-1} - n_{b,j-1-b}, j = b+1, \ldots$$

Lemma 3.3. $n_{b,j+1} \geq n_{b,j} \geq 1$ for $b \geq 2$ and $j > 0$.

Proof. For any fixed integer $b \geq 2$, prove by induction on j. The lemma is true when $j = 0, 1, \ldots, b$, as in these cases $n_{b,j} = 2^j$. Assume that it is true for all integers less than j and $j \geq b$. It implies that $n_{b,s} \geq n_{b,t} \geq n_{b,0} = 1$ for all s, t such that $j > s \geq t \geq 0$. Then we have $n_{b,j+1} = 2n_{b,j} - n_{b,j-b} = n_{b,j} + (n_{b,j} - n_{b,j-b}) \geq n_{b,j} \geq 1$. Therefore, the lemma is true for all integers $j \geq 0$ by induction. ∎

Lemma 3.4. $n_{b,j} \leq 2^j$ for all integers $b \geq 2$ and $j \geq 0$.

Proof. For any fixed integer $b \geq 2$, prove by induction on j. Since $n_{b,j} = 2^j$ when $j = 0, 1, \ldots, b$, the lemma is true for $j = 0, 1, \ldots, b$. Assume that it is true for all positive integers less than j and $j \geq b+1$, then $n_{b,j} = 2n_{b,j-1} - n_{b,j-1-b} \leq 2n_{b,j-1} \leq 2 \times 2^{j-1} = 2^j$. Therefore, the lemma is true for all integers $j \geq 0$ by induction. ∎

Lemma 3.5. Let $\delta_{b,j} = n_{b,j} - n_{b-1,j}$ for all $b \geq 3, j \geq 0$. Then $\delta_{b,j} \geq \delta_{b,j-1} \geq 0$ for $b \geq 3$ and $j \geq 1$.

Proof. For any fixed $b \geq 3$ we prove by induction on j. For $j = 1, \ldots, b-1$, $\delta_{b,j} = n_{b,j} - n_{b-1,j} = 2^j - 2^j = 0$, so we have $\delta_{b,j} \geq \delta_{b,j-1} \geq 0$. When $j = b$, $\delta_{b,b} = n_{b,b} - n_{b-1,b} = 2^b - (2n_{b-1,b-1} - n_{b-1,0}) = 2^b - (2 \times 2^{b-1} - 1) = 1$, so that $\delta_{b,b} \geq \delta_{b,b-1} = 0$. Hence the lemma is true when $j = b$.

When $j = b+1$,

$$\delta_{b,b+1} = n_{b,b+1} - n_{b-1,b+1}$$
$$= (2n_{b,b} - n_{b,0}) - (2n_{b-1,b} - n_{b-1,1})$$
$$= (2 \times 2^b - 1) - (2(2n_{b-1,b-1} - n_{b-1,0}) - n_{b-1,1})$$
$$= (2^{b+1} - 1) - (2(2 \times 2^{b-1} - 1) - 2)$$
$$= (2^{b+1} - 1) - (2^{b+1} - 2 - 2)$$
$$= 3$$
$$\geq 1 = \delta_{b,b} \geq 0.$$

Hence, the lemma is true when $j = b+1$. Assume that the lemma is true for all positive integers less than j and $j \geq b+2$, we prove it is true for j.

$$\delta_{b,j} = n_{b,j} - n_{b-1,j}$$
$$= (2n_{b,j-1} - n_{b,j-b-1}) - (2n_{b-1,j-1} - n_{b-1,j-b})$$
$$= 2(n_{b,j-1} - n_{b-1,j-1}) - n_{b,j-b-1} + n_{b-1,j-b}$$
$$= 2\delta_{b,j-1} - n_{b,j-b-1} + 2n_{b-1,j-b-1} - n_{b-1,j-2b}$$
$$= 2\delta_{b,j-1} - (n_{b,j-b-1} - n_{b-1,j-b-1}) + (n_{b-1,j-b-1} - n_{b-1,j-2b})$$
$$= 2\delta_{b,j-1} - \delta_{b,j-b-1} + (n_{b-1,j-b-1} - n_{b-1,j-2b})$$
$$= \delta_{b,j-1} + (\delta_{b,j-1} - \delta_{b,j-b-1}) + (n_{b-1,j-b-1} - n_{b-1,j-2b})$$

Then by the induction hypothesis and Lemma 3.3, we have

$$\delta_{b,j} - \delta_{b,j-1} = (\delta_{b,j-1} - \delta_{b,j-b-1}) + (n_{b-1,j-b-1} - n_{b-1,j-2b}) \geq 0,$$

and hence $\delta_{b,j} \geq \delta_{b,j-1} \geq 0$. Therefore the lemma is true for all $j \geq 1$ by induction. ∎

Lemma 3.6. $1.618^j - 2 \leq n_{b,j} \leq n_{b+1,j} \leq 2^j$ for $b \geq 2$ and $j \geq 0$.

Proof. By Lemmas 3.4 and 3.5, we have $n_{2,j} \leq n_{b,j} \leq n_{b+1,j} \leq 2^j$. We next show that $n_{2,j} \geq 1.618^j$.

Clearly, $n_{2,0} = 1, n_{2,1} = 2, n_{2,2} = 4, n_{2,j} = 2n_{2,j-1} - n_{2,j-3}$. Solve the recursive equation, we have

$$n_{2,j} = -1 + (2/5\sqrt{5} + 1)(1/2 + 1/2\sqrt{5})^j + (-2/5\sqrt{5} + 1)(1/2 - 1/2\sqrt{5})^j$$
$$\geq -1 + 1 \times (1/2 + 1/2\sqrt{5})^j + 0.1 \times (-0.6180339880)^j$$
$$\geq -1 + 1 \times (1/2 + 1/2\sqrt{5})^j + (-1)$$
$$\geq 1.618^j - 2.$$

This completes the proof of the lemma. ∎

Lemma 3.7. *Along $T_b(n)$ the file distribution can be done in time $\Theta(\log n)$ with EFDP.*

Proof. Suppose $n_{b,j} = n$, then by Lemma 3.6 we have $1.618^j \leq n_{2,j} + 2 \leq n_{b,j} = en + 2 \leq 2^j + 2$. Then, $j \log_2(1.618) \leq \log_2(n + 2) \leq \log_2(2^j + 2) \leq j + 1$. This implies that $j = \Theta(\log(n))$. ∎

Finally, if the file transfer delays between all pairs of nodes are not the same, the EFDP topology algorithm returns a tree T. We extend the delay time on edges of T to $M = \max\{t(p_i, p_j) : i, j = 0, \ldots, n - 1\}$. Using constant delay, we obtain the best ordering of children at each node. Then with this ordering, the file distribution time is at most $\Theta(M \log n)$ or $\Theta(\log n)$ if we take M as a constant. This completes the proof of Theorem 2.3.

4   Conclusions

In this paper, we proposed and studied a new optimal spanning tree problem. The problem came from the application of file distribution along a structured overlay network with the edge-based file distribution protocol. The file distribution time was formulated and used as the objective function. We showed that the problem is NP-hard in general and then presented a heuristic algorithm. The tree derived by the algorithm is of file distribution time $\Theta(\log n)$ and it is optimal when the file transfer delays between all pairs of nodes are same. Similar optimal overlay network design problems exist for other structured file distribution protocols.

References

1. Bermond, J.-C., Hell, P., Liestman, A.L., Peters, J.G.: Broadcasting in bounded degree graphs. SIAM J. Discret. Math. 5(1), 10–24 (1992)
2. Cohen, B.: Incentives build robustness in bittorrent. Technical report (2003)
3. Fan, B., Lui, J.C.S., Chiu, D.-M.: The design trade-offs of bittorrent-like file sharing protocols. IEEE/ACM Trans. Netw. 17(2), 365–376 (2009)
4. Ge, Z., Figueiredo, D.R., Jaiswal, S., Kurose, J.F., Towsley, D.F.: Modeling peer-peer file sharing systems. In: INFOCOM (2003)
5. Harutyunyan, H.A., Liestman, A.L.: More broadcast graphs. Discrete Applied Mathematics 98(1-2), 81–102 (1999)
6. Harutyunyan, H.A., Liestman, A.L.: k-broadcasting in trees. Networks 38(3), 163–168 (2001)
7. Li, Q., Lui, J.C.-S.: On modeling clustering indexes of bt-like systems. In: ICC, pp. 1–6 (2009)
8. Liu, J., Rao, S.G., Li, B., Zhang, H.: Opportunities and challenges of peer-to-peer internet video broadcast. In: Proceedings of the IEEE, Special Issue on Recent Advances in Distributed Multimedia Communications (2007)

Optimal Balancing of Satellite Queues in Packet Transmission to Ground Stations

Evangelos Kranakis[1], Danny Krizanc[2], Ioannis Lambadaris[3],
Lata Narayanan[4], and Jaroslav Opatrny[4]

[1] School of Computer Science, Carleton University, Ottawa, ON, K1S 5B6, Canada.
Supported in part by NSERC and MITACS grants
[2] Department of Mathematics and Computer Science, Wesleyan University,
Middletown CT 06459, USA
[3] Department of Systems and Computer Engineering, Carleton University, Ottawa,
ON, K1S 5B6, Canada. Supported in part by NSERC and MITACS grants
[4] Department of Computer Science, Concordia University, Montréal, QC, H3G 1M8,
Canada. Supported in part by NSERC grant

Abstract. Satellites collecting data store packets in queues and transmit these packets to ground stations within their view. In a given time step, a ground station may see several satellites, but can receive a packet from only one of them. A satellite can send each packet from its queue to a ground station in its view. We consider the problem of finding an assignment of satellites to ground stations that results in all ground stations receiving a packet while optimally balancing the sizes of remaining queues at satellites. We give a polynomial time algorithm for solving this problem which requires $O((m+n)^3 n)$ arithmetic operations, where m is the number of satellite queues and n is the number of ground stations.

Keywords and Phrases: Bipartite Graph, Matching, Optimal Balancing, Servers, Queues.

1 Introduction

The transmission of data gathered in outer space (e.g., weather and surveillance images) to earth involves relaying information from satellites to ground stations placed at different geographical locations and within view of the satellite. Choosing a ground station to transmit the information depends on available resources along the route, the utilization of the route, as well as weather conditions that may affect the quality of the link. In this setting important problems arising include 1) transmission scheduling (to which beam should the data be transmitted; and allocating the satellite's transmitters to the different downlink beams), 2) optimal throughput (select the requests to be served that maximize overall throughput), and 3) efficient protocol design (see [10,11] for additional details on this model concerning communication efficiency in satellite data networks).

In this paper we consider the problem of choosing ground stations to transmit data while balancing the sizes of remaining queues at satellites. More specifically, consider a group of orbiting satellites communicating with ground stations. Each

W. Wu and O. Daescu (Eds.): COCOA 2010, Part II, LNCS 6509, pp. 303–316, 2010.
© Springer-Verlag Berlin Heidelberg 2010

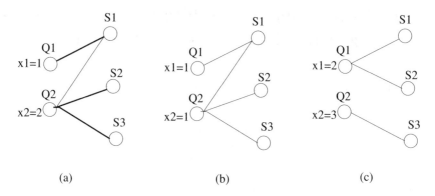

Fig. 1. (a) A satellite-ground station system. A saturated assignment is shown with bold edges. (b) A saturated assignment is impossible. (b) The minimum value of balance factor possible is 2.

satellite holds a queue of packets. In any given time step, a ground station can see several satellites, but can receive or download a packet from only a single satellite within its view. On the other hand, a satellite can send one packet from its queue each to any ground station within its view, provided it has enough packets in its queue. An assignment pairs a ground station with a satellite with available packets in its view. To optimize throughput, we need to find an assignment that contains as many ground stations as possible. Clearly, the assignment is constrained both by the connectivity of satellites to ground stations and the number of packets available at satellites. Naturally, once the transmission of packets from satellites to ground stations as specified by the assignment takes place, the number of packets remaining at the satellite queues changes. In this paper, we are interested in optimizing throughput while balancing the sizes of the satellite queues. The problem we consider is to find an assignment that serves all or a maximum possible number of ground stations in such a way that the number of packets remaining in the satellite queues are as balanced as possible.

Observe that the satellite-ground station communication system can be modelled as a bipartite graph (V_1, V_2, E) where V_1 represents the satellite queues, V_2 represents the ground stations and $\{u, v\} \in E$ if the satellite with queue u and the ground station v can see each other. Each node $u_i \in V_1$ is associated with an integer x_i, the number of packets in the queue. Given a subset E' of E, define the $deg_{E'}(v)$ of a node $v \in V_1 \cup V_2$ to be the number of edges in E' with v as an endpoint. An assignment then is a set $A \subseteq E$ such that

1. $\forall u_i \in V_1, deg_A(u_i) \leq x_i$ and
2. $\forall v \in V_2, deg_A(v) \leq 1$.

The *size* of an assignment is the number of edges in it, and is clearly related to the throughput of the system. We call an assignment *saturated* if every ground station is paired with a queue. Observe that the size of a saturated assignment is n. Figure 1(a) shows the bipartite graph model of a satellite-ground station system with the bold edges showing a saturated assignment. In the example shown in Figure 1(b), a saturated assignment does not exist. The *balance factor* of an assignment is the difference between the size of the longest and shortest queues after performing the assignment. This is motivated from the fact that over time, optimally balanced assignments not only lead to equally distributed (satellite) queue utilization but may also avert undesirable buffer overflows when queues are close to capacity

Given a satellite-ground station system, we are interested in several questions. Does a saturated assignment exist? If so, what is the smallest balance factor possible for a saturated assignment? If not, what is the largest possible assignment? What is the smallest balance factor that can be achieved for a given system? For the example shown in Figure 1(c), the minimum achievable balance factor is 2. Given an achievable value of balance factor, what is the largest possible assignment?

1.1   Outline and Results of the Paper

We give polynomial time algorithms to answer all the questions mentioned above. We show a relationship between the existence of assignments and network flow. The latter can be used to obtain a valid but not necessarily maximum-sized assignment. Next, given a particular value of balance factor, we use the technique of augmenting paths (see [4,13]), whereby starting from an assignment, we produce a sequence of assignments that provide successive improvements in terms of size, and result in an optimal final solution for the given balance factor. The main result of the paper is a polynomial time algorithm for computing a saturated assignment (when one exists) with optimal balance factor which requires $O((m+n)^3 n)$ arithmetic operations, where m is the number of queues and n is the number of stations.

In Section 2, we provide some preliminary definitions and notation that will be used throughout the paper. Related work is described in Section 3. Details of the main algorithm and its analysis can be found in Section 4.

2   Preliminaries and Notation

Given a satellite queue and ground station system with m queues and n stations, we model it as a bipartite graph (V_1, V_2, E) where $V_1 = \{Q_1, Q_2, \ldots, Q_m\}$ represents the satellite queues, $V_2 = \{s_1, s_2, \ldots s_n\}$ represents the ground stations and $\{Q_i, s_j\} \in E$ if the satellite with queue Q_i and the ground station s_j can see each other. Each node $Q_i \in V_1$ is associated with an integer x_i, the number of packets in the queue. Given a subset E' of E, define the $deg_{E'}(v)$ of a node in $v \in V_1 \cup V_2$ to be the number of edges in E' with v as an endpoint. An assignment then is a set $A \subseteq E$ such that

1. $\forall Q_i \in V_1, deg_A(Q_i) \le x_i$ and
2. $\forall s_j \in V_2, deg_A(s_j) \le 1$.

In a saturated assignment, we have $deg_A(s_j) = 1$ for all stations s_j.

For a given assignment A, the queue sizes are updated as follows

$$x_i(A) \leftarrow x_i - deg_A(Q_i),$$

where $i = 1, 2, \ldots, m$. We are interested in the following problem: Given a bipartite graph of queues and stations as above provide an algorithm for computing an assignment A satisfying the following min-max optimization condition

$$\min_A \max_{1 \leq i, i' \leq m} |x_i(A) - x_{i'}(A)|, \tag{1}$$

where A ranges over all possible assignments.

We denote $\min(A) = \min_{1 \leq i \leq m} x_i(A)$ and $\max(A) = \max_{1 \leq i \leq m} x_i(A)$ to be the minimum and maximum sizes respectively of any queue after the assignment A is applied. The min-max optimization problem can be restated as finding the assignment A that minimizes $\max(A) - \min(A)$. In order to study the min-max optimization problem it will be necessary to introduce the concept of a k-balanced assignment.

Definition 1. *For a given non-negative integer k, an assignment A is called k-balanced if $\max_{i,i'} |x_i(A) - x_{i'}(A)| \leq k$, where $i, i' = 1, 2, \ldots, m$ or equivalently $\max(A) - \min(A) \leq k$.*

Finally, we define the notion of a station that is not assigned a queue under a given (partial) assignment that is used throughout the proofs in this paper.

Definition 2. *A station s_j is called A-free with respect to assignment A if $deg_A(s_j) = 0$.*

Auxiliary flow network. The solution to the proposed problem will be related to flows in networks [7]. We use the bipartite graph in Figure 1 to construct a new "flow network" from a source S to a sink T by joining S to each of the queues and each of the stations to the sink T. Each of the edges of the original bipartite graph is assigned capacities equal to 1. The i-th edge of the flow network from S to the queue Q_i is assigned an integer capacity y_i, for $i = 1, 2, \ldots, m$. In the sequel, $y_1, y_2, \ldots, y_m$ will be called the *input* capacities. Note that the input capacities will vary depending upon the situation being considered. All other edges have capacity 1.

Given the original bipartite graph, a key technique used in this paper is to construct an auxiliary flow network with suitably chosen input capacities $y_1, y_2, \ldots, y_m$ and then determining whether a given net flow f can be attained, where

$$f \leq \sum_{i=1}^m y_i. \tag{2}$$

The *integral flow* theorem states that if each edge in a flow network has integral capacity, then there exists an integral maximal flow [7].

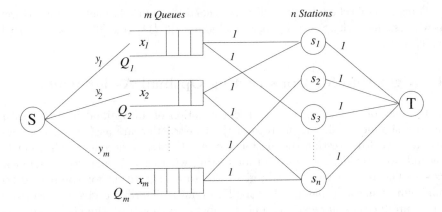

Fig. 2. Auxiliary flow network from source S to a sink T

Observe that if $y_i \leq x_i$ for all $i = 1, \ldots, m$, then the unit capacities on the edges from the stations to node T imply that an integral flow in the auxiliary flow network corresponds directly to an assignment in the corresponding original bipartite graph; all edges with flow 1 between the queues and stations belong to the assignment. Furthermore, the net flow into node T is equal to the size of the assignment.

3   Related Work

Despite the existence of extensive literature on network flows and bipartite matchings, to the best of our knowledge, the problem of constructing saturated assignments between stations and queues satisfying the balance condition above has not been considered before. Existing work includes dynamic station allocation, as well as fair scheduling and transmission scheduling policies.

The papers [1], [2], [3], and [14] consider dynamic server allocation with randomly varying connectivity. In the first paper the authors look at stationary dynamics, in the second additional routing constraints are considered (including random connectivities, random accessibilities and multiple classes of flows), the processing system considered in the third paper is comprised of several parallel queues and a processor, which operates in a time-varying environment that fluctuates between various states or modes, and in the fourth paper the authors propose queue allocation policies suitable for networks with changing topology (such as wireless networks).

Several papers consider fair scheduling policies. For example, [6] considers the problem of allocating resources (time slots, frequency, power, etc.) at a base station to many the channel capacity region is assumed to be known. The competing flows, where each flow is intended for a different re-scheduler is allowed to know the current queue-lengths and ceiver. In addition, [9] proposes a new model for wireless fair scheduling based on an adaptation of fluid fair queueing (FFQ) to handle location-dependent error bursts.

Transmission scheduling policies are crucial for multi-channel satellite as well as wireless networks. Several papers, including [12], [11], and [8], investigate such policies.

4 Algorithm for Constructing Optimal Assignments

In this section, we detail the main construction of an optimal balanced saturated assignment. The main steps of the construction are as follows. First, we give an algorithm testing the existence of a saturated assignment (Theorem 1). Second, we provide an algorithm for testing whether a k-balanced assignment exists (Theorem 2) which easily gives rise to an algorithm for computing the minimum achievable balance factor (Theorem 3). Third, we give an algorithm which produces the largest (i.e., max number of stations) possible k-balanced assignment for the minimum achievable k (Theorem 4). This leads to an algorithm which given k produces a largest possible k-balanced assignment (Theorem 5). The final result is given in Theorem 6 and provides an algorithm for constructing a saturated assignment with minimum balance factor, if one exists. All the algorithms given require a polynomial number of arithmetic operations and the detailed complexity analysis is provided in Subsection 4.6.

4.1 Saturated Assignments

We can test for the existence of a saturated assignment in polynomial time using an algorith for integral maximum flow.

Theorem 1. *There is an $O((m + n)^3)$ algorithm for testing whether there is a saturated assignment in an m-satellite and n-ground station system.*

Proof. Consider the auxiliary flow network depicted in Figure 2 obtained by adding a source S which is connected to all the queues and a sink T which is connected to all the stations. Let the input capacities be defined so that $y_i = x_i$, for all $i = 1, 2, \ldots, m$. As observed in Section 2, an integer net flow in the auxiliary network corresponds to an assignment. Therefore, an integer net flow $f = n$ can be realized if and only if the corresponding assignment of stations to queues is saturated. This completes the proof of Theorem 1.   □

Observe that the proof generalizes to testing for an assignment of any size, and also to finding the largest possible assignment.

Clearly, the assignment resulting from the proof of Theorem 1 cannot be guaranteed to be balanced, in the sense that $\max_{i,i'} |x_i(A) - x_{i'}(A)|$, where $i, i' = 1, 2, \ldots, m$, is minimized over all possible assignments. A different technique will be necessary to find balanced assignments.

4.2 Balanced Assignments

Recall that in a k-balanced assignment, we have $|x_i(A) - x_{i'}(A)| \leq k$, for all $i, i' = 1, 2, \ldots, m$.

Theorem 2. *There is an $O((m + n)^3)$ algorithm for testing whether there is a k-balanced assignment in an m-satellite and n-ground station system.*

Proof. Without loss of generality we may assume that the original queue sizes are sorted so as to satisfy $x_1 \geq x_2 \geq \cdots \geq x_m$. Consider the auxiliary flow network depicted in Figure 2. Given a non-negative integer k, let the input capacities be $y_i = \max\{0, x_i - x_m - k\}$, for $i = 1, 2, \ldots, m$. Observe that $y_i = 0$ if and only if $x_m + k \geq x_i$ (so clearly, $y_m = 0$). In this case we test whether there is a net flow f, where f is defined by

$$f := \sum_{i=1}^{m} y_i. \tag{3}$$

If there is a solution to the flow problem above then there is a k-balanced assignment. Indeed, assume that the flow above is achievable. Observe that in view of Equation 3, the link from the source S to the i-th queue must be used to capacity, that is, its flow must be equal to $\max\{0, x_i - x_m - k\}$. Therefore the number $x_i(A)$ of packets remaining in the i-th queue after the assignment must satisfy

$$x_i(A) = x_i - y_i = x_i - \max\{0, x_i - x_m - k\} \leq x_m + k.$$

Also, since $\max\{0, x_m - x_m - k\} = 0$ the flow at the m-th queue remains at 0. Therefore the balance factor of this assignment will be at most k.

Conversely, suppose there is a k-balanced assignment A such that $x_i(A) = x_i - z_i$, for $i = 1, 2, \ldots, m$. We claim that $z_i \geq y_i$. If $y_i = 0$ then the claim is trivial. Otherwise, assume on the contrary that $z_i < y_i = x_i - x_m - k$. It follows that

$$x_i(A) = x_i - z_i > x_m + k \geq x_m - z_m + k = x_m(A) + k.$$

Therefore the assignment A is not k-balanced, which is a contradiction. This proves the claim.

There exists a solution to the auxiliary flow problem with input capacities z_i and net flow $\sum_{i=1}^{m} z_i$. It follows from the claim above that there is also a solution to the auxiliary net flow problem with input capacities y_i and corresponding net flow $\sum_{i=1}^{m} y_i$. This completes the proof of Theorem 2.   □

Using binary search between the smallest and largest value of k guaranteed by Theorem 1, it is now possible to compute the minimum achievable balance factor. As a corollary we obtain the following result.

Theorem 3. *There is an $O((m + n)^3 \log n)$ algorithm for computing the minimum achievable balance factor for an m-satellite and n-ground station system.*

Proof. This is an immediate consequence of Theorem 2 using binary search. This completes the proof of Theorem 3.   □

4.3   Largest Assignment with Minimum Achievable Balance Factor

We start with the following interesting property of an assignment with minimum achievable balance factor that will be useful in the proof of optimality of our constructions for any achievable balance factor.

Lemma 1. *Let A be an assignment that achieves the minimum possible balance factor. Then $\max(A') \geq \max(A)$ for any other assignment A'.*

Proof. Let A' be an assignment with $\max(A') < \max(A)$. Then by taking edges away from the assignment A', it is possible to ensure that $\min(A') \geq \min(A)$. But this means the balance factor achieved by A' is less than that achieved by A, a contradiction.

A basic ingredient in the proofs for constructing assignments is the concept of an augmenting path [4]. Let $x_1(A), x_2(A), \ldots, x_m(A)$ be the queue sizes resulting from an assignment A. There will be two instantiations of the concept of augmenting path that will be used in the sequel both of which result in a new assignment A' as follows.

1. An A-augmenting path of Type 1 for assignment A, starts at an A-free station and uses edges alternating between edges in A and not in A and ending at a queue Q_i such that $x_i(A) > \min(A)$. By swapping the edges in A and out of A in the path from the assignment A, we obtain a new assignment A' that does not increase the balance factor of A.
2. An A-augmenting path of Type 2 starts at an A-free station and uses edges alternating between edges in A and not in A and ends at a queue with $x_i(A) = \min(A)$. The effect of the new assignment A' differs from that of assignment of Type 1 in that it increases the previous balance factor $\max_{1 \leq i, i' \leq m} |x_i(A) - x_{i'}(A)|$ by 1.

If the assignment A is easily understood from the context then we simply use the term augmenting path without mentioning A explicitly. Both types of augmenting paths will be used in the proofs of Theorems 4 and 5 below.

To illustrate the concept of path of Type 1, consider the augmenting path depicted in Figure 3 concerning a set

$$Q_{i_1}, Q_{i_2}, \ldots, Q_{i_k}$$

of queues and a set

$$s_{j_1}, s_{j_2}, \ldots, s_{j_k}$$

of stations numbered from top to bottom. For the given assignment A, assume that station s_{j_r} is assigned the queue $Q_{i_{r+1}}$, for all $r < k$ (see dashed edges). Moreover, for all $r \leq k$, station s_{j_r} is not assigned the queue Q_{i_r} (see solid edges). We now define a new assignment A' which differs from A only in that all edges $\{Q_{i_{r+1}}, s_{j_r}\}$, for all $r < k$, are removed from A and edges $\{Q_{i_r}, s_{j_r}\}$, for all $r \leq k$, are added to A'. Observe that A' has one edge more than A. Moreover, A' is a legal assignment since the sizes of queues Q_{i_r} remain the same in A' for all $r > k$, while the first queue which is reduced by 1 packet can participate in the new assignment without increasing the balance factor because it contains more packets than $\min(A)$.

A similar illustration is easy to give for augmenting paths of Type 2. Here, we increase by 1 the balance factor $\max(A) - \min(A)$ corresponding to the assignment A.

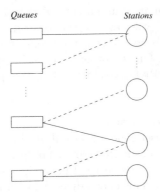

Fig. 3. An augmenting path. Dashed edges are in the assignment A while solid edges are not.

Now that we have explained the concept of augmenting path we are in a position to state and prove the following theorem.

Theorem 4. *Given an m-satellite and n-ground station system, there is an $O((m+n)^3)$ algorithm which when given the minimum achievable balance factor, say k, produces a k-balanced assignment satisfying the largest number of stations.*

Proof. Consider the auxiliary flow network depicted in Figure 2. Given a non-negative integer k, let the input capacities be $y_i = \max\{0, x_i - x_m - k\}$, for $i = 1, 2, \ldots, m$. Solve the max flow problem for these capacities to obtain an initial assignment A.

Use the method of augmenting paths (of Type 1) to augment the assignment A. Each time an augmenting path is found there results a new assignment which increases the number of assigned stations. Continue until you can no longer find an augmenting path and let the resulting final assignment be denoted by R (Red). Clearly there is no R-augmenting path of type 1. It follows from the fact that $y_m = 0$ in the auxiliary flow problem and the definition of augmenting paths that $\min(R) = x_m$ and from Lemma 1 that $\max(R) = \max(A)$.

The main argument is now contained in the following claim.

Claim 1. *Assignment R is an optimal k-balanced assignment.*

Proof. Suppose on the contrary there is a k-balanced assignment satisfying a larger number of stations, denoted by B (Blue). Without loss of generality let B be the optimal assignment that has the maximum-sized intersection with the assignment R in terms of the stations that are being served by the two assignments, respectively.

To this end label the edges R, B or R/B depending on whether they are part of the Red, Blue or both assignments, respectively. Similarly, color stations R, B or R/B depending on the corresponding color of the edges incident on the station. (Notice that some edges and stations may have no color.)

Start with a Blue station and build a path of alternating Blue and Red edges. Such a path can terminate either at a station or at a queue. If it terminates at a station it is because there is no Blue edge incident on the station, in which case the station is colored Red. On the other hand, if it terminates at a queue it is because there are no more Red edges incident on the queue. Therefore there are only three cases to consider.

Case 1. Path stops at a Red station.
In this case from the Blue assignment take out all the Blue edges from the above mentioned path and add in all the Red edges. This gives rise to a new assignment of the same size and balance factor as the Blue assignment but which differs from the Red by one station less. This contradicts the maximality of the intersection between the Red and Blue assignments.

Case 2. Path stops at a queue Q_j such that $x_j(R) > x_m$.
In this case there is an R-augmenting path and R would not be the final assignment computed by the algorithm, which is a contradiction.

Case 3. Path stops at a queue Q_j such that $x_j(R) = x_m$.
We know from Lemma 1 that $\max(B) \geq \max(A) = \max(R)$. Since there are no more red edges incident on Q_j, it follows that $deg_B(Q_j) > deg_R(Q_j)$ which implies that $x_j(B) < x_j(R) = x_m$. Consequently, $\min(B) < \min(R)$ contradicting the fact that Blue is a k-balanced assignment. This completes the proof of the claim.

Clearly, Claim 1 implies that the assignment R previously constructed is optimal in the number of stations being assigned to queues which also completes the proof of Theorem 4.   □

4.4   Largest Assignment with Given Balance Factor

Theorem 4 gives the construction of an optimal assignment for the minimum possible achievable balance factor. It remains to provide the construction for any possible balance factor. We give an inductive construction which starts with an assignment for the minimum achievable balance factor.

Theorem 5. *Given an m-satellite and n-ground station system, there is an $O(n(m+n)^3)$ algorithm which when given a non-negative integer $k \geq k_0$ (where k_0 is the minimum achievable balance factor), produces a k-balanced assignment satisfying the largest possible number of stations.*

Proof. The proof is by induction on k. For the initial step, Theorem 4 gives an assignment A satisfying the largest number of stations for the minimum possible balance factor. Further assume inductively that we have an assignment satisfying the largest number of stations for balance factor $k - 1 \geq k_0$. Call this assignment R (Red) and denote $M := \min(R)$. Clearly $\max(R) \geq \max(A)$ from

Lemma 1, and assume inductively that $\max(R) = \max(A)$. We will show how to augment this assignment to obtain the assignment satisfying the largest number of stations with balance factor equal to k.

Build an augmenting path of type 2: Start with an R-free station and alternate edges in and out of the assignment R and ending at a queue Q_j such that $x_j(R) = M$. If an augmenting path exists then a swap of the edges results in an assignment R' which is one edge larger and which has balance factor k. Now continue by finding augmenting paths of type 1 (thereby not increasing the balance factor) and stop when it is no longer possible to find an augmenting path. Call the final assignment B (Blue). Observe that $\min(B) = M - 1$ and by Lemma 1 and by the construction, $\max(B) = \max(A)$, that is, the Blue assignment has balance factor k.

The main claim is the following.

Claim 2. B *is an optimal k-balanced assignment.*

Proof. Assume to the contrary that there is a k-balanced assignment, say G (Green), which satisfies a larger number of stations than B. Without loss of generality we can take such an assignment having the maximum number of stations in common with the Blue assignment. As before color edges and stations with $B, G, B/G$ depending on the corresponding assignments. Construct a path starting at a Green station alternating Green and Blue edges. Using a similar argument as before it is easy to see that there are three possible cases to consider.

Case 1. Path stops at a Blue station.
In this case from the Green assignment take out all the Green edges from the above mentioned path and add in all the Blue edges. This gives rise to a new assignment of the same size and balance factor as the Green assignment but which differs from the Blue by one station less. This contradicts the maximality of the intersection between the Blue and Green assignments.

Case 2. Path stops at a queue Q_j with $x_j(B) \geq M$.
In this case we obtain a B-augmenting path, which is a contradiction.

Case 4. Path stops at a queue Q_j with $x_j(B) = M - 1$.

In this case, we first observe that $\min(G) < \min(B)$ since there are more green edges incident on Q_j than blue edges which implies $x_j(G) < x_j(B) = \min(B)$. It follows from Lemma 1 that $\max(G) \geq \max(A) = \max(B)$, which contradicts the assertion that Green is a k-balanced assignment.

This completes the proof of Claim 2.   □

Claim 2 gives the main ingredient required for the inductive construction and is thus sufficient to complete the proof of Theorem 5.   □

4.5 Optimal Saturated Assignments

It is now easy to construct a saturated assignment with optimal balance factor using the previous theorems.

Theorem 6. *Given an m-satellite and n-ground station system, there is an algorithm for constructing a saturated assignment (if such an assignment exists) and attains the optimal value*

$$\min_{A} \max_{1 \leq i,i' \leq m} |x_i(A) - x_{i'}(A)|,$$

where A ranges over all possible assignments between queues and stations. The algorithm requires $O((m+n)^3 n)$ arithmetic operations.

Proof. First use Theorem 1 in order to test if there is a saturated assignment. If the answer is yes, then next start with the minimum possible balance factor, say k_0, which ensures the existence of a feasible assignment (see Theorem 3) and use the following algorithm.

1. Set $k \leftarrow k_0$; with input capacities z_i and netflow $\sum_{i=1}^{m} z_i$
2. Construct the k-balanced assignment satisfying the largest number of stations;
3. **if** the resulting assignment is saturated then stop;
4. **Else** set $k \leftarrow k + 1$ and **goto** Step 2

Observe that Step 2 is justified using Theorem 5. The algorithm halts in Step 3 the first time a saturated assignment is found. Details of the running time of the algorithm is given in Subsection 4.6. This completes the proof of Theorem 6. □

4.6   Complexity Analysis of the Algorithms

There are two main tools used in the proofs: maximum net flow algorithms in order to determine feasible solutions to the assignment problem and augmenting paths that improve the assignments, if possible. The auxiliary flow network depicted in Figure 2 and which was used in the proof of the main theorem has $m + n$ vertices and at most $mn + m + n$ edges. Running the Edmonds-Karp algorithm [5] (see also [13]) therefore takes $O((m+n)^3)$ arithmetic operations. An augmenting path operation involves n stations and producing the path involves the construction of a BFS tree of size $O(nm)$ (the number of edges of the bipartite graph, i.e. $O(n^2 m)$ time and there are $O(n)$ such operations. It follows that all the algorithms in Theorems 1, 2, and 4, take $O((m+n)^3)$ arithmetic operations. As described, the recursive algorithm in Theorem 5 has an additional multiplicative factor of $k - k_0$. Similarly, the algorithm in Theorem 4 finds the first k that has a saturated k-balanced assignment. If the queue sizes in sorted order are $x_1 \geq x_2 \geq \cdots \geq x_m$. the maximum possible balance factor of a saturated assignment, or indeed any assignment is $x_1 - x_m + n$, which implies an extra multiplicative factor equal to $x_1 - x_m + n$ for the number of required iterations of both algorithms. In addition, the algorithm of Theorem 3 is based on binary search and therefore requires an additional multiplicative factor of only $\log(x_1 - x_m + n)$.

As presented above, the additional multiplicative factors $x_1 - x_m + n$ and $\log(x_1 - x_m + n)$ depend on the sizes of the queues. However, this is not necessary as explained in the following argument.

Let $x_1, x_2, \ldots, x_m$ be the given queue sizes and assume that $x_1 - x_m > 2n + 2$. Let i be the largest integer such that $x_1 - x_i \leq n$ and j be the smallest integer such that $x_j - x_m \leq n$. Notice that in any optimal assignment only the queues $Q_1, Q_2, \ldots, Q_i$ can be the queues whose size would become the largest and similarly only the queues $Q_j, Q_{j+1}, \ldots, Q_m$ can be the queues whose sizes can become the smallest, since the size of any queue can change at most by n.

Consider a modified instance with queue sizes x' as follows:

1. $1 \leq k \leq i \Rightarrow x'_i = x_i - x_1 + x_m + 2n + 2$
2. $i + 1 \leq k \leq j - 1 \Rightarrow x'_i = x_m + n + 1$
3. $j \leq k \leq m \Rightarrow x'_i = x_i$

The following two claims establish the relationship between assignments for the original and modified instances.

Claim 3. *A is an assignment for the original instance if and only if it is an assignment for the modified instance.*

Proof. Assume A is an assignment for the original instance. Since in the modified instance all queue sizes are either the same as in the original instance or are greater than n, the number of stations, it is clear that A is also an assignmet for the modified instance. The other direction follows from the fact that queue sizes in the original instance are at least the corresponding queue sizes in the modified instance.

Claim 4. *Let A be an assignment for the original and modified instances. Then A achieves a balance factor of k for the original instance if and only if it achieves a balance factor of $k - x_1 + x_m + 2n + 2$ for the modified instance.*

Proof. Let k_1 and k_2 be the indices of the largest and smallest queues respectively after applying the assignment A to the original instance. Clearly, $1 \leq k_1 \leq i$ and $j \leq k_2 \leq m$. Applying the assignment A to the modified instance reduces the values of $x'_1, x'_2, \ldots x'_i$ by the same amounts as the corresponding values of $x_1, x_2, \ldots x_i$ and thus k_1 remains the index of the largest queue size, and for the same reason k_2 remains the index of the smallest queue size in the modified instance. Since by definition, $x'_{k_1} = x_{k_1} - x_1 + x_m + 2n + 2$ and $x'_{k_2} = x_{k_2}$, the lemma follows. The proof of the other direction is similar.

Suppose A is an optimal solution for the modified instance achieving a balance factor of k. We claim that A is also an optimal solution for the original instance. Suppose instead that A' achieves a better balance factor than A on the original instance. Then it follows from Claim 4 that A' achieves a better balance factor than A for the modified instance, contradicting the optimality of A for the modified instance.

It follows from the previous argument that we can assume that in the original queues the number of packets is such that the difference in the numbers of packets between the largest and smallest queues is $\leq 2n + 2$. As a consequence, the algorithm of Theorem 3 requires $O((m + n)^3 \log n)$ and the algorithm of Theorem 6, $O((m + n)^3 n)$ arithmetic operations.

5   Conclusion

In this paper we have provided an algorithm for constructing saturated assignments between stations and queues with optimal balance factor. In addition to the possibility of improved trade-offs for various types of flows and bipartite graphs, several interesting problems remain open and worth investigating. One such problem concerns the study of weighted versions of the balanced, saturated assignment problem for bipartite graphs, and another the problem of balanced integer flows in arbitrary flow networks.

References

1. Bambos, N., Michailidis, G.: On the stationary dynamics of parallel queues with random serverconnectivities. In: Proceedings of the 34th IEEE Conference on Decision and Control 1995, vol. 4 (1995)
2. Bambos, N., Michailidis, G.: On parallel queuing with random server connectivity and routing constraints. Probability in the Engineering and Informational Sciences 16(02), 185–203 (2002)
3. Bambos, N., Michailidis, G.: Queueing and scheduling in random environments. Advances in Applied Probability 36(1), 293–317 (2004)
4. Bondy, J.A., Murty, U.S.R.: Graph theory with applications. Macmillan, London (1976)
5. Edmonds, J., Karp, R.M.: Theoretical Improvements in Algorithmic Efficiency for Network Flow Problems. Journal of the ACM (JACM) 19(2), 248–264 (1972)
6. Eryilmaz, A., Srikant, R.: Fair resource allocation in wireless networks using queue-length-based scheduling and congestion control. In: Proceedings IEEE INFOCOM 2005. 24th Annual Joint Conference of the IEEE Computer and Communications Societies, vol. 3 (2005)
7. Ford, L.R., Fulkerson, D.R.: Maximal flow through a network. Canadian Journal of Mathematics 8(3), 399–404 (1956)
8. Ganti, A., Modiano, E., Tsitsiklis, J.N.: Transmission Scheduling for Multi-Channel Satellite and Wireless Networks. In: Proceedings of the Annual Allerton Conference on Communication Control and Computing, vol. 40, pp. 1319–1328 (2002)
9. Lu, S., Bharghavan, V., Srikant, R.: Fair scheduling in wireless packet networks. IEEE/ACM Transactions on Networking (TON) 7(4), 473–489 (1999)
10. Modiano, E.: Satellite data networks. AIAA Journal on Aerospace Computing, Information and Communication 1, 395–398 (2004)
11. Neely, M.J.: Dynamic Power Allocation and Routing for Satellite and Wireless Networks with Time Varying Channels. PhD thesis, Massachusetts Institute of Technology (2003)
12. Neely, M.J., Modiano, E., Rohrs, C.E.: Power allocation and routing in multi-beam satellites with time-varying channels. IEEE/ACM Transactions on Networking (TON) 11(1), 152 (2003)
13. Papadimitriou, C.H., Steiglitz, K.: Combinatorial optimization: algorithms and complexity. Dover Publications, New York (1998)
14. Tassiulas, L., Ephremides, A.: Dynamic server allocation to parallel queues with randomly varying connectivity. IEEE Transactions on Information Theory 39(2), 466–478 (1993)

The Networked Common Goods Game

Jinsong Tan

Department of Computer and Information Science
University of Pennsylvania, Philadelphia, PA 19104
jinsong@seas.upenn.edu

Abstract. We introduce a new class of games called the *networked common goods game* (NCGG), which generalizes the well-known *common goods game* [12]. We focus on a fairly general subclass of the game where each agent's utility functions are the same across all goods the agent is entitled to and satisfy certain natural properties (diminishing return and smoothness). We give a comprehensive set of technical results listed as follows.

- We show the optimization problem faced by a single agent can be solved efficiently in this subclass. The discrete version of the problem is however NP-hard but admits a *fully polynomial time approximation scheme* (FPTAS).
- We show uniqueness results of pure strategy Nash equilibrium of NCGG, and that the equilibrium is fully characterized by the structure of the network and independent of the choices and combinations of agent utility functions.
- We show NCGG is a *potential game*, and give an implementation of best/better response Nash dynamics that lead to fast convergence to an ϵ-approximate pure strategy Nash equilibrium.
- Lastly, we show the *price of anarchy* of NCGG can be as large as $\Omega(n^{1-\epsilon})$ (for any $\epsilon > 0$), which means selfish behavior in NCGG can lead to extremely inefficient social outcomes.

1 Introduction

A collection of members belong to various communities. Each member belongs to one or more communities to which she can make contributions, either monetary or in terms of service but subject to a budget, and in turn benefits from contributions made by other members of the communities. The extent to which a member benefits from a community is a function of the collective contributions made by the members of this community.

A collection of collaborators are collaborating on various projects. Each collaborator is collaborating on one or more projects and each project has one or more collaborators. Each collaborator comes with certain endowment of resources, in terms of skills, time and energy, that she can allocate across the projects on which she is collaborating. The extent to which a project is successful is a function of the resources collectively allocated to it by its collaborators, and each of its collaborator in turn derives a utility from the successfulness of the project.

W. Wu and O. Daescu (Eds.): COCOA 2010, Part II, LNCS 6509, pp. 317–331, 2010.
© Springer-Verlag Berlin Heidelberg 2010

A collection of friends interact with each other, and friendships are reinforced through mutual interactions or weakened due to the lack of them. The more time and effort mutually devoted by two friends in their friendship, the stronger the friendship is; the stronger the friendship is, the more each benefit from it. However, each friend is constrained by her time and energy and has to decide how much to devote to each of her friends.

Suppose the community members, the collaborators and the friends (which we collectively call *agents*) are all self-interested and interested in allocating their limited resources in a way that maximizes their own total utility derived from the communities, projects, and mutual friendships (which we collectively call *goods*) that they have access to. Interesting computational and economics questions abound: Can the agents efficiently find optimal ways to allocate their resources? Viewed as a game played by the agents over a bipartite network, how does the network structure affect the game? In particular, does there exist a pure strategy Nash equilibrium? Is it unique and will myopic and selfish behaviors of the agents lead to a pure strategy Nash equilibrium? And how costly are these myopic and selfish behaviors?

In this paper we address these questions by first proposing a model that naturally captures these strategic interactions, and then giving a comprehensive set of results to the scenario where there is only one resource to be allocated by the agents, and the utility an agent derives from a good to which she is entitled is a concave and smooth function of the total resource allocated to that good. We start by giving our model that we call the *networked common goods game* (NCGG).

The Model. The *networked common goods game* is played on a bipartite graph $G = (P, A, E)$, where $P = \{p_1, p_2, ..., p_n\}$ is a set of *goods* and $A = \{a_1, a_2, ..., a_m\}$ is a set of *agents*. If there is an edge $(p_i, a_j) \in E$, then agent a_j is entitled to good p_i. There is a single kind of divisible resource of which each agent is endowed with one unit (we note this is not a loss of generality as our results generalize easily to the case where different agents start with different amounts of resource). Moreover, we can assume Nature has endowed each common good p_i with α_i amount of resource that we call the *ground level*; this can be viewed as modelling p_i as having access to some external sources of contributions.

Let $\mathcal{N}(v)$ denote the set of neighbors of a node $v \in P \cup A$, $x_{ij} \in [0,1]$ the amount of resource agent a_j contribute to good p_i and $\omega_i = \alpha_i + \sum_{a_k \in \mathcal{N}(p_i)} x_{ik}$ the total amount of resource allocated to good p_i. Each agent a_j derives certain utility $\mathcal{U}_j(\omega_i)$ from each p_i of which she is a member. We always assume $\mathcal{U}_j(0) = 0$ and for the most part of the paper, we consider the case where $\mathcal{U}_j(\cdot)$ is increasing, concave and differentiable. Being self-interested, agent a_j is interested in allocating her resources across the goods to which she is entitled in a way that maximizes her total utility $\sum_{p_i \in \mathcal{N}(a_j)} \mathcal{U}_j(\omega_i)$.

Our Results. We first consider the optimization problem faced by a single agent: Given the resources already allocated to the goods to which agent a_j is entitled, find a way to allocate resource so that a_j's total utility is maximized. We call this the *common goods problem* (CGP) and consider both continuous and discrete versions, where the agent's resource is either infinitely divisible or atomic.

- We show that for the continuous version, if $\mathcal{U}_j(\cdot)$ is assumed to be increasing, concave and differentiable, then CGP has an analytical solution. On the other hand, the discrete version of CGP is NP-hard but admits an FPTAS.

We then turn to investigate the existence and uniqueness of pure strategy Nash equilibrium of NCGG[1]. We consider two concepts of uniqueness of equilibrium, among which *strong uniqueness* is the standard concept of equilibrium uniqueness whereas *weak uniqueness* is defined as follows: For any two equilibria $\mathcal{E}$ and $\mathcal{E}'$ of the game and for any good $p_i \in P$, the total amount of resource allocated to p_i is the same under both $\mathcal{E}$ and $\mathcal{E}'$. We have the following results.

- We show for any NCGG instance, a Nash equilibrium always exists. And we show that this Nash equilibrium is weakly unique not only in a particular NCGG instance, but across all NCGG instances played on the same network as long as the utility function of each agent is increasing, concave and differentiable. And if in addition the underlying graph is a tree, the equilibrium is strongly unique. Our results do not assume that different agents have the same utility function; this demonstrates that Nash equilibrium in NCGG is completely characterized by network structure.

We also consider the convergence of Nash dynamics of the game, and its *price of anarchy*: The worst-case ratio between the social welfare of an optimal allocation of resources and that of a Nash equilibrium [17].

- We show that NCGG is a *potential game*, a concept introduced in [16], therefore any (better/best response) Nash dynamics always converge to the (unique) pure strategy Nash equilibrium. We then propose a particular implementation of Nash dynamics that leads to fast convergence to a state that is an additive ϵ-approximation of the pure strategy Nash equilibrium of NCGG. The convergence takes $O(Kmn)$ time, where $K = \max_j \mathcal{U}_j^{-1}(\epsilon/n)$, which for most reasonable choices of $\mathcal{U}_j$ is a polynomial of n and m. (For example, for $\mathcal{U}_j(x) = x^p$ where $p \in (0,1)$ is a constant, it is sufficient to set $K = (n/\epsilon)^{1/p}$, which is a polynomial in n.)
- We show the price of anarchy of the game is $\Omega(n^{1-\epsilon})$ (for any $\epsilon > 0$), which means selfish behavior in this game can lead to extremely inefficient social outcomes, for a reason that echoes the phenomenon of *tragedy of the commons* [11].

We note that NCGG introduced in this paper has the particularly nice property that very little is assumed about agents' utility functions. Unlike most economic

---

[1] Since we concern ourself with only pure strategy Nash equilibrium in this paper, we use it interchangeably with Nash equilibrium.

models considered in the literature where not only a particular form of utility function is assumed about a particular agent, but very often the same utility function is imposed across all agents, so that the model remains mathematically tractable, our model do not assume more than the following: 1) $\mathcal{U}_j(0) = 0$; 2) $\mathcal{U}_j(\cdot)$ has diminishing return (increasing and convex); 3) $\mathcal{U}_j(\cdot)$ is smooth (differentiable). In particular, we do not need to assume different agents share a common utility function for our results to go through.

Related Work. The networked common goods game we consider is a natural generalization of the well-known common goods game [12]. Bramoullé and Kranton considered a different generalization of the common goods game to networks [3]. In their formulation a (general, non-bipartite) network is given where each node represents an agent a_i, who can exert certain amount of effort $e_i \in [0, +\infty)$ towards certain common good and such effort incurs a cost of ce_i on the part of the agent, for some constant c. a_i's effort directly benefits another agent a_j iff they are directed connected in the network, and the utility of a_i is defined as $\mathcal{U}_i(e_i + \sum_{a_j \in \mathcal{N}(i)} e_j) - ce_i$. Bramoullé and Kranton then analyze this model to yield the following interesting insights: First, in every network there is an equilibrium where some individuals contribute whereas others free ride. Second, specialization can be socially beneficial. And lastly, a new link in the network can reduce social welfare as it can provide opportunities to free ride and thus reduce individual incentives to contribute. We note both the model and the research perspectives are very different from those considered in this paper.

A more closely related model is that studied by Fol'gardt [8,9]. The author considered a resource allocation game played on a bipartite graph that is similar to our setting. In Fol'gardt's model, each agent has certain amount of discrete resources, each of unit volume, that she can allocate across the 'sites' that she has access to. Each site generates certain utility for the agent, depending on the resources jointly allocated to it by all its adjacent agents. In Fol'gardt's formulation, each agent is interested in maximizing the minimum utility obtained from a single site she has access to. The analysis of Fol'gardt's resource allocation game is limited to very specific and small graphs [8,9].

A variety of other models proposed and studied in the literature bear similarities to the networked common goods game considered here. These include Fisher's model of economy [7], the bipartite exchange economy [13,4], the fixed budget resource allocation game [6,18], the Pari-Mutuel betting as a method of aggregating subjective probabilities [5], and the market share game [10]. However these model all differ significantly in the ways allocations yield utility.

2   The Common Goods Problem

Recall that CGP is the optimization problem faced by a single agent: An agent has access to n goods, each good p_i has already been allocated $\alpha_i \geq 0$ resources. The agent has certain amount of resource to allocate across the n goods. Denote by x_i ($i = 1, ..., n$) the amount of resource the agent allocates to goods i, she receives a total utility of $\sum_{i=1}^{n} \mathcal{U}(\alpha_i + x_i)$. In this section, we consider two versions of this optimization problem, where the resource is either infinitely divisible or discrete.

2.1 Infinitely Divisible Resource

Without loss of generality, assume the agent has access to one unit of resource. In the infinitely divisible case, CGP is a convex optimization problem captured by the following convex program.

$$
\begin{aligned}
\text{maximize} \quad & \sum_{i=1}^{n} \mathcal{U}(\alpha_i + x_i) \\
\text{subject to} \quad & \sum_{i=1}^{n} x_i = 1 \\
& x_i \geq 0 \qquad (i = 1, 2, ..., n)
\end{aligned}
\tag{1}
$$

where the constraint $\sum_{i=1}^{n} x_i = 1$ comes from the observation that $\mathcal{U}(\cdot)$ is an increasing function so an optimal solution must have allocated the entire unit of resource.

As it turns out, as long as $\mathcal{U}(\cdot)$ is increasing, concave and differentiable, the above convex program admits exactly the same unique solution regardless of the particular choice of $\mathcal{U}(\cdot)$. And we note this solution coincides with what is known in the literature as the *water-filling* algorithm [2]. This is summarized in the following theorem. The proof relies on the above program being convex to apply the well-known *Karush-Kuhn-Tucker* (KKT) optimality condition [2], and is relegated to the appendix.

Theorem 1. *For any utility function $\mathcal{U}$ that is concave and differentiable, the convex program admits a unique analytical solution. Moreover, the solution is unique across all choices of $\mathcal{U}(\cdot)$ as long as it is increasing, concave and differentiable.*

Therefore the unique optimal way to allocate resources across the goods is independent of the agent's utility function as long as it is differentiable and has diminishing return, which is a very reasonable assumption. We note this is a particularly nice property of the model as it frees us from imposing any particular form of utility function, which can often be arbitrary, and the risk of observing artifacts thus introduced. In NCGG considered later, this property frees us from making the assumption that each agent has the same utility function, which is standard of most economic models whose absence would often render the underlying model intractable.

2.2 Discrete Resource

In the discrete case, the agent has access to a set of atomic resources, each of integral volume. We show in the next two theorems that although the discrete CGP is NP-hard even in a rather special case, the general problem always admits an FPTAS.

Theorem 2. *The discrete common goods problem is NP-hard even when each atomic resource is of unit volume and $\mathcal{U}(\cdot)$ is increasing.*

Proof. We prove the hardness result by giving a reduction from the NP-hard *unbounded knapsack problem* [15].

UNBOUNDED KNAPSACK PROBLEM (UKP)
INSTANCE: A finite set $U = \{1, 2, ..., n\}$ of items, each item i has value $v_i \in \mathbb{Z}^+$, weight $w_i \in \mathbb{Z}^+$ and unbounded supply, a positive integer $B \in \mathbb{Z}^+$.
QUESTION: Find a multi-subset U' of U such that $\sum_{i \in U'} v_i$ is maximized and $\sum_{i \in U'} w_i \leq B$.

Since supply is unlimited we can assume without loss of generality that no two items are of the same weight and no item is strictly dominated by any other item, i.e. $w_i > w_j$ implies $v_i > v_j$. Now create n goods, $p_1, ..., p_n$, where p_i corresponds to item i and has a ground level $(i-1)B$. Let the agent have access to a total of B atomic resource, each of unit volume. Define the utility function $\mathcal{U}(\cdot)$ as follows: $\mathcal{U}(\omega) = \sum_{i=1}^{\mu(\omega)-1} \lfloor \frac{B}{w_i} \rfloor v_i + \lfloor \frac{\nu(\omega)}{w_{\mu(\omega)}} \rfloor v_{\mu(\omega)} + \frac{\omega}{(n^2-n+2)B^2}$ where $\mu(\omega) = \lceil \omega/B \rceil$ and $\nu(\omega) = \omega \bmod B$.

Clearly, $\mathcal{U}(\cdot)$ is a strictly increasing function, and thus we only concern ourselves with those CGP solutions that allocate all B atomic units of resources. One can then verify that there is a solution of total value K to the UKP instance iff there is a solution of total utility $\sum_{j=1}^{n} \sum_{i=1}^{j-1} \lfloor \frac{B}{w_i} \rfloor v_i + \frac{1}{2B} + K$ to the corresponding CGP instance. Therefore the discrete common goods problem is NP-hard.

Theorem 3. *The discrete common goods problem always admits an FPTAS.*

Proof. The discrete common goods problem can be reduced to the *multiple-choice knapsack problem*

MULTIPLE-CHOICE KNAPSACK PROBLEM (MCKP)
INSTANCE: A finite set $U = \{1, 2, ..., k\}$ of items, each item i has value v_i, weight w_i and belongs to one of n classes, a capacity $B > 0$.
QUESTION: Find a subset U' of U such that $\sum_{i \in U'} v_i$ is maximized, $\sum_{i \in U'} w_i \leq B$, and at most one item is chosen from each of the n classes.

The reduction goes as follows. For a general CGP instance, where there are B atomic unit-volume resources, and goods $\{p_1, ..., p_n\}$ such that good p_i has ground level α_i, create a MCKP instance such that there are n classes $c_1, ..., c_n$. Class c_i corresponds to good p_i and has B items of weight j and value $\mathcal{U}(\alpha_i + j)$, for $j = 1, ..., B$. The knapsack is of total capacity B.

It is not hard to see that there is a solution of total utility K to the CGP instance if and only if there is a solution of total value K to the MCKP instance. Therefore, any approximation algorithm for the latter translates into one for the former with the same approximation guarantee. Since an FPTAS is known for MCKP [1,14], CGP also admits an FPTAS.

3   Pure Strategy Nash Equilibrium

We consider in this section the existence and uniqueness of Nash equilibrium in NCGG.

3.1   The Existence of Nash Equilibrium

First we show a Nash equilibrium always exists in NCGG when the utility functions satisfy certain niceness properties.

Theorem 4. *For any NCGG instance, a pure strategy Nash equilibrium always exists as long as $\mathcal{U}_j$ is increasing, concave and differentiable for any agent a_j.*

Proof. Let $\deg(a_i)$ be the degree of agent a_i and $D = \sum_{a_i \in A} \deg(a_i)$. Let $s \in [0,1]^D$ be the state vector that corresponds to how the m agents have allocated their resources, where the $(\sum_{k=1}^{i-1} \deg(a_k))$th to the $(\sum_{k=1}^{i} \deg(a_k))$th dimension of s correspond to a_i's allocation of her resource on the $\deg(a_i)$ goods she is connected to (assume an arbitrary but fixed order of the goods a_i is connected to). Define function $f : [0,1]^D \to [0,1]^D$ such that $f(s)$ maps to the *best response* state s', where $\left(s'_{\sum_{k=1}^{i-1} \deg(a_k)}, ..., s'_{\sum_{k=1}^{i} \deg(a_k)} \right)$ corresponds to a_i's best response. Note s' is unique because each agent a_i's best response is unique by Theorem 1, therefore $f(s)$ is well-defined.

It is clear that $[0,1]^D$ is compact (i.e. closed and bounded) and convex, and f is continuous. Therefore, applying Brouwer's fixed point theorem shows that f has a fixed point, which implies NCGG has a Nash equilibrium.

We note on the other hand, it is easy to see that if $\mathcal{U}_j$ is allowed to be convex, then a pure strategy Nash equilibrium may not exist in NCGG.

3.2   The Uniqueness of Nash Equilibrium

We next establish uniqueness results of Nash equilibrium of NCGG in the next two theorems. Apparently, NCGG played on a general graph does not have a unique Nash equilibrium in the standard sense: Consider for example the 2×2 complete bipartite graph where $P = \{p_1, p_2\}$ and $A = \{a_1, a_2\}$, for any $0 \leq \delta \leq 1$, a_1 (resp. a_2) allocating δ (resp. $1 - \delta$) resource on p_1 and $1 - \delta$ (resp. δ) resource on p_2 constitutes a pure strategy Nash equilibrium and therefore there are uncountably infinite many of them. However, all these equilibria can still be considered as equivalent to each other in the sense that they all allocate exactly the same amount of resource to each good. And the reader is encouraged to verify as an exercise that any Nash equilibrium in the above NCGG instance belongs to this equivalence class. Therefore, the Nash equilibrium is still unique, albeit in a weaker sense.

To capture this, we thus consider two concepts of uniqueness of equilibrium: We say an NCGG instance has a *weakly unique* equilibrium if all its equilibria allocate exactly the same amount of resource on each good p_i. And if an NCGG instance has an equilibrium that is unique in the standard sense, we call it *strongly unique*. We note the concept of weak uniqueness is a useful one as it implies the uniqueness of each agent's utility in equilibrium, which is really what we ultimately care about.

We show two uniqueness results in this section. The first one establishes that NCGG has a strongly unique Nash equilibrium if the underlying graph is a tree.

The second one indicates that it is not a coincidence that the example shown above has a weakly unique equilibrium — in fact, we show *any* NCGG instance has a weakly unique Nash equilibrium. Furthermore, our results indicate that the equilibrium is a function of the structure of the underlying graph only, and independent of the particular forms and combinations of agents' utility functions, as long as these functions are increasing, concave and differentiable.

Theorem 5. *The Nash equilibrium of NCGG is* weakly unique *across all networked common goods games played on a given bipartite graph* $G = (P, A, E)$, *as long as* $\mathcal{U}_j$ *is increasing, convex and differentiable for any agent* a_j.

Proof. Suppose otherwise that there are two equilibria $\mathcal{E}$ and $\mathcal{E}'$ that have different amount of resource ω_i and ω_i' allocated to some good p_i (throughout the rest of the paper whenever it is clear from the context, for any good p_x we denote by ω_x and ω_x' the amount of resource allocated to p_x in $\mathcal{E}$ and $\mathcal{E}'$, respectively). Without loss of generality assume $\omega_i' < \omega_i$. Then there must exists some agent $a_j \in \mathcal{N}(p_i)$ who is allocating less resource on p_i in $\mathcal{E}'$ than in $\mathcal{E}$, and as a result, a_j must be allocating more resource on some good $p_k \in \mathcal{N}(a_j) \backslash \{p_i\}$ in $\mathcal{E}'$ because in equilibrium each agent allocates all of its resources. The fact that a_j is allocating nonzero resource on p_i in $\mathcal{E}$ implies $\omega_i \leq \omega_k$, and for the same reason $\omega_k' \leq \omega_i'$. Therefore we have $\omega_k - \omega_k' \geq \omega_i - \omega_i' > 0$.

Now consider the following process: Starting from set $S_0 = \{p_i\}$, add goods to S_0 that share an agent with p_i and whose total resource have decreased by at least $\omega_i - \omega_i'$ in $\mathcal{E}'$; let the new set be S_1. Then grow the set further by adding goods that share an agent with some good in S_1 and whose total resource are reduced by at least $\omega_i - \omega_i'$ in $\mathcal{E}'$. Continue this process until no more goods can be added and let the resulting set be S. By construction every good in S has its total resource decreased by at least $\omega_i - \omega_i'$ in $\mathcal{E}'$ than in $\mathcal{E}$; in fact, it can be shown that the decrease is exactly $\omega_i - \omega_i'$ for each good in S.

If $S = P$, then we have a contradiction immediately because if each good in P has its total resource decreased by a positive amount in $\mathcal{E}'$ then it implies the agents collectively have a positive amount of resources not allocated, contradicting the fact that $\mathcal{E}'$ is a Nash equilibrium.

We now claim that indeed $S = P$. Suppose otherwise $P = S \cup T$ and $T \neq \emptyset$. Then, $\mathcal{N}(S)$, the neighboring agents of S are collectively spending less resources on S in $\mathcal{E}'$ than in $\mathcal{E}$, which implies there exists an agent $a \in \mathcal{N}(S)$ who is allocating more resources to a good $p_t \in T$ in $\mathcal{E}'$ than in $\mathcal{E}$ and less resources to a good $p_s \in S$ in $\mathcal{E}'$ than in $\mathcal{E}$. By an argument similar to one given above, we have $\omega_s \leq \omega_t$ and $\omega_t' \leq \omega_s'$, and thus $\omega_t - \omega_t' \geq \omega_s - \omega_s' \geq \omega_i - \omega_i'$. This implies that p_t should be in S rather than T; so we must have $T = \emptyset$ or $S = P$.

Therefore $\mathcal{E}$ and $\mathcal{E}'$ must be equivalent in the sense that for any good $p_i \in P$, $\omega_i = \omega_i'$; this allows us to conclude that the Nash equilibrium of NCGG on any graph is *weakly unique*.

Next, we move to establish the strong uniqueness result on trees. We need the following lemma before we proceed to the main theorem of the section.

Lemma 1. *For any instance of NCGG on a tree* $G = (P, A, E)$, *let* $\mathcal{E}$ *be a Nash equilibrium of this game,* α_i *the ground level of* $p_i \in P$ *and* ω_i *the total resource allocated on* p_i *in* $\mathcal{E}$. *For any other instance of NCGG where everything is the same except that* α_i *is increased, if* $\mathcal{E}'$ *is an equilibrium of this new instance and* ω_i' *is total resource allocated to* p_i *in* $\mathcal{E}'$, *then* $\omega_i' \geq \omega_i$.

Proof. Without loss of generality assume all leafs of the tree are goods (because a leaf agent has no choice but to allocate all her resources to the unique good she is connected to) and root the tree at p_i. Suppose $\omega_i' < \omega_i$. Since $\alpha_i' > \alpha_i$, it must be the case that there exists some agent $a_j \in \mathcal{N}(p_i)$ who is allocating less resource on p_i in $\mathcal{E}'$ than in $\mathcal{E}$, this in turn implies that a_j is allocating more resource to some good $p_k \in \mathcal{N} \backslash \{p_i\}$ in $\mathcal{E}'$ than in $\mathcal{E}$. Therefore we have $\omega_i \leq \omega_k$ and $\omega_k' \leq \omega_i'$ and thus $\omega_k - \omega_k' \geq \omega_i - \omega_i' > 0$. If k is a leaf then this is obviously a contradiction. Otherwise, we can continue the above reasoning recursively and eventually we will reach a contradiction by having a leaf good whose total resource decreases in $\mathcal{E}'$ whereas at the same time its unique neighboring agent is allocating more resources to it.

Theorem 6. *The Nash equilibrium is strongly unique across all NCGG played on a given tree* $G = (P, A, E)$, *as long as* $\mathcal{U}_j$ *is increasing, convex and differentiable for any agent* a_j.

Proof. Again without loss of generality assume leafs are all goods. We have the following claim.

<u>Claim.</u> *For any NCGG instance on a tree* $G = (A, P, E)$, *if there is an equilibrium* $\mathcal{E}$ *where total resource allocated is the same across all goods, then* $\mathcal{E}$ *is the strongly unique Nash equilibrium.*

<u>Proof.</u> Suppose $\mathcal{E}$ is not strongly unique. Let $\mathcal{E}'$ be a different Nash equilibrium. By Theorem 5 $\mathcal{E}'$ can only be weakly different from $\mathcal{E}$. Since $\mathcal{E}$ and $\mathcal{E}'$ are weakly different there must exist edge (p_i, a_j) such that a_j is allocating different amount of resource in $\mathcal{E}$ and $\mathcal{E}'$; without loss of generality, assume a_j is allocating less resource in $\mathcal{E}'$ than in $\mathcal{E}$. Root the tree at p_i, then a_j must be allocating more resource in $\mathcal{E}'$ to one of its child $p_k \in \mathcal{N}(a_j) \backslash \{p_i\}$. Note given the amount of resource allocated by a_j on p_k, the game played at the subtree rooted at p_k can be viewed as independent of the game played in the rest of the tree, by viewing the resource allocated by a_j on p_k as part of the ground level of p_k. Now that the ground level has increased, by Lemma 1 any equilibrium on the subtree rooted at p_k must not have the total resource allocated on p_k decreased, so we have $\omega_k' \geq \omega_k$. If $\omega_k' > \omega_k$, then this is a contradiction to weak uniqueness. If $\omega_k' = \omega_k$, then one of p_k's child must be allocating less resource to p_k in $\mathcal{E}'$ than in $\mathcal{E}$ and we can repeat the above reasoning recursively. Continue this process until we either reach the conclusion that $\mathcal{E}$ and $\mathcal{E}'$ are strongly different, which is a contradiction, or reach a leaf good whose allocated resource in $\mathcal{E}'$ is the same as that in $\mathcal{E}$ even when his unique neighboring agent is allocating more resource to it in $\mathcal{E}'$, which is again a contradiction.

<u>Resume Proof of Theorem</u>. We prove this theorem by giving an induction on the size of the tree $N = |A| + |P|$. First note the equilibrium is unique when $N \leq 2$ (in the trivial case where either $E = \emptyset$, the claim is vacuously true). Assume the theorem is true for any tree of size $N \leq K$, consider the case $N = K + 1$.

For any instance G_{K+1} with $N = K + 1$, let $\mathcal{E}$ be a Nash equilibrium (whose existence is implied by Theorem 4). We want to show that $\mathcal{E}$ is strongly unique. Let

$$E(\mathcal{E}) = \{(p_i, a_j) \mid \omega_i > \omega_k \text{ and } x_{jk} > 0 \text{ in } \mathcal{E}\}$$

If $E(\mathcal{E}) = \emptyset$ then it must be the case that the total resource allocated is the same across all goods, and by the above claim $\mathcal{E}$ is thus strongly unique and we are through. Otherwise, partition G into subtrees by removing $E(\mathcal{E})$ from E. Note the size of each subtree thus resulted is at most N, so by induction they each has a strongly unique equilibrium; this implies that if we can prove $E(\mathcal{E}') = E(\mathcal{E})$ for any equilibrium $\mathcal{E}'$, then $\mathcal{E}' = \mathcal{E}$ and we are again through. To this end, suppose G_{K+1} has a weakly different equilibrium $\mathcal{E}'$ such that $(p_i, a_j) \in E(\mathcal{E})$ and $(p_i, a_j) \notin E(\mathcal{E}')$ and consider the following two cases.

CASE I: a_j IS ALLOCATING RESOURCE TO p_i IN $\mathcal{E}'$. Consider the game played on the subtree of G_{K+1} rooted at p_i and not containing a_j. Since a_j allocates more resource on p_i in $\mathcal{E}'$ than in $\mathcal{E}$, by Lemma 1 $\omega_i' \geq \omega_i$. On the other hand, a_j must be allocating less resource to some other good p_k in $\mathcal{E}'$ than in $\mathcal{E}$, so again by Lemma 1 $\omega_k \geq \omega_k'$. Note we also have $\omega_i > \omega_k$ and thus conclude that $\omega_i' > \omega_k'$; since a_j allocates non-zero resource to p_i in $\mathcal{E}'$, she is not acting optimally and this gives a contradiction to the fact that $\mathcal{E}'$ is an equilibrium.

CASE II: a_j IS NOT ALLOCATING RESOURCE TO p_i IN $\mathcal{E}'$. Since the subtree rooted at p_i and not containing a_j is of size at most $N - 1$, by induction we have $\omega_i = \omega_i'$. Since a_j is allocating the same total amount of resource to $\mathcal{N}(a_j) \backslash p_i$, there exists p_k on which a_j is allocating nonzero resource in $\mathcal{E}$ and not allocating strictly more resource in $\mathcal{E}'$ than in $\mathcal{E}$; by Lemma 1 this implies $\omega_k' \leq \omega_k$. Note we also have $\omega_i > \omega_k$ because $(p_i, a_j) \in E(\mathcal{E})$, and thus we have $\omega_i' > \omega_k'$. Consider the following two cases: CASE 1) If a_j allocates nonzero resource to p_k in $\mathcal{E}'$ then $\omega_i' = \omega_k'$ because $(p_i, a_j) \notin E(\mathcal{E}')$; but this is a contradiction. CASE 2) If a_j allocates zero resource to p_k then there exists good $p_l \in \mathcal{N}(a_j) \backslash \{p_i, p_k\}$ on which a_j is allocating strictly more resource in $\mathcal{E}'$ than in $\mathcal{E}$. The fact that $(p_i, a_j) \notin E(\mathcal{E}')$ implies $\omega_i' = \omega_l'$, so we have $\omega_l' > \omega_k'$; but this is a contradiction to the fact that $\mathcal{E}'$ is an equilibrium.

Now we conclude that $E(\mathcal{E}) = E(\mathcal{E}')$ and this completes the proof.

4   Nash Dynamics

Pick any utility function that is increasing, concave and differentiable, say $\mathcal{U}(x) = \sqrt{x}$, and define potential function $\Psi(\omega_1, ..., \omega_n) = \sum_{i=1}^{n} \sqrt{\omega_i}$. It is clear that for any agent a_j, whenever a_j updates her allocation such that increases her total utility, the potential increases as well. This proves the following theorem.

Theorem 7. *NCGG is a potential game.*

Therefore, better/best response Nash dynamics always converge. However it is not clear how fast the convergence is as the increment in a_j's total utility can be either larger or smaller than the increment of the potential, depending both on $\mathcal{U}_j(\cdot)$ and the amount of resources already allocated to a_j's neighboring goods. In the rest of the section, we present a particular Nash dynamics where we can show fast convergence to an ϵ-approximate Nash equilibrium. We only give details for the best response Nash dynamics (Algorithm 1), and it is easy to see the same convergence result holds for the corresponding better response Nash dynamics as well. To this end we consider K-discretized version of the game, where each agent has access to a total of K identical *atomic* resources, each of volume $1/K$. We start by giving the following two lemmas.

Lemma 2. *A solution to the K-discretized CGP is optimal iff the following two conditions are satisfied: 1) the agent has allocated all of its K atomic units of resource; 2) for any two goods $p_i, p_j \in P$, $\omega_i - \omega_j > 1/K$ (where $\omega_i = \alpha_i + x_i$ and $\omega_j = \alpha_j + x_j$) implies $x_i = 0$.*

Proof. First we prove the 'only if' direction. It is obvious that an optimal solution must have allocated all of its K atomic units of resource because the utility function is increasing, so we focus on the proof of the second condition. Suppose otherwise we have $p_i, p_j \in P$ with $\omega_i - \omega_j > 1/k$, where $\omega_i = \alpha_i + x_i$, $\omega_j = \alpha_j + x_j$ and $x_i > 0$. Construct another solution by moving one atomic unit of resource from good p_i to p_j gives a new solution of total utility strictly higher because the utility function is increasing and concave. Therefore we have a contradiction.

Next we prove the 'if' direction of the lemma. Suppose the solution x is not optimal. Let ω_k and ω'_k (where $p_k \in P$) denote the total resource induced by this 'suboptimal' solution and a true optimal solution x', respectively. Since an optimal solution must have allocated all of its K units of atomic resource among the goods, it must be true that there exist $p_i, p_j \in P$ such that $\omega_i - \omega'_i \geq 1/K$ and $\omega'_j - \omega_j \geq 1/K$, and if both inequality holds in equality, then $\omega'_i \neq \omega_j$ (because otherwise x and x' are essentially the same, which means x is already optimal). Note $\omega'_j - \omega_j \geq 1/K$ implies that good j' has resource allocated to it in the optimal solution (i.e. $x'_j \geq 1/K$), so by the 'only if' part of proof above, we must have $\omega'_i \geq \omega'_j - 1/K$. Now we show that $\omega_i - \omega_j > 1/K$ by considering the following two cases:

CASE I: $(\omega_i - \omega'_i) + (\omega'_j - \omega_j) > 2/K$. In this case, it is easily checked that $\omega_i - \omega_j > 1/K$.

CASE II: $\omega_i - \omega'_i = 1/K$ AND $\omega'_j - \omega_j = 1/K$. As discussed above, we must not have $\omega'_i = \omega_j$. In fact, we must have $\omega'_i > \omega_j$ because otherwise we will have $\omega'_j = \omega_j + 1/K > \omega'_i + 1/K$, which is a contradiction to optimality because $x'_j \geq 1/K$. Therefore, again we have reached the conclusion that $\omega_i - \omega_j > 1/K$.

Now note $\omega_i - \omega'_i \geq 1/K$ implies $x_i \geq 1/K$, but this is a contradiction to $\omega_i - \omega_j > 1/K$, which by assumption implies $x_i = 0$. Therefore, x must itself be an optimal solution.

Lemma 3. *For any $\epsilon > 0$, an optimal solution to the K-discretized common goods problem, where $K = 1/\mathcal{U}^{-1}(\epsilon/n)$, is an ϵ-approximation to the optimal solution in the continuous common goods problem.*

Proof. Denote by OPT and OPT_K the optimal utility attained by an optimal solution in the continuous version and the K-discretized version, respectively; denote by $\mathcal{W}^*$ and $\mathcal{W}_K^*$ the set of goods to which non-zero resource is allocated in the two optimal solutions, respectively. By Lemma 2, any two goods in $\mathcal{W}_K^*$ must have their total resources allocated differ by at most $1/K$, i.e. $\omega_{max} - \omega_{min} \leq 1/K$, where $\omega_{min} = \min\{\omega_i \mid p_i \in \mathcal{W}_K^*\}$ and $\omega_{max} = \max\{\omega_i \mid p_i \in \mathcal{W}_K^*\}$. Since the agent has access to n goods, it must be the case that $\omega_{min} \geq 1/n - 1/K$ because otherwise $\omega_{max} < 1/n$. Now consider the set $\mathcal{W} = \mathcal{W}_K^* \cup \{p_i \notin \mathcal{W}_K^* \mid \alpha_i \leq \omega_{max}\}$ of goods whose total resource is at most ω_{max}, it is clear that: 1) $\mathcal{W}^*$, the optimal solution to the continuous version of the problem, forms a subset of $\mathcal{W}$; 2) $\max\{\omega_i \mid i \in \mathcal{W}^*\} \leq \omega_{max}$.

Now suppose we have access to an additional of $|\mathcal{W}|$ atomic units of resource, each of volume $1/K$, construct a new allocation by doing the following: Start with an allocation same as $\mathcal{W}_K^*$, then assign one atomic unit of resource to each good in $\mathcal{W} \supseteq \mathcal{W}_K^*$. It is clear from the above discussion that for any good $p_i \in \mathcal{W}$, its total resource under the new allocation is at least that of the total resource allocated under $\mathcal{W}^*$, which means the utility that we obtain under the new allocation, OPT'', is at least OPT. Therefore $OPT - OPT_K \leq OPT'' - OPT_K \leq n\mathcal{U}(1/K)$; so to upper bound $OPT - OPT_K$ by ϵ, it is sufficient to set $K = 1/\mathcal{U}^{-1}(\epsilon/n)$.

Note for most reasonable choices of $\mathcal{U}$ (e.g. $\mathcal{U}(x) = x^p$ where $p \in (0,1)$), K is polynomial in n. We have the following theorem.

Theorem 8. *For any $\epsilon > 0$, Algorithm 1 converges to an ϵ-approximate Nash equilibrium in $O(Kmn)$ time, where $K = \max_{j \in [m]} \mathcal{U}_j^{-1}(\epsilon/n))$, for any updating schedule σ.[2]*

Proof. First note according to the characterization of Lemma 2, the response of each agent $a_{\sigma(t)}$ in Algorithm 1 is a K-discretized best response. The rest of this proof is to define a potential function[3] whose range are positive integers that span an interval no greater than Kmn, and to show each time an agent updates his allocation with a best response, the value of this potential function strictly decreases.

For simplicity of exposition, we write p_i in place of $p_{\pi(i)}$ in the rest of the proof. Let $p_1, p_2, ..., p_n$ be the n goods arranged in non-increasing order of total resource allocated, that is, $\omega_1 \geq \omega_2 \geq ... \geq \omega_n$. Define potential function $\Phi(\omega_1, \omega_2, ..., \omega_n) = \sum_{i=1}^{n}(n-i) \cdot \omega_i$. Apparently, $\Phi(\cdot)$ is a positive integer valued function and the difference between the greatest and smallest function value is

---

[2] σ is assumed to at any time only pick an agent whose state is not already a best-response.

[3] This potential function is different from the one given in the proof of Theorem 7; this new potential function is convenient in upper bounding the convergence time.

Algorithm 1. K-discretized Best Response Nash Dynamics

1: // INPUT: G, $\alpha \succeq 0$, $\epsilon > 0$, and schedule σ
2: // OUTPUT: An ϵ-approximate Nash Equilibrium
3: Start by setting $K = \max_{j \in [m]} \mathcal{U}_j^{-1}(\epsilon/n))$
4: // Set an arbitrary initial state $s = (s_1, s_2, ..., s_m)$
5: **for** $j = 1$ to m **do**
6:     a_j discretizes his one unit of resource into $2K$ atomic units, each of volume $1/2K$;
        arbitrarily assigns them to her adjacent goods, resulting in s_j
7: **end for**
8: // Sort in non-increasing order of total resource allocated
9: Arrange goods in the order $p_{\pi(1)}, p_{\pi(2)}, ..., p_{\pi(n)}$ s.t. $\omega_{\pi(i)} \geq \omega_\pi(j)$ if $1 \leq i < j \leq n$
10: // Best response Nash Dynamics
11: **for** $t = 1$ to T **do**
12:     Let $a_{\sigma(t)}$ be the agent *active* in round t;
13:     **while** $\exists\, 1 \leq i < j \leq n$ s.t. $\omega_{\pi(i)} - \omega_{\pi(j)} \geq 1/K$ and $x_{\sigma(t)\pi(i)} > 0$ **do**
14:         $x_{\sigma(t)\pi(i)} = x_{\sigma(t)\pi(i)} - 1/2K$; $x_{\sigma(t)\pi(j)} = x_{\sigma(t)\pi(j)} + 1/2K$
15:         If necessary, re-define π to maintain total resource allocated in non-increasing
            order.
16:     **end while**
17: **end for**

---

upper bounded by Kmn. We are done if we can show that for any node $a_{\sigma(t)}$, the computation that $a_{\sigma(t)}$ does on line 13-17 of Algorithm 1 results in a strict decrease in the potential.

On line 14-15 of Algorithm 1, an atomic unit of resource of volume $1/2K$ is moved from good p_i to p_j. In doing so, the goods may no longer be sorted in non-increasing order of total resource, and in this case we restore it on line 16 of Algorithm 1, which without loss of generality can be thought of as moving p_i to the right for some $\mu \geq 0$ positions (with μ being the minimum necessary), and moving p_j to the left in the ordering for some $\nu \geq 0$ positions (again with ν being the minimum necessary). This results in the new ordering of the goods:

$$p_1, ..., p_{i-1}, p_{i+1}, ..., p_{i+\mu}, p_i, ..., p_j, p_{j-\nu}, ..., p_{j-1}, p_{j+1}, ..., p_n$$

Note p_i still precedes p_j (i.e. $i + \mu < j - \nu$) in this ordering because prior to line 14-15 of Algorithm 1, $\omega_i - \omega_j \geq 1/K$, therefore, the total resource of p_i is still at least that of p_j after a $1/2K$ amount of resource has been moved from p_i to p_j. With this observation, we can analyze the change in potential by looking at the changes of potential on $\{p_i, ..., p_{i+\mu}\}$ and $\{p_{j-\nu}, ..., p_j\}$ separately, and ignore the rest of the goods, whose contribution to potential remain unchanged. Clearly, the contribution to potential from $\{p_i, ..., p_{i+\mu}\}$ decreases, and by an amount of $\Delta\Phi_\downarrow = (\omega_i - \omega_{i+1}) \cdot (n - i) + (\omega_{i+1} - \omega_{i+2}) \cdot (n - i - 1)\ \ + ... +\ (\omega_{i+\mu} - \omega_i + 1/2K) \cdot (n - i - \mu) \geq (n - i - \mu)/2K$. Similarly, the contribution to potential from $\{p_{j-\nu}, ..., p_j\}$ increases by $\Delta\Phi_\uparrow = (\omega_{j-1} - \omega_j) \cdot (n - j) + (\omega_{j-2} - \omega_{j-1}) \cdot (n - j + 1)\ \ + ... + (\omega_j + 1/2K - \omega_{j-\nu}) \cdot (n - j + \nu) \leq (n - j + \nu)/2K$. Since $i + \mu < j - \nu$, we have $\Delta\Phi_\downarrow > \Delta\Phi_\uparrow$, which means the potential decrease by at least 1. Therefore, in at most Kmn steps Algorithm 1 converges to a

Nash equilibrium in the K-discretized game. By Lemma 3, this constitutes an ϵ-approximate Nash equilibrium to the original game.

5   Price of Anarchy of the Game

We show in this section the *price of anarchy* of NCGG is unbounded, and it is for a reason that echoes the well-known phenomenon called *tragedy of the commons* [11].

Theorem 9. *The* price of anarchy *of NCGG is* $\Omega(n^{1-\epsilon})$, *for any* $\epsilon > 0$.

Proof. Consider the bipartite graph $G = (P, A, E)$ where $P = \{p_c, p_1, ..., p_n\}$, $A = \{a_1, ..., a_n\}$ and $E = \{(p_j, a_j), (p_c, a_j) \mid j \in [n]\}$ so that all agents share the 'common' good p_c and each agent a_j has a 'private' good p_j to himself. Assume each agent a_j has the same utility function $\mathcal{U}$, $\alpha_i = 0 \; \forall \; a_i \in \{p_1, ..., p_n\}$ and $\alpha_c = 1$.

It is clear that it is a Nash equilibrium for every agent a_j to allocate her entire unit of resource to her private good p_j. And in this case the social welfare is $2n \cdot \mathcal{U}(1)$. On the other hand, if every agent devotes her entire unit of resource to the common good, then the social welfare is $n \cdot \mathcal{U}(n + 1)$. Therefore the price of anarchy of this particular example is at least $\frac{\mathcal{U}(n+1)}{2\mathcal{U}(1)} = O(\mathcal{U}(n + 1))$. Since $\mathcal{U}(\cdot)$ is concave, we can set $\mathcal{U}(x) = x^{1-\epsilon}$; therefore the theorem follows.

References

1. Bansal, M.S., Venkaiah, V.C.: Improved fully polynomial time approximation scheme for the 0-1 multiple-choice knapsack problem. In: Proc. of SIAM Conference on Discrete Mathematics (2004)
2. Boyd, S.P., Vandenberghe, L.: Convex optimization. Cambridge University Press, Cambridge (2004)
3. Bramoullé, Y., Kranton, R.E.: Public goods in networks. Journal of Economic Theory 135(1), 478–494 (2007)
4. Even-Dar, E., Kearns, M., Suri, S.: A network formation game for bipartite exchange economies. In: Proc. of ACM SODA 2007, pp. 697–706 (2007)
5. Eisenberg, E., Gale, D.: Consensus of subjective probabilities: The Pari-Mutuel method. Annals of Mathematical Statistics 30, 165–168 (1959)
6. Feldman, M., Lai, K., Zhang, L.: A price-anticipating resource allocation mechanism for distributed shared clusters. In: Proc. of ACM EC 2005, pp. 127–136 (2005)
7. Fisher, I.: PhD thesis. Yale University, New Haven (1891)
8. Fol'gardt, A.V.: Solution of a resource allocation game. Computational Mathematics and Modeling 4(3), 273–274 (1993)
9. Fol'gardt, A.V.: Games with allocation of discrete resources to several sites. Computational Mathematics and Modeling 6(3), 172–176 (1995)
10. Goemans, M.X., Li, E.L., Mirrokni, V.S., Thottan, M.: Market sharing games applied to content distribution in ad-hoc networks. In: Proc. of MobiHoc 2004, pp. 55–66 (2004)

11. Hardin, G.: Tragedy of the commons. Science 162, 1243–1248 (1968)
12. Kagel, J.H., Roth, A.E. (eds.): The handbook of experimental economics. Princeton University Press, Princeton (1995)
13. Kakade, S.M., Kearns, M.J., Ortiz, L.E., Pemantle, R., Suri, S.: Economic properties of social networks. In: Proc. of NIPS 2004 (2004)
14. Kellerer, H., Pferschy, U., Pisinger, D.: Knapsack problems. Springer, Heidelberg (2004)
15. Martello, S., Toth, P.: Knapsack problems: Algorithms and computer implementation. John Wiley and Sons, Chichester (1990)
16. Monderer, D., Shapley, L.S.: Potential games. Games and Economic Behavior 14, 124–143 (1996)
17. Papadimitriou, C.: Algorithms, games, and the Internet. In: Proc. of STOC 2001, pp. 749–753 (2001)
18. Zhang, L.: The efficiency and fairness of a fixed budget resource allocation game. In: Caires, L., Italiano, G.F., Monteiro, L., Palamidessi, C., Yung, M. (eds.) ICALP 2005. LNCS, vol. 3580, pp. 485–496. Springer, Heidelberg (2005)

A   Proof of Theorem 1

Proof. Let λ_i $(i = 1, 2, ..., n)$ be the *Lagrange multiplier* associated with the inequality constraint $x_i \geq 0$ and ν the *Lagrange multiplier* associated with the equality constraint $\sum_{i=1}^{n} x_i = 1$. Since the above program is convex, the following KKT optimality conditions,

$$\lambda_i^* \geq 0, \ x_i^* \geq 0 \quad (i \in [n]) \tag{2a}$$

$$\sum_{i=1}^{n} x_i^* = 1 \tag{2b}$$

$$\lambda_i^* x_i^* = 0 \quad (i \in [n]) \tag{2c}$$

$$-\frac{d}{dx_i}\mathcal{U}(\alpha_i + x_i^*) - \lambda_i^* + \nu^* = 0 \quad (i \in [n]) \tag{2d}$$

are sufficient and necessary for x^* to be the optimal solution to the (primal) convex program (1) and (λ^*, ν^*) the optimal solution to the associated dual program.

Let $\mathcal{V}$ be the inverse function of $\frac{d}{dt}\mathcal{U}$. Note equation (2c) and (2d) implies $(-\frac{d}{dx_i}\mathcal{U}(\alpha_i + x_i^*) + \nu^*)x_i^* = 0$; equation (2a) and (2d) implies $-\frac{d}{dx_i}\mathcal{U}(\alpha_i + x_i^*) + \nu^* \geq 0$, which combing with the fact that $\mathcal{U}(\cdot)$ is convex implies $\mathcal{V}(\nu^*) \leq \alpha_i + x_i^*$. If $\alpha_i < \mathcal{V}(\nu^*)$, then $x_i^* > 0$ and thus $-\frac{d}{dx_i}\mathcal{U}(\alpha_i + x_i^*) + \nu^* = 0$, i.e. $x_i^* = \mathcal{V}(\nu^*) - \alpha_i$. On the other hand, if $\alpha_i \geq \mathcal{V}(\nu^*)$, then we must have $x_i^* = 0$. To see why this is true, suppose otherwise $x_i^* > 0$; this leads to $x_i^* = \mathcal{V}(\nu^*) - \alpha_i \leq 0$, which is a contradiction. Therefore $x_i^* = \mathcal{V}(\nu^*) - \alpha_i$ if $\alpha_i < \mathcal{V}(\nu^*)$, and $x_i^* = 0$ otherwise, where ν^* is a solution to $\sum_{i=1}^{n} \max\{0, \mathcal{V}(\nu^*) - \alpha_i\} = 1$.

It is easy to see that $\sum_{i=1}^{n} \max\{0, \mathcal{V}(\nu^*) - \alpha_i\} = 1$ admits a unique solution if we treat $\mathcal{V}(\nu^*)$ as the variable, i.e. different utility functions only leads to different solutions of the Lagrange multiplier ν^* but $\mathcal{V}(\nu^*)$ remains invariant. Therefore the optimal solution x^* is unique not only of a particular choice of $\mathcal{U}(\cdot)$, but across all utility functions that are increasing, concave and differentiable.

A Novel Branching Strategy for Parameterized Graph Modification Problems*

James Nastos and Yong Gao**

Department of Computer Science, Irving K. Barber School of Arts and Sciences,
University of British Columbia Okanagan, Kelowna, Canada V1V 1V7
jnastos@interchange.ubc.ca, yong.gao@ubc.ca

Abstract. Many *fixed-parameter tractable* algorithms using a bounded
search tree have been repeatedly improved, often by describing a larger
number of branching rules involving an increasingly complex case analysis. We introduce a novel and general branching strategy that branches
on the forbidden subgraphs of a relaxed class of graphs. By using the
class of P_4-sparse graphs as the relaxed graph class, we obtain efficient
bounded-search tree algorithms for several parameterized deletion problems. For the cograph edge-deletion problem and the trivially perfect
edge-deletion problem, the branching strategy yields the first non-trivial
bounded-search tree algorithms. For the cograph vertex deletion problem, the running time of our simple bounded search algorithm matches
those previously designed with the help of complicated case distinctions
and non-trivial running time analysis [16] and computer-aided branching
rules [7].

Keywords: Fixed-parameter tractability; edge-deletion; graph modification; cographs; trivially perfect graphs; quasi-threshold graphs; bounded
search tree.

1 Introduction

A graph is a *cograph* [18] if it has no induced subgraph isomorphic to a P_4, an
induced path on four vertices. The name originates from *complement reducible
graphs* as cographs are also characterized as being those graphs G which are
either disconnected or else its complement $\overline{G}$ is disconnected [18]. They are a
well-studied class of graph and many NP-complete problems on graphs have
been shown to have polynomial time solutions when the input is a cograph [4].

A *graph modification problem* is a general term for a problem that takes a
graph as input and asks how the graph can be modified to arrive at a new
graph with a desired property. Usually, graph modifications are edge additions
or deletions, vertex additions or deletions, or combinations of these. Our work
on the following problems originally stems from studying social networks from
which edge removals are made to reveal underlying structures in the network.

* Supported in part by NSERC Discovery Grant RGPIN 327587-09.
** We thank Dr. Donovan Hare for our discussions on these results.

W. Wu and O. Daescu (Eds.): COCOA 2010, Part II, LNCS 6509, pp. 332–346, 2010.
© Springer-Verlag Berlin Heidelberg 2010

This paper concerns the problem of determining when a graph $G = (V, E)$ has a set S of at most k edges which can be removed in order to make $G_2 = (V, E \backslash S)$ a cograph. This problem is known to be NP-complete [14] and also known to be *fixed parameter tractable* [2]. We show how our method for cograph edge-deletion can also solve the vertex-deletion problem for cographs and to the edge-deletion problem to *trivially perfect* graphs.

We note that since the class of cographs is self-complementary, an algorithm solving k-edge-deletion problem also serves as a solution to the problem of k-edge-addition to cographs and the k-edge-addition problem to co-trivially perfect graphs.

This paper is structured as follows: Section 2 summarizes previous results related to cograph modification problems and gives some background on the class of P_4-sparse graphs; Section 3 gives the cograph edge-deletion algorithms and its runtime analysis; Section 4 modifies the algorithm to solve the *vertex deletion* problem for cographs in $O(3.30^k(m + n))$ time and the *trivially perfect* edge deletion problem in $O(2.45^k(m + n))$ time, where m is the number of edges and n is the number of vertices in the input graph; Section 5 summarizes and discusses these results and mentions how cograph edge deletion can be solved in $O(2.415^k(m + n))$ time through the use of *semi-P_4-sparse graphs*.

2   Previous Results and Background

2.1   Previous Fixed-Parameter Tractability Results

While cographs can be recognized in linear time [4], it is also known that it is NP-complete to decide whether a graph is a cograph with k extra edges [14]. Graph modification problems have been studied extensively: Yannakakis shows that vertex-deletion problems to many types of structures is NP-hard [21]. Elmallah and and Colbourn give hardness results for many edge-deletion problems [14].

Recently, a lot of research has been devoted to finding fixed-parameter tractable algorithms for graph modification problems: Guo [8] studied edge deletion to split graphs, chain graphs, threshold graphs and co-trivially perfect graphs; Kaplan et al. [11] studied edge-addition problems to chordal graphs, strongly chordal graphs and proper interval graphs; Cai [2] showed fixed-parameter tractability for the edge deletion, edge addition, and edge editing problem to any class of graphs defined by a finite set of forbidden induced subgraphs. The constructive proof implies that k-edge-deletion problems to a class of graphs defined by a finite number of forbidden subgraphs is $O(M^k p(m + n))$ where p is some polynomial and M is the maximum over the number of edges in each of the forbidden induced subgraphs defining that graph class in question. For k-edge-deletions to P_4-free graphs in particular, Cai's result implies an algorithm running in $O(3^k(m + n))$ time. This algorithm would work by finding a P_4: $a - b - c - d$ in a graph and branching on the 3 possible ways of removing an edge in order to destroy the P_4 (that is, removing either the edge $\{a, b\}$ or $\{b, c\}$ or $\{c, d\}$).

Nikolopoulos and Palios study the edge-deletion to cograph problem for a graph $G - xy$ where G is a cograph and xy is some edge of G [17]. Lokshtanov et al.

study cograph edge-deletion sets to determine whether they are *minimal*, but not a minimum edge-deletion set [13]. To the best of our knowledge, ours is the first study that specifically addresses the edge-deletion problem to cographs. We present a bounded search tree algorithm that solves k-edge-deletion to cographs in $O(2.562^k(m+n))$ time by performing a search until we arrive at a P_4-sparse graph and then optimally solving the remainder of the search space using the structure of P_4-sparse graphs.

Graph modification problems can also be regarded as a type of graph recognition problem. Following the notation of Cai [2], for any class of graphs $\mathcal{C}$, we call $\mathcal{C} + ke$ the set of all graphs which are formed from a graph of class $\mathcal{C}$ with k extra edges. Similarly, $\mathcal{C} - ke$ is the set of graphs which are formed from a graph of class $\mathcal{C}$ with k edge removals. Replacing 'edges' by 'vertices' in these definitions gives analogous classes for $\mathcal{C} + kv$ and $\mathcal{C} - kv$. A k-edge-deletion problem to a class of graphs $\mathcal{C}$ can thusly be restated as a recognition problem for the class of $\mathcal{C} + ke$ graphs. Our results on cographs here can be restated as recognition algorithms for the classes: Cograph+ke, Cograph-ke, Cograph+kv, Trivially Perfect+ke.

2.2 Background Information: P_4-sparse Graphs

One generalization to the class of cographs is formed by allowing P_4s to exist in a graph but in restricted amounts. Hoáng [9] introduced P_4-sparse graphs to be those for which every induced subgraph on five vertices induces at most one P_4. This immediately implies a forbidden induced subgraph characterization which restricts any subgraph of five vertices inducing two or more P_4s. We include these graphs in Figure 1.

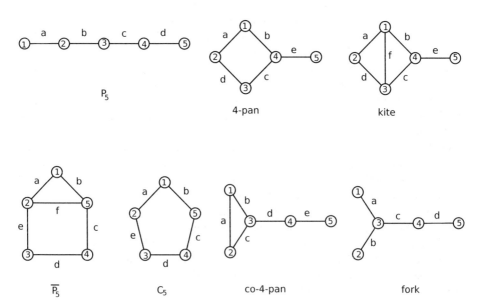

Fig. 1. The forbidden induced subgraphs for P_4-sparse graphs

A special graph structure called a *spider* [10] commonly occurs in graph classes of bounded cliquewidth. We define two types of spiders here:

Definition 1. *A graph $G = (V, E)$ is a* thin spider *if V can be partitioned into K, S and R such that:*

i) K is a clique, S is a stable set, and $|K| = |S| \geq 2$.
ii) every vertex in R is adjacent to every vertex of K and to no vertex in S
iii) each vertex in S has a unique neighbour in K, that is: there exists a bijection $f : S \to K$ such that every vertex $k \in K$ is adjacent to $f(k) \in S$ and to no other vertex in S.

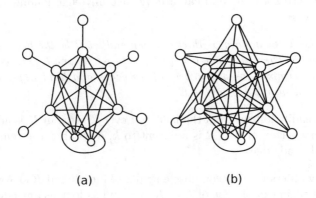

(a)   (b)

Fig. 2. A thin (a) spider and a thick (b) spider with $|K| = |S| = 5$ and $|R| = 2$

A graph G is called a *thick spider* if $\overline{G}$ is a thin spider. Note that the vertex sets K and S swap roles under graph complementation, that condition (i) and (ii) hold for thick spiders, and that statement (iii) changes to saying that every vertex in S has a unique *non-neighbour* in K. The sets K, S and R are called the *body*, *feet* and *head* of the spider, respectively. The edges with one endpoint in S are called *thin legs* or *thick legs* for thin spiders or thick spiders, respectively. Examples of spiders are given in Figure 2.

Hoàng [9] defined a graph G to be P_4-sparse if every induced subgraph with exactly five vertices contains at most one P_4. The following decomposition theorem for P_4-sparse graphs was proven in [10]:

Lemma 1. *[10] Let G be a P_4-sparse graph. Then at least one of the following is true:*

i) G is disconnected
ii) $\overline{G}$ is disconnected
iii) G is a thin spider
iv) G is a thick spider

We note here that for the purposes of an edge-deletion problem, if G is disconnected then the edge-deletion problem on G decomposes into separate edge-deletion problems on the connected components of G, and their individual solutions combine to solve the edge-deletion problem on G. We show that for an edge-deletion or vertex-deletion problem in which the goal is to break P_4s, we can decompose the deletion problem into separate smaller problems. Solving these deletion problems on spiders requires separate algorithms for the thin and thick cases, and these algorithms are presented in the next section and used as subroutines for the main cograph deletion algorithm.

Decomposing the deletion problems into subproblems requires the observation that no edge joining a vertex in R with a vertex in K is in any P_4, even after some edge-deletions of a certain type. Call a *leg edge* any edge joining a vertex $s \in S$ with a vertex $k \in K$, and call a *head edge* any edge joining some $r_1 \in R$ with some $r_2 \in R$.

Lemma 2. *Let G be a (thin or thick) spider and let R be the head of G and K be the body of G. Then any edge $e = \{r, k\}$ with $r \in R$ and $k \in K$ is not in any P_4 in G. Furthermore, for any subset of leg edges and head edges E' the edge $e = \{r, k\}$ is not in any P_4 in $G - E'$.*

Proof. This readily follows from the fact that every vertex in K is adjacent to r and that every vertex in $K \cup R$ is adjacent to k, even after the removal of any leg edges and head edges.

This property allows us to decompose a spider G into R and $K \cup S$ where every P_4 in G must be in exactly one of R or $K \cup S$, and as long as any edge deletions are leg edges or head edges, we can solve the deletion problem for $K \cup S$ and R separately.

3   Edge-Deletion Algorithm

In this section, we discuss the algorithm for the cograph edge-deletion problem defined as follows:

Problem 1. Cograph Deletion (G, k):
Given graph $G = (V, E)$, does there exist a set S of k edges such that $(V, E \setminus S)$ is a cograph?

The idea of the algorithm is to focus on the forbidden subgraphs of P_4-sparse graphs so that efficient branching rules can be designed systematically. This depends critically on whether the cograph deletion problem can be solved polynomially on P_4-sparse graphs. Therefore, we first show how to solve the problem on P_4-sparse graphs in linear time.

3.1   Finding Cograph Edge-Deletion Sets in P_4-Sparse Graphs

Our algorithm to find cograph edge-deletion sets in P_4-sparse graphs is presented in Algorithm 1. Its correctness depends on the structural decomposition of a P_4-sparse graph discussed in Section 2 and the following two lemmas.

Lemma 3. *Let G be a thin spider with body $K = \{k_1, \ldots, k_{|K|}\}$ and legs $S = \{s_1, \ldots, s_{|K|}\}$, and $\{s_i, k_j\}$ is an edge if and only if $i = j$. Then a minimum cograph edge-deletion set for $K \cup S$ is $\{\{s_i, k_i\}, i = 1..|K| - 1\}$.*

Proof. Since K is a clique and S is stable, every P_4 in $K \cup S$ has its endpoints in S. Furthermore, every pair of vertices in S are in a unique P_4. Deleting any $|S| - 1$ thin legs will clearly destroy all of the P_4s, so this edge-deletion set is indeed a cograph edge-deletion set. To see that it is of minimum size, assume there is a deletion set of size $|K| - 2$ or less in which two legs are not part of the deletion set. Let these two legs be $\{s_1, k_1\}$ and $\{s_2, k_2\}$ and call them "permanent" in this case. Since $\{s_1, k_1, k_2, s_2\}$ is a P_4 and the edges $\{s_1, k_1\}$ and $\{s_2, k_2\}$ are not in the deletion-set, it must be that $\{k_1, k_2\}$ is in the deletion set. There at most $|K| - 3$ other edges in the deletion set. Now $\{s_1, k_1, k_j, k_2\}$ induces a P_4 for every $j = 3 \ldots |K|$. This means that the permanent edge $\{s_1, k_1\}$ is still in $|K| - 2$ P_4s and every pair of these P_4s have distinct edges aside from $\{s_1, k_1\}$. Thus it is impossible to destroy all of these remaining P_4s with only $|K| - 3$ additional deletions or less.

Lemma 4. *Let G be a thick spider with body $K = \{k_1, \ldots, k_{|K|}\}$ and feet $S = \{s_1, \ldots, s_{|K|}\}$, and $\{s_i, k_j\}$ is an edge if and only if $i \neq j$. Then a minimum cograph edge-deletion set for $K \cup S$ is $\{\{k_i, s_j\}, i < j\}$.*

Proof. Every edge in $K \cup S$ is in exactly one P_4: an edge $\{k_i, k_j\}$ is only in the P_4 $\{s_j, k_i, k_j, s_i\}$ and any edge $\{s_i, k_j\}$ is only in the P_4 $\{s_i, k_j, k_i, s_j\}$ so the number of P_4s in $K \cup S$ is $\binom{|S|}{2}$, and since no two of these P_4s share an edge, at least $\binom{|S|}{2}$ deletions are required. Consider the edge set $T = \{\{k_i, s_j\}, i < j\}$. When deleting T from $K \cup S$, K is still a clique and S is still a stable set, and so if there is any P_4 in $(K \cup S) \setminus T$, its endpoints must still be in S. But after deletion of T, we have that the neighbourhood of s_i is $N(s_i) = \{k_{i+1}, \ldots, k_{|K|}\}$ which means that $N(s_i) \subset N(s_j)$ for all $i > j$, and so no two vertices in S can be the endpoints of a P_4. So T indeed destroys all the P_4s in $K \cup S$ and since $|T| = \binom{|S|}{2}$, this is a minimum set.

Theorem 1. *Algorithm 1 correctly solves the cograph edge-deletion problem for P_4-sparse graphs and can be implemented in $O(m + n)$ time.*

Proof. Lemma 3 and Lemma 4 show that there are optimal edge-deletion sets from $K \cup S$ that remove only leg edges. We can then use Lemma 2 to combine the edge deletions from $K \cup S$ with any edge deletions from R for a complete solution to the cograph edge-deletion problem on P_4-sparse graphs.

Algorithm 1 can be implemented in linear time, as the spider structure of P_4-sparse graphs can be identified in linear time [10]. Identifying the connected or co-connected components can also be done in linear time, as these types of vertex partitions are special cases of the more general notion of a *homogeneous set* or *module*, and there are a number of modular decomposition algorithms running in linear time [15], [5].

Algorithm SPIDER(G):
Input: A P_4-Sparse Graph $G = (V, E)$
Output: A set $S \subset E$
if G *(or $\overline{G}$) is disconnected* **then**
  Let $C_1, \ldots, C_p$ be the connected components of G (or $\overline{G}$);
  Recurse on each C_i and add SPIDER(C_i) to the solution set S;
end
G is a spider with $K = \{k_1, \ldots, k_{|K|}\}$ and $S = \{s_1, \ldots, s_{|K|}\}$;
if G *is a thin spider* **then**
  Notation: k_i adjacent to s_j if and only if $i = j$;
  Add edge $\{k_i, s_i\}$ to solution set S for every $i = 1, \ldots, |K| - 1$;
end
if G *is a thick spider* **then**
  Notation: k_i adjacent to s_j if and only if $i \neq j$;
  Add edge $\{k_i, s_j\}$ to solution set S for every pair $i < j$;
end
Recurse on the head R of the spider. Return $S \cup$ SPIDER(R);

Algorithm 1. Cograph edge-deletion algorithm for P_4-sparse graphs

3.2 Cograph Edge-Deletion in General Graphs

The algorithm that follows takes any general graph as input and uses Algorithm 1 from the previous section as a subroutine.

Jamison and Olariu [10] give a linear time recognition algorithm for P_4-sparse graphs. In the case that the graph being tested is not P_4-sparse, the algorithm terminates upon finding a 5-set of vertices isomorphic to one of the forbidden subgraphs shown in Figure 1. In $O(m + n)$ time on a general graph, we can find one of the subgraphs in Figure 1 or else assert that our graph is P_4-sparse.

Algorithm 2 finds 5-vertex subsets that induce at least 2 P_4s, branches on the possible ways of destroying the P_4s, and then finally arrives at a P_4-sparse graph and calls Algorithm 1. This algorithm either terminates with a call to the subroutine (in the case that a spider structure is encountered) or detects a cograph structure early, or else its integer parameter k has been reduced to 0 or less in which case the number of allowed edge-deletions has been exhausted without reaching a cograph.

Refer to Figure 1 for the possible subgraphs the general search algorithm may encounter. We refer to specific edges as they are labeled in Figure 1 for each subgraph. The pseudocode description of the general search algorithm uses these branching rules.

Let H be one of the forbidden subgraphs. The possible edge-deletion sets for that subgraph are:

$$H = \begin{cases} \begin{array}{c|c} \text{Subgraph} & \text{Minimal Edge Deletion Sets} \\ \hline C_5 & \{a,c\},\ \{a,d\},\ \{b,d\},\ \{b,e\},\ \{c,e\} \\ P_5 & \{a,d\},\ \{b\},\ \{c\} \\ \overline{P}_5 & \{a,b\},\ \{e,c\},\ \{d,e\},\ \{c,d\},\ \{a,d,f\},\ \{a,c,f\},\ \{b,d,f\},\ \{b,e,f\} \\ \text{4-pan} & \{a,d\},\ \{a,c\},\ \{b,c\},\ \{b,d\},\ \{e\} \\ \text{co-4-pan} & \{b,c\},\ \{d\},\ \{e\} \\ \text{fork} & \{a,b\},\ \{c\},\ \{d\} \\ \text{kite} & \{a,d\},\ \{a,c,f\},\ \{b,d,f\},\ \{b,c\},\ \{e\} \end{array} \end{cases}$$

Algorithm COGRAPHDELETION(G, k)
Input: A Graph $G = (V, E)$ and a positive integer k
Output: A set S of edges of G with $|S| \le k$ where $(V, E \setminus S)$ is a cograph
         if it exists, otherwise No
Initialize $S = \emptyset$;
if G *is a cograph* **then**
| Return S;
end
if $k \le 0$ **then**
| Return No;
end
Apply a P_4-sparse recognition algorithm;
if G *is P_4-sparse* **then**
| $S \leftarrow S \cup$ SPIDER(G);
| If $|S| \le k$, return S; Otherwise, return No.
end
else
| A forbidden graph H from Figure 1 exists;
| **foreach** *minimal edge-deletion set E' for H* **do**
| | Create a branch with edges from E' deleted;
| | Add the edges E' to the solution set S;
| | Apply COGRAPHDELETION to $G - E'$, with parameter k reduced
| | according to the number of edges removed in the branch;
| **end**
end

Algorithm 2. Bounded search tree algorithm finding a cograph edge-deletion set

It is routine to verify that any edge-deletion set from each of the 7 induced subgraph cases must contain one of the deletion set cases given in the table. Since every P_4 in the graph must be destroyed with an edge deletion, these rules will eventually enumerate all possible cograph edge-deletion sets.

The runtime of the algorithm is dominated by the branching steps. The spider structure can be identified in linear time, while finding the co-connected components takes the same time as computing graph complementation. We find the runtime $T(k)$ of the algorithm with parameter k from each branch rule separately:

1. C_5: five branches, each reducing the parameter by 2 gives $T(k) = 5T(k-2)$ and so $T(k) \le 2.237^k$
2. P_5: $T(k) = 2T(k-1) + T(k-2)$ giving $T(k) \le 2.415^k$
3. $\overline{P}_5$: $T(k) = 4T(k-2) + 4T(k-3)$ giving $T(k) \le 2.383^k$
4. 4-pan: $T(k) = T(k-1) + 4T(k-2)$ giving $T(k) \le 2.562^k$
5. co-4-pan: $T(k) = 2T(k-1) + T(k-2)$ giving $T(k) \le 2.415^k$
6. fork: $T(k) = 2T(k-1) + T(k-2)$ giving $T(k) \le 2.415^k$
7. kite: $T(k) = T(k-1) + 2T(k-2) + 2T(k-3)$ giving $T(k) \le 2.270^k$

The branching process is thus upper-bounded by the worst case of deleting P_4s in a 4-pan: $T(k) \le 2.562^k$.

Theorem 2. *Algorithm 2 correctly solves the cograph k-edge-deletion problem in $O(2.562^k(n+m))$ time.*

4   Cograph Vertex-Deletion and Trivially Perfect Edge-Deletion Problems

4.1   Vertex-Deletion to Cographs

Since removing a vertex set S from a graph $G = (V, E)$ is equivalent to taking the induced subgraph on the vertex set $V \setminus S$, these problems are also often named *maximum induced subgraph* problems. In our case of asking if there is a vertex set of size at most k that can be removed to leave behind a cograph, this is equivalent to asking if there is an induced cograph subgraph of size at least $|V|-k$. Removing a vertex from G can never create a new induced subgraph in G, and so deleting vertices to destroy induced subgraphs is commonly modeled as a HITTING SET problem. In this case in which each P_4 maps to a 4-set in a HITTING SET instance, we have the restricted problem of a 4-HITTING SET. Algorithms for such vertex-deletion problems should always be compared against the state-of-the-art algorithms of d−HITTING SET if not anything else. d-HITTING SET is a well-studied NP-complete problem which admits fixed-parameter tractable algorithms, the fastest of which (for $d = 4$) runs in $O(3.30^k)$ time [16].

The simple spider structure of P_4 sparse graphs allows us to describe a simple algorithm for the vertex-deletion problem to cographs. The runtime of this simple algorithm matches that of [7] and of [16]. The algorithm in [7] was developed by an automated search, where branching rules were made to delete vertices breaking the P_4s in every subgraph of size t. Testing various values of t deduced that rules based on subgraphs of size 7 yielded the optimal runtime of an algorithm of this sort, with runtime $O(3.30^k)$. The automated algorithm builds branching rules from 447 graphs of size 7, while our algorithm only involves seven graphs on 5 vertices (Figure 1.)

We omit a pseudocode description of the subroutine SPIDER VERTEX-DELETION to save space. The algorithm works in the same way as algorithm 1, taking as input a P_4-sparse graph and returning the optimal number of vertices to remove in order to break all P_4s in the graph. For thin spiders, every pair of feet is the end-pair of a P_4, and removing any $|S| - 1$ vertices from S will destroy all the P_4s in the body and legs.

Algorithm COGRAPHVERTEXDELETION(G, k)
Input: A Graph $G = (V, E)$ and a positive integer k
Output: A set S of vertices of G with $|S| \leq k$ where $(V \setminus S, E)$ is a
            cograph if it exists, otherwise NO
Initialize $S = \emptyset$;
if *G is a cograph* **then**
  | Return S;
end
if $k \leq 0$ **then**
  | Return NO;
end
Apply a P_4-sparse recognition algorithm;
if *G is P_4-sparse* **then**
  | $S \leftarrow S \cup$ SPIDER VERTEX-DELETION(G);
  | If $|S| \leq k$, return S; Otherwise, return NO.
end
else
  | A forbidden graph H from Figure 1 exists;
  | **foreach** *minimal vertex-deletion set S' for H* **do**
  | | Create a branch with vertices from S' deleted;
  | | Add the vertices S' to the solution set S;
  | | Apply COGRAPHVERTEXDELETION to $G - S'$, with parameter k
  | | reduced according to the number of vertices removed in the branch;
  | **end**
end

Algorithm 3. Bounded search tree algorithm finding a cograph vertex-deletion set

---

Since a set of 4 vertices induces a P_4 in a graph G if and only if they induce a P_4 in $\overline{G}$, deleting any $|K| - 1$ vertices from K in a thick spider will destroy all the P_4s in $K \cup S$. In either the thin or thick spider case, the subroutine is then applied to head R. The branching rules for the vertex deletions are given in a table as before:

$$H = \begin{cases} \begin{array}{c|c} \text{Subgraph} & \text{Minimal Vertex Deletion Sets} \\ \hline C_5 & \{1,2\}, \{1,3\}, \{1,4\}, \{1,5\}, \{2,3\}, \{2,4\}, \{2,5\}, \{3,4\}, \{3,5\}, \{4,5\} \\ P_5 & \{1,5\}, \{2\}, \{3\}, \{4\} \\ \overline{P_5} & \{1\}, \{3\}, \{4\}, \{2,5\} \\ \text{4-pan} & \{2\}, \{4\}, \{5\}, \{1,3\} \\ \text{co-4-pan} & \{3\}, \{4\}, \{5\}, \{1,2\} \\ \text{fork} & \{3\}, \{4\}, \{5\}, \{1,2\} \\ \text{kite} & \{2\}, \{4\}, \{5\}, \{1,3\} \end{array} \end{cases}$$

The runtime of the algorithm is dominated by the branching steps. The runtime $T(k)$ for the C_5 case depends on 10 branches, while each of the other cases have equivalent runtime analysis.

1. C_5: ten branches, each reducing the parameter by 2 gives $T(k) = 10T(k-2)$ and so $T(k) \leq 3.163^k$
2. All others: $T(k) = 3T(k-1) + T(k-2)$ giving $T(k) \leq 3.303^k$

The runtime of this vertex-deletion algorithm is bounded by $O(3.303^k(m+n))$, matching the best known FTP algorithm for 4-HITTING SET [16] and matching the algorithm developed with the automated search [7].

Theorem 3. *Algorithm 3 solves the vertex-deletion problem for cographs in* $O(3.303^k(m+n))$ *time.*

4.2   Edge-Deletion to Trivially Perfect Graphs

A graph is *trivially perfect* if it has no induced subgraphs isomorphic to a P_4 or a C_4 [20]. In [8], Guo studied the edge-deletion problem for *complements* of trivially perfect graphs. We know of no prior study of this specific problem. A naïve solution would find a subgraph isomorphic to either a P_4 or a C_4 and then branch on the possible ways of deleting an edge from that subgraph, resulting in a worst-case search tree of size $O(4^k)$. A minor observation that deleting any one edge from a C_4 always results in the other forbidden subgraph, P_4, allows us to branch on the 6 possible ways of deleting any 2 edges from a C_4. This results in a worst-case search tree of size $O(3^k)$ due to the 3 edges in a P_4.

We show that our edge-deletion algorithm for cographs can be adapted to solve the edge-deletion problem for trivially perfect graphs, again by finding any P_4-sparse forbidden subgraph and branching on it, or else by solving the problem optimally on a smaller portion of the graph (a connected component or the head of a spider). The branching rules become simpler in that only 5 of the 7 graphs in Figure 1 need consideration. In particular, the *4-pan* that caused the bottleneck of Algorithm 2, is no longer considered and this changes the runtime of the process from $O(2.562^k)$ to $O(2.450^k)$.

Algorithm TRIVIALLYPERFECTEDGEDELETION(G, k)
Input: A Graph $G = (V, E)$ and a positive integer k
Output: A set S of edges of G with $|S| \leq k$ where $(V, E \setminus S)$ is trivially
        perfect if it exists, otherwise No
Initialize $S = \emptyset$;
if *G is a trivially perfect* **then**
 | Return S;
end
if $k \leq 0$ **then**
 | Return No;
end
while *There exists H isomorphic to C_4* **do**
 | Create 6 branches corresponding to the possible ways of removing any
 | 2 edges in H
end
Apply a P_4-sparse recognition algorithm;
if *G is P_4-sparse* **then**
 | $S \leftarrow S \cup$ SPIDER VERTEX-DELETION(G);
 | If $|S| \leq k$, return S; Otherwise, return No.
end
else
 | A forbidden graph H from Figure 1 exists;
 | **foreach** *minimal vertex-deletion set S' for H* **do**
 | | Create a branch with vertices from S' deleted;
 | | Add the vertices S' to the solution set S;
 | | Apply TRIVIALLYPERFECTEDGEDELETION to $G - S'$, with
 | | parameter k reduced according to the number of vertices removed
 | | in the branch;
 | **end**
end

Algorithm 4. Bounded search tree algorithm finding a trivially perfect edge-deletion set

One main difference in this algorithm from Algorithm 2 is that C_4s are found and destroyed first, and after any of the P_4-sparse deletions are made, the process restarts with looking for C_4s to destroy again. Once the C_4s are destroyed and the resulting graph is P_4-sparse, we proceed with removing edges with edge-deletion algorithm for thin or thick spiders (Algorithm 1).

The correctness of decomposing the edge-deletion problem into separate problems on $K \cup S$ and R depends a lemma similar to Lemma 2.

Lemma 5. *Let G be a (thin or thick) spider and let R be the head of G and K be the body of G. Then any edge $e = \{r, k\}$ with $r \in R$ and $k \in K$ is not in any C_4 in G. Furthermore, for any subset of leg edges and head edges E' the edge $e = \{r, k\}$ is not in any P_4 in $G - E'$.*

Proof. Notice that no C_4 can include a vertex s from S in a spider even after removals of leg edges and head edges since the neighbourhood of s induces a clique. Since K is a clique, and every $k \in K$ is adjacent to every $r \in R$, it is clear that there can not exist a C_4 in $K \cup R$ unless the C_4 is completely contained in R. So no C_4 contains an edge from R to K.

This lemma shows us that since all the C_4s are destroyed in the branching stage, once we arrive at a C_4-free spider, we are free to delete leg edges without worry that a new C_4 will be created. When recursing on the head R and its spider structure is found, moving those edges will also not create C_4s since they are leg edges in the sub-spider and head edges in the original graph.

One source of worry may be that if R contains two co-connected components C_1 and C_2 (co-connected here means that every $c \in C_1$ is adjacent with every $d \in C_2$) an edge removal from each of these co-components will create a C_4. One does not need to worry about this:

Lemma 6. *If a graph G is C_4-free and its complement is disconnected, then G has exactly two co-components C_1, C_2, one of which is a clique and one that is not.*

Proof. If C_1 is not a clique, it has two non-adjacent vertices u and v. Now, u and v are each adjacent to everything in $G \setminus C_1$, so if there was a non-edge $\{w, x\} \in G \setminus C_1$ then $\{u, v, w, x\}$ would induce a C_4. Hence $G \setminus C_1$ is a clique.

Corollary 1. *If a C_4-free P_4-sparse graph has more than one co-connected component, then it is a spider with a head R that is a clique.*

This puts an end to the worry that removing edges from separate co-connected components of R during the SPIDER(G) subroutine may create any C_4s.

The runtime of Algorithm 4 is dominated by the branching rules once again. Encountering a C_4 results in 6 branches which delete 2 edges each. The resulting recurrence is $T(k) = 6T(k-2)$ and so $T(k) \leq 2.450^k$. Having deleted all the C_4s, we no longer include the $\overline{P_5}$ or the *4-pan* cases in our analysis. The runtime analysis for the rest remain unchanged: $C_5 : 2.237^k, P_5 : 2.415^k$, co-4-pan: 2.415^k, fork: 2.415^k, kite: 2.270^k. The search tree is thus bounded by the C_4 case of size $O(2.450^k)$. Finding a C_4 directly is a problem that is currently best-achieved using matrix multiplication [12], so this entire process *as described* runs in $O(2.450^k n^\alpha)$ where $O(n^\alpha)$ is the time required for matrix multiplication ($\alpha \leq 2.376$ [3]).

We can, in fact, modify the algorithm to run linearly in n and m by observing that a graph is P_4-free and C_4-free if and only if it is a chordal cograph. By first running a certifying chordal recognition algorithm [19], we can either deduce that there is no C_4 or else find a C_4 or a C_5 or a larger induced cycle (and thus a P_5) and branch on these subgraphs according to the rules we gave, and if the graph is chordal then we apply a P_4-sparse recognition algorithm to find one of the other forbidden induced subgraph, branch on it, and then re-apply the chordal recognition process.

Theorem 4. *Finding a trivially perfect k-edge-deletion set can be solved in* $O(2.450^k(n+m))$ *time.*

5   Conclusions and Future Work

We presented a framework for solving a variety of graph modification problems by branching on the forbidden subgraphs of a superclass of graphs. The algorithms presented here depend on the fact that deleting to a cograph or a trivially-perfect graph can be solved in linear time from a P_4-sparse graph. We gave the first non-trivial algorithm for the cograph edge-deletion problem and trivially-perfect edge-deletion problem, and matched existing algorithms for the cograph vertex-deletion problem while only using a small number of branching rules. Furthermore, by applying our edge-deletion algorithms to the complement of the input graph, the edge-deletion problem for cographs also serves as an edge-completion problem to cographs and the edge-deletion problem to trivially perfect also serves as an edge-completion problem to co-trivially perfect graphs.

In a future edition of this work, we also show how to further improve the runtime of the cograph edge-deletion problem (from $O(2.562^k)$ to $O(2.415^k)$) by considering a class of graphs more general than P_4-sparse called the *semi-P_4-sparse* graphs [6]. These are $(P_5, \overline{P}_5, \text{kite})$-free graphs, and they also decompose into manageable structures. In fact, a restricted form of this graph class suffices. By using the same process as we have here for P_4-sparse graphs, and not including the *4-pan* cases in the branching process, it reduces the bottleneck of the algorithm. More work is involved in handling the special structures that result because they are more general than spider.

We are currently working on applying the same idea to the edge-deletion problem for P_3-free graphs (the well-studied CLUSTER DELETION problem,) using trivially perfect graphs as a superclass to P_3-free graphs. Initial results are promising.

The literature on graph classes is extensive [1], and many of these classes admit polynomial time solutions to many NP-complete problems. We suspect that new and fast fixed-parameter tractable algorithms will soon develop through the use of superclasses as we have used in this paper.

References

1. Brandstädt, A., Le, V.B., Spinrad, J.P.: Graph classes: a survey. Society for Industrial and Applied Mathematics, Philadelphia (1999)
2. Cai, L.: Fixed-parameter tractability of graph modification problems for hereditary properties. Inf. Process. Lett. 58(4), 171–176 (1996)
3. Coppersmith, D., Winograd, S.: Matrix multiplication via arithmetic progressions. J. Symb. Comput. 9(3), 251–280 (1990)
4. Corneil, D.G., Perl, Y., Stewart, L.K.: A linear recognition algorithm for cographs. SIAM J. Comput. 14, 926–934 (1985)
5. Cournier, A., Habib, M.: A new linear algorithm for modular decomposition. In: Tison, S. (ed.) CAAP 1994. LNCS, vol. 787, pp. 68–84. Springer, Heidelberg (1994)

6. Fouquet, J.-L., Giakoumakis, V.: On semi-P_4-sparse graphs. Discrete Mathematics 165-166, 277–300 (1997)
7. Gramm, J., Guo, J., Hüffner, F., Niedermeier, R.: Automated generation of search tree algorithms for hard graph modification problems. Algorithmica 39(4), 321–347 (2004)
8. Guo, J.: Problem kernels for NP-complete edge deletion problems. In: Tokuyama, T. (ed.) ISAAC 2007. LNCS, vol. 4835, pp. 915–926. Springer, Heidelberg (2007)
9. Hoàng, C.T.: Perfect graphs (Ph.D. thesis). School of Computer Science, McGill University Montreal (1985)
10. Jamison, B., Olariu, S.: Recognizing P_4-sparse graphs in linear time. SIAM J. Comput. 21(2), 381–406 (1992)
11. Kaplan, H., Shamir, R., Tarjan, R.E.: Tractability of parameterized completion problems on chordal, strongly chordal, and proper interval graphs. SIAM J. Comput. 28(5), 1906–1922 (1999)
12. Kratsch, D., Spinrad, J.: Between O(nm) and O(n^α). SIAM J. Comput. 36(2), 310–325 (2006)
13. Lokshtanov, D., Mancini, F., Papadopoulos, C.: Characterizing and computing minimal cograph completions. Discrete Appl. Math. 158(7), 755–764 (2010)
14. El Mallah, E.S., Colbourn, C.J.: Edge deletion problems: properties defined by weakly connected forbidden subgraphs. In: Proc. Eighteenth Southeastern Conference on Combinatorics, Graph Theory, and Computing, Congressus Numerantium, vol. 61, pp. 275–285 (1988)
15. McConnell, R.M., Spinrad, J.: Modular decomposition and transitive orientation. Discrete Mathematics 201(1-3), 189–241 (1999)
16. Niedermeier, R., Rossmanith, P.: An efficient fixed-parameter algorithm for 3-hitting set. J. Discrete Algorithms 1(1), 89–102 (2003)
17. Nikolopoulos, S.D., Palios, L.: Adding an edge in a cograph. In: Kratsch, D. (ed.) WG 2005. LNCS, vol. 3787, pp. 214–226. Springer, Heidelberg (2005)
18. Seinsche, D.: On a property of the class of n-colorable graphs. J. Combin. Theory (B) 16(2), 191–193 (1974)
19. Tarjan, R.E., Yannakakis, M.: Addendum: Simple linear-time algorithms to test chordality of graphs, test acyclicity of hypergraphs, and selectively reduce acyclic hypergraphs. SIAM J. Comput. 14(1), 254–255 (1985)
20. Yan, J.-H., Chen, J.-J., Chang, G.J.: Quasi-threshold graphs. Discrete Applied Mathematics 69(3), 247–255 (1996)
21. Yannakakis, M.: The effect of a connectivity requirement on the complexity of maximum subgraph problems. J. ACM 26(4), 618–630 (1979)

Listing Triconnected Rooted Plane Graphs

Bingbing Zhuang and Hiroshi Nagamochi

Graduate School of Informatics, Kyoto University
{zbb,nag}@amp.i.kyoto-u.ac.jp

Abstract. A plane graph is a drawing of a planar graph in the plane such that no two edges cross each other. A rooted plane graph has a designated outer vertex. For given positive integers $n \geq 1$ and $g \geq 3$, let $\mathcal{G}_3(n, g)$ denote the set of all triconnected rooted plane graphs with exactly n vertices such that the size of each inner face is at most g. In this paper, we give an algorithm that enumerates all plane graphs in $\mathcal{G}_3(n, g)$. The algorithm runs in constant time per each by outputting the difference from the previous output.

1 Introduction

The problem of enumerating (i.e., listing) all graphs in particular classes of graphs is one of the most fundamental and important issues in graph theory [10]. Cataloguing graphs, i.e., making the complete of graphs in a particular class can be used in a various way: search for a possible counterexample to a mathematical conjecture; choosing the best graph among all candidate graphs; and experiment for measuring the average performance of a graph algorithm over all possible input graphs.

The common idea behind most of the recent efficient enumeration algorithms is to define a parent-child relationship among all graphs in a given class in order to induce a rooted tree that connects all graphs in the class, called the *family tree* $\mathcal{F}$, where each node in $\mathcal{F}$ corresponds to a graph in the class. Then all graphs in the class will be generated one by one according to the depth-first traversal of the family tree $\mathcal{F}$. Time delay of an enumeration algorithm is a time bound between two consecutive outputs. Enumerating graphs with a polynomial time delay would be rather easy since we can examine the whole structure of the current graph anytime. However, algorithms with a constant time delay in the worst case is a hard target to achieve without a full understanding of the structure of graphs to be enumerated, because not only the difference between two consecutive outputs is required to be $O(1)$, but also any operation for examining symmetry and identifying the edges/vertices to be modified to get the next output needs to be executable in $O(1)$ time.

Enumeration for a particular class of graphs also has practical applications in various fields such as the inference of structures of chemical compounds [5], virtual exploration of chemical universe, and reconstruction of molecular structures from their signatures. It is known that 94.3% of chemical compounds in NCI chemical database have planar structures [3]. Hence planar graphs is an

W. Wu and O. Daescu (Eds.): COCOA 2010, Part II, LNCS 6509, pp. 347–361, 2010.
© Springer-Verlag Berlin Heidelberg 2010

important class to be investigated. Our research group has been developing algorithms for enumerating chemical graphs that satisfy given various constraints [1,4,5]. We have designed efficient branch-and-bound algorithms for enumerating tree-like chemical graphs [1,5], which are based on the tree enumeration algorithm [9], and implementations of these algorithms are available on our web server[1]. Currently we aim to provide efficient algorithms for enumerating chemical graphs for a wider class of graphs than trees such as cacti and outerplanar graphs in our web server. To facilitate such development of algorithms, the authors recently proposed a general enumeration scheme for classes $\mathcal{H}$ of "rooted graphs with a reflective block structure" [18]. A *reflective block* means a rooted biconnected component which may admit reflective symmetry around its root.

In particular, triconnected planar graphs is a mathematically important class of graphs in the sense that every triconnected planar graph has a unique embedding on a sphere only up to its reversal. Furthermore, Tutte [11] proved that *triconnected* plane graphs is the class of plane graphs that admit convex drawings in the plane for any prescribed polygonal boundary, where a convex drawing is an embedding in a plane graph such that all the edges of the graph are drawn as straight-line segments and every facial cycle is drawn as a convex polygon. Steinitz [12] proved that *triconnected* planar graphs is the class of vertex-edge graphs of three-dimensional convex polyhedra.

Yamanaka and Nakano [13] gave an algorithm for generating all connected rooted plane graphs with at most m edges, where an outer edge with an orientation is designated as the root of each plane graph. The algorithm uses $O(m)$ space and generates such graphs in $O(1)$ time per graph on average without duplications. Li and Nakano [6] presented an efficient algorithm that enumerates all biconnected rooted triangulated plane graphs in constant time per each. Nakano [8] presented an algorithm with the same time complexity to generate all triconnected rooted triangulated plane graphs. Recently, in our companion papers [14,15], we gave efficient enumeration algorithms under reflective symmetry for classes of biconnected rooted triangulations and biconnected rooted outerplanar graphs, respectively, where a planar graph designates an outer vertex v and two outer edges incident to v, and the algorithm does not generate plane graphs which are reflectively symmetric along the root v. In our companion paper, we also gave an efficient enumeration algorithm for the class $\mathcal{G}_2(n, g)$ of biconnected rooted plane graphs [16,17]. The algorithms generate biconnected rooted plane graphs with exactly n vertices such that the size of each inner face is at most g, where $n \geq 1$ and $g \geq 3$ are prescribed integers. The algorithm runs in $O(n)$ space and in $O(1)$ time per graph in the worst case.

In this paper, we consider the class $\mathcal{G}_3(n, g)$ of all triconnected rooted plane graphs with exactly n vertices such that the size of each inner face is at most g. The structure of triconnected plane graphs is more complicated than those of biconnected plane graphs. We present an algorithm that enumerates all plane graphs in $\mathcal{G}_3(n, g)$ in $O(n)$ space and in $O(1)$ time per graph in the worst case. However, our algorithm does not exploit any dynamic data structure that

---

[1] http://sunflower.kuicr.kyoto-u.ac.jp/tools/enumol/

represents 4-connected components to test triconnectivity of possible candidates for graphs to be generated, because an $O(1)$ time maintenance of such a data structure required to achieve an $O(1)$-time delay seems extremely difficult. Our algorithm also yields an $O(n^3)$-time delay algorithm for generating all triconnected *unrooted* plane graphs with exactly n vertices such that the size of each inner face is at most g.

Recently we used the results obtained in this paper to design an $O(1)$-time delay enumeration algorithm for internally triconnected rooted plane graphs [19].

The rest of the paper is organized as follows. After introducing basic notations in Section 2, Sections 3 and 4 examine the structure of triconnected graphs and triconnected plane graphs, respectively. Section 5 introduces the parent of each triconnected rooted plane graph, and Section 6 characterizes the children of a triconnected rooted plane graph. Section 7 describes an algorithm for enumerating all triconnected rooted plane graphs, and analyzes the time and space complexities of the algorithm. Section 8 makes some concluding remarks. The proofs omitted due to the space limitation can be found in a full version of the paper [20].

2   Preliminaries

Throughout the paper, a graph stands for a simple undirected graph unless stated otherwise. A graph is denoted by a pair $G = (V, E)$ of a vertex set V and an edge set E. The set of vertices and the set of edges of a given graph G are denoted by $V(G)$ and $E(G)$, respectively.

For a subset $E' \subseteq E(G)$, $G - E'$ denotes the graph obtained from a graph G by removing the edges in E'. Let X be a subset of $V(G)$. We denote by $G - X$ the graph obtained from G by removing the vertices in X together with the edges incident with a vertex in X. Let $deg(v; G)$ denote the degree of a vertex v in a graph G. For a vertex $v \in V$ in a graph G, let $\Gamma(v; G)$ denote the set of *neighbours* of v (i.e., vertices adjacent to v). For a subset $X \subseteq V$, let $\Gamma(X; G) = \cup_{v \in X} \Gamma(v; G) - X$. The *vertex-connectivity* of G is denoted by $\kappa(G)$. A *fan* is the graph obtained from a path P with at least one vertex by adding a new vertex v together with an edge incident to each vertex in the path, where the vertex v is called the *center* of a fan. A fan with n vertices is denoted by F_n. The graph W_n $(n \geq 4)$ obtained from F_n by joining the two vertices of degree 2 by a new edge is called a *wheel*. Note that F_n $(n \geq 2)$ and W_n $(n \geq 4)$ are biconnected and triconnected, respectively.

A graph is called *planar* if its vertices and edges can be drawn as points and curves on the plane so that no two curves intersect except for their endpoints. A planar graph with such a fixed embedding is called a *plane* graph, where a face is designated as the *outer face* and all other faces are called *inner faces*. Let $F(G)$ denote the set of faces in a plane graph G. For a face f in a plane graph G, let $V(f)$ and $E(f)$ denote the sets of vertices and edges on the facial cycle of f, and define the size $|f|$ of face f to be $|V(f)|$. For a vertex v and an edge e, let $F(v)$ (or $F(v; G)$) denote the set of inner faces f with $v \in V(f)$ and $F(e)$ (or $F(e; G)$) denote the set of inner faces f with $e \in E(f)$. An inner face $f \in F(v)$ is called a *v-face*.

A *rooted* plane graph is a plane graph which has a designated outer edge (u, r) with orientation from u to r, where r is called the *root*. Two rooted plane graphs G_1 and G_2 are *equivalent* if their vertex sets admit a bijection by which the designated directed edge and the incidence-relation between edges and vertices/faces in G_1 correspond to those in G_2.

3   Triconnected Graphs

In this section, we introduce a transformation that preserves the triconnectivity of a graph, which is not necessarily planar. The next lemma is the key property that enables us to design several graph transformations such that triconnectivity of the resulting new graph can be tested by checking the degree of a constant number of vertices.

Lemma 1. *Let G be a triconnected graph, and let $e = (u_1, u_2)$ be an edge such that a vertex v of degree 3 is incident to both end vertices u_1 and u_2 in G. If $G - e$ is not triconnected, then any vertex cut X with $|X| \leq 2$ is given by $X = \Gamma(u_1; G - e)$ or $\Gamma(u_2; G - e)$.*

Let G be a triconnected graph. For a vertex u of degree 3 in a graph G, let G/u denote the graph obtained from G by removing the vertex u and adding a new edge between any two nonadjacent neighbours of u. See Fig. 1(b)-(c). For a cycle $C = (v_1, v_2, v_3)$ of length 3 in a graph G, let G/C denote the graph obtained from G by adding a new vertex u together with three new edges (u, v_i), $i = 1, 2, 3$.

Lemma 2. *Let G be a triconnected graph.*
(i) For a vertex u of degree 3 in G, the graph G/u remains simple and triconnected.
(ii) For a cycle $C = (v_1, v_2, v_3)$ of length 3 in G, the graph G/C remains simple and triconnected. For a subset F of edges in C, the graph $G/C - F$ remains triconnected if and only if the degree of each v_i is at least three.

For a cycle C of length 3 in a graph G, let $\mathcal{G}_3(G/C)$ denote the set of all graphs obtained from G/C by removing edges in C so that the degree of each vertex v_i is at least three, where $|\mathcal{G}_3(G/C)| \leq 2^3$.

4   Triconnected Rooted Plane Graphs

In this section, let G be a triconnected rooted plane graph, where the root r is an outer vertex. Denote the vertices along the boundary of G in the clockwise order by $u_1 = r, u_2, \ldots, u_B$, and let e_i denote the outer edge (u_i, u_{i+1}), where we let $e_B = (u_B, u_1)$. For each outer edge e, let $f(e)$ denote the unique inner face which contains e on its facial cycle. For two outer vertices u and v in G, let $\beta[u, v]$ denote the path obtained by traversing the boundary of G from u to v in the clockwise order.

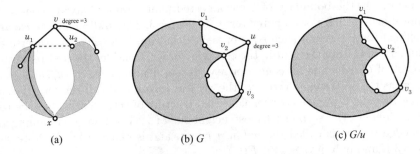

Fig. 1. (a) A vertex cut X containing a vertex v of degree 3; (b) a triconnected graph G; (c) graph G/u

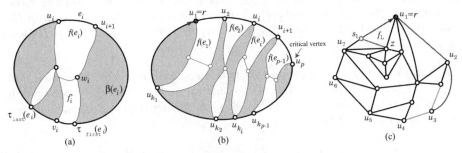

Fig. 2. (a) $\tau_{\text{first}}(e_i)$ and $\tau_{\text{last}}(e_i)$ for an outer edge e_i; (b) active path $\beta[r, u_p]$ (c) a triconnected graph G with $\psi(G) = \{s_1\}$

In a triconnected plane graph G, an edge e is called *removable* if $G - e$ remains triconnected, and is called *irremovable* otherwise. We characterize removability of edges. Two inner faces are *internally adjacent* in G if they share at least one inner vertex. For each inner face f in a plane graph G, let $\Gamma(f; G)$ denote the set of all inner faces internally adjacent to f, and let $W^o(f)$ denote the set of all outer vertices in $\cup_{f' \in \Gamma(f;G)} V(f') - V(f)$.

Lemma 3. *Let $e_i = (u_i, u_{i+1})$ be an outer edge in a triconnected plane graph G. If $W^o(f(e_i)) \neq \emptyset$, then for each vertex $v \in W^o(f(e_i))$, $G - e_i$ has a vertex-cut X with $|X| = 2$ and $v \in X$. Furthermore e_i is irremovable if and only if $W^o(f(e_i)) \neq \emptyset$.*

For each outer edge $e_i = (u_i, u_{i+1})$, let $\tau_{\text{first}}(e_i)$ (resp., $\tau_{\text{last}}(e_i)$) be the vertex in $W^o(f(e_i))$ that appears first (resp., last) when we traverse the boundary of G from u_i in the clockwise order (recall that $\{u_i, u_{i+1}\} \cap W^o(f(e_i)) = \emptyset$). See Fig. 2(a). For each irremovable outer edge $e_i = (u_i, u_{i+1})$, define $\beta(e_1)$ to be the path $\beta[u_{i+1}, \tau_{\text{first}}(e_i)]$.

Lemma 4. *Let $e_i = (u_i, u_{i+1})$ be an irremovable outer edge in a triconnected plane graph G with $n \geq 4$ vertices. Then $\beta(e_i)$ contains a removable outer edge or an outer vertex of degree 3.*

By Lemma 4, the boundary of a triconnected plane graph G always contains a removable outer edge or an outer vertex of degree 3. The *first removable element* is defined to be the first such edge or vertex that appears when we traverse the boundary of G from the root in the clockwise order. Consider the first removable element of G, which is an edge e or a vertex u, denote by $e_p = (u_p, u_{p+1})$ and u_p, respectively. We call vertex u_p the *critical vertex*, and call the path $\beta[r, u_p]$ *active*. See Fig. 2(b). We now consider the structure of active paths. For the root r, let z denote the second leftmost neighbour of r. The r-face containing the root edge (u_B, r) is called the *leftmost r-face*, and is denoted by f_L. Note that $|f_L| = 3$ if and only if $(u_B, z) \in E(G)$.

The *fan factor* of G is defined to be the maximal sequence $u_{B+1-t}, u_{B+2-t}, \ldots, u_B$ of outer vertices such that each $s_i = u_{B-t+i}$, $1 \le i \le t$ is of degree 3 and z is adjacent to each u_{B-t+i}, $0 \le i \le t$ (hence s_i is shared by two inner faces of length 3). See Fig. 2(c), where $t = 1$. Let $\psi(G)$ denote the fan factor of G. Note that G is W_n if and only if $|\psi(G)| = n - 2$. Let $s_0 = u_B$ and $k_0 = B$ for a notational convenience.

Lemma 5. *Let G be a triconnected plane graph with $n \ge 4$ vertices such that $|f_L| = 3$. Let $t = |\psi(G)|$, and u_p be the critical vertex. Then the index k_i with $u_{k_i} = \tau_{\text{last}}(e_i)$ for each edge $e_i = (u_i, u_{i+1})$ in the active path $\beta[u_1, u_p]$ satisfies $k_i \in [i + 2, B]$, and it holds*

$$p + 1 \le k_{p-1} \le \cdots \le k_2 \le k_1 \le k_0 = B. \tag{1}$$

Moreover if the vertex u_p is the first removable element, then it holds $2 \le p \le B - t - 1$.

Let $G \oplus s_{t+1}$ denote the graph obtained from G with $t = |\psi(G)|$ by introducing a new vertex s_{t+1} together with a new edge (s_{t+1}, z), replacing (s_t, r) with two edges (s_t, s_{t+1}) and (s_{t+1}, r). We say that $\psi(G)$ is *augmented by 1* when we construct $G \oplus s_{t+1}$ from G. Note that $G \oplus s_{t+1}$ remains triconnected. Conversely, for the last vertex $s_t \in \psi(G)$ in G with $t = |\psi(G)| \ge 1$, let $G \ominus s_t$ denote the graph obtained by removing the vertex s_t and rejoining the two outer vertices s_{t-1} and r incident to s_t with an outer edge; i.e., G is obtained from $G \ominus s_t$ by inserting vertex s_t on edge (s_{t-1}, r) and joining s_t and z with a new edge. We say that $\psi(G)$ is *reduced by 1* when we construct $G \ominus s_t$ from G. We see that $G \ominus s_t$ remains triconnected by applying Lemma 2(i) to $G \ominus s_t = G/s_t$.

Let f' be the u_B-face adjacent to the leftmost r-face f_L, and z be the second leftmost neighbour of r. An r-face $f \in F(r; G) - \{f_L\}$ is called *separating* if f and f' share shares an inner vertex other than z, as shown in Fig. 3(b).

Lemma 6. *Let G be a triconnected plane graph rooted at edge $(u_B, r = u_1)$ such that $|f_L| = 3$, i.e., $e = (u_B, z) \in E(G)$. Then $G - e$ is triconnected if and only if $\deg(u_B; G) \ge 4$ and G has no separating r-face.*

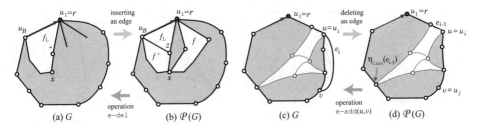

Fig. 3. (a) A plane graph G with $(u_B, z) \notin E(G)$; (b) the parent $\mathcal{P}(G)$ of G in (a); (c) a plane graph G such that the first removable element is an edge e_i; (d) the parent $\mathcal{P}(G)$ of G in (c)

5   Parents of Triconnected Rooted Plane Graphs

A triconnected plane graph with $n \leq 4$ vertices is unique. In what follows, we assume that $n \geq 5$ and $g \geq 3$, and treat wheel W_n as a rooted plane graph such that the center is drawn as an inner vertex and a non-center is chosen as the root r. Let G be a triconnected rooted plane graph with $n \geq 5$ such that $G \neq W_n$. We define the *parent* $\mathcal{P}(G)$ of G to be the following graph with n vertices.

P1. The length $|f_L|$ of the leftmost r-face f_L is at least 4 (see Fig. 3(a)): Define $\mathcal{P}(G)$ to be the graph obtained from G by inserting a new inner edge between the leftmost and second leftmost neighbours of r (see Fig. 3(b)).

P2. $|f_L| = 3$, $B = 3$ and the first removable element is vertex u_2 (where $(u_3, u_1) \in E(G)$): Construct G/u_2 from G, where no new edge joining vertices $u_1, u_3 \in V(G)$ is introduced, and edges (u_1, w) and (u_3, w) for the neighbour $w \in \Gamma(u_2; G) - \{u_1, u_3\}$ are outer edges e_1 and e_2 in $\mathcal{P}(G)$. Then $\mathcal{P}(G)$ is defined to be the graph obtained by augmenting $\psi(G/u_2)$ by 1. See Fig. 4(c) and (d).

P3. $G[v_h, v_k] \neq W_n$, $|f_L| = 3$, $B \geq 4$ and the first removable element is a vertex u_{i+1}, $1 \leq i \leq B - 1$: Construct G/u_{i+1} from G, where the new edge joining vertices $u_i, u_{i+2} \in V(G)$ becomes the ith outer edge e_i in $\mathcal{P}(G)$, and edges (u_i, w) and (u_{i+2}, w) for the neighbour $w \in \Gamma(u_{i+1}; G) - \{u_i, u_{i+2}\}$ are inner edges in $\mathcal{P}(G)$. Then $\mathcal{P}(G)$ is defined to be the graph obtained by augmenting $\psi(G/u_{i+1})$ by 1. See Fig. 4(a) and (b).

P4. $|f_L| = 3$ and the first removable element is an edge e: $\mathcal{P}(G)$ is defined to be $G - e$. See Fig. 3(c) and (d).

For a given integer $n \geq 1$, let $\mathcal{G}_3(n)$ denote the set of all triconnected rooted plane graphs with exactly n vertices. Define function $\Phi(G) = |\psi(G)| + |B| - \min\{|f_L| - 3, 1\}$, where f_L is the leftmost r-face and B is the length of the boundary of G.

Lemma 7. *For any graph $G \in \mathcal{G}_3(n) - \{W_n\}$ with $n \geq 5$, the leftmost r-face of its parent $\mathcal{P}(G)$ is of length 3, the maximum size of inner faces in $\mathcal{P}(G)$ never exceeds that of G, and it holds $3 \leq \Phi(G) < \Phi(\mathcal{P}(G)) \leq 2n - 2$.*

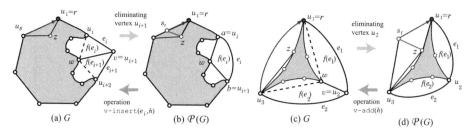

Fig. 4. (a) A plane graph G such that the first removable element is a vertex u_{i+1}; (b) the parent $\mathcal{P}(G)$ of G in (a); (c) a plane graph G such that $B = deg(u_2; G) = 3$; (d) the parent $\mathcal{P}(G)$ of G in (c).

Let $\mathcal{P}^0(G) = G$ and $\mathcal{P}^i(G) = \mathcal{P}(\mathcal{P}^{i-1}(G))$ for integers $i \geq 1$. Lemma 7 implies that, for any graph $G \in \mathcal{G}_3(n)$ with $n \geq 5$, there is an integer $i \in [0, 2n - 5]$ such that $\mathcal{P}^i(G) = W_n$.

6   Children of Triconnected Rooted Plane Graphs

Let G be a triconnected rooted plane graph with $n \geq 5$ vertices. A rooted plane graph G' is called a *child* of G if $G = \mathcal{P}(G')$. Let $\mathcal{C}(G)$ denote the set of all children of G, and let $\mathcal{C}^i(G)$, $i = 1, 2, 3, 4$ denote the set of all children G' of G such that $\mathcal{P}(G')$ is given by definition Pi of parents. In order to generate all children of G, we introduce the following four operations, e-del, v-add, v-insert and e-add.

O1. Assume $|f_L| = 3$. Operation e-del removes the edge between the the leftmost and second leftmost neighbours of r. See Fig. 3(a) and (b).

O2. Assume $B = 4$ and $\psi(G) \neq \emptyset$. Define four subsets of $\{e_1, e_2\}$ by $E_0 = \emptyset$, $E_1 = \{e_1\}$, $E_2 = \{e_2\}$, and $E_3 = \{e_1, e_2\}$. Then operation v-add(h), $h \in \{0, 1, 2, 3\}$ reduces $\psi(G)$ by 1 by removing $s_t \in \psi(G)$, and adds a new outer vertex v together with three edges (v, u_1), (v, u_2) and (v, u_3), deleting an edge set E_h. In the resulting graph, v serves as the second outer vertex u_2. See Fig. 4(c) and (d).

O3. Assume $B \geq 4$ and $\psi(G) \neq \emptyset$, and let $e_i = (a = u_i, b = u_{i+1})$ be an outer edge such that $|f(e_i)| = 3$ and $V(f(e_i)) = \{a, b, w\}$. Define four subsets of $E(f(e_i))$ by $E_0 = \emptyset$, $E_1 = \{(a, w)\}$, $E_2 = \{(b, w)\}$, and $E_3 = \{(a, w), (b, w)\}$. Then operation v-insert(e_i, h), $h \in \{0, 1, 2, 3\}$ reduces $\psi(G)$ by 1 by removing $s_t \in \psi(G)$, replaces e_i with three edges (a, v), (v, b) and (v, w), introducing a new vertex v, and delete edge set E_h. In the resulting graph, v serves as the $(i + 1)$st outer vertex u_{i+1}. See Fig. 4(a) and (b).

O4. Assume $B \geq 4$ and $|f_L| = 3$. For two outer vertices u_i and u_j ($1 \leq i+2 \leq j \leq \min\{B, B - 2 + i\}$) which are not adjacent (i.e., they does not consecutively appear along the outer boundary) in G, operation e-add(u_i, u_j) adds a new edge (u_i, u_j) as the ith outer edge e_i. See Fig. 3(c) and (d). Clearly operation e-add(u_i, u_j) preserves the triconnectivity of G.

Since the above operations are the reverse of the operations that define the parents, the sets of all graphs G' that can be constructed from G by e-del, v-add, v-insert and e-add contains $\mathcal{C}^1(G)$, $\mathcal{C}^2(G)$, $\mathcal{C}^3(G)$ and $\mathcal{C}^4(G)$, respectively. Note that such a graph G' is a child of G if and only if $\kappa(G') \geq 3$ and the edge/vertex introduced by the operation is the first removable element of G' (if any). Based on this, $\mathcal{C}(G)$ is characterized as follows. Let f_L denote the leftmost r-face in G. Let u_p be the critical vertex of G, and k_i denote the index k of such vertex $u_k = \tau_{\text{last}}(e_i)$ in G.

Lemma 8. *Let G^1 be the plane graph obtained from G by operation* e-del. *Then $G^1 \in \mathcal{C}^1(G)$ if and only if $|f_L| = 3$, $deg(u_B; G) \geq 4$ and G has no separating r-face.*

Lemma 9. *Let G_h^2, $0 \leq h \leq 3$, be the plane graph obtained from G by operation* v-add(h) *for the second outer vertex u_2, and let v denote the new second outer vertex in G_h^2. Then $G_h^2 \in \mathcal{C}^2(G)$ if and only if $B = 4$, $|f_L| = 3$, $|\psi(G)| \geq 1$ and $deg(x; G_h^2) \geq 3 \ \forall x \in \Gamma(v; G_h^2)$.*

Lemma 10. *Let $G_{i,h}^3$ be the plane graph obtained from G by operation* v-insert (e_i, h) *for an outer edge $e_i = (u_i, u_{i+1})$, $i \in [1, B]$ and let v denote the new $(i+1)$st outer vertex in $G_{i,h}^3$. If the first removable element is an edge $e_p = (u_p, u_{p+1})$ (resp., a vertex u_p), then $G_{i,h}^3 \in \mathcal{C}^3(G) \Leftrightarrow G \neq W_n$, $B \geq 4$, $|f_L| = |f(e_i)| = 3$, $|\psi(G)| \geq 1$, $i \leq p$ (resp., $i \leq p - 1$), $deg(x; G_{i,h}^3) \geq 3 \ \forall x \in \Gamma(v; G_{i,h}^3)$ and $deg(u_i; G_{i,h}^3) \geq 4$ (where $deg(u_1; G_{i,h}^3) = 3$ is allowed).*

Lemma 11. *Let $G_{i,j}^4$ be the plane graph obtained from G by operation* e-add(u, v) *for two outer vertices $u = u_i$ and $v = u_j$, where $i \in [1, B-2]$ and $j \in [i+2, B]$. Then $G_{i,j}^4 \in \mathcal{C}^4(G)$ if and only if $B \geq 4$, $|f_L| = 3$, $i \leq p$ and $j \leq k_{i-1}$.*

7   Algorithm

This section describes an algorithm for generating children of a given graph $G \in \mathcal{G}_3(n)$ based on the characterization of children in Lemmas 8- 11.

To generate all plane graphs $G' \in \mathcal{C}(G) \cap \mathcal{G}_3(n, g)$, we generate only those $G' \in \mathcal{C}(G)$ such that the new face introduced by e-add and the face $f(e_i)$ and/or $f(e_{i+1})$ enlarged by v-insert or v-add are of length at most g. To generate all triconnected rooted plane graphs in $\mathcal{G}_3(n, g)$, we set $G := W_n$, and execute the following procedure GEN($G, \varepsilon = u_2$), where the second argument ε stands for the first removable element in the first argument G.

In GEN(G, ε), we first generate G^1 from G, and children $G_h^2 \in \mathcal{C}(G)$ if $B = 4$, and then generate children $G_{i,h}^3, G_{i,i+\Delta}^4 \in \mathcal{C}(G)$ for all vertices u_i, $i = 1, 2, \ldots, p$ in the active path of G by increasing step size $\Delta \geq 2$ by 1. We also generate G^1 from the children $G' = G_h^2$ obtained from G, without using a recursive call.

Procedure GEN(G, ε)
Input: A triconnected rooted plane graph $G \in \mathcal{G}_3(n, g)$ with $|f_L| = 3$ and the

first removable element ε of G, where ε is either an edge $e_p = (u_p, u_{p+1})$ or a vertex u_p.

Output: All descendants $G' \in \mathcal{G}_3(n, g)$ of G.

begin

  if the depth of the current recursive call is odd **then** Output G **endif**;

  /* Let $(u_1 = r, u_2, \ldots, u_B)$ denote the boundary of G in the clockwise
    order, and $e_i = (u_i, u_{i+1})$ denote the edge between u_i and u_{i+1} */

  if $e = (u_{B'}, z') \in E(G)$, $\kappa(G - e) \geq 3$ and $|f| < g$ for the $u_{B'}$-face f
  adjacent to f_L **then** Output $G' := G - e$ /* G' has no child */
  endif;

  if $B = 4$ and $|\psi(G)| \geq 1$ **then**

  for $h = 0, 1, 2, 3$ **do**

    Let G' be the graph G_h^2 obtained from G by v-add(h), and
    let v be the newly introduced vertex;

    if $deg(x; G') \geq 3$ for all $x \in \Gamma(v; G')$ and $|f| \leq g$
    for all faces $f \in F(v; G')$ **then**

      Output G';

      if $\kappa(G' - e) \geq 3$ and $|f| < g$ for the edge $e = (u_B, z)$ and
      the u_B-face f adjacent to the leftmost r-face in G' **then**

        Output $G'' := G' - e$ /* G'' has no child */

    endif endif endfor

  endif;

  if $G \neq W_n$, $B \geq 4$ and $|\psi(G)| \geq 1$ **then**

  for $i = 1, 2, \ldots, q$ **do**

    if $|f(e_i)| = 3$, and "$i < p$" or "$i = p$ and ε is an edge" **then**

    for $h = 0, 1, 2, 3$ **do**

      Let G' be the graph $G_{i,h}^3$ obtained from G by v-insert(e_i, h);

      Let v be the newly introduced vertex;

      if $deg(u_i; G') \geq 4$ when $i > 1$, $deg(x; G') \geq 3$ for all $x \in \Gamma(v; G')$
      and $|f| \leq g$ for all faces $f \in F(v; G')$ **then** GEN(G', v) **endif**

    endfor endif endfor

  endif;

  for $\Delta = 2, 3, \ldots, \min\{B - 1, g - 1\}$ **do**

  $i := 1$;

  while $i + \Delta \leq k_{i-1}$ and $i \leq p$ **do**

    /* k_{i-1} be the index $k \in [i + 1, B]$ of $u_k = \tau_{\text{last}}(e_{i-1})$ and $k_0 := B$ */

    $j := i + \Delta$;

    Let G' be the graph $G_{i,j}^4$ obtained from G by e-add(u_i, u_j);

    GEN$(G', e_i = (u_i, u_j))$;

    $i := i + 1$

  endwhile

  endfor;

  if the depth of the current recursive call is even **then** Output G **endif**;

  Return

end.

Note that the while-loop terminates once it holds $i + \Delta > k_{i-1}$ for some i without executing an iteration for $i' > i$. If $i + \Delta > k_{i-1}$ holds for some i, then it also holds $i' + \Delta > k_{i'-1}$ for any $i' \in [i,p]$ since $k_{i-1} \geq k_{i'-1}$ by (1). Therefore, $\mathrm{GEN}(G, \varepsilon)$ inspects all possible cases that can generate a child $G' \in \mathcal{C}(G) \cap \mathcal{G}_3(n,g)$.

We first show that each line of $\mathrm{GEN}(G, \varepsilon)$ can be executed in $O(1)$ time and $O(n)$ space. Since it is easy to maintain data for the size $|f|$ of each inner face f in $O(1)$ per change on an inner face, it suffices to show that $\tau_{\mathrm{last}}(e)$ for each edge in the active path can be found in $O(1)$ time and that whether $\kappa(G-e) \geq 3$ in e-del or not, (i.e., G has no separating r-face or not) can be tested in $O(1)$ time.

Lemma 12. *Let G be a triconnected plane graph rooted at $r = u_1$ such that $B \geq 4$.*
(i) For each edge e in the active path of G, $\tau_{\mathrm{last}}(e)$ can be found in $O(1)$ time and $O(n)$ space.
(ii) For the first edge $e = (r = u_1, u_2)$, whether G has no separating r-face or not can be tested in $O(1)$ time and $O(n)$ space.

Proof. (i) To facilitate computation of $\tau_{\mathrm{last}}(e)$, we introduce several notions. Denote the vertices along the boundary of G in the clockwise order by $u_1 = r, u_2, \ldots, u_B$, where $u_{B+1} = r$. Recall that, for each outer edge $e_i = (u_i, u_{i+1})$, $\tau_{\mathrm{last}}(e_i)$ denotes the vertex in $W^o(f(e_i))$ that appears first last when we traverse the boundary of G from u_i in the clockwise order, where $\{u_i, u_{i+1}\} \cap W^o(f(e_i)) = \emptyset$.

For an inner face f which contains at least one outer vertex, let $\tau(f)$ denote the outer vertex $u_j \in V(f')$ with the largest index $j \leq B+1$ for an inner face f' which shares a vertex w with f (where f' is not necessarily an inner face internally adjacent to f; i.e., w may be an outer vertex).

For an outer vertex in the current graph G, let $f_1(u_i)$ and $f_2(u_i)$ (resp., $f_4(u_i)$ and $f_3(u_i)$) denote the rightmost and second rightmost (resp., the leftmost and second leftmost) v-faces when we regard (u_{i-1}, u_i) and (u_i, u_{i+1}) as the leftmost and rightmost edges incident to u_i. Define

$$\tau_j(u_i) = \tau(f_j(u_i)), \quad j = 1, 2, 3, 4 \text{ (see Fig. 5(a)-(b))}.$$

We first show that each outer edge $e_i = (u_i, u_{i+1})$ in the active path satisfies

$$\tau_1(u_i) = \tau_{\mathrm{last}}(e_i).$$

Let $e_i = (u_i, u_{i+1})$ be an outer edge in the active path. Since all edges in the active path are irremovable, $\tau_{\mathrm{last}}(e_i)$ is a vertex u_j with $i < j \leq B$ by the property (1). Thus, the vertex $u_j = \tau_{\mathrm{last}}(e_i)$ is one of the candidates to define $\tau_1(u_i)$, and no other outer vertex $u_{j'}$ with $j' > j$ can be chosen as $\tau_1(u_i)$ because otherwise such a vertex $u_{j'}$ would be $\tau_{\mathrm{last}}(e_i)$ (note that in this case the face f' adjacent to f that attains $u_{j'}$ must also be internally adjacent to f).

We next show how to update the values of τ_j, $j = 1, 2, 3, 4$. For this, we define data η as follows. Let w be a vertex in the current graph G such that a w-face f_w

contains an outer vertex. Then $\eta(w)$ is defined to be the outer vertex $u_j \in V(f_w)$ with the largest index $j \leq B + 1$ among all such w-faces f_w (see Fig. 5(d)).

In what follows, we show how to update η, τ_1, τ_2, τ_3 and τ_4 when one of operations e-add and v-insert is applied to generate a child G' from the current graph G so that $\eta(u)$, $\tau_1(u)$ and $\tau_2(u)$, $\tau_3(u)$ and $\tau_4(u)$ for outer vertices u in the active path take the correct value. Note that, after operation v-add is applied, only e-del is applicable and there is no need to update τ_i to obtain τ_{last}.

(1) Initialization: For $G = W_n$, where $B = n - 1$, we set

$\eta(u_i) := u_{i+1}$ and $\tau_1(u_i) := \tau_2(u_i) := \tau_3(u_i) := \tau_4(u_i) := r$ for all $i = 1, 2, \ldots$, $n - 1$, and $\eta(z) := r$ for the center z of W_n (see Fig. 5(c)).

(2) Operation e-add(u, v) is applied (see Fig. 5(a) and (b)): We update as follows.

$\eta(u) := v$; $\tau_2(u) := \tau_1(u)$; $\tau_1(u) := \eta(v)$; $\tau_3(v) := \tau_4(v)$; $\tau_4(v) := \eta(v)$.

(3) Operation v-insert(e_i, h), $h \in \{0, 1, 2, 3\}$ is applied: We update η by $\eta(v) := b$; $\eta(a) := v$ and τ_i, $i = 1, 2, 3, 4$ as follows.

1. v-insert$(e_i, 0)$: See Fig 5(a).
   $\tau_1(v) := \tau_3(v) := \tau_4(b)$; $\tau_2(v) := \tau_4(v) := \tau_1(a)$; $\tau_1(a) := \tau_4(b) := \eta(w)$.
2. v-insert$(e_i, 1)$: See Fig 5(b).
   $\tau_1(v) := \tau_3(v) := \tau_4(b)$; $\tau_2(v) := \tau_4(v) := \tau_2(a)$; $\tau_1(a) := \tau_2(a)$.
3. v-insert$(e_i, 2)$: See Fig. 5(c).
   $\tau_1(v) := \tau_3(v) := \tau_3(b)$; $\tau_2(v) := \tau_4(v) := \eta(w)$; $\tau_4(b) := \tau_3(b)$.
4. v-insert$(e_i, 3)$: See Fig. 5(d).
   $\tau_1(v) := \tau_3(v) := \tau_3(b)$; $\tau_2(v) := \tau_4(v) := \tau_2(a)$; $\tau_4(b) := \tau_3(b)$; $\tau_1(a) := \tau_2(a)$.

It is easy to see that η and τ_i can be updated in $O(1)$ time per operation using $O(n)$ space.

We here show that $\eta(w)$ is correctly updated. For $w = z$ in $G = W_n$, $\eta(w) = r$ remains unchanged in any descendant of G. For other w, we see that the value $\eta(w)$ is correctly updated by the above procedure for e-add when w is still an outer vertex. When an outer vertex w becomes an inner vertex by e-add, the latest value u_h for $\eta(w)$ will never be changed by any of the above procedures. We show that $\eta(w) = u_h$ remains valid as long as a w-face contains an outer vertex. When an outer vertex w becomes an inner vertex in a graph $\hat{G}$, e-add is applied and the newly introduced edge (u, v) is the first removable element ε in the resulting graph G', where ε is situated in $\beta[r, u_h]$. See Fig. 5(d). If u_h becomes an inner vertex and a w-face f'_w still has an outer vertex u_i in a graph G'', then the newly introduced element ε' to G'' would not be the first removable element in G'', since $\beta[r, u_h]$ contains a removable element. This proves that $\eta(w)$ stores the correct value.

Supposing that $\eta(w)$ for all vertices w store the correct values, it is a simple matter to see that the values τ_i for outer vertices in the active path are correctly updated by the above procedures.

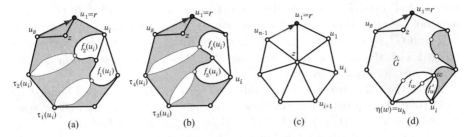

Fig. 5. (a) Inner faces f_1 and f_2 and $\tau_1(u_i)$ and $\tau_2(u_i)$; (b) inner faces f_3 and f_4 and $\tau_3(u_i)$ and $\tau_4(u_i)$; (c) initial graph $G = W_n$; (d) $\eta(w) = u_h$ and w-faces f_w and f'_w

(ii) Let f_{s_0} denote the s_0-face such that $(s_0, z) \in E(f_{s_0})$ and $\psi(G) \cap E(f_{s_0}) = \emptyset$. We call an r-face $f \in F(r; G) - \{f_L\}$ s_0-*separating* if f and f_{s_0} share an inner vertex other than the second leftmost neighbour z of r. Let δ be the existent function of s_0-separating r-face, i.e., $\delta = 1$ if there is a s_0-separating r-face, and $\delta = 0$ otherwise. Note that an s_0-separating r-face is a separating r-face when $\psi(G) = \emptyset$ and $s_0 = u_B$. Hence it suffices to show how to compute δ in $O(n)$ space and $O(1)$ time per operation.

Since there is no s_0-separating r-face in wheel $G = W_n$, we initialize $\delta := 0$. Since both v-add and v-insert do not change the set of inner faces internally adjacent to r-faces, they never create or eliminate any s_0-separating r-face. Hence δ shall only be updated when e-add is performed. Note that applying e-add may change the positions of s_0 and f_{s_0}. When e-add(u, v) is performed to generate a child G' of G, we update δ as follows:

(1) $u \neq r$ and $v = s_i$, $i > 0$: Let $\delta := 0$, since s_0 in G becomes an inner vertex in the child G', the previous s_0-separating r-face is no longer s_0-separating in G', and no new r-face becomes s_0-separating for the new s_0.

(2) $u = r$ and $v = s_i$, $i > 0$: Let $\delta := 1$, since s_i becomes a new s_0 in G' and the newly created rightmost r-face f is internally adjacent with the new f_{s_0}.

(3) $u = r$, $v = s_0$, $deg(s_0; G) = 3$: Let $\delta := 1$ analogously with (2).

(4) otherwise: δ remains unchanged, since no s_0-separating r-face is created or eliminated in this case.

Therefore, before e-del is applied, it holds $\psi(G) = \emptyset$ and δ tells whether the current G has a separating r-face or not. □

Finally we show that $\text{GEN}(W_n, \varepsilon = u_2)$ can be implemented to run in $O(|\mathcal{G}_3(n, g)|)$ time. For this, it suffices to show that the time complexity $T(G)$ of $\text{GEN}(G, \varepsilon)$ without including the computation time for recursive calls of $\text{GEN}(G', \varepsilon)$ is $O(|\mathcal{C}(G) \cap \mathcal{G}_3(n, g)|)$. Constructing each of graphs G^1, G_h^2, $G_{i,h}^3$ and $G_{i,j}^4$ can be done in $O(1)$ time during an execution of $\text{GEN}(G, \varepsilon)$. We next show that the delay spent to generate the next child G' during an execution of $\text{GEN}(G, \varepsilon)$ is $O(1)$ time, which implies $T(G) = O(|\mathcal{C}(G) \cap \mathcal{G}_3(n, g)|)$. The time for computing all G_h^2, $h = 0, 1, 2, 3$ is $O(1)$. The time delay to generate the next child $G' = G_{i,j}^4$ is $O(1)$ time during the last for-loop, since any constructed graph $G' = G_{i,j}^4$

belongs to $\mathcal{C}(G) \cap \mathcal{G}_3(n, g)$. The delay spent to generate the next child $G' = G_{i,h}^3$ may not be $O(1)$ time. This, however, can be easily amortized by generating $G_{i,h}^3$ and $G_{i,i+2}^4$ alternately; i.e., the iteration for $\Delta = 2$ in the last for-loop is merged with the for-loop for generating $G_{i,h}^3$. The current description of GEN avoids such a complicated for-loop.

Hence the delay between two children G' in GEN(G, ε) can be bounded by $O(1)$ time. This proves that the time complexity $T(G)$ of GEN(G, ε) without recursive calls GEN(G', ε) is $O(|\mathcal{C}(G) \cap \mathcal{G}_3(n, g)|)$, and that GEN$(W_n, \varepsilon = u_2)$ runs in $O(|\mathcal{G}_3(n, g)|)$ time. Furthermore, by outputting a child G' before calling GEN(G', ε) if the current depth of recursive call is odd and after calling GEN(G', ε) if the current depth of recursive call is even, the delay between two outputs in the entire execution is $O(1)$ in the worst case. It is easy to see that the entire algorithm GEN$(W_n, \varepsilon = u_2)$ can be implemented in $O(n)$ space.

Theorem 1. *For integers $n \geq 1$ and $g \geq 3$, all triconnected rooted plane graphs with exactly n vertices such that each inner face is of length at most g can be enumerated without duplication in $O(n)$ space by an algorithm that outputs the difference between two consecutive outputs in $O(1)$ time in a series of all outputs after an $O(n)$ time preprocessing.*

We can use our algorithm for generating unrooted plane graphs. During an execution of GEN$(W_n, \varepsilon = u_2)$, we check in $O(n^2)$ time whether a newly generated rooted graph G is the representative among rooted graphs with the same plane graphs or not by computing its signature [2].

Corollary 1. *For a given integer $n \geq 1$, all triconnected planar graphs with exactly n vertices can be enumerated without duplication in $O(n)$ space by an algorithm that outputs the difference between two consecutive outputs in $O(n^3)$ time in average in a series of all outputs.*

8   Concluding Remarks

In this paper, we gave an $O(1)$-time delay enumeration algorithm for the class of triconnected rooted plane graphs with exactly n vertices and an inner face size bounded by g. The $O(1)$-time delay in $O(n)$ space is attained mainly by introducing graph transformations to define parents based on the property in Lemma 1, which allows us to test triconnectivity of a whole graph by degree by checking the degree of a constant number of vertices.

It is our future work to design enumeration algorithms for rooted plane graphs with a higher vertex-connectivity and to take into account the reflective symmetry around the root, as studied in our companion paper [14].

References

1. Fujiwara, H., Wang, J., Zhao, L., Nagamochi, H., Akutsu, T.: Enumerating tree-like chemical graphs with given path frequency. Journal of Chemical Information and Modeling 48, 1345–1357 (2008)

2. Hopcroft, J.E., Wong, J.K.: Linear time algorithm for isomorphism of planar graphs. In: STOC 1974, pp. 172–184 (1974)
3. Horváth, T., Ramon, J., Wrobel, S.: Frequent subgraph mining in outerplanar graphs. In: Proc. 16th ACM SIGKDD International Conference on Knowledge Discovery and Data Mining, pp. 197–206 (2006)
4. Imada, T., Ota, S., Nagamochi, H., Akutsu, T.: Enumerating stereoisomers of tree structured molecules using dynamic programming. In: Dong, Y., Du, D.-Z., Ibarra, O. (eds.) ISAAC 2009. LNCS, vol. 5878, pp. 14–23. Springer, Heidelberg (2009)
5. Ishida, Y., Zhao, L., Nagamochi, H., Akutsu, T.: Improved algorithm for enumerating tree-like chemical graphs. In: Genome Informatics, GIW 2008, vol. 21, pp. 53–64 (2008)
6. Li, Z., Nakano, S.: Efficient generation of plane triangulations without repetitions. In: Orejas, F., Spirakis, P.G., van Leeuwen, J. (eds.) ICALP 2001. LNCS, vol. 2076, pp. 433–443. Springer, Heidelberg (2001)
7. Nakano, S.: Efficient generation of plane trees. IPL 84, 167–172 (2002)
8. Nakano, S.: Efficient generation of triconnected plane triangulations. In: Computational Geometry Theory and Applications, vol. 27(2), pp. 109–122 (2004)
9. Nakano, S., Uno, T.: Efficient generation of rooted trees, NII Technical Report, NII-2003-005 (2003)
10. Read, R.C.: How to avoid isomorphism search when cataloguing combinatorial configurations. Annals of Discrete Mathematics 2, 107–120 (1978)
11. Tutte, W.T.: Convex representations of graphs. Proc. of London Math. Soc. 10(3), 304–320 (1960)
12. Steinitz, E.: Polyeder und Raumeinteilungen. Encyclopädie der mathematischen Wissenschaften, Band 3 (Geometrie), Teil 3AB12, 1–139 (1922)
13. Yamanaka, K., Nakano, S.: Listing all plane graphs. In: Nakano, S.-i., Rahman, M. S. (eds.) WALCOM 2008. LNCS, vol. 4921, pp. 210–221. Springer, Heidelberg (2008)
14. Zhuang, B., Nagamochi, H.: Enumerating rooted biconnected planar graphs with internally triangulated faces, Kyoto University, Technical Report 2009-018 (2009), http://www-or.amp.i.kyoto-u.ac.jp/members/nag/Technical_report/TR2009-018.pdf
15. Zhuang, B., Nagamochi, H.: Efficient generation of symmetric and asymmetric biconnected rooted outerplanar graphs. In: AAAC 2010, p. 21 (2010)
16. Zhuang, B., Nagamochi, H.: Enumerating biconnected rooted plane graphs, Kyoto University, Technical Report 2010-001 (2010), http://www-or.amp.i.kyoto-u.ac.jp/members/nag/Technical_report/TR2010-001.pdf
17. Zhuang, B., Nagamochi, H.: Constant time generation of biconnected rooted plane graphs. In: Lee, D.-T., Chen, D.Z., Ying, S. (eds.) FAW 2010. LNCS, vol. 6213, pp. 113–123. Springer, Heidelberg (2010)
18. Zhuang, B., Nagamochi, H.: Enumerating rooted graphs with reflectional block structures. In: Calamoneri, T., Diaz, J. (eds.) CIAC 2010. LNCS, vol. 6078, pp. 49–60. Springer, Heidelberg (2010)
19. Zhuang, B., Nagamochi, H.: Generating internally triconnected rooted plane graphs. In: Kratochvíl, J., Li, A., Fiala, J., Kolman, P. (eds.) TAMC 2010. LNCS, vol. 6108, pp. 467–478. Springer, Heidelberg (2010)
20. Zhuang, B., Nagamochi, H.: Listing triconnected rooted plane graphs, Kyoto University, Technical Report 2010-002 (2010), http://www-or.amp.i.kyoto-u.ac.jp/members/nag/Technical_report/TR2010-002.pdf

Bipartite Permutation Graphs Are Reconstructible

Masashi Kiyomi[1], Toshiki Saitoh[2], and Ryuhei Uehara[1]

[1] School of Information Science, JAIST, 1-1, Asahidai,
Nomi, Ishikawa 923-1292, Japan
{mkiyomi,uehara}@jaist.ac.jp
[2] ERATO MINATO Discrete Structure Manipulation System Project,
JST, North 14, West 9, Sapporo, Hokkaido 060-0814, Japan
t-saitoh@erato.ist.hokudai.ac.jp

Abstract. The graph reconstruction conjecture is a long-standing open problem in graph theory. The conjecture has been verified for all graphs with at most 11 vertices. Further, the conjecture has been verified for regular graphs, trees, disconnected graphs, unit interval graphs, separable graphs with no pendant vertex, outer-planar graphs, and unicyclic graphs. We extend the list of graph classes for which the conjecture holds. We give a proof that bipartite permutation graphs are reconstructible.

Keywords: the graph reconstruction conjecture, bipartite permutation graphs.

1 Introduction

The graph reconstruction conjecture proposed by Ulam and Kelly[1] has been studied by many researchers intensively. In order to state the conjecture, we first introduce some terms. A graph G' is called a *card* of a graph $G = (V, E)$, if G' is isomorphic to $G - v$ for some $v \in V$, where $G - v$ is a graph obtained from G by removing v and incident edges. A multi-set of n graphs with $n - 1$ vertices for some positive integer n is called a *deck*. Especially, the multi-set of the $|V|$ cards of G, each of which is isomorphic to $G - v$ for each $v \in V$, is a deck of G. A graph G is a *preimage* of a deck D, if D is a deck of G. We also say that a graph G is a preimage of the n graphs if each card of G is isomorphic to each of them. The graph reconstruction conjecture is that there is at most one preimage of given n graphs ($n \geq 3$). No one has given a positive nor a negative proof of this conjecture, while there is a positive proof for small graphs [11].

The graph reconstruction conjecture has been verified for some graph classes. Kelly showed that the conjecture is true on regular graphs, trees, and disconnected graphs. Other classes proven to be reconstructible[2] are unit interval

---

[1] Determining the first person who proposed the graph reconstruction conjecture is difficult, actually. See [7] for the detail.
[2] A graph is reconstructible, if its deck has only one preimage.

W. Wu and O. Daescu (Eds.): COCOA 2010, Part II, LNCS 6509, pp. 362–373, 2010.
© Springer-Verlag Berlin Heidelberg 2010

graphs [12], separable graphs with no pendant vertex [2], outer-planar graphs [5], unicyclic graphs [10], etc. We extend the list of graph classes for which the conjecture holds. We give a proof that bipartite permutation graphs are reconstructible.

Rimscha showed that unit interval graphs are reconstructible [12]. Unit interval graphs have somewhat path-like structures, and so do bipartite permutation graphs. Further, the representation of a unit interval graph is unique, similar to that of a bipartite permutation graph. Thus, we first thought that we can easily prove that bipartite permutation graphs are reconstructible. There are two differences between the two classes, that make it difficult to prove that bipartite permutation graphs are reconstructible. One is that, bipartite permutation graphs are bipartite. Therefore, we have to determine from which vertex set a vertex was removed for cards in a deck. The second difference is that, in the case of unit interval graphs, there is no disconnected card obtained by removing a vertex laying at the end of the path structure. In a deck of a bipartite permutation graph, there can be a disconnected card that is obtained by removing a polar vertex which lays at the end of the path structure. we will define a polar vertex later. Therefore, we had to consider many exceptional cases.

Kelly showed the following lemma.

Lemma 1 (Kelley's Lemma [9]). *Let G be any preimage of the given deck, and let H be a graph whose number of vertices is smaller than that of G. Then we can uniquely determine the number of subgraphs in G isomorphic to H from the deck.*

Greenwell and Hemminger extended this lemma to a more general form [6]. We can determine the degree sequence of a preimage of the given deck from these lemmas. Moreover, given a deck of a graph, we can determine the degree of removed vertex for each card in the deck. Note that $\sum_{v:\ \text{vertex}} \deg(v) = 2 \times (\#$ of edges$)$. Thus, we can easily show for example that cycles are reconstructible, since a graph is a cycle if and only if it is connected, and all its vertices have degree exactly two.

Tutte proved that the dichromatic rank and Tutte polynomials are reconstructible (i.e. looking at the deck, they are uniquely determined) [14]. Bollobás showed that almost all graphs are reconstructible from three well-chosen graphs in its deck [1]. About permutation graphs, Rimscha showed that permutation graphs are recognizable in the sense that looking at the deck of G one can determine whether or not G belongs to permutation graphs [12]. To be precise Rimscha showed in the paper that comparability graphs are recognizable. Even's result [4] directly gives a proof in the case of permutation graphs. Rimscha also showed in the same paper that many subclasses of perfect graphs including perfect graphs themselves are recognizable, and moreover, some of subclasses, such as unit interval graphs, are reconstructible. There are a lot of papers about the conjecture, and many good surveys about this conjecture. See for example [3,7].

We explain about bipartite permutation graphs in the next section. Then, we prove the statement in Section 3. The proof has two subsections. In the first subsection, we give the main idea of the proof. In the second subsection, we consider some exceptional cases. The proof uses some lemmas on bipartite

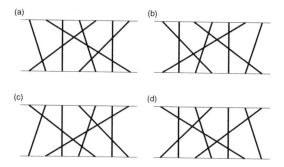

Fig. 1. (a) is an example of a permutation diagram. (b), (c), and (d) are permutation diagrams obtained from (a) by reversing horizontally, reversing vertically, and rotating 180°, respectively. They represent permutations (2,7,3,5,1,6,4), (4,2,7,3,5,1,6), (5,1,3,7,4,6,2), and (6,2,4,1,5,7,3), respectively.

permutation graphs. Since we think that checking these lemmas one by one may make readers lose the way, we write the proofs of some of them in Section 4.

2 Bipartite Permutation Graphs

All the graphs in this paper are simple unless stated otherwise.

2.1 Permutation Diagram

Let $\pi = (\pi_1, \pi_2, \ldots, \pi_n)$ be a permutation of $1, \ldots, n$. We denote $(\pi_n, \pi_{n-1}, \ldots, \pi_1)$ by $\overline{\pi}$.

We call a set L of line segments connecting two horizontal parallel lines on Euclidean plane a *permutation diagram*. A permutation diagram represents a permutation. Let $l_1, l_2, \ldots, l_{|L|}$ be the line segments in L. We assume that the end-points of them appear in this order from left to right on the upper horizontal line. Then, the permutation represented by L is $(\pi_1, \pi_2, \ldots, \pi_{|L|})$, such that the end-points of $l_1, l_2, \ldots, l_{|L|}$ appear in the order of $\pi_1, \ldots, \pi_n$ on the lower horizontal line. Equivalently, the ith left-most end-point among those of the segments in L is that of $l_{\pi_i^{-1}}$, on the lower horizontal line, for each $i \in \{1, 2, \ldots, |L|\}$. See Fig. 1(a) for example. The permutation diagram represents $(2, 7, 3, 5, 1, 6, 4) = (5, 1, 3, 7, 4, 6, 2)^{-1}$.

Let P be a permutation diagram. We denote by P^{H} a permutation diagram obtained by reversing P horizontally. See Fig. 1(b) for example. The permutation diagram is obtained by reversing (a) horizontally. Similarly, we denote by P^{V} and P^{R} permutation diagrams obtained by reversing P vertically, and by rotating P 180°, respectively.[3]

---

[3] Let P be a permutation diagram representing a permutation π. For those who want concrete expressions, it is not difficult to check that P^{V} represents $\pi^{\mathrm{V}} = \pi^{-1}$, P^{H} represents $\pi^{\mathrm{H}} = \overline{\overline{\pi}^{-1}}^{-1}$, and P^{R} represents $\pi^{\mathrm{R}} = \overline{\overline{\pi}^{-1}}$.

Fig. 2. Forbidden graphs of bipartite permutation graphs are these graphs, K_3, and cycles of length more than four

2.2 Bipartite Permutation Graphs

Let π be a permutation of the numbers $1, 2, \ldots, n$. $G_\pi = (V_\pi, E_\pi)$ is a graph satisfying that

- $V_\pi = \{1, \ldots, n\}$, and
- $\{i, j\} \in E_\pi \Leftrightarrow (i - j)(\pi_i^{-1} - \pi_j^{-1}) < 0$.

A graph G is called a *permutation graph* if there exists a permutation π such that G is isomorphic to G_π. Equivalently, a graph G is a permutation graph if there exists a permutation π such that G is an intersection model of the permutation diagram of π. We say that π (or, sometimes, the permutation diagram of π) represents G. If a permutation graph G is bipartite, we call G a *bipartite permutation graph*.

There is a good characterization for bipartite permutation graphs.

Theorem 1 (P. Hell and J. Huang [8]). *A graph G is a bipartite permutation graph if and only if G has neither the graphs in Fig. 2, nor K_3, nor cycles of length more than four as an induced subgraph.*

It is known that a connected bipartite permutation graph has at most four representing permutation diagrams. If a permutation diagram P is representing a connected bipartite permutation graph G, the other representing permutation diagram of G must be one of P^H, P^V, and P^R [13]. Thus a permutation diagram representing a connected bipartite permutation graph is essentially unique. Note that a disconnected bipartite permutation graph may have more than four representing permutation diagrams. Together with the fact that cards in a deck of a connected graph can be disconnected, this is the reason why our proof is not very simple.

Let P be a permutation diagram representing a connected bipartite permutation graph G. There are two left-most segments in P, and there are two right-most segments in P. Here, we say that a segment is the *left-most(right-most)* if it is the left-most (right-most) among the segments not intersecting with it. We call the vertices that can correspond to the left-most or right-most segments *polar vertices*. Note that there are at least four polar vertices in a bipartite permutation graph. The number of polar vertices may be more than four, since there may exist some isomorphic polar vertices.[4] See Fig. 3 for an example.

By repeatedly removing degree one polar vertices from a connected bipartite permutation graph G, we obtain a connected bipartite permutation graph G'. We call the graph G' *trunk* of G, and we denote the trunk by $\text{Tr}(G)$. The vertex

---

[4] Vertices v and u are isomorphic, if the neighbors of them are identical.

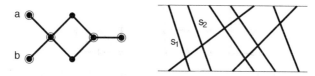

Fig. 3. An bipartite permutation graph and its representation. The polar vertices are circled. Vertices a and b are isomorphic, and can correspond to both the segments s_1 and s_2. Thus, both a and b are polar vertices.

in $\mathrm{Tr}(G)$ nearest from a degree one polar vertex v of G is called the *root* of v. The path in G whose ends are v and v's root is called a *limb*.

It is clear that every card G' of a bipartite permutation graph G is a bipartite permutation graph, since we can obtain a representing permutation diagram of G' by removing a line segment from a representing permutation diagram of G.

3   Main Proof

The main idea of our proof is simple. However, if there is a degree one polar vertex, there are many exceptional cases, and the proof gets complex. Therefore, we first show the simple case, and then prove the exceptional cases.

3.1   No Degree One Polar Vertex Case

We show an algorithm which reconstructs G from its deck. The algorithm directly shows the uniqueness of the preimage. However, the proof of the uniqueness uses a bunch of bipartite permutation graph specific properties. We are afraid that checking the properties one by one makes the readers lose the way in the mainline of the proof. Therefore, we leave some of the proofs in Section 4.

We need the two lemmas below to keep the main proof simple.

Lemma 2. *All the preimages of the deck of a bipartite graph G are bipartite.*

Proof. Immediate from the fact that the chromatic number of G is reconstructible [15]. □

Lemma 3. *All the preimages of a deck of a bipartite permutation graph are bipartite permutation graphs.*

Proof. Immediate from Lemma 2 and the fact that permutation graphs are recognizable [12]. □

We can easily check the following lemma.

Lemma 4. *A card obtained from a connected bipartite graph G by removing its polar vertex is connected, if every polar vertex of G has degree more than one.*

Our main proof assumes that a bipartite permutation graph $G = (X, Y, E)$ does not have a vertex in X whose degree is $|Y|$. Therefore, we need the following lemma.

Lemma 5. *Let $G = (X, Y, E)$ be a connected bipartite permutation graph with a vertex $x \in X$ whose degree is $|Y|$. Then G is reconstructible.*

We leave the proof in Section 4. Now, we make assumption below.

Assumption 1. *In a connected bipartite permutation graph $G = (X, Y, E)$, there is no vertex $x \in X$ such that $\deg(x) = |Y|$, and there is no vertex $y \in Y$ such that $\deg(y) = |X|$.*

Note that this is equivalent that a polar vertex on the left-end cannot be adjacent to any polar vertex on the right-end. Moreover, we can assume that $|X|, |Y| \geq 2$, since if $|X| = 1$, then $x \in X$ must be adjacent to every vertex in Y.

Now, we explain the main idea. Let $G = (X, Y, E)$ be a bipartite permutation graph satisfying Assumption 1, and every polar vertex of G has degree more than one.

We state that $|\tilde{X}|$ and $|\tilde{Y}|$ are reconstructible for any connected bipartite permutation graph $\tilde{G} = (\tilde{X}, \tilde{Y}, \tilde{E})$. Since the proof of this fact becomes a bit long, we leave it in Section 4.

We first consider the case that $|X| \neq |Y|$. We assume without loss of generality that $|X| > |Y|$. There are two polar vertices x_1 and x_r in X, such that x_1 corresponds to the left-most line segment, and x_r corresponds to the right-most line segment in a permutation diagram representing G. We denote the degrees of x_1 and x_r by p and q. In a similar fashion, we denote by r and s the degrees of polar vertices y_1 and y_r in Y. We assume without loss of generality that $p \leq q$ holds.

Let $G_1 = (X_1, Y_1, R_1)$ and $G_r = (X_r, Y_r, R_r)$ be cards of G obtained by removing y_1 and y_r from G, respectively. By Lemma 4, G_1 and G_r are connected.

We denote by D_Y the set of G's connected cards that are obtained by removing a vertex belonging to Y. Clearly, G_1 and G_r are in D_Y. Consider a connected bipartite permutation graph $G' = (X', Y', E')$ in D_Y. We assume without loss of generality that $|X'| \geq |Y'|$ holds. Then, since $|X| > |Y|$ holds, $|X| = |X'| > |Y'| = |Y| - 1$ holds. Therefore, we can choose all the cards that belong to D_Y from the deck of G, and we can determine which vertex set of each card corresponds to X. Now, consider the degrees of the polar vertices in X'. If G' is G_1, the degrees of the polar vertices in X' are $\{p - 1, q\}$. If G' is G_r, the degrees of the polar vertices in X' are $\{p, q - 1\}$. Otherwise, the degrees of the polar vertices in X' are either $\{p - 1, q\}$, $\{p, q - 1\}$, or $\{p, q\}$. We call G' *good*, if the degrees of the polar vertices in X' are $\{p - 1, q\}$.

Let $\{G'_1, \ldots, G'_k\}$ be the set of good graphs in D_Y. Let d_i be the degree of the vertex v_i such that G'_i is obtained by removing v_i from G. Note that we can determine d_i for each card, since the degree sequence is reconstructible. It is clear that $r = \min_{i=1,\ldots,k} d_i$ by Lemma 12 which we state in Section 4. Thus, we can uniquely reconstruct the preimage from the deck. (We only have to add a degree r polar vertex adjacent to the polar vertex of degree $p - 1$. This can be done deterministically on the permutation diagram by Lemma 13 which we prove in Section 4.)

Now we consider the case that $|X| = |Y|$. In this case, every connected card is in the form $G_i = (X_i, Y_i, E)$ such that $|X_i| = |Y_i| - 1$. Thus we cannot determine which vertex set corresponds to which vertex set of G. However, we know that G_i is obtained by removing a vertex in Y_i's side. That is, polar vertices of G_i in X_i are also polar vertices of a preimage. Therefore, the minimum degree $\overline{p'}$ of polar vertices in X_i among all the connected cards is equal to $p' - 1$, where p' is the minimum degree of polar vertices of a preimage. Moreover, a card that has a polar vertex of degree $\overline{p'}$ in X_i is obtained by removing a vertex adjacent to a polar vertex of degree p' from a preimage. Thus, we can uniquely reconstruct a preimage in the same fashion above (using Lemmas 12,13).

Therefore, we have the theorem below.

Theorem 2. *A connected bipartite permutation graph* $G = (X, Y, E)$ *satisfying Assumption 1 is reconstructible, if every polar vertex of G has degree more than one.*

3.2   Polar Vertices with Degree One

We can determine if a preimage G has a polar vertex of degree one, by Lemma 14 in Section 4. In this subsection, we consider the case that G has a polar vertex of degree one.

First, we show the fundamental lemmas.

Lemma 6. *Let P be a permutation diagram of a connected bipartite permutation graph G having at least one cycle. Any two limbs of the same side (the left-side or the right-side) of P have the same root.*

Proof. If not, G cannot be connected.   □

Lemma 7. *If a connected bipartite permutation graph G has a polar vertex of degree one, $Tr(G)$ is reconstructible.*

Proof. Let G' be a card obtained by removing a polar vertex of degree one from G. Then, $Tr(G')$ and $Tr(G)$ are clearly isomorphic. Let G'' be a connected card obtained by removing a vertex that is not a polar vertex of degree one. Then, $|V(Tr(G''))| < |V(Tr(G))|$ holds. Thus, we can reconstruct $Tr(G)$ by choosing the $Tr(G')$ whose number of vertices is the maximum.   □

Now we prove the reconstructivity, one by one.

Lemma 8. *A connected bipartite permutation graph $G = (X, Y, E)$ with a limb whose length is more than one is reconstructible.*

Proof. Let L be a permutation diagram of G. Let $\{G'_1, \ldots, G'_k\}$ be the multi-set of connected cards of G that satisfy $Tr(G'_i) = Tr(G)$. If there are more than one limbs having the same root, and both of them have the lengths more than one, G contains the left forbidden graph in Fig 2. Thus, only one limb can have the length more than one among limbs of the same root. We concentrate the limbs of the maximum length, on the both sides.

First we consider the case that there are two limbs, one is on the left-side of L having the maximum length among limbs on the left-side, and the other is on the right-side having the maximum length among limbs on the right-side. If both the two limbs have the lengths more than one, we can easily reconstruct G, since we can determine the lengths p, q of the two limbs from $\{G'_1, \ldots, G'_k\}$. Note that each G'_i has limbs of lengths $p-1, q$, or p, q, or $p, q-1$. Hence, we consider the case that the left-side limb has the length exactly one. The right-side limb has the length q more than one. Even in this case, we can determine that the maximum lengths of limbs on the both sides of any preimage are one and q. The remaining problem is how to reconstruct G from $\{G'_1, \ldots, G'_k\}$. If q is more than two, the reconstruction is easy. Only find the limb of length $q-1$, and add a degree one vertex to it. Thus, we consider the case that q is equal to two. In this case, there is a card in $\{G'_1, \ldots, G'_k\}$ that has length one limbs on the both side. Thus, we can determine if the roots of the two limbs of a preimage belongs to the same vertex set. And, there is a card G' in $\{G'_1, \ldots, G'_k\}$ that has a limb l of the length of two. Thus, we can reconstruct G uniquely by adding a degree one vertex to the opposite side to l.

Next, we consider the case that G has limbs only on the left-side. It is easy to reconstruct G in this case, since finding the connected card that has limbs most, and adding a degree one vertex to the longest limb (the other limbs have length one), we have G. □

Lemma 9. *A bipartite permutation graph $G = (X, Y, E)$ with two limbs of different roots is reconstructible.*

Proof. From Lemma 8, we only have to prove the case that every limb has length exactly one. In this case, we can determine if the two roots belong to the same vertex set, since we can reconstruct $\{|X|, |Y|\}$. Thus we can reconstruct G uniquely. □

Lemma 10. *A bipartite permutation graph $G = (X, Y, E)$ whose limbs have the same root is reconstructible.*

Proof. If there are more than one limbs, it is easy to reconstruct G. Let G' be a connected card satisfying $\mathrm{Tr}(G') = \mathrm{Tr}(G)$. Find a limb in G' and add a degree one vertex to its root. Therefore, we consider the case that G has only one limb, and the length is one. Assume that the limb is on the left-side of L, where L is a permutation diagram of G. Then two polar vertices on the right-side have degrees p, q larger than one. If both of p and q are larger than two, we can reconstruct G, since connected cards of G with degree one polar vertex on the one side of their permutation diagram have polar vertices of degree p, q, or $p-1, q$, or $p, q-1$, on the opposite side.

Now we consider the case that p is exactly two, and q is also equal to two. There is a connected card of G whose polar vertices on the same side have degrees one and two. Since the limb of G has length exactly one, the polar vertices of the same side having degree one and two cannot be the degree one vertex of G. Therefore we can reconstruct G uniquely.

Lastly, we consider the case that p is exactly two, and q is larger than two. Let $\{G'_1, \ldots, G'_k\}$ be connected cards of G obtained by removing a vertex whose degree is larger than one. Looking $\{G'_1, \ldots, G'_k\}$, we can determine the value p and q. Hence we can reconstruct G uniquely. □

From above lemmas, we have the theorem below.

Theorem 3. *A connected bipartite permutation graph with a polar vertex of degree one is reconstructible.*

4   Miscellaneous Proofs

We prove Lemmas not proved yet, in this section.

Lemma 11. *The numbers of vertices in X and Y are reconstructible for a connected bipartite permutation graph $G = (X, Y, E)$.*

Proof. Let D be the deck of G. There are at least two connected cards in a deck of a connected graph with more than two vertices. Let $G_1 = (X_1, Y_1, E_1), G_2 = (X_2, Y_2, E_2), \ldots, G_k = (X_k, Y_k, E_k)$ be connected bipartite graphs in D. The following cases can occur.

1. $\{|X_i|, |Y_i|\} = \{p_1, q_1\}$ for some $i \in \{1, \ldots, k\}$, and $\{|X_i|, |Y_i|\} = \{p_2, q_2\}$ for other $i \in \{1, \ldots, k\}$, where $\{p_1, q_1\} \neq \{p_2, q_2\}$ holds.
2. $\{|X_i|, |Y_i|\}$ is the identical set $\{p, q\}$ for every $i \in \{1, \ldots, k\}$.

First we consider the case 1. It is clear that $\max\{p_1, p_2, q_1, q_2\}$ is equal to $\max\{|X|, |Y|\}$, and $\min\{p_1, q_1, p_2, q_2\}$ is equal to $\min\{|X|, |Y|\} - 1$. Therefore, we can easily reconstruct $\{|X|, |Y|\}$, in this case.

Now, we consider the case 2. There are two more detailed cases. One case is that $|X| = |Y|$ (case 2a), and the other case is that every connected card is obtained by removing a vertex from one vertex set (case 2b).

In the case 2a, $\max\{p, q\} = \min\{p, q\} + 1 = |X| = |Y|$ holds. Thus, we can determine $\{|X|, |Y|\}$, if we can realize that the case is 2a, not 2b. We explain how to distinguish the case 2a from the case 2b later.

In the case 2b, let T be a spanning tree of G. Since a graph obtained from G by removing a leaf of T is connected, all the leaves of T belong to the same vertex set X or Y. We can assume without loss of generality that the vertex set is X. Then apparently $|X| > |Y|$ holds. Thus $\{|X|, |Y|\}$ is $\{\max\{p, q\} + 1, \min\{p, q\}\}$.

The remaining problem is how to distinguish the case 2a from the case 2b. In the case 2a, $|p - q| = 1$ always holds. Therefore, we consider the case that $|p - q| = 1$ holds in the case 2b. In this case, $|X| = |Y| + 2$ must hold.

Let L be a permutation diagram of G. Let x_l and x_r be polar vertices in X that correspond to the left-most segment, and the right-most segment of L, respectively. Let P be the shortest path in G from x_l to x_r. Let y_l be the vertex adjacent to x_l in P, and let y_r be the vertex adjacent to x_r in P. Consider the number of vertices in P. Since every vertex $y \in Y$ is a cut-vertex of G, every path from x_l to x_r passes y. Therefore, all the vertices in Y are in P. Hence, there exist $|Y| + 1$ X-vertices in P. Note that the graph induced from G by these X-vertices and vertices in Y is exactly P, since otherwise some vertex in Y is not a cut-vertex of G. Since we here consider the case that $|X| = |Y| + 2$ holds, there is only one vertex remaining not in P. Thus, the degree in G of at least one of y_l and y_r is exactly two. Assume that the degree of y_l is two. Removing

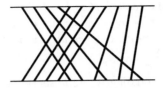

Fig. 4. An example of permutation diagram of a connected bipartite graph $G = (X, Y, E)$ with a vertex $x \in X$ whose degree is $|Y|$

y_1 from G results in two connected graphs. One is a graph with only one vertex x_1, and the other is the graph induced from G by $X \cup Y \setminus \{x_1, y_1\}$. The latter is a bipartite graph, and the difference of the numbers of vertices in the two vertex sets is exactly two. On the other hand, if there is a card consists of an isolated vertex and a connected component in the case 2a, the size of each vertex set of the connected component must be the same. Therefore, we can distinguish the two cases. □

Proof (Proof of Lemma 5). If $\min\{|X|, |Y|\} = 1$, G is a tree, and is thus reconstructible. Therefore we assume that $\min\{|X|, |Y|\} \geq 2$. Since x is adjacent to every vertex in Y, x is a polar vertex of G. Let L be a permutation diagram representing G. Assume without loss of generality that x corresponds to a line segment s in L whose lower-end is the left-most among all the lower-ends of the segments in L. Then each vertex in $X \setminus \{x\}$ corresponds to each segment that lays on the right-side of s in L. On the other hand, a segment s' corresponding to a vertex in Y must intersect with s. Therefore, the upper-end of s' must be on the left-side of that of s. Consider the segment s'' whose upper-end is the right-most among segments corresponding to vertices in Y. Then lower-end of s'' must be the right-most, since otherwise G cannot be connected. Hence, the vertex corresponding to s'' is adjacent to all the vertices in X. This means that if a bipartite permutation graph $G = (X, Y, E)$ has a vertex $x \in X$ satisfying $\deg(x) = |Y|$, then G also has a vertex $y \in Y$ satisfying $\deg(y) = |X|$. See Fig. 4 for an illustration.

   Now, we know that there are two special polar vertices x and y in G. There must be two other polar vertices. One corresponds to the segment in L whose upper-end is the left-most, and the other corresponds to the segment whose upper-end is the right-most. We denote the vertices by v and w, respectively. We assume without loss of generality $|X| \geq |Y|$. Removing $v(\in Y)$ results in a connected bipartite graph G'. The size of the vertex sets of G is $|X|$ and $|Y| - 1$. Thus, there is at least one connected card whose vertex sets have sizes $|X|$ and $|Y| - 1$. Moreover, since $|X| > |Y| - 1$, we can find such a card from the deck of G. Let G'' be a card of G whose vertex sets have sizes $|X|$ and $|Y| - 1$. Since the degree sequence of G is reconstructible, we can determine the degree of the vertex z, where G'' is obtained by removing z from G. Hence, we can determine the preimage uniquely. □

Lemma 12. *Given a connected bipartite permutation graph $G = (X, Y, E)$ satisfying Assumption 1, let x be a polar vertex in X, and let Y' be the set of vertices adjacent to x. A vertex $y \in Y'$ is a polar vertex of G if and only if y's degree is the minimum in Y'.*

Proof. Let L be a permutation diagram representing G. Since x is a polar vertex of G, we can assume without loss of generality that the line segment s in L whose upper-end is the left-most corresponds to x. Then, all the line segments corresponding to the vertices in $X \setminus \{x\}$ are at the right-side of s.

Let y^* be a polar vertex in Y'. Since G satisfies Assumption 1, a line segment s' in L corresponding to y^* is the left-most in those corresponding to vertices in Y'. Therefore, the degree of y^* is the minimum among the vertices in Y'.   □

Lemma 13. *Let $G' = (X', Y', E')$ be a connected bipartite permutation graph. Let x be a polar vertex in X'. Then, graph $G = (X, Y, E)$ that is obtained by adding a degree $k \in \{1, \ldots, |X'|\}$ vertex y to Y' is uniquely determined, under the conditions that G is a bipartite permutation graph, y is a polar vertex of G, and y is adjacent to x in G.*

Proof. Let L' be a permutation diagram representing G', and let L be a permutation diagram representing G. It is clear that L can be obtained by adding to L' a line segment s_y corresponding to y.

Since x is a polar vertex of G', we can assume without loss of generality that the line segment s_x in L' and L corresponding to x is the left-most among those corresponding to vertices in X. We can assume without loss of generality that the upper-end of s_x is the left-most among the upper-ends of all the segments in L. Since y is a polar vertex in G, s_y in L corresponding to y is the left-most among those corresponding to vertices in Y. That is, the lower-end of the s_y is the left-most among the lower-ends of all the segments in L. Then, we can determine the position of the upper-end of s_y uniquely, since s_y must intersect to exactly k segments in X.   □

Lemma 14. *The number of polar vertices whose degree is one is reconstructible for a connected bipartite permutation graph $G = (X, Y, E)$ satisfying Assumption 1.*

Proof. If a polar vertex v of G has degree one, the polar vertex adjacent to v has degree more than one, since otherwise G is disconnected.

When we remove a polar vertex that is adjacent to another polar vertex of degree one from G, we obtain a graph consisting some isolated vertices and a connected component. Conversely, if there is a graph consisting some isolated vertices and a connected component in the deck of G, this graph must be obtained from some preimage by removing a polar vertex adjacent to an other polar vertex of degree one. Otherwise, there must be at least two connected components. Thus, the number of polar vertices whose degree is one equal to the number of cards consisting some isolated vertices and a connected component.   □

5 Concluding Remarks

Combining Theorems 2 and 3, we have the main theorem. Note that since disconnected graphs are reconstructible, disconnected bipartite permutation graphs are of course reconstructible.

Theorem 4. *Bipartite permutation graphs are reconstructible.*

References

1. Bollobás, B.: Almost every graph has reconstruction number three. Journal of Graph Theory 14, 1–4 (1990)
2. Bondy, J.A.: On Ulam's conjecture for separable graphs. Pacific Journal of Mathematics 31, 281–288 (1969)
3. Bondy, J.A.: A graph reconstructor's manual. In: Surveys in Combinatorics. London Mathematical Society Lecture Note Series, vol. 166, pp. 221–252 (1991)
4. Even, S.: Algorithmic Combinatorics. Macmillan, New York (1973)
5. Giles, W.B.: The reconstruction of outerplanar graphs. Journal of Combinatorial Theory, Series B 16, 215–226 (1974)
6. Greenwell, D.L., Hemminger, R.L.: Reconstructing the n-connected components of a graph. Aequationes Mathematicae 9, 19–22 (1973)
7. Harary, F.: A survey of the reconstruction conjecture. In: Graphs and Combinatorics. Lecture Notes in Mathematics, vol. 406, pp. 18–28 (1974)
8. Hell, P., Huang, J.: Certifying LexBFS recognition algorithms for proper interval graphs and proper interval bigraphs. SIAM Journal on Discrete Mathematics 18, 554–570 (2005)
9. Kelly, P.J.: A congruence theorem for trees. Pacific Journal of Mathematics 7, 961–968 (1957)
10. Manvel, B.: Reconstruction of unicyclic graphs. In: Proof Techniques in Graph Theory. Academic Press, New York (1969)
11. McKay, B.D.: Small graphs are reconstructible. Australasian Journal of Combinatorics 15, 123–126 (1997)
12. von Rimscha, M.: Reconstructibility and perfect graphs. Discrete Mathematics 47, 283–291 (1983)
13. Saitoh, T., Otachi, Y., Yamanaka, K., Uehara, R.: Random generation and enumeration of bipartite permutation graphs. In: Zaroliagis, C. (ed.) ISAAC 2009. LNCS, vol. 5868, pp. 1104–1113. Springer, Heidelberg (2009)
14. Tutte, W.T.: On dichromatic polynomials. Journal of Combinatorial Theory 2, 310–320 (1967)
15. Tutte, W.T.: All king's horses. A guide to reconstruction. In: Graph Theory and Related Topics, pp. 15–33. Academic Press, New York (1979)

A Transformation from PPTL to S1S*

Cong Tian and Zhenhua Duan**

Institute of Computing Theory and Technology, and ISN Laboratory,
Xidian University, Xi'an, 710071, P.R. China

Abstract. A transformation from Propositional Projection Temporal Logic (PPTL) as well as Propositional Interval Temporal Logic (PITL) with infinite models to monadic second order logic with one successor (S1S) is presented in this paper. To this end, intervals where PPTL and PITL formulas are interpreted over are represented as $\mathfrak{T}$-structures. Further, the semantics of PPTL and PITL formulas are redefined over $\mathfrak{T}$-structures. Moreover, according to $\mathfrak{T}$-structure semantics, a PPTL or PITL formula is translated to a formula in S1S. As a result, many mature theoretical and technical results, such as decidability etc. for S1S can be easily inherited by PPTL and PITL.

Keywords: Propositional Projection Temporal Logic, S1S, Propositional Interval Temporal Logic, Decidability, Verification.

1 Introduction

Linear-time temporal logic has been developed into two categories, state based and interval based logics. The most famous state based linear-time temporal logic is Linear Temporal Logic (LTL) [1] and its variations while the most extensively investigated interval based temporal logic is Interval Temporal Logic (ITL) [4] and Projection Temporal Logic (PTL) [6]. Currently, LTL has been widely used in the specification and verification of concurrent systems. Particularly, it has been an important property specification language with model checking [10]. This is benefited from the early researching work on the theoretical aspects of Propositional LTL (PLTL) [13,9,14], especially the decidability result. Nevertheless, for interval-based temporal logics, even though it is more expressive and convenient for the specification, it has not been broadly used in the specification and verification of concurrent systems because of the decidability problem.

Within the community of interval-based temporal logics, several researchers have investigated the decidability of the logic with various kinds of extensions. Rosner and Pnueli proved the decidability of Choppy Logic [3] which is an extension of PLTL with chop operator. Halpern and Moszkowski [4] proved that satisfiability of Propositional Interval Temporal Logic (PITL) is undecidable, while Quantifier Propositional ITL (QPITL), a subset of PITL, over finite time is decidable. Kono presented a tableaux-based decision procedure for QPITL with projection [15]. Bowman and Thompson [12]

* This research is supported by the NSFC Grant No. 61003078, 60873018, 60910004, 60433010 and 61003079, 973 Program Grant No. 2010CB328102 and SRFDP Grant No. 200807010012.
** Corresponding author.

W. Wu and O. Daescu (Eds.): COCOA 2010, Part II, LNCS 6509, pp. 374–386, 2010.
© Springer-Verlag Berlin Heidelberg 2010

presented the first tableaux-based decision procedure for quantifier-free propositional ITL (PITL) over finite intervals with projection. Later, Gomez and Bowman [16] gave another executable MONA-based decision procedure for PITL with finite models. Decidability of interval based temporal logics confined in finite models is not enough for verification since many reactive systems are designed not to terminate. Accordingly, to verify those systems with interval based temporal logics, especially using model checking, decidability of these logics with infinite models are required. Motivated by this, Duan, Tian and Zhang proved the decidability of Propositional Projection Temporal Logic (PPTL) with infinite models [7]. It is well known that complementing infinite objects are difficult [18]. So the decision procedure is significant since complementing infinite words is explicitly involved in.

It has been proved that PLTL is equivalent to first order logic (FOL) [19,20]. Thus, some theoretical results of FOL can automatically be inherited by PLTL. Naturally, we can ask a question whether or not there is a well extensively investigated logic which could be equivalent PPTL? The answer is yes. Monadic second order logic with one successor (S1S, for short) is a well known second order logic with expressiveness of full regular expressions. The decidability result of S1S was given by Büchi by transforming S1S formulas to the notation of Büchi automata [17]. Theory of S1S have well been established. Therefore, if any PPTL formula can be equally transformed to an S1S formula, some matured theory and technique results, such as decision procedure etc. for S1S can be inherited by PPTL and PITL. To do so, intervals are first represented as $\mathfrak{T}$-structures. Further, the semantics of PITL or PPTL formulas are redefined over $\mathfrak{T}$-structures. Moreover, according to $\mathfrak{T}$-structure semantics, a PITL or PPTL formula is equivalently translated to a formula in S1S.

The rest parts of this paper are organized as follows. In the next section, the syntax and semantics of PPTL with infinite models are presented. S1S logic is briefly introduced in Section 3. In Section 4, we show how a PPTL formula can be translated to a formula in S1S. Finally, the conclusions are drawn in Section 5.

2   Propositional Projection Temporal Logic

Formula P of PPTL over a countable set of atomic propositions $Prop$ is inductively defined by the following grammar,

$$P ::= p \mid \neg P \mid P \vee Q \mid \bigcirc P \mid P^* \mid (P_1, ..., P_m) prj Q$$

where $p \in Prop$; $\bigcirc$, $*$ and prj are basic temporal operators.

Following the definition of Kripke's structure [2], we define a state s over $Prop$ to be a mapping from $Prop$ to $B = \{true, false\}$, $s : Prop \longrightarrow B$. We will use $s[p]$ to denote the valuation of p at state s. An interval σ is a non-empty sequence of states, which can be finite or infinite. The length, $|\sigma|$, of σ is ω if σ is infinite, and the number of states minus 1 if σ is finite. To have a uniform notation for both finite and infinite intervals, we will use extended integers as indices. That is, we consider the set N_0 of non-negative integers and ω, $N_\omega = N_0 \cup \{\omega\}$, and extend the comparison operators, $=, <, \leq$, to N_ω by considering $\omega = \omega$, and for all $i \in N_0$, $i < \omega$. Moreover, we define $\leq$ as $\leq -\{(\omega, \omega)\}$. To simplify definitions, we will denote σ by $< s_0, ..., s_{|\sigma|} >$, where

$s_{|\sigma|}$ is undefined if σ is infinite. With such a notation, $\sigma_{(i..j)}$ $(0 \le i \le j \le |\sigma|)$ denotes the sub-interval $< s_i, ..., s_j >$ and $\sigma^{(k)}$ $(0 \le k \le |\sigma|)$ denotes $< s_k, ..., s_{|\sigma|} >$. Further, the concatenation of a finite σ with another interval (or empty string) σ' is denoted by $\sigma \cdot \sigma'$. Let $\sigma = < s_0, s_1, ..., s_{|\sigma|} >$ be an interval and $r_1, ..., r_h$ be integers $(h \ge 1)$ such that $0 \le r_1 \le r_2 \le ... \le r_h \le |\sigma|$. The projection of σ onto $r_1, ..., r_h$ is the interval (namely projected interval)

$$\sigma \downarrow (r_1, ..., r_h) = < s_{t_1}, s_{t_2}, ..., s_{t_l} >$$

where $t_1, ..., t_l$ is obtained from $r_1, ..., r_h$ by deleting all duplicates. That is, $t_1, ..., t_l$ is the longest strictly increasing subsequence of $r_1, ..., r_h$. For instance,

$$< s_0, s_1, s_2, s_3, s_4 > \downarrow (0, 0, 2, 2, 2, 3) = < s_0, s_2, s_3 >$$

This is convenient to define an interval obtained by taking the endpoints (rendezvous points) of the intervals over which $P_1, ..., P_m$ are interpreted in the projection construct.

An interpretation is a tuple $I = (\sigma, k, j)$, where σ is an interval, k is an integer, and j an integer or ω such that $k \le j \le |\sigma|$. We use the notation $(\sigma, k, j) \models P$ to denote that formula P is interpreted and satisfied over the subinterval $< s_k, ..., s_j >$ of σ with the current state being s_k. The satisfaction relation ($\models$) is inductively defined as follows:

$(\sigma, k, |\sigma|) \models p$ iff $s_k[p] = true$, for any proposition p
$(\sigma, k, |\sigma|) \models \neg P$ iff $I \not\models P$
$(\sigma, k, |\sigma|) \models P \vee Q$ iff $I \models P$ or $I \models Q$
$(\sigma, k, |\sigma|) \models \bigcirc P$ iff $k < j$ and $(\sigma, k+1, |\sigma|) \models P$
$(\sigma, k, |\sigma|) \models P^*$ iff $|\sigma| = 0$ or there exists $k_0 = 0 \le k_1 \le ... \le k_m = |\sigma|$ such that
$\qquad (\sigma, k_i, k_{i+1}) \models P$ for all $i, 0 \le i < m$
$(\sigma, k, |\sigma|) \models (P_1, ..., P_m) \, prj \, Q$ if there exist integers $k = r_0 \le r_1 \le ... \le r_m \le j$ such
$\qquad$ that $(\sigma, r_0, r_1) \models P_1$, $(\sigma, r_{l-1}, r_l) \models P_l$, $1 < l \le m$, and $(\sigma', 0, |\sigma'|) \models Q$
$\qquad$ for one of the following σ' :
$\qquad$ (a) $r_m < j, \sigma' = \sigma \downarrow (r_0, ..., r_m) \cdot \sigma_{(r_m+1..j)}$
$\qquad$ (b) $r_m = j, \sigma' = \sigma \downarrow (r_0, ..., r_h)$ for some $0 \le h \le m$

The abbreviations $true, false, \wedge, \rightarrow$ and $\leftrightarrow$ are defined as usual. In particular, $true \stackrel{def}{=} P \vee \neg P$ and $false \stackrel{def}{=} P \wedge \neg P$ for any formula P. Also we have the following derived formulas:

$$empty \stackrel{def}{=} \neg \bigcirc true \qquad\qquad more \stackrel{def}{=} \neg empty$$
$$\bigcirc^0 P \stackrel{def}{=} P \qquad\qquad \bigcirc^n P \stackrel{def}{=} \bigcirc(\bigcirc^{n-1} P)$$
$$len(n) \stackrel{def}{=} \bigcirc^n empty \qquad\qquad skip \stackrel{def}{=} len(1)$$
$$\bigodot P \stackrel{def}{=} empty \vee \bigcirc P \qquad\qquad P; Q \stackrel{def}{=} (P, Q) \, prj \, empty$$
$$\Diamond P \stackrel{def}{=} true; P \qquad\qquad \Box P \stackrel{def}{=} \neg \Diamond \neg P$$

where $\Box$ (always), $\Diamond$ (sometimes) and ; (chop) are derived temporal operators; $empty$ denotes an interval with zero length; and $more$ means the current state is not the final one over an interval.

A formula P is satisfied by an interval σ, denoted by $\sigma \models P$, if $(\sigma, 0, |\sigma|) \models P$. A formula P is satisfiable if $\sigma \models P$ for some σ. A formula P is valid, denoted by $\models P$, if $\sigma \models P$ for all σ.

Note that the subset of PPTL without *projection* (but *chop* is included as a basic operator) is propositional interval temporal logic (PITL) given in 1995 [5] by extending the original PITL confined in finite models to infinite models after the appearance of PPTL in 1994 [11].

3 Monadic Second Order Logic with One Successor

S1S is a monadic second order logic with one successor. Let $V_1 = \{x, y, ...\}$ be a countable set of first-order variables, and $V_2 = \{X, Y, ...\}$ be a countable set of second-order variables. Term t and formula φ can inductively be defined as follows,

$$t ::= 0 \mid x \mid Suc(t)$$
$$\varphi ::= t \in X \mid t_1 = t_2 \mid \neg \varphi \mid \varphi_0 \vee \varphi_1 \mid \exists x.\varphi \mid \exists X.\varphi$$

where $Suc(t) = t + 1$. A first-order variable x is interpreted over N_ω, $I_1 : V_1 \to N_\omega$, while a second-order variable is interpreted as a subset of N_ω, $I_2 : V_2 \to 2^{N_\omega}$.

An interpretation for a term or a formula is a $\mathcal{I}_{I_1,I_2} = < m_1, ..., m_k, M_1, ..., M_l >$, where $m_i \in N_\omega$, $1 \le i \le k$, and $M_j \in 2^{N_\omega}$, $1 \le j \le l$. We use $\mathcal{I}_{I_1,I_2}$ to mean that a term or a formula is interpreted under the interpretation $I_1(x_i) = m_i$ and $I_2(X_j) = M_j$ for the first and second-order variables appearing in the term or formula. For every term t, the evaluation of t relative to interpretation $\mathcal{I}_{I_1,I_2}$ is defined as $\mathcal{I}_{I_1}[t]$ by induction on the structure of term,

$$\mathcal{I}_{I_1}[0] = 0$$
$$\mathcal{I}_{I_1}[x] = I_1(x)$$
$$\mathcal{I}_{I_1}[Suc(t)] = \mathcal{I}_{I_1}[t] + 1$$

The satisfaction relation for formulas $\models$ is inductively defined as follows,

$\mathcal{I}_{I_1,I_2} \models t \in X$ iff $\mathcal{I}_{I_1}[t] \in I_2(X)$

$\mathcal{I}_{I_1,I_2} \models t_1 = t_2$ iff $\mathcal{I}_{I_1}[t_1] = \mathcal{I}_{I_1}[t_2]$

$\mathcal{I}_{I_1,I_2} \models \neg\varphi$ iff $\mathcal{I}_{I_1,I_2} \not\models \varphi$

$\mathcal{I}_{I_1,I_2} \models \varphi_1 \vee \varphi_2$ iff $\mathcal{I}_{I_1,I_2} \models \varphi_1$ or $\mathcal{I}_{I_1,I_2} \models \varphi_2$

$\mathcal{I}_{I_1,I_2} \models \exists x.\varphi$ iff there is an $a \in N_\omega$ such that $I_1'(y) = \begin{cases} I_1(y), \text{ if } y \ne x \\ a, \text{ otherwise} \end{cases}$ and $\mathcal{I}_{I_1',I_2} \models \varphi$.

$\mathcal{I}_{I_1,I_2} \models \exists X.\varphi$ iff there is an $A \in 2^{N_\omega}$ such that $I_2'(Y) = \begin{cases} I_2(Y), \text{ if } Y \ne X \\ A, \text{ otherwise} \end{cases}$ and $\mathcal{I}_{I_1,I_2'} \models \varphi$.

The abbreviations $true$, $false$, $\wedge$, $\to$ and $\leftrightarrow$ can be derived as usual. In addition, many other useful abbreviations can be derived. The following abbreviations are used in this paper.

$$\forall X.\varphi \stackrel{def}{=} \neg\exists X.\neg\varphi$$
$$x \notin Y \stackrel{def}{=} \neg(x \in Y)$$
$$x \ne y \stackrel{def}{=} \neg(x = y)$$
$$X \subseteq Y \stackrel{def}{=} \forall z.(z \in X \to z \in Y)$$

$$X = Y \overset{def}{=} X \subseteq Y \wedge Y \subseteq X$$
$$Suff(X) \overset{def}{=} \forall y.(y \in X \rightarrow Suc(y) \in X)$$
$$x \leq y \overset{def}{=} \forall Z.(x \in Z \wedge Suff(Z) \rightarrow y \in Z)$$
$$Min(X) = x \overset{def}{=} x \in X \wedge \neg \exists y.(y \in X \wedge y < x))$$
$$Max(X) = x \overset{def}{=} x \in X \wedge \neg \exists y.(y \in X \wedge y > x))$$
$$x \leq y \overset{def}{=} (y \neq \omega \rightarrow x \leq y) \wedge (y = \omega \rightarrow x < y)$$
$$Con(K, k_m, k_n) \overset{def}{=} K \subseteq N_\omega \wedge (k_m, k_n \in K) \wedge k_m \leq k_n \wedge \neg \exists k_l \in K \wedge (k_m < k_l < k_n)$$

The meaning of most of the derived constructs is intuitive, e.g. $\forall X.\varphi$, $x \notin Y$, $x \neq y$, $X \subseteq Y$, $X = Y$, $x \leq y$, $Suff(X)$, $Min(X)$ and $Max(X)$. For $x \leq y$, it has the same meaning as the one given in section 2. $Con(K, k_m, k_n)$ means that k_n is the smallest successor of k_m in set $K \subseteq N_\omega$ or k_n equals to k_m.

An S1S formula $\varphi(x_1, ..., x_k, X_1, ..., X_l)$ is satisfiable by $\mathcal{I}_{l_1, l_2}$, denoted by $\mathcal{I}_{l_1, l_2} \models \varphi(x_1, ..., x_k, X_1, ..., X_l)$, if $\varphi(m_1, ..., m_k, M_1, ..., M_l)$ is $true$. A formula φ is satisfiable if $\mathcal{I}_{l_1, l_2} \models \varphi(x_1, ..., x_k, X_1, ..., X_l)$ for some $\mathcal{I}_{l_1, l_2}$. A formula P is valid, denoted by $\models \varphi$, if $\mathcal{I}_{l_1, l_2} \models \varphi(x_1, ..., x_k, X_1, ..., X_l)$ for all $\mathcal{I}_{l_1, l_2}$.

4   Transformation from PPTL to S1S

4.1   From Intervals to $\mathfrak{T}$-Structures

As mentioned before, an interval is defined as a sequence of states where each state is a set of propositions which hold at the state. From another point of view, a sequence of states can be denoted by a set N consisting all the subscripts of the states in the sequence. And a proposition can be viewed as a set consisting of subscripts of the states at which it holds over an interval. Formally, for a given interval $\sigma =< s_0, s_1, ... >$ and a proposition p we define,

$$N = \begin{cases} \{0, 1, ..., i\}, & \text{if } |\sigma| \neq \omega \text{ and the last state in } \sigma \text{ is } s_i. \\ N_\omega, & \text{otherwise} \end{cases}$$

and

$$Set(p, \sigma) = \begin{cases} \{i \mid s_i[p] = true, i \in N_0\}, & \text{if } |\sigma| \neq \omega \\ \{i \mid s_i[p] = true, i \in N_0\} \cup \{\omega\}, & \text{otherwise} \end{cases}$$

Note that for $\sigma = \sigma_1 \cdot \sigma_2$, $N = N_1 \cup N_2$.

Example 1. $\sigma_1 =< \{p\}, \{p, q\}, \{p, q\}, \{q\}, \{q\}, \{p\}, \{p\}, \{p\}, \{p\}, ... >$, where for $i = 0, 1, 2$ and $i \geq 5$, $i \in N_0$, $s_i[p] = true$, and for $i = 1, 2, 3$ and 4, $s_i[q] = true$ and $|\sigma_1| = \omega$. Therefore, $Set(p, \sigma_1) = \{\omega, 0, 1, 2, 5, 6, 7, 8, ...\}$ and $Set(q, \sigma_1) = \{1, 2, 3, 4\}$. Equivalently, we can denote the valuation of p at state s_i by 1 if $s_i[p] = true$ otherwise 0, thus interval σ_1 can be shown as follows,

p	1,	1,	1,	0,	0,	1,	1,	1,	1,	...
q	0,	1,	1,	1,	1,	0,	0,	0,	0,	...
	s_0,	s_1,	s_2,	s_3,	s_4,	s_5,	s_6,	s_7,	s_8,	...

In this way, we add i to $Set(p, \sigma)$ if the value of p is 1 at the ith state s_i. Accordingly, since proposition p holds on states $s_0, s_1, s_2, s_5, s_6, s_7, s_8, ...,$ and q only holds on states s_1, s_2, s_3 and s_4. So, we can also obtain $Set(p, \sigma) = \{\omega, 0, 1, 2, 5, 6, 7, 8, ...\}$ and $Set(q, \sigma) = \{1, 2, 3, 4\}$. □

Now, we represent an interval as a so called $\mathcal{I}$-structure, $\mathcal{I} =< i, j, N, \mathcal{P} >$, where $i, j \in N_\omega$, $N \subseteq N_\omega$, and $\mathcal{P} \subseteq 2^N$. Note that $\mathcal{I} =< i, j, N, \mathcal{P} >$ is a special case of $I_{l_1, l_2} =< m_1, ..., m_k, M_1, ..., M_l >$ with i and j being first order variables while $N \cup \mathcal{P}$ being the set of second order variables. Let Σ be the set of all intervals, and $\mathbb{T}$ be the set of all $\mathcal{I}$-structures. Function $\Theta : \Sigma \to \mathbb{T}$ is defined to map an interval σ to a $\mathcal{I}$-structure. For an interval $\sigma_{(i..j)}$,

$$\Theta(\sigma_{(i..j)}) =< i, j, N, \mathcal{P} >$$

where $\mathcal{P}$ is a set of second-order variables which are renamed by propositions appearing in the interval $\sigma_{(i..j)}$. Function $\theta : Prop \to \mathcal{P}$ is employed for renaming a proposition $p \in Prop$ as a second-order variable in S1S,

$$\theta(p) = X_p$$

Therefore,

$$\mathcal{P} = \{X_p \mid p \in Prop, \text{ and appears in } \sigma\}$$

and for each X_p, $I_2(X_p) = Set(\sigma, p)$.

Example 2. For $\sigma =< \{p\}, \{p, q\}, \{p, q\}, \{q\}, \{q\}, \{p\}, \{p\}, \{p\}, \{p\}, ... >$, where for $i = 0, 1, 2$ and $i \geq 5$, $i \in N_0$, $s_i[p] = true$, and for $i = 1, 2, 3$ and 4, $s_i[q] = true$.

$$\begin{aligned} &\Theta(\sigma_{(0..\omega)}) \\ &= < 0, \omega, N = N_\omega, \{X_p = \{\omega, 0, 1, 2, 5, 6, 7, 8, ...\}, \\ &\quad X_q = \{1, 2, 3, 4\}\} > \end{aligned}$$

□

Further, for projected interval $\sigma \downarrow (k_0, ..., k_m)$, $\Theta(\sigma \downarrow (k_0, ..., k_m)) =< k_0, k_m, N = \{k_0, ..., k_m\}, \mathcal{P} >$.

Example 3. For $\sigma_{(0..\omega)} \models (P_1, P_2, P_3) \, prj \, Q$ as depicted in Fig.1, $\Theta(\sigma_{(0..\omega)}) =< 0, \omega, N_\omega, \mathcal{P}_1 >$, $\mathcal{P}_1 = \{X_p \mid p \in Prop, \text{ and appears in } \sigma_{0..\omega}\}$. $\Theta((\sigma_{(0..\omega)} \downarrow \{0, 5, 7, 13\}) \cdot \sigma_{(14..\omega)}) =< 0, \omega, N, \mathcal{P}_2 >$ where $N = \{0, 5, 7, 13, 14, 15, ...\} \cup \{\omega\}$, $\mathcal{P}_2 = \{X_p \mid p \in Prop, \text{ and appears in } < s_0, s_5, s_7, s_{13}, s_{14}, s_{15}, ... >\}$. □

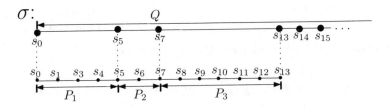

Fig. 1. $\sigma \models (P_1, P_2, P_3) \, prj \, Q$

4.2   From PPTL to S1S

Accordingly, PPTL formulas can be interpreted over $\mathfrak{I}$-structures as follows,

$< i, j, N, \mathcal{P} > \models_{\mathfrak{I}} p$ iff $\theta(p) \in \mathcal{P}$ and $i \in Set(p, \sigma_{(i..j)})$.

$< i, j, N, \mathcal{P} > \models_{\mathfrak{I}} \neg P$ iff $< i, j, N, \mathcal{P} > \not\models_{\mathfrak{I}} P$.

$< i, j, N, \mathcal{P} > \models_{\mathfrak{I}} P \vee Q$ iff $< i, j, N, \mathcal{P} > \models_{\mathfrak{I}} P$ or $< i, j, N, \mathcal{P} > \models_{\mathfrak{I}} Q$.

$< i, j, N, \mathcal{P} > \models_{\mathfrak{I}} \bigcirc P$ iff $i < j$ and $< i+1, j, N \setminus \{i\}, \mathcal{P} > \models_{\mathfrak{I}} P$.

$< i, j, N, \mathcal{P} > \models_{\mathfrak{I}} P; Q$ iff there exists $k \in N, i \le k \le j$ such that
$< i, k, \{i, ..., k\}, \mathcal{P} > \models_{\mathfrak{I}} P$ and $< k, j, \{k, ..., j\}, \mathcal{P} > \models_{\mathfrak{I}} Q$.

$< i, j, N, \mathcal{P} > \models_{\mathfrak{I}} P^*$ iff $i = j$ or there exist $k_0, k_1, ..., k_m \in N$, such that
$i = k_0 \le k_1 \le ... \le k_m = j$ and $< k_l, k_{l+1}, N, \mathcal{P} > \models_{\mathfrak{I}} P$
for all $0 \le l < m$.

$< i, j, N, \mathcal{P} > \models_{\mathfrak{I}} (P_1, ..., P_m) \ prj \ Q$ iff there exists $M = \{k_0, ..., k_m\} \subseteq N$, such that
$i = k_0 \le k_1 \le ... \le k_m \le j, < k_{l-1}, k_l, \{k_{l-1}, ..., k_l\},$
$\mathcal{P} > \models_{\mathfrak{I}} P_l$ for all $0 < l \le m$, and
$(a) < k_0, j, N', \mathcal{P} > \models_{\mathfrak{I}} Q$ and $N' = M \cup \{k_m + 1, k_m + 2, ..., j\}$, if $k_m < j$.
$(b) < k_0, k_m, N', \mathcal{P} > \models_{\mathfrak{I}} Q$ and $N' = M$, if $k_m = j$.

Satisfaction of PPTL formulas over $\mathfrak{I}$-structures can be shown to be necessary and sufficient w.r.t satisfaction over intervals, for both describe exactly the same models.

Theorem 1. For any $\sigma_{(i..j)} \in \Sigma$, and for any PPTL formula P, $\Theta(\sigma_{(i..j)}) = < i, j, N, \mathcal{P} > \models_{\mathfrak{I}}$ P iff $\sigma_{(i..j)} \models P$.

Proof. The proof proceeds by induction on structures of PPTL formulas.
Base case: $< i, j, N, \mathcal{P} > \models_{\mathfrak{I}} p$ iff $\sigma_{(i..j)} \models p$

$< i, j, N, \mathcal{P} > \models_{\mathfrak{I}} p$
iff $\theta(p) \in \mathcal{P}$ and $i \in \theta(p)$ Definition of $\models_{\mathfrak{I}}$
iff $s_i[p] = true$ Definition of $\mathcal{P}$
iff $\sigma_{(i..j)} \models p$ Definition of $\models$

Inductive step: Suppose, for any formula P and Q in PPTL, $< i, j, N, \mathcal{P} > \models_{\mathfrak{I}} P$ iff $\sigma_{(i..j)} \models P$, and $< i, j, N, \mathcal{P} > \models_{\mathfrak{I}} Q$ iff $\sigma_{(i..j)} \models Q$.

1. **Negation:** $< i, j, N, \mathcal{P} > \models_{\mathfrak{I}} \neg P$ iff $\sigma_{(i..j)} \models \neg P$

$< i, j, N, \mathcal{P} > \models_{\mathfrak{I}} \neg P$
iff $< i, j, N, \mathcal{P} > \not\models_{\mathfrak{I}} P$ Definition of $\models_{\mathfrak{I}}$
iff $\sigma_{(i..j)} \not\models P$ Hypothesis
iff $\sigma_{(i..j)} \models \neg P$ Definition of $\models$

2. **Disjunction:** $< i, j, N, \mathcal{P} > \models_{\mathfrak{I}} P \vee Q$ iff $\sigma_{(i..j)} \models P \vee Q$

$< i, j, N, \mathcal{P} > \models_{\mathfrak{I}} P \vee Q$
iff $< i, j, N, \mathcal{P} > \models_{\mathfrak{I}} P$ or $< i, j, N, \mathcal{P} > \models_{\mathfrak{I}}$ Definition of $\models_{\mathfrak{I}}$
iff $\sigma_{(i..j)} \models P$ or $\sigma_{(i..j)} \models Q$ Hypothesis
iff $\sigma_{(i..j)} \models P \vee Q$ Definition of $\models$

3. **Next:** $< i, j, N, \mathcal{P} > \models_{\mathfrak{I}} \bigcirc P$ iff $\sigma_{(i..j)} \models \bigcirc P$

$$< i, j, N, \mathcal{P} > \models_{\mathfrak{I}} \bigcirc P$$

iff $i < j$ and $< i + 1, j, N \setminus \{i\}, \mathcal{P} > \models_{\mathfrak{I}} P$	Definition of $\models_{\mathfrak{I}}$
iff $\sigma_{(i+1..j)} \models P$	Hypothesis
iff $\sigma_{(i..j)} \models \bigcirc P$	Definition of $\models$

4. **Chop:** $< i, j, N, \mathcal{P} > \models_{\mathfrak{I}} P; Q$ iff $\sigma_{(i..j)} \models P; Q$

$$< i, j, N, \mathcal{P} > \models_{\mathfrak{I}} P; Q$$

iff there exists $k \in N_0, k \leq j$ such that

$< i, k, \{i, ..., k\}, \mathcal{P} > \models_{\mathfrak{I}} P$ and $< k, j, \{k, ..., j\}, \mathcal{P} > \models_{\mathfrak{I}} Q$	Definition of $\models_{\mathfrak{I}}$

iff there exists $k \in N_0, k \leq j$ such that

$\sigma_{(i..k)} \models P$ and $\sigma_{(k..j)} \models Q$	Hypothesis
iff $\sigma_{(i..j)} \models P; Q$	Definition of $\models$

5. **Chop star:** $< i, j, N, \mathcal{P} > \models_{\mathfrak{I}} P^*$ iff $\sigma_{(i..j)} \models P^*$

$$< i, j, N, \mathcal{P} > \models_{\mathfrak{I}} P^*$$

iff $i = j$ or there exists $k_0, k_1, ..., k_m \in N_\omega$, such that $i = k_0 \leq k_1 \leq ... \leq k_m = j$ and $< k_l, k_{l+1}, \{k_l, ..., k_{l+1}\}, \mathcal{P} > \models_{\mathfrak{I}} P$ for all $0 \leq l < m$.	Definition of $\models_{\mathfrak{I}}$
iff $i = j$ or there exists $k_0, k_1, ..., k_m \in N_\omega$, such that $i = k_0 \leq k_1 \leq ... \leq k_m = j$ and $\sigma_{(k_l..k_{l+1})} \models P$ for all $0 \leq l < m$.	Hypothesis
iff $< i, j, N, \mathcal{P} > \models P^*$	Definition of $\models$

6. **Projection:** $< i, j, N, \mathcal{P} > \models_{\mathfrak{I}} (P_1, ..., P_m) \, prj \, Q$ iff $\sigma_{(i..j)} \models (P_1, ..., P_m) \, prj \, Q$

$$< i, j, N, \mathcal{P} > \models_{\mathfrak{I}} (P_1, ..., P_m) \, prj \, Q$$

iff there exist $M = \{k_0, ..., k_m\} \subseteq N$, such that $i = k_0 \leq k_1 \leq ... \leq k_m = j, < k_{l-1}, k_l, \{k_{l-1}, ..., k_l\}, \mathcal{P}_l > \models_{\mathfrak{I}} P_l$ for all $0 < l \leq m$, and $(a) < k_0, k_m, N', \mathcal{P}' > \models_{\mathfrak{I}} Q$ and $N' = M$, if $k_m = j$ or $(b) < k_0, j, N', \mathcal{P}' > \models_{\mathfrak{I}} Q$ and $N' = M \cup \{k_m + 1, k_m + 2, ..., j\}$, if $K_m < j$.	Definition of $\models_{\mathfrak{I}}$
iff there exist $M = \{k_0, ..., k_m\} \subseteq N$, such that $i = k_0 \leq k_1 \leq ... \leq k_m = j, \sigma_{(k_{l-1}..k_l)} \models P_l$ for all $0 < l \leq m$, and $(a) \sigma_{(i..j)} \downarrow (k_0, ..., k_m) \models Q$, if $k + m = j$ or $(b) \sigma_{(i..j)} \downarrow (k_0, ..., k_m) \cdot \sigma_{(k_m+1..j)} \models Q$, if $k_m < j$.	Hypothesis
iff $\sigma_{(i..j)} \models (P_1, ..., P_m) \, prj \, Q$	Definition of $\models$

So the theorem holds. $\square$

Now, the function for transforming a PPTL formula P interpreted over an interval $\sigma_{i..j}$, $i \in N_0$, $j \in N_\omega$, $i \leq j$, to an S1S formula is presented as follows.

$$toS\,1S\,(i, j, N, p)$$
$$= N \subseteq N_\omega \wedge i = Min(N) \wedge j = Max(N) \wedge \exists X_p.(X_p \subseteq N \wedge i \in X_p))$$

$$toS\,1S\,(i, j, N, \neg P)$$
$$= \neg toS\,1S\,(i, j, N, P)$$

$$toS\,1S\,(i, j, N, P \vee Q)$$
$$= toS\,1S\,(i, j, N, P) \vee toS\,1S\,(i, j, N, Q)$$

$$toS\,1S\,(i, j, N, \bigcirc P)$$
$$= N \subseteq N_\omega \wedge i = Min(N) \wedge j = Max(N) \wedge i < j \wedge toS\,1S\,(i + 1, j, N \setminus \{i\}, P)$$

$$toS\,1S\,(i, j, N, P; Q)$$
$$= N \subseteq N_\omega \wedge i = Min(N) \wedge j = Max(N) \wedge \exists k.(k \in N \wedge k \geq i \wedge k \leq j \wedge$$
$$toS\,1S\,(i, k, N \setminus \{k + 1, ..., j\}, P) \wedge toS\,1S\,(k, j, N \setminus \{i, ..., k - 1\}, Q))$$

$$toS\,1S\,(i, j, N, P^*)$$
$$= N \subseteq N_\omega \wedge i = Min(N) \wedge j = Max(N) \wedge$$
$$\exists K.(K \subseteq N \wedge j = Max(K) \wedge i = Min(K) \wedge ((i = j) \vee$$
$$\forall k_m, k_n \in K.(Con(K, k_m, k_n) \rightarrow toS\,1S\,(k_m, k_n, \{k_m, ..., k_n\}, P)))$$

$$toS\,1S\,(i, j, N, (P_1, ..., P_m)\,prj\,Q)$$
$$= N \subseteq N_\omega \wedge i = Min(N) \wedge j = Max(N) \wedge$$
$$\exists K.(K \subseteq N \wedge Max(K) \leq j \wedge i = Min(K) \wedge$$
$$\forall k_{l-1}, k_l \in K.(Con(K, k_{l-1}, k_l) \rightarrow (toS\,1S\,(k_{l-1}, k_l, \{k_{l-1}...k_l\}, P_l)) \wedge$$
$$((Max(K) < j) \rightarrow (toS\,1S\,(i, j, K \cup \{Max(K) + 1...j\}, Q))) \wedge$$
$$((Max(K) = j) \rightarrow (toS\,1S\,(i, Max(K), K, Q)))))$$

Theorem 2. For any formula P, $\sigma_{(i..j)} \models P$ iff $< i, j, \{i, ..., j\}, \mathcal{P} > \models toS\,1S\,(i, j, \{i, ..., j\}, P)$.

Proof. By Theorem 1, it has $\sigma_{(i..j)} \models_{\mathfrak{I}} P$ iff $< i, j, \{i, ..., j\}, \mathcal{P} > \models_{\mathfrak{I}} P$. So, we need to further prove that $< i, j, \{i, ..., j\}, \mathcal{P} > \models_{\mathfrak{I}} P$ iff $< i, j, \{i, ..., j\}, \mathcal{P} > \models toS\,1S\,(i, j, \{i, ..., j\}, P)$. The proof bases on the semantics of PPTL formulas interpreted over $\mathfrak{I}$-structures and proceeds on the induction of PPTL formulas.

Base case: $< i, j, N, \mathcal{P} > \models_{\mathfrak{I}} p$ iff $< i, j, N, \mathcal{P} > \models toS\,1S\,(i, j, N, p)$

$< i, j, N, \mathcal{P} > \models_{\mathfrak{I}} p$	
iff $X_p \in \mathcal{P}$ and $i \in X_p$	Definition of $\models_{\mathfrak{I}}$
iff $N \subseteq N_\omega \wedge i = Min(N) \wedge j = Max(N) \wedge$	
$\exists X_p.(X_p \in 2^N \wedge i \in X_p)$	Definition of $\Theta(\sigma_{(i..j)})$
iff $N \subseteq N_\omega \wedge i = Min(N) \wedge j = Max(N) \wedge$	
$\exists X_p.(X_p \subseteq N \wedge i \in X_p)$	
iff $< i, j, N, \mathcal{P} > \models toS\,1S\,(i, j, N, p)$	Definition of $toS\,1S\,(i, j, N, P)$

Inductive step: Suppose, for any formula P and Q in PPTL, $< i, j, N, \mathcal{P} > \models_{\mathfrak{I}} P$ iff $< i, j, N, \mathcal{P} > \models toS\,1S\,(i, j, N, P)$, and $< i, j, N, \mathcal{P} > \models_{\mathfrak{I}} Q$ iff $< i, j, N, \mathcal{P} > \models toS\,1S\,(i, j, N, Q)$.

1. **Negation:** $< i, j, N, \mathcal{P} > \models_{\mathfrak{I}} \neg P$ iff $< i, j, N, \mathcal{P} > \models toS\,1S\,(i, j, N, P)$

$< i, j, N, \mathcal{P} > \models_{\mathfrak{I}} \neg P$	
iff $< i, j, N, \mathcal{P} > \not\models_{\mathfrak{I}} P$	Definition of $\models_{\mathfrak{I}}$
iff $< i, j, N, \mathcal{P} > \not\models toS\,1S\,(i, j, N, P)$	Hypothesis
iff $< i, j, N, \mathcal{P} > \models toS\,1S\,(i, j, N, \neg P)$	Definition of $toS\,1S\,(i, j, N, P)$

2. **Disjunction:** $< i, j, N, \mathcal{P} > \models_{\mathfrak{I}} P \vee Q$ iff $< i, j, N, \mathcal{P} > \models toS1S(i, j, N, P \vee Q)$

$< i, j, N, \mathcal{P} > \models_{\mathfrak{I}} P \vee Q$

iff $< i, j, N, \mathcal{P} > \models_{\mathfrak{I}} P$ or $< i, j, N, \mathcal{P} > \models_{\mathfrak{I}} Q$ Definition of $\models_{\mathfrak{I}}$

iff $< i, j, N, \mathcal{P} > \models toS1S(i, j, N, P)$ or

$< i, j, N, \mathcal{P} > \models toS1S(i, j, N, Q)$ Hypothesis

iff $< i, j, N, \mathcal{P} > \models toS1S(i, j, N, P \vee Q)$ Definition of $toS1S(i, j, N, P)$

3. **Next:** $< i, j, N, \mathcal{P} > \models_{\mathfrak{I}} \bigcirc P$ iff $< i, j, N, \mathcal{P} > \models toS1S(i, j, N, \bigcirc P)$

$< i, j, N, \mathcal{P} > \models_{\mathfrak{I}} \bigcirc P$

iff $i < j$ and $< i+1, j, N \setminus \{i\}, \mathcal{P} > \models_{\mathfrak{I}} P$.   Definition of $\models_{\mathfrak{I}}$

iff $N \subseteq N_\omega \wedge i = Min(N) \wedge j = Max(N) \wedge i < j \wedge$

$< i+1, j, N, \mathcal{P} > \models_{\mathfrak{I}} P$ Definition of $\Theta(\sigma_{(i..j)})$

iff $N \subseteq N_\omega \wedge i = Min(N) \wedge j = Max(N) \wedge i < j \wedge$

$toS1S(i+1, j, N \setminus \{i\}, P)$ Hypothesis

iff $< i, j, N, \mathcal{P} > \models toS1S(i, j, N, \bigcirc P)$ Definition of $toS1S(i, j, N, P)$

4. **Chop:** $< i, j, N, \mathcal{P} > \models_{\mathfrak{I}} P; Q$ iff $< i, j, N, \mathcal{P} > \models toS1S(i, j, N, P; Q)$

$< i, j, N, \mathcal{P} > \models_{\mathfrak{I}} P; Q$

iff there exists $k \in N, i \leq k \leq j$ such that

$< i, k, \{i, ..., k\}, \mathcal{P} > \models_{\mathfrak{I}} P$

and $< k, j, \{k, ..., j\}, \mathcal{P} > \models_{\mathfrak{I}} Q$.   Definition of $\models_{\mathfrak{I}}$

iff $N \subseteq N_\omega \wedge i = Min(N) \wedge j = Max(N) \wedge \exists k.$

$(k \geq i \wedge k \leq j \wedge < i, k, \{i, ..., k\}, \mathcal{P} > \models_{\mathfrak{I}} P \wedge$

$< k, j, \{k, ..., j\}, \mathcal{P} > \models_{\mathfrak{I}} Q)$ Definition of $\Theta(\sigma_{(i..j)})$

iff $N \subseteq N_\omega \wedge i = Min(N) \wedge j = Max(N) \wedge \exists k.(k \geq i \wedge$

$k \leq j \wedge < i, k, \{i, ..., k\}, \mathcal{P} > \models toS1S(i, k, \{i, ..., k\}, P)$

$\wedge < k, j, \{k, ..., j\}, \mathcal{P} > \models toS1S(k, j, \{k, ..., j\}, Q)$ Hypothesis

iff $< i, j, N, \mathcal{P} > \models toS1S(i, j, N, P; Q)$ Definition of $toS1S(i,j,N,P)$

5. **Chop star:** $< i, j, N, \mathcal{P} > \models_{\mathfrak{I}} P^*$ iff $< i, j, N, \mathcal{P} > \models toS1S(i, j, N, P^*)$

$< i, j, N, \mathcal{P} > \models_{\mathfrak{I}} P^*$

iff $i = j$ or there exist $k_0, k_1, ..., k_m \in N$, such that

$i = k_0 \leq k_1 \leq ... \leq k_m = j$ and

$< k_l, k_{l+1}, \{k_l, ..., k_{l+1}\}, \mathcal{P} > \models_{\mathfrak{I}} P$ for all $0 \leq l < m$. Definition of $\models_{\mathfrak{I}}$

iff $N \subseteq N_\omega \wedge i = Min(N) \wedge j = Max(N) \wedge$

$\exists K.(K \subseteq N \wedge j = Max(K) \wedge i = Min(K) \wedge$

$((i = j) \vee \forall k_m, k_n \in K.(Con(K, k_m, k_n) \rightarrow$

$< k_m, k_n, \{k_m, ..., k_n\}, \mathcal{P} > \models P)))$ Definition of $\Theta(\sigma_{(i..j)})$

iff $N \subseteq N_\omega \wedge i = Min(N) \wedge j = Max(N) \wedge$

$\exists K.(K \subseteq N \wedge j = Max(K) \wedge i = Min(K) \wedge$

$((i = j) \vee \forall k_m, k_n \in K.(Con(K, k_m, k_n) \rightarrow$

$toS1S(k_m, k_n, \{k_m, ..., k_n\}, P)))$ Hypothesis

iff $< i, j, N, \mathcal{P} > \models toS1S(i, j, N, P^*)$ Definition of $toS1S(i, j, N, P)$

6. **Projection:** $< i, j, N, \mathcal{P} >\models_{\mathfrak{I}} (P_1, ..., P_m)prjQ$ iff $< i, j, N, \mathcal{P} >\models toS1S(i, j, N,$ $(P_1, ..., P_m)prjQ)$

$< i, j, N, \mathcal{P} >\models_{\mathfrak{I}} (P_1, ..., P_m)prjQ$
iff there exists $M = \{k_0, ..., k_m\} \subseteq N$, such that $i = k_0 \leq$
$k_1 \leq ... \leq k_m \leq j, < k_{l-1}, k_l, \{k_{l-1}, ..., k_l\}, \mathcal{P} >\models_{\mathfrak{I}} P_l$
for all $0 < l \leq m$, and
$(a) < k_0, j, N', \mathcal{P} >\models_{\mathfrak{I}} Q$ and $N' = M \cup \{k_m + 1,$
$\quad k_m + 2, ..., j\}$, if $k_m < j$.
$(b) < k_0, k_m, N', \mathcal{P} >\models_{\mathfrak{I}} Q$ and $N' = M$, if $k_m = j$.   Definition of $\models_{\mathfrak{I}}$
iff $N \subseteq N_\omega \wedge i = Min(N) \wedge j = Max(N) \wedge$
$\exists K.(K \subseteq N \wedge Max(K) \leq j \wedge i = Min(K) \wedge$
$\forall k_{l-1}, k_l \in K.(Con(K, k_{l-1}, k_l) \rightarrow$
$(toS1S(k_{l-1}, k_l, \{k_{l-1}, ..., k_l\}, P_l)) \wedge$
$((Max(K) < j) \rightarrow (toS1S(i, j, K \cup \{Max(K) + 1, ..., j\}, Q)))$
$((Max(K) = j) \rightarrow (toS1S(i, Max(K), K, Q)))))$ Definition of $\Theta_{(i..j)}$
iff $N \subseteq N_\omega \wedge i = Min(N) \wedge j = Max(N) \wedge$
$\exists K.(K \subseteq N \wedge Max(K) \leq j \wedge i = Min(K) \wedge$
$\forall k_{l-1}, k_l \in K.(Con(K, k_{l-1}, k_l) \rightarrow$
$(toS1S(k_{l-1}, k_l, \{k_{l-1}, ..., k_l\}, P_l)) \wedge$
$((Max(K) < j) \rightarrow (toS1S(i, j, K \cup \{Max(K) + 1, ..., j\}, Q)))$
$((Max(K) = j) \rightarrow (toS1S(i, Max(K), K, Q)))))$ Hypothesis
iff $< i, j, N, \mathcal{P} >\models toS1S(i, j, N, (P_1, ..., P_m)prjQ)$ Definition of $toS1S(i, j, N, P)$
So the theorem holds.   □

Accordingly, an arbitrary PPTL formula P can be transformed to a S1S formula by,

$$Tr(P) = \exists i, j, N. ((i \in N_0 \wedge j \in N_\omega \wedge N \subseteq N_\omega) \wedge toS1S(i, j, N, P))$$

Example 4. Transform $\bigcirc p; q$ to an S1S formula.

$Tr(\bigcirc p; q) = \exists i, j, N. ((i \in N_0 \wedge j \in N_\omega \wedge N \subseteq N_\omega) \wedge toS1S(i, j, N, \bigcirc p; q))$
$\quad = \exists i, j, N. ((i \in N_0 \wedge j \in N_\omega \wedge N \subseteq N_\omega) \wedge$
$\quad\quad (i = Min(N) \wedge j = Max(N) \wedge \exists k.(k \in N \wedge k \geq i \wedge k \leq j \wedge$
$\quad\quad toS1S(i, k, \{i, ..., k\}, \bigcirc p) \wedge toS1S(k, j, \{k, ..., j\}, q))))$
$\quad = \exists i, j, N. ((i \in N_0 \wedge j \in N_\omega \wedge N \subseteq N_\omega) \wedge$
$\quad\quad (i = Min(N) \wedge j = Max(N) \wedge \exists k.(k \in N \wedge k \geq i \wedge k \leq j \wedge$
$\quad\quad (\{i, ..., k\} \subseteq N_\omega \wedge i = Min(\{i, ..., k\}) \wedge k = Max(\{i, ..., k\}) \wedge i < k \wedge$
$\quad\quad toS1S(i + 1, k, \{i, ..., k\} \setminus \{i\}, p)) \wedge$
$\quad\quad (\{k, ..., j\} \subseteq N_\omega \wedge k = Min(\{k, ..., j\}) \wedge j = Max(\{k, ..., j\}) \wedge$
$\quad\quad \exists X_p.(X_p \subseteq \{k, ..., j\} \wedge k \in X_p)))))$
$\quad = \exists i, j, N. ((i \in N_0 \wedge j \in N_\omega \wedge N \subseteq N_\omega) \wedge$
$\quad\quad (i = Min(N) \wedge j = Max(N) \wedge \exists k.(k \in N \wedge k \geq i \wedge k \leq j \wedge$
$\quad\quad (\{i, ..., k\} \subseteq N_\omega \wedge i = Min(\{i, ..., k\}) \wedge k = Max(\{i, ..., k\}) \wedge i < k \wedge$
$\quad\quad (\{i + 1, ..., k\} \subseteq N_\omega \wedge i + 1 = Min(\{i + 1, ..., k\}) \wedge k = Max(\{i + 1, ..., k\}) \wedge$
$\quad\quad \exists X_p.(X_p \subseteq \{i + 1, ..., k\} \wedge i + 1 \in X_p)))) \wedge$
$\quad\quad (\{k, ..., j\} \subseteq N_\omega \wedge k = Min(\{k, ..., j\}) \wedge j = Max(\{k, ..., j\}) \wedge$
$\quad\quad \exists X_p.(X_p \subseteq \{k, ..., j\} \wedge k \in X_p))))$ □

5   Other Transformations

To directly obtain a PPTL formula from an S1S formula seems to be difficult. How-ever, an indirect transformation exists with Büchi automata and omega regular expressions being the bridges. Büchi provided a transformation from S1S formulas to Büchi automata [17]. In [21], further transformations from Büchi automata to omega regular expressions, and omega regular expressions to PPTL formulas were presented. So, given an S1S formula, an equivalent PPTL formula can also be obtained. Currently, the related transformations available are illustrated in Fig.2.

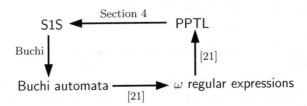

Fig. 2. Existing transformations

6   Conclusions

As presented in this paper, for any formula in PITL and PPTL, it can be equivalently translated to an S1S formula. Thus, PITL and PPTL are decidable since S1S is de-cidable. The decision procedure of S1S formulas mainly is based on a transforming procedure from S1S formulas to Büchi automata. So, for PITL and PPTL, a decision procedure given by translating formulas in PITL and PPTL to S1S formulas, then trans-form the S1S formulas to Büchi automata.

Decidability of PPTL and PITL is significant in practice since it makes it possible to model check interval based temporal logics. This is useful in practice since PPTL and PITL with infinite models have the expressiveness of full omega regular expressions, and can express the typical properties such as "even(p)", which cannot be specified by both PLTL and CTL.

In [21], a transformation from omega regular expressions to PPTL formulas was give. So PPTL has the same expressiveness with S1S. However, it can be easily observed that the resulting S1S formula of $Tr(P)$ is much longer than the original PPTL formula P. Thus, PPTL is more succinct than S1S.

References

1. Pnueli, A.: The temporal logic of programs. In: Proceedings of the 18th IEEE Symposium on Foundations of Computer Science, pp. 46–67.2. IEEE, New York (1977)
2. Kripke, S.A.: Semantical analysis of modal logic I: normal propositional calculi. Z. Math. Logik Grund. Math. 9, 67–96 (1963)
3. Rosner, R., Pnueli, A.: A choppy logic. In: First Annual IEEE Symposium on Logic In Computer Science, LICS, pp. 306–314 (1986)

4. Moszkowski, B.: Reasoning about digital circuits. Ph.D Thesis, Department of Computer Science, Stanford University. TRSTAN-CS-83-970 (1983)
5. Moszkowski, B.C.: Compositional reasoning about projected and infinite time. In: Proceeding of the First IEEE Int'l Conf. on Enginneering of Complex Computer Systems (ICECCS 1995), pp. 238–245. IEEE Computer Society Press, Los Alamitos (1995)
6. Duan, Z.: An Extended Interval Temporal Logic and A Framing Technique for Temporal Logic Programming. PhD thesis, University of Newcastle Upon Tyne (May 1996)
7. Duan, Z., Tian, C., Zhang, L.: A Decision Procedure for Propositional Projection Temporal Logic with Infinite Models. Acta Informatica 45(1), 43–78 (2008)
8. Tian, C., Duan, Z.: Model Checking Propositional Projection Temporal Logic Based on SPIN. In: Butler, M., Hinchey, M.G., Larrondo-Petrie, M.M. (eds.) ICFEM 2007. LNCS, vol. 4789, pp. 246–265. Springer, Heidelberg (2007)
9. Wolper, P.L.: Temporal logic can be more expressive. Information and Control 56, 72–99 (1983)
10. Holzmann, G.J.: The Model Checker Spin. IEEE Trans. on Software Engineering 23(5), 279–295 (1997)
11. Duan, Z., Koutny, M., Holt, C.: Projection in temporal logic programming. In: Pfenning, F. (ed.) LPAR 1994. LNCS, vol. 822, pp. 333–344. Springer, Heidelberg (1994)
12. Bowman, H., Thompson, S.: A decision procedure and complete axiomatization of interval temporal logic with projection. Journal of logic and Computation 13(2), 195–239 (2003)
13. Vardi, M., Wolper, P.: Reasoning about infinite computations. Information and Computation 115(1), 1–37 (1994)
14. Kesten, Y., Manna, Z., McGuire, H., Pnueli, A.: A decision algorithm for full propositional temporal logic. In: Courcoubetis, C. (ed.) CAV 1993. LNCS, vol. 697, Springer, Heidelberg (1993)
15. Kono, S.: A combination of clausal and non-clausal temporal logic programs. In: Fisher, M., Owens, R. (eds.) IJCAI-WS 1993. LNCS (LNAI), vol. 897, pp. 40–57. Springer, Heidelberg (1995)
16. Gomez, R., Bowman, H.: A MONA-based Decision Procedure for Propositional Interval Temporal Logic. In: Workshop of Interval Temporal Logics and Duration Calculi (part of the 15th European Summer School in Logic, Language and Information (August 2003)
17. Büchi, J.R.: On a decision method in restricted second order arithmetic. In: Nagel, E. (ed.) Proceedings of the 1960 International Congress on Logic, Methodology and Philosophy of Science, pp. 1–12. Stanford University Press, Stanford (1960)
18. Vardi, M.Y.: The Büchi Complementation Saga. In: Thomas, W., Weil, P. (eds.) STACS 2007. LNCS, vol. 4393, pp. 12–22. Springer, Heidelberg (2007)
19. Gabbay, D., Pnueli, A., Shelah, S., Stavi, J.: On the temporal analysis of fairness. In: Proceedings of the 7th ACM SIGPLAN-SIGACT Symposium on Principles of Programming Languages, POPL 1980, pp. 163–173. ACM Press, New York (1980)
20. McNaughton, R., Papert, S.A.: Counter-Free Automata. M.I.T research monograph, vol. 65. The MIT Press, Cambridge (1971)
21. Tian, C., Duan, Z.: Propositional Projection Temporal Logic, Büchi Automata and ω-Expressions. In: Agrawal, M., Du, D.-Z., Duan, Z., Li, A. (eds.) TAMC 2008. LNCS, vol. 4978, pp. 47–58. Springer, Heidelberg (2008)

Exact and Parameterized Algorithms for Edge Dominating Set in 3-Degree Graphs

Mingyu Xiao*

School of Computer Science and Engineering
University of Electronic Science and Technology of China
Chengdu 610054, P.R. China
myxiao@gmail.com

Abstract. Given a graph $G = (V, E)$, the *edge dominating set* problem is to find a minimum set $M \subseteq E$ such that each edge in $E - M$ has at least one common endpoint with an edge in M. The edge dominating set problem is an important graph problem and has been extensively studied. It is well known that the problem is NP-hard, even when the graph is restricted to a planar or bipartite graph with maximum degree 3. In this paper, we show that the edge dominating set problem in graphs with maximum degree 3 can be solved in $O^*(1.2721^n)$ time and polynomial space, where n is the number of vertices in the graph. We also show that there is an $O^*(2.2306^k)$-time polynomial-space algorithm to decide whether a graph with maximum degree 3 has an edge dominating set of size k or not. Above two results improve previously known results on exact and parameterized algorithms for this problem.

Keywords: Edge Dominating Set, Exact Algorithm, Parameterized Algorithm, Cubic Graph.

1 Introduction

Since we donot know whether $P = NP$ or not, currently the best we can exactly solve NP-complete problems is super-polynomial time algorithms. Although most NP-complete problems have exhaustive search algorithms, the forbiddingly large running time of the search algorithms makes them impractical even on instances of fairly small size. People wonder whether we can design algorithms that are significantly faster than trivial exhaustive search, though they are still not polynomial-time. Research on exponential-time algorithms for some natural and basic problems, such as independent set [1,2], coloring [3], exact satisfiability [4] and so on, has a long history. Recently, some other basic graph problems, such as dominating set [5], edge dominating set [6] and feedback set [7], also draw much attention in this line of research. Furthermore, to get more understanding of the structural properties of NP-complete problems, people also have

---

* This work was supported in part by National Natural Science Foundation of China Grant No. 60903007.

W. Wu and O. Daescu (Eds.): COCOA 2010, Part II, LNCS 6509, pp. 387–400, 2010.

interests in exactly solving problems in sparse and low-degree graphs. The independent set problem in 3-degree graphs can be solved in $O^*(1.0854^n)$ time [8], the k-vertex cover problem in 3-degree graphs can be solved in $O^*(1.1616^k)$ time [9], and TSP in cubic graphs can be solved in $O^*(1.251^n)$ time [10]. More fresh results on problems in sparse and low-degree graphs can be found in the literature [11,12,13,14]. To a certain extent, some graph problems in low-degree graphs are the bottleneck of improving the algorithms for the problems in general graphs. Motivated by those, in this paper, we study the classic edge dominating set problem in 3-degree graphs and present some improved algorithms for it. Except exact algorithms, we also study *parameterized algorithms* for this problem. In parameterized algorithms, we first pick up a parameter of the problem (the parameter can be the size of the solution, number of the vertices of the input graph, treewidth of the input graph, and so on), and try to design algorithms such that the exponential part of the running time is only related to the parameter (but not the whole input size). We can regard parameterized algorithms as a kind of exact algorithms. Parameterized algorithms for some basic graph problems, including the edge dominating set problem, have been extensively studied recently. For more details about parameterized algorithms, readers can refer to recent monographs [15]. In this paper, we will take k, the size of the solution to the edge dominating set problem, as the parameter to study the parameterized algorithms.

The edge dominating set problem is a basic problem introduced in Garey and Johnson's work [16] on NP-completeness. Yannakakis and Gavril [17] proved that the edge dominating set problem is NP-hard even in planar or bipartite graphs of maximum degree 3. Randerath and Schiermeyer [18] designed the first nontrivial exact algorithm for the minimum edge dominating set problem, which runs in $O^*(1.4423^m)$ time, where m is the number of edges in the graph. Later Raman *et al.* [19] improved the result to $O^*(1.4423^n)$ and Fomin *et al.* [20] further improved to $O^*(1.4082^n)$. Currently, the best result is Rooij and Bodlaender's $O^*(1.3226^n)$-time algorithm [6], which is analyzed by using 'measure and conquer'. In terms of parameterized algorithms with parameter k being the size of the solution, Fernau [21] gave an $O^*(2.6181^k)$-time algorithm and Fomin *et al.* [20] improved the result to $O^*(2.4181^k)$.

Almost all above algorithms are based on the idea of enumerating minimal vertex covers. We first find out a set C of vertices, which is the vertex set of a minimum edge dominating set, and then find a maximum matching in the induced graph $G[C]$. To get vertex cover C, usually we search in the following way: picking up a vertex of highest degree and branching into two branches by including it into the vertex cover or excluding it from the vertex cover. This method is also an effective way to solve the independent set and vertex cover problems. Although we have many useful techniques for the independent set and vertex cover problems in low-degree graphs and can solve these two problems in 3-degree graphs very fast, few of the techniques can be used in the edge dominating set problem. All previous algorithms for the edge dominating set problem in general graphs can not be improved directly when we restrict the graphs to 3-degree graphs. In this paper, we design faster exact and parameterized algorithms for the edge dominating set problem in 3-degree graphs. Our exact algorithm is the first effective algorithm that are not based on enumerating minimal vertex covers and the algorithm is analyzed by measuring the number of degree-3 vertices instead of the number of vertices or edges. We also show that the exact algorithm can be used to derive an improved parameterized algorithm for our parameterized problem.

The rest of the paper is organized as follows: Section 2 gives the basic definitions and our notation system. Section 3 gives some reduction rules which will be used as preprocesses to simplified the input graphs. Section 4 gives our exact algorithm for the edge dominating set problem in 3-degree graphs. Section 5 designs the algorithm for the parameterized version of this problem. Finally Section 6 makes the conclusion.

2   Preliminaries

Let $G = (V, E)$ be a graph with $n = |V|$ vertices and $m = |E|$ edges. A graph is called a 3-*degree graph* if any vertex in the graph is of degree ≤ 3. We may use p to denote the number of degree-3 vertices in a 3-degree graph. Let $G[V']$ be the subgraph induced by a subset $V' \subseteq V$, and $V(G)$ be the vertex set of graph G. A subset $S \subseteq E$ is called an *edge dominating set* of the graph if every edge in $E - S$ has at least a common endpoint with an edge in S. The *edge dominating set problem* is to find an edge dominating set of minimum size. We also define the *annotated edge dominating set problem*, in which, given a graph G with a subset $V_0 \subseteq V$, we are asked to find a minimum edge dominating set M such that $V_0 \subseteq V(M)$. We will also call the vertices in the given subset V_0 *annotated vertices*. For a vertex v, we will use $N(v)$ to denote the neighbor set of v, i.e., the set of vertices adjacent to v. Let $N[v] = N(v) \cup \{v\}$ denote the *closed neighbor set* of v. An ordered list of distinct vertices $v_1 v_2 \cdots v_i$ is called a *path* of length i, if for all $2 \leq j \leq i$ there is an edge between v_{j-1} and v_j. We may use (i)-*path* to denote a path of length i. A path $v_1 v_2 \cdots v_i$ is called an *inner path*, if $v_2, v_3, \cdots, v_{i-1}$ are degree-2 vertices and v_1 and v_i are not degree-1 vertices. Specially, if v_1 and v_i are vertices of degree ≥ 3, we also call such inner path a *pure path*. We also say that pure path $v_1 v_2 \cdots v_i$ is *incident on* v_1 (or v_i). In this paper, we will use a modified O notation that suppresses all polynomially bounded factors. For two functions f and g, we write $f(n) = O^*(g(n))$ if $f(n) = g(n)n^{O(1)}$.

The algorithms presented in this paper are based on the branch-and-reduce paradigm. We first apply some reduction rules to reduce the size of the input. Then we apply some branching rules to branch on the graph by including some vertices (or edges) into the solution set or excluding some vertices (or edges) from the solution set. In each branch, we will delete some vertices and get an annotated edge dominating set problem in a graph with a smaller measure (the measure can be the number of vertices or edges or others). When a vertex is removed, we also think that all the edges incident on it are removed. Assume that we branch on graph G with measure p into several graphs $G_1, G_2, \cdots, G_r$ and the measure of graph G_i is p_i ($i = 1, 2, \cdots, r$). Let $C(p)$ be the worst size of the search tree in the algorithm when the graph has measure p, then we get the recurrence relation $C(p) \leq \sum_{i=1}^{r} C(p_i)$. Solving the recurrence, we get $C(p) = [\alpha(p_1, p_2, \cdots, p_r)]^p$, where $\alpha(p_1, p_2, \cdots, p_r)$ is the largest root of the function $f(x) = 1 - \sum_{i=1}^{r} x^{-p_i}$. In the next section, we first introduce our reduction rules.

3   Reduction Rules

Reduction rules are frequently used as preprocesses to reduce the input size. We can apply the rules to transform an instance to an equivalent smaller instance (an instance with smaller measure) in polynomial time. There are several nice reduction rules for the vertex cover problem to deal with degree-1 and degree-2 vertices, by using which we can design fast algorithms for the vertex cover problem in 3-degree graphs and general graphs. However, few of the rules can be extended to the edge dominating set problem in spite of the similarity between the two problems. In this section, we will present some reduction rules to deal with some special cases of the degree-1 and degree-2 vertices for the annotated edge dominating set problem. Recall that annotated vertices are required to be endpoints of edges in the edge dominating set. Our reduction rules are given as follows.

Rule 1: Folding degree-1 vertices
Let v be a degree-1 vertex in the graph and u the unique neighbor of it.

Case 1: *If v is an annotated vertex, we put edge vu into the edge dominating set and remove v and u (also all the edges incident on them and degree-0 vertices) from the graph;*
Case 2: *If v is not an annotated vertex, we remove vertex v from the graph and annotate vertex u.*

It is easy to see the correctness of Rule 1. To deal with degree-2 vertices, we have the following two reduction rules.

Rule 2: Folding some special triangles
Let v be a degree-2 vertex in the graph, and u and w the two neighbors of it. If v is not annotated and u and w are adjacent, then we remove v from the graph and annotate u and w.
Proof. If vu or vw is in the minimum edge dominating set S, we can simply replace it with uw in S to get another minimum edge dominating set. Then we can assume that $vu, vw \notin S$ and $u, w \in V(S)$. Therefore, we can delete v and annotate u and w in the graph.   ∎

Rule 3: Folding inner paths of length 4
Let $abcde$ be an inner path in the graph. We can get the following reduction rules that depend on b, c, d annotated or not (please see Fig. 1 for the illustration):

Case 1: *If none of b, c, d is an annotated vertex, we delete b, c, d, and introduce a new edge ae if there is not an edge between a and e;*
Case 2: *If only b (resp. d) is an annotated vertex, we delete b, c, d, and introduce a new edge ae if there is not an edge between a and e, and annotate e (resp. a);*

Case 3: *If only b and d are annotated vertices, we put ab and de into the edge dominating set and remove a, b, c, d, e from the graph;*
Case 4: *If all of b, c, d are annotated vertices, we remove b, c, d and introduce a new vertex v that is adjacent to a and e and annotate v.*

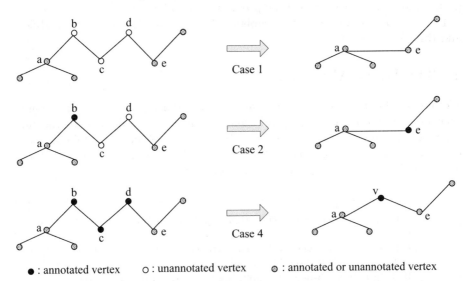

● : annotated vertex   ○ : unannotated vertex   ◉ : annotated or unannotated vertex

Fig. 1. Illustration of Reduction Rule 3

Proof. First, we prove Case 1. Let G be the original graph and G' the graph after replacing path $abcde$ with edge ae in G. We prove that G has an edge dominating set of size k if and only if G' has an edge dominating set of size $k - 1$.

For the first direction, we assume that G has an edge dominating set S of size k and show that G' will have an edge dominating set of size $k - 1$. We consider three cases for S: $ab \in S$, $ab \notin S$ but $bc \in S$, and $ab, bc \notin S$ but $\{a, c\} \subset V(S)$. If $ab \in S$, we can assume that $de \in S$, otherwise bc or cd will be in S and we can simply replace bc or cd with de in S. It is easy to see that $(S - \{ab, de\}) \cup \{ae\}$ is an edge dominating set of size $k - 1$ for G'. If $bc \in S$ and $ab \notin S$, we can assume that $cd, de \notin S$, otherwise we can simply replace cd or de with ef in S, where f is a neighbor of e other than d. Therefore, $S - \{bc\}$ will be an edge dominating set of size $k - 1$ for G'. If $ab, bc \notin S$ and $\{a, c\} \subset V(S)$, then $cd \in S$ and we can assume that $de \notin S$. For this case, $S - \{cd\}$ is still an edge dominating set of size $k - 1$ for G'.

For the other direction, we assume that S' is an edge dominating set of size $k - 1$ for G' and consider three cases for S': $a, e \in V(S)$, $a \in V(S)$ but $e \notin V(S)$, and $e \in V(S)$ but $a \notin V(S)$. For the case $a, e \in V(S)$, if $ae \in S$, then $(S' - \{ae\}) \cup \{ab, de\}$ is an edge dominating set of size k for G; if $ae \notin S$, then $S' \cup \{bc\}$ is an edge dominating set of size k for G. If $a \in V(S)$ but $e \notin V(S)$, then $S' \cup \{cd\}$ is an edge dominating set of size k for G. If $e \in V(S)$ but $a \notin V(S)$, then $S' \cup \{bc\}$ is an edge dominating set of size k for G.

We have finished the proof of Case 1. In the same way, we can prove Case 2-4. ∎

Note that Rule 3 can not be used to reduce all inner paths of length 4, because the four cases do not cover all cases. But we can use Rule 3 to reduce all inner paths of length ≥ 6 (please see the proof of Lemma 3).

For convenience, we will call a graph a *reduced graph*, if none of Rule 1, Rule 2 and Rule 3 can be applied.

4   The Exact Algorithm

We present the main steps of our exact algorithm in Fig. 2. The description of each branching operation (Step 3-7) and the analysis are delayed to the subsections following the algorithm. We will use S^* to denote a minimum edge dominating set that contains all the edges in the current solution set in our algorithm if such kind of minimum edge dominating sets exist.

---

1. If the graph is not a reduced graph, apply Reduction Rule 1-3 to reduce the graph.
2. If there is a component of less than 20 vertices, find the optimal solution to this component directly.
3. If the graph is a 3-regular graph, branch on the 3-regular graph.
4. If there is a pure path abc of length 2 such that b is annotated, branch on it.
5. If there is a pure path of length 3, branch on it.
6. If there is a pure path $abcde$ of length 4 such that b and c are annotated, branch on it.
7. Else pick up a degree-3 vertex adjacent to at least a degree-2 vertex and branch on it.
8. Return a solution.

Fig. 2. The exact algorithm

To analyze the running time of our algorithm, we need to analyze the size of the searching tree of the algorithm. Traditionally, we may use the number of vertices or edges in the graph to measure the size. Some recent references [22,14] used the number of degree-3 vertices as the measure to analyze the running time for the independent set and vertex cover problems in 3-degree graphs and got significant improvements. We will also use this technique to analyze our algorithm and get the improved running time bound. Recall that we use p to denote the number of degree-3 vertices in the graph. First of all, we give the following two lemmas.

Lemma 1. *Let G be a connected graph with maximum degree 3. If G has x degree-1 vertices and at least x degree-3 vertices, then after iteratively applying Reduction Rule 1 until the graph has no degree-1 vertex, we can reduce at least x degree-3 vertices from the graph.*

This lemma for the independent set and vertex cover problems has been proved in [14] (Lemma 8). We can simply modify the proof in [14] to get a similar proof for this lemma.

Lemma 2. *After applying Rule 2-3, the number of degree-3 vertices in the graph will not increase.*

Now we are ready to describe and analyze the branching operations in Fig. 2.

4.1   Branching on 3-Regular Graphs (Step 3)

Assume the graph is a 3-regular graph, we consider two cases: whether there is an annotated vertex in the graph or not. If all the vertices are not annotated vertices, we arbitrarily select a vertex a (assume that b, c, d are the three neighbors of a), and branch into four branches by either including ab or ac or ad into the edge dominating set or excluding all the three edges from the graph. When ab (or ac or ad) is included into the edge dominating set, we delete ab (or ac or ad) from the graph. After deleting ab (or ac or ad), we will iteratively apply Reduction Rule 1 to reduce degree-1 vertices in Step 1 if some degree-1 vertices are created. By Lemma 1, we know that together at least 6 degree-3 vertices will be reduced in this branch. When none of the three edges is included into the edge dominating set, we can delete a from the graph and annotate b, c, d. Then we will reduce 4 degree-3 vertices. Let $C(p)$ be the worst size of the searching tree when the graph has at most p degree-3 vertices, we get the following recurrence

$$C(p) \le 3C(p-6) + C(p-4), \tag{1}$$

which solves to $C(p) = O(1.2930^p)$.

If the graph has an annotated vertex a (with three neighbors b, c, d), we only need to branch into three branches by either including ab or ac or ad into the edge dominating set, then we get the following recurrence

$$C(p) \le 3C(p-6), \tag{2}$$

which is covered by (1).

4.2   Branching Operation in Step 4

Assume the graph has a pure 2-path abc such that b is an annotated vertex. We branch on abc by including ab into the edge dominating set or including bc into the edge dominating set. After deleting ab (or bc) from the graph and iteratively applying Reduction Rule 1, we can reduce at least 4 degree-3 vertices in each branch (note that abc is a pure path and then a and c are degree-3 vertices). Then we get the following recurrence

$$C(p) \le 2C(p-4), \tag{3}$$

which solves to $C(p) = O(1.1893^p)$.

4.3   Branching Operation in Step 5

Assume the graph has a pure path $abcd$ of length 3. We will consider whether b and c are annotated or not.

If neither b or c is annotated, we branch into three branches by including either ab or bc or cd into the edge dominating set. Recall that S^* stands for a minimum edge dominating set that contains the current solution set. Note that in the branch where bc is included into the edge dominating set, we can assume that $a, d \notin V(S^*)$, because if $a \in V(S^*)$ (or $d \in V(S^*)$) we can replace bc with cd (or replace bc with ab) in S^* to get another solution that does not contain bc. Then in this branch, we annotate all neighbors of a and d and delete a, b, c, d from the graph and include bc into the edge dominating set. Totally we can reduce p by at least 6. In the other two branches, we delete ab or cd from the graph, and can reduce p by at least 4. We get

$$C(p) \leq C(p - 6) + 2C(p - 4), \tag{4}$$

which solves to $C(p) = O(1.2721^p)$.

If only b is an annotated vertex, we branch by including ab or bc into the edge dominating set. Note that in the branch where bc is included into the edge dominating set, we can assume that $d \notin V(S^*)$, because if $d \in V(S^*)$ we can replace bc with ab in S^* to get another solution that does not contain bc. Then in this branch, we annotate all neighbors of d and delete b, c, d from the graph and include bc into the edge dominating set. Totally we can reduce p by at least 4. In the branch where ab is included into the edge dominating set, we will delete a and b from the graph and reduce p by at least 4. We will get a recurrence as (3).

Similarly, we can branch with (3) by the same analysis, if only c is annotated.

If both b and c are annotated, we branch by including bc into the edge dominating set or excluding it from the edge dominating set. In the first branch, we will delete bc from the graph, and reduce p by 2. In the second branch, then ab and cd will be included into the edge dominating set. We will delete a, b, c, d from the graph and reduce p by at least 6. We get

$$C(p) \leq C(p - 6) + C(p - 2), \tag{5}$$

which solves to $C(p) = O(1.2107^p)$.

4.4   Branching Operation in Step 6

Assume $abcde$ is a pure path of length 4 such that b and c are annotated vertices and d is not an annotated vertex (If d is also an annotated vertex, the path will be reduced in Step 1). We look at vertex e. If e is not an endpoint of any edge in the minimum edge dominating set, we can simply assume that ab and cd are in the edge dominating set. Then we can annotate all neighbors of e and delete a, b, c, d, e from the graph and reduce p by 6. Else e is an endpoint of an edge in the minimum edge dominating set, we can simply assume that bc is in the edge dominating set. Then we can delete b, c, d from the graph and annotate e and reduce p by 2. Therefore, we can branch as above and get a recurrence as (5).

4.5   Branching Operation in Step 7

This case is the most complicated case, but it is not the bottleneck of the algorithm. First of all, we prove some properties of the graph in Step 7.

Lemma 3. *The graph in Step 7 has only three kinds of pure paths besides pure path of length* 1 *(please see Fig. 3 for the illustration):*
(a) pure path abc of length 2 *such that b is not an annotated vertex;*
(b) pure path abcde of length 4 *such that c is an annotated vertex and b and d are not annotated vertices;*
(c) pure path abcdef of length 5 *such that c and d are annotated vertices and b and e are not annotated vertices.*

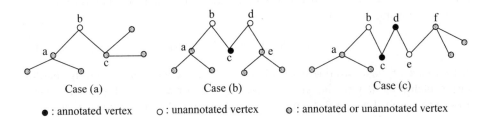

Case (a)   Case (b)   Case (c)

● : annotated vertex   ○ : unannotated vertex   ⊚ : annotated or unannotated vertex

Fig. 3. The three kinds of pure paths in Lemma 3

Proof. Clearly, if the graph has a pure path of length 2 or 3 or 4 other than Case (*a*) or (*b*) in the lemma, the path will be reduced in Step 1-6. Next, we assume that there is an inner path *abcdef* of length 5 (which will appear in all pure paths of length ≥ 5). We prove that this path must be a pure path of Case (*c*). Since *abcde* and *bcdef* can not apply Reduction Rule 3, the only case of *abcdef* is that *c* and *d* are annotated vertices and *b* and *e* are not annotated vertices. Furthermore, path *abcdef* can not be a subpath of a pure path of length > 5, otherwise Reduction Rule 3 still can be applied. Therefore, we know that *abcdef* can only be a pure path of Case (*c*).   ∎

Lemma 4. *Let abcde be a pure path of length* 4 *such that c is an annotated vertex and b and d are not annotated vertices (Case (b) in Lemma 3). Then there is a minimum edge dominating set that does not contain edge ab.*

Proof. If there is a minimum edge dominating set that contains *ab*, we can simply replace *ab* with *ab'* to get another minimum edge dominating set, where *b'* is another neighbor of *a*.   ∎

In the same way, we can prove

Lemma 5. *Let abcdef be a pure path of length* 5 *such that c and d are annotated vertices and b and e are not annotated vertices (Case (c) in Lemma 3). Then there is a minimum edge dominating set that does not contain edge ab.*

We are ready to introduce the branching operation in this step now. We pick up a degree-3 vertex a (assume that b, c, d are the three neighbors of a) that is adjacent to at least a degree-2 vertex in the graph. We distinguish the following three cases according to the number of pure 2-paths incident on v.

Case 1: No pure 2-path is incident on v. If there is only one degree-2 neighbor of a, say b, we will branch by either including ac or ad into the edge dominating set or excluding all the three edges from the graph. The correctness of this operation follows from this observation: b is a degree-2 vertex and ab must be on a pure path of Case (b) or Case (c) in Lemma 3. By Lemma 4 and Lemma 5, we know that we do not need to consider the case of including ab. In the branch where ac or ad is included into the edge dominating set, we can reduce p by 6, and in the branch where all the three edges are excluded from the edge dominating set, we can reduce p by 4 (just as we analyzed in Section 4.1). Therefore, we can get the following recurrence

$$C(p) \leq 2C(p-6) + C(p-4). \tag{6}$$

If there are at least two degree-2 neighbors of a, say c and d, we will branch by either including ab into the edge dominating set or excluding all the three edges from the graph. Lemma 3, Lemma 4 and Lemma 5 also show the correctness of this operation. We only need to pay attention to the case where all the three neighbors of a are degree-2 vertices. For this case, we can not simply exclude all the three edge from the graph without branching, because we may miss the case $a \in V(S^*)$ if we do that. Therefore, we can get

$$C(p) \leq C(p-6) + C(p-4). \tag{7}$$

Case 2: Only one pure 2-path is incident on v, say abb'. We look at the other two neighbors c and d of a (except b). If both c and d are degree-3 vertices, we will branch on abb' into three branches. We consider a and b'. If $a \notin V(S^*)$, then $bb' \in S^*$. For this case, we can annotate all neighbors of a and delete a, b, b' from the graph and include bb' into the edge dominating set. We can reduce 6 degree-3 vertices (a, c, d, b' and two from the other two neighbors of b'). If $b' \notin V(S^*)$, we can annotate all neighbors of b' and delete a, b, b' from the graph and include ab into the edge dominating set. In this branch we still can reduce p by 6. If $a, b' \in V(S^*)$, we can simply assume that $ab, bb' \notin V(S^*)$ (if $ab \in S$, we can replace ab with ac in S to get another solution), and annotate a and b', and delete ab and bb' from the graph. In the third branch, we can reduce two degree-3 vertices a and b'. Then we can branch into three branches as above and get the recurrence

$$C(p) \leq 2C(p-6) + C(p-2). \tag{8}$$

Note that in the third branch, cad is a pure 2-path ($b' \neq c$ and $b' \neq d$, otherwise Reduction Rule 2 can be applied, and then c and d are still degree-3 vertices after removing ab and bb') and a is annotated, which implies that the condition of Step 4 of the algorithm holds. Then we can further branch with (3) at least in this branch. By putting these together, we get

$$C(p) \leq 2C(p-6) + 2C(p-2-4) = 4C(p-6),\tag{9}$$

which solves to $C(p) = O(1.2600^p)$.

If at least one of c and d, say c, is a degree-2 vertex, then ac is on a pure path of Case (b) or Case (c) in Lemma 3. We will branch on v into three branches: including ab into the edge dominating set, including ad into the edge dominating set, and excluding ab, ac, ad from the edge dominating set. By Lemma 4 and Lemma 5, we know that we can ignore the case of including ac into the edge dominating set. In the branch where ab or ad is included into the edge dominating set, we can reduce p by 4 at least. In the last branch, we will delete a and annotate b, c, d, and then we can directly put edge bb' into the edge dominating set according to Reduction Rule 1. Note that b' is a degree-3 vertex, and then we can reduce at least 6 degree-3 vertices (a, b', c, d and two from the other two neighbors of b') in this branch. Therefore, we can get a recurrence as (4).

Case 3: At least two pure 2-paths are incident on v, say abb' and acc'. For this case, we will pick up a pure 2-path, say abb', and branch on it into three branches. We consider a and b'. If $a \notin V(S^*)$, we can annotate all neighbors of a and delete a from the graph. Note that in the branch b and c become annotated degree-1 vertices and we can further include bb' and cc' into the edge dominating set. Totally, we can reduce 8 degree-3 vertices (a, b', c, d and two from the other two neighbors of b' and two from the other two neighbors of c'). If $b' \notin V(S^*)$, we can annotate all neighbors of b' and delete a, b, b' from the graph and include ab into the edge dominating set. In this branch we can reduce p by 6. If $a, b' \in V(S^*)$, we can simply assume that $ab, bb' \notin V(S^*)$ (if $ab \in S^*$, we can replace ab with aa' in S^* to get another solution, where a' is another neighbor of a), and annotate a and b' and remove edges ab and bb' from the graph. In the third branch, we can reduce two degree-3 vertices a and b'. Then we get

$$C(p) \leq C(p-8) + C(p-6) + C(p-2),\tag{10}$$

which solves to $C(p) = O(1.2721^p)$.

4.6   Putting All Together

Among all above cases, the worst case is that we branch with (1) in Section 4.1. In fact, the worst case will not always happen. It is easy to see that in each subbranch after branching on a 3-regular graph, we will not get a 3-regular graph again (after applying the reduction rules). In the branch where all the three edges are excluded from the edge dominating set, the graph will have three annotated degree-2 vertices. No matter whether the three degree-2 vertices are adjacent or not, we will branch with (3) at least. Combining with that, we get the following recurrence

$$C(p) \leq 2C(p-4-4) + 3C(p-6) = 2C(p-8) + 3C(p-6),\tag{11}$$

which solves to $C(p) = O(1.2721^p)$.

Since (11) is still one of the worst cases among all the cases, we get

Theorem 1. *The edge dominating set problem in graphs with maximum degree 3 can be solved in $O^*(1.2721^p)$ time and polynomial space, where $p \leq n$ is the number of degree-3 vertices in the graph.*

Our result improves the best previous result of $O^*(1.3226^n)$ by Rooij and Bodlaender [6] (their result also holds in general graphs).

5   The Parameterized Algorithm

We show that our exact algorithm presented in Section 4 can be used to get an improved algorithm for the parameterized edge dominating set problem, in which we are asked to decide whether a 3-degree graph has an edge dominating set of size k or not. The following property is crucial for our algorithm.

Lemma 6. *Let G be a graph with maximum degree 3. If G has p degree-3 vertices and an edge dominating set of size k, then*

$$k \geq \frac{3}{10}p. \tag{12}$$

Proof. Let $C = V(S)$ be the vertex size of the edge dominating set S, where $|S| = k$. We consider the set E_0 of edges with one endpoint in C and the other in $V - C$.

Assume that there are p' degree-3 vertices in C (it is easy to see that $p' \leq 2k$). All other vertices in C are of degree ≤ 2 and the total number is not greater than $2k - p'$. Then we have

$$|E_0| \leq 3p' + 2(2k - p') - 2k. \tag{13}$$

On the other hand, there are at least $p - p'$ degree-3 vertices in $V - C$. Note that $V - C$ is an independent set. We have

$$|E_0| \geq 3(p - p'). \tag{14}$$

By combining (13) with (14), we get

$$3p' + 2(2k - p') - 2k \geq 3(p - p')$$

and then

$$3p - 2k \leq 4p' \leq 8k.$$

Therefore, we get (12).   ∎

Based on Lemma 6, we can solve the parameterized edge dominating set problem in 3-degree graphs in the following way: We first count the number p of degree-3 vertices in the input graph. If $k < \frac{3}{10}p$, we report that the graph has no edge dominating set of size k. Else we use our exact algorithm in Section 4 to find a minimum edge dominating set in $O^*(1.2721^p) = O^*(1.2721^{\frac{10}{3}k}) = O^*(2.2306^k)$ time. Then we get

Theorem 2. *We can use $O^*(2.2306^k)$ time and polynomial space to decide whether a graph with maximum degree 3 has an edge dominating set of size k or not.*

Our algorithm improves the $O^*(2.6181^k)$-time polynomial-space algorithm by Fernau [21] and $O^*(2.4181^k)$-time exponential-space algorithm by Fomin *et al.* [20]. Note that the algorithms in [21] and [20] also work for the problem in general graphs.

6   Concluding Remarks

In this paper, we have presented an $O^*(1.2721^n)$-time exact algorithm and an $O^*(2.2306^k)$-time parameterized algorithm for the edge dominating set problem in degree-3 graphs. Currently, these two algorithms are the fastest algorithms for the two corresponding problems. Together with our algorithms, we have presented some data reduction rules for the (annotated) edge dominating set problem, which can be used to reduce the input size of the graph.

Many branch-and-search algorithms for graph problems have good performance when the graph has some high-degree vertices. So fast algorithms for the problems in low-degree graphs may directly lead to the improvements on the algorithms for the problems in general graphs. This situation holds for independent set, vertex cover, edge dominating set and some other basic problems. It would be interesting to know whether there are faster algorithms for some basic graph problems when the graph is restricted to a low-degree graph.

References

1. Tarjan, R., Trojanowski, A.: Finding a maximum independent set. SIAM Journal on Computing 6(3), 537–546 (1977)
2. Fomin, F.V., Grandoni, F., Kratsch, D.: Measure and conquer: a simple $O(2^{0.288n})$ independent set algorithm. In: SODA, pp. 18–25. ACM Press, New York (2006)
3. Beigel, R., Eppstein, D.: 3-coloring in time $O(1.3289^n)$. J. Algorithms 54(2), 168–204 (2005)
4. Bjorklund, A., Husfeldt, T.: Exact algorithms for exact satisfiability and number of perfect matchings. Algorithmica 52(2), 226–249 (2008)
5. Fomin, F.V., Grandoni, F., Kratsch, D.: Measure and conquer: Domination - a case study. In: Caires, L., Italiano, G.F., Monteiro, L., Palamidessi, C., Yung, M. (eds.) ICALP 2005. LNCS, vol. 3580, pp. 191–203. Springer, Heidelberg (2005)
6. Rooij, J.M., Bodlaender, H.L.: Exact algorithms for edge domination. In: Grohe, M., Niedermeier, R. (eds.) IWPEC 2008. LNCS, vol. 5018, pp. 214–225. Springer, Heidelberg (2008)
7. Razgon, I.: Exact computation of maximum induced forest. In: Arge, L., Freivalds, R. (eds.) SWAT 2006. LNCS, vol. 4059, pp. 160–171. Springer, Heidelberg (2006)
8. Bourgeois, N., Escoffier, B., Paschos, V.T., Rooij, J.M.M.: Maximum independent set in graphs of average degree at most three in $O(1.08537^n)$. In: Kratochvíl, J., Li, A., Fiala, J., Kolman, P. (eds.) Theory and Applications of Models of Computation. LNCS, vol. 6108, pp. 373–384. Springer, Heidelberg (2010)

9. Xiao, M.: A note on vertex cover in graphs with maximum degree 3. In: Thai, M.T., Sahni, S. (eds.) COCOON 2010. LNCS, vol. 6196, pp. 150–159. Springer, Heidelberg (2010)

10. Iwama, K., Nakashima, T.: An improved exact algorithm for cubic graph tsp. In: Lin, G. (ed.) COCOON 2007. LNCS, vol. 4598, pp. 108–117. Springer, Heidelberg (2007)

11. Chen, J., Kanj, I.A., Xia, G.: Labeled search trees and amortized analysis: Improved upper bounds for NP-hard problems. Algorithmica 43(4), 245–273 (2005); A preliminary version appeared in ISAAC 2003

12. Fomin, F.V., Høie, K.: Pathwidth of cubic graphs and exact algorithms. Inf. Process. Lett. 97(5), 191–196 (2006)

13. Razgon, I.: Faster computation of maximum independent set and parameterized vertex cover for graphs with maximum degree 3. J. of Discrete Algorithms 7(2), 191–212 (2009)

14. Xiao, M.: A simple and fast algorithm for maximum independent set in 3-degree graphs. In: Rahman, M. S., Fujita, S. (eds.) WALCOM 2010. LNCS, vol. 5942, pp. 281–292. Springer, Heidelberg (2010)

15. Niedermeier, R.: Invitation to Fixed-Parameter Algorithms. Oxford University Press, Oxford (2006)

16. Garey, M.R., Johnson, D.S.: Computers and intractability: A guide to the theory of NP-completeness. Freeman, San Francisco (1979)

17. Yannakakis, M., Gavril, F.: Edge dominating sets in graphs. SIAM J. Appl. Math. 38(3), 364–372 (1980)

18. Randerath, B., Schiermeyer, I.: Exact algorithms for minimum dominating set. Technical Report zaik 2005-501, Universitat zu Koln, Cologne, Germany (2005)

19. Raman, V., Saurabh, S., Sikdar, S.: Efficient exact algorithms through enumerating maximal independent sets and other techniques. Theory of Computing Systems 42(3), 563–587 (2007)

20. Fomin, F., Gaspers, S., Saurabh, S., Stepanov, A.: On two techniques of combining branching and treewidth. Algorithmica 54(2), 181–207 (2009)

21. Fernau, H.: Edge dominating set: Efficient enumeration-based exact algorithms. In: Bodlaender, H., Langston, M. (eds.) IWPEC 2006. LNCS, vol. 4169, pp. 142–153. Springer, Heidelberg (2006)

22. Razgon, I.: A faster solving of the maximum independent set problem for graphs with maximal degree 3. In: Broersma, H., Dantchev, S.S., Johnson, M., Szeider, S. (eds.) ACiD. Texts in Algorithmics, vol. 7, pp. 131–142. King's College, London (2006)

Approximate Ellipsoid in the Streaming Model

Asish Mukhopadhyay, Animesh Sarker, and Tom Switzer

School of Computer Science, University of Windsor, ON N9B 3P4, Canada
{asishm,sarke1a,switzec}@uwindsor.ca

Abstract. In this paper we consider the problem of computing an approximate ellipsoid in the streaming model of computation, motivated by a 3/2-factor approximation algorithm for computing approximate balls. Our contribution is twofold: first, we show how to compute an approximate ellipsoid as done in the approximate ball algorithm, and second, construct an input to show that the approximation factor can be unbounded, unlike the algorithm for computing approxinmate balls. Though the ratio of volumes can become unbounded, we show that there exists a direction in which the ratio of widths is bounded by a factor of 2.

1 Introduction

Of late there has been a burgeoning interest in geometric optimzation problems over data streams [5,1,6], inspired by Feigenbaum, Kannan and Zhang's result on computing the diameter of a point set approximately in the streaming and sliding-window models [3].

In [9], Zarrabi-Zadeh and Chan proposed a very simple algorithm for computing an approximate spanning ball of a set of n points $P = \{p_1, p_2, p_3, ..., p_n\}$ in the streaming model and showed by an elegant analysis showing that the radius of the approximate ball is to within 3/2 of the exact one. Mukhopadhyay et al [7] studied this problem for computing an approximate ellipse, extending the ball algorithm in a non-trivial way and also constructing an example input to show that the approximation ratio can be unbounded. In this paper, we show that the results extend to d dimensional space.

The rest of the paper has four sections. In the next, we define our streaming model. The following section contains our algorithm for computing an approximate ellipsoid with appropriate recipes and analysis. In the fourth section we construct an example to show that the approximation ratio is unbounded. We conclude in the last section.

2 Streaming Model

The streaming model was first proposed in [4], providing computational geometers with an impetus to examine problems involving geometric data streams. In this model, we are allowed to make only one pass over the data, the amount of storage we are allowed is much smaller than the input data size and updates

W. Wu and O. Daescu (Eds.): COCOA 2010, Part II, LNCS 6509, pp. 401–413, 2010.

must be done fast. These are real-time constraints when we attempt to do computation involving an infinite data stream.

For our problem, at iteration $i + 1$ of the algorithm, we are given a point p_{i+1}, and also some previous ellipsoid E_i, whose description we store, requiring $O(d)$ space. We know that E_i is an approximate minimum enclosing ellipsoid for some unknown set S_i. We then want a new ellipsoid E_{i+1} that is an approximate minimum enclosing ellipsoid for $S_{i+1} = S_i \bigcup \{s_{i+1}\}$. Though it can be set up as a convex programming optimization problem [2], here we adopt a more direct approach that we have invented for this rather special case.

3   Approximation Algorithm for Ellipsoid

Our algorithm is based on a solution to the following problem:

Problem: Given an ellipsoid E_A be of full volume in k-dimensional space ($k \leq d$) and a and an input point p that lies outside E_A, find an ellipsoid of minimum volume that spans E_A and p.

We solve this problem by finding an elliptic transformation (nomenclature due to Post [8]), T, that transforms E_A to a unit ball; the same transformation is applied to p. Then, in the transformed space, we solve the easier problem of finding a minimum spanning ellipsoid of the unit ball and $T(p)$. Finally, we apply an inverse transformation $T^{-1}(.)$ to the minimum spanning ellipsoid in the transformed space to find the desired minimum spanning ellipsoid in the original space.

That the transformed ellipsoid in the original space is of minimum volume is due to the fact that the elliptic transformation T preserves relative volumes. We will prove this. The details are in the next two subsections.

3.1   Finding T(.)

$T(.)$ is required to have the following properties:

P1 For any ellipsoid E, $T(E)$, as well as its inverse image $T^{-1}(E)$, is an ellipsoid;
P2 $T(.)$ preserves relative volumes, that is, $volume(E_1) \leq volume(E_2) \Leftrightarrow volume(T(E_1)) \leq volume(T(E_2))$.

If $T(.)$ is a rotation or translation then it obviously has both properties. The following lemma shows this is also true when $T(.)$ is a scaling.

Lemma 1. *Let $\alpha > 0$. Then the scaling $T(.) = \begin{bmatrix} \alpha & 0 & 0 & 0 & \cdots & 0 \\ 0 & 1 & 0 & 0 & \cdots & 0 \\ 0 & 0 & 1 & 0 & \cdots & 0 \\ & & & \vdots & & \\ 0 & 0 & 0 & 0 & \cdots & 1 \end{bmatrix}$*

along the x_1-axis satisfies P1 and P2.

Proof. Let

$$1 = [p - p_0]^T A [p - p_0]$$

be the matrix equation of an ellipsoid E with center at p_0, where

$$A = \begin{bmatrix} a_{11} & a_{12} & a_{13} & \cdots & a_{1d} \\ a_{21} & a_{22} & a_{23} & \cdots & a_{2d} \\ a_{31} & a_{32} & a_{33} & \cdots & a_{3d} \\ & & \vdots & & \\ a_{d1} & a_{d2} & a_{d3} & \cdots & a_{dd} \end{bmatrix}$$

is a positive definite matrix $(det(A) > 0)$.

Let $p' = T(p)$ for some p on E and $p'_0 = T(p_0)$. Since T^{-1} is well-defined, the matrix equation of the transformed ellipsoid is

$$\left[T^{-1}p' - T^{-1}p'_0 \right]^T A \left[T^{-1}p' - T^{-1}p'_0 \right] = 1$$

or, on a little simplification

$$[p' - p'_0]^T T^{-1} A T^{-1} [p' - p'_0] = 1,$$

where

$$T^{-1}AT^{-1} = \begin{bmatrix} \frac{1}{\alpha} & 0 & 0 & 0 & \cdots & 0 \\ 0 & 1 & 0 & 0 & \cdots & 0 \\ 0 & 0 & 1 & 0 & \cdots & 0 \\ & & \vdots & & & \\ 0 & 0 & 0 & 0 & \cdots & 1 \end{bmatrix} \begin{bmatrix} \frac{a_{11}}{\alpha} & a_{12} & a_{13} & \cdots & a_{1d} \\ \frac{a_{21}}{\alpha} & a_{22} & a_{23} & \cdots & a_{2d} \\ \frac{a_{31}}{\alpha} & a_{32} & a_{33} & \cdots & a_{3d} \\ & & \vdots & & \\ \frac{a_{d1}}{\alpha} & a_{d2} & a_{d3} & \cdots & a_{dd} \end{bmatrix}$$

$$= \begin{bmatrix} \frac{a_{11}}{\alpha^2} & \frac{a_{12}}{\alpha} & \frac{a_{13}}{\alpha} & \cdots & \frac{a_{1d}}{\alpha} \\ \frac{a_{21}}{\alpha} & a_{22} & a_{23} & \cdots & a_{2d} \\ \frac{a_{31}}{\alpha} & a_{32} & a_{33} & \cdots & a_{3d} \\ & & \vdots & & \\ \frac{a_{d1}}{\alpha} & a_{d2} & a_{d3} & \cdots & a_{dd} \end{bmatrix}$$

Hence $T(E)$ is an ellipsoid.

Since the volume of E is given by $\dfrac{K}{\sqrt{det(A)}}$, where K is a constant that depends on the dimension d, the volume of $T(E) = \dfrac{K\alpha^2}{\sqrt{det(A)}}$. From this and the fact that $T^{-1}(E)$ is an ellipsoid because T^{-1} is also an x_1-scaling, it follows that T satisfies $P1$ and $P2$. $\qquad\square$

3.2 Finding E'_A

Let $x_i = (x_{1i}, x_{2i}, \ldots, x_{di})$ represent a point in d-dimensional space.

Lemma 2. *Suppose* $\dfrac{(x_1 - x_0)^2}{a_1^2} + \dfrac{x_2^2}{a_2^2} = 1$ *is an ellipse that encloses the unit circle* $x_1^2 + x_2^2 = 1$. *Then the ellipsoid* $\dfrac{(x_1 - x_0)^2}{a_1^2} + \dfrac{x_2^2}{a_2^2} + \cdots + \dfrac{x_d^2}{a_2^2} = 1$ *encloses the unit ball* $x_1^2 + x_2^2 + \cdots + x_d^2 = 1$

Proof. Let (x_{11}, x_{21}) be a point in $x_1^2 + x_2^2 \leq 1$. Then $x_{11}^2 + x_{21}^2 = r^2$ for some $r \leq 1$. Since $\dfrac{(x_1 - x_0)^2}{a_1^2} + \dfrac{x_2^2}{a_2^2} = 1$ encloses $x_1^2 + x_2^2 = 1$, we can write $\dfrac{(x_{11} - x_0)^2}{a_1^2} + \dfrac{x_{21}^2}{a_2^2} \leq 1$ i.e.

$$\frac{(x_{11} - x_0)^2}{a_1^2} + \frac{r^2 - x_{11}^2}{a_2^2} \leq 1$$

Let $(x_{11}, x_{21}, \cdots, x_{d1})$ be a point in $x_1^2 + x_2^2 + \cdots + x_d^2 \leq 1$. Then $x_{11}^2 + x_{21}^2 + \cdots + x_{d1}^2 = s^2$ for some $s \leq 1$. Now,

$$\frac{(x_{11} - x_0)^2}{a_1^2} + \frac{x_{21}^2}{a_2^2} + \cdots + \frac{x_{d1}^2}{a_2^2} = \frac{(x_{11} - x_0)^2}{a_1^2} + \frac{x_{21}^2 + \cdots + x_{d1}^2}{a_2^2}$$

$$= \frac{(x_{11} - x_0)^2}{a_1^2} + \frac{s^2 - x_{11}^2}{a_2^2}$$

$$\leq 1$$

This proves the lemma.   □

Lemma 3. *If* $\dfrac{(x_1 - x_0)^2}{a_1^2} + \dfrac{x_2^2}{a_2^2} + \cdots + \dfrac{x_d^2}{a_d^2} = 1$ *is the equation of an ellipsoid E with minimum volume in an d-dimensional space that encloses the unit ball B centered at the origin and a point $(s, 0, 0, \cdots, 0)$ then $a_2 = a_3 = \cdots = a_d$.*

Proof. Suppose $a_m = \min\{a_2, a_3, \cdots a_d\}$ where $2 \leq m \leq d$. Consider the projection of the ellipsoid and the unit ball on the two dimensional plane $x_1 x_m$. These are respectively the ellipse $\dfrac{(x_1 - x_0)^2}{a_1^2} + \dfrac{x_m^2}{a_m^2} = 1$ and the unit circle; the ellipse encloses the unit circle and passes through the point $(s, 0)$. By Lemma 2, the ellipsoid of revolution E': $\dfrac{(x_1 - x_0)^2}{a_1^2} + \dfrac{x_2^2}{a_m^2} + \cdots + \dfrac{x_d^2}{a_m^2} = 1$, obtained from this ellipse encloses the unit ball B and passes through the point $(s, 0, 0, \cdots, 0)$. Since the volume of E' is less than or equal to the volume of E, we conclude that if E is the ellipsoid with minimum volume then $a_2 = a_3 = \cdots = a_d$.   □

Lemma 4. *If E:* $\dfrac{(x_1 - p_1)^2}{a_1^2} + \dfrac{(x_2 - p_2)^2}{a_2^2} + \cdots + \dfrac{(x_d - p_d)^2}{a_d^2} = 1$ *is an ellipsoid of minimum volume that encloses the unit ball B and is incident on the point $(s, 0, 0, \cdots, 0)$ then $p_2 = \cdots = p_d = 0$.*

Proof. Since $\dfrac{(x_1 - p_1)^2}{a_1^2} + \dfrac{(x_2 - p_2)^2}{a_2^2} + \cdots + \dfrac{(x_n - p_d)^2}{a_d^2} = 1$ is incident on the

point $(s, 0, 0, \cdots, 0)$ and encloses the unit ball, we can verify that $\dfrac{(x_1 - p_1)^2}{a_1^2} +$

$\dfrac{(x_2 + p_2)^2}{a_2^2} + \cdots + \dfrac{(x_d + p_d)^2}{a_d^2} = 1$ also encloses the unit ball and is incident on

the point $(s, 0, 0, \cdots, 0)$. Adding the equations of the two ellipsoids we get the
following equation:

$$\frac{(x_1 - p_1)^2}{a_1^2} + \frac{x_2^2}{a_2^2} + \cdots + \frac{x_d^2}{a_d^2} = 1 - \frac{p_2^2}{a_2^2} - \frac{p_2^2}{a_2^2} - \cdots - \frac{p_d^2}{a_d^2}$$

This represents an ellipsoid whose volume is smaller than that of the original

ellipsoid. Thus if $E : \dfrac{(x_1 - p_1)^2}{a_1^2} + \dfrac{(x_2 - p_2)^2}{a_2^2} + \cdots + \dfrac{(x_d - p_d)^2}{a_d^2} = 1$ is an ellipsoid

minimum volume then $p_2 = \cdots = p_d = 0$. $\qquad\square$

Lemma 3 and Lemma 4 completely chracterize an ellipsoid of minimum volume
that encloses the unit ball B and is incident on the point $(s, 0, 0, \cdots, 0)$.

Lemma 5. *If* $E : \dfrac{(x_1 - x_0)^2}{a_1^2} + \dfrac{x_2^2}{a_2^2} + \dfrac{x_3^2}{a_2^2} + \cdots + \dfrac{x_d^2}{a_2^2} = 1$ *is an ellipsoid of minimum*
volume that encloses the unit ball B *centered at the origin and is incident on the*
point $p' = (s, 0, 0, \cdots, 0)$ *then* $a_1^2 = (\beta - x_0)(\beta^{-1} - x_0)$ *and* $a_2^2 = 1 - x_0\beta$ *where*

$\beta = \dfrac{s - \sqrt{s^2 + \frac{4d}{(d-1)^2}}}{\frac{2d}{d-1}}$ *and* $x_0 = \dfrac{s^2 - 1}{2s - \beta - \beta^{-1}}$.

Proof. Since p' is on E,

$$\frac{(s - x_0)^2}{a_1^2} = 1 \tag{1}$$

Let α be the angle made by the x-axis, the origin, and the point at which $E'(t)$
touches the upper half of the unit circle, where $E'(t)$ is the projection of E on
$x_1 x_2$. (See Figures 1.) The tangent to the circle at $(\cos\alpha, \sin\alpha)$ is

$$x_1 \cos\alpha + x_2 \sin\alpha = 1 \tag{2}$$

The tangent to the ellipse at $(\cos\alpha, \sin\alpha)$ is

$$\frac{(x_1 - x_0)(\cos\alpha - x_0)}{a_1^2} + \frac{x_2 \sin\alpha}{a_2^2} = 1$$

$$\frac{x_1(\cos\alpha - x_0)}{a_1^2} + \frac{x_2 \sin\alpha}{a_2^2} = 1 + \frac{x_0(\cos\alpha - x_0)}{a_1^2} \tag{3}$$

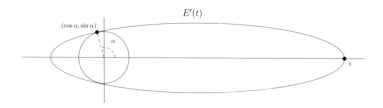

Fig. 1. An ellipse containing and tangent to the unit circle, and passing through p'

Since (2) and (3) are the same line, we know $\dfrac{a_1^2 \cos \alpha}{\cos \alpha - x_0} = a_2^2 = \dfrac{1}{1 + \frac{x_0(\cos \alpha - x_0)}{a_1^2}}$.

So

$$\frac{(\cos \alpha - x_0)}{a_1^2} = \cos \alpha + \frac{x_0 \cos \alpha (\cos \alpha - x_0)}{a_1^2}$$

$$\frac{(\cos \alpha - x_0 - x_0 \cos \alpha (\cos \alpha - x_0))}{a_1^2} = \cos \alpha$$

$$\frac{(\cos \alpha - x_0)(1 - x_0 \cos \alpha)}{a_1^2} = \cos \alpha$$

$$a_1^2 = \frac{(\cos \alpha - x_0)(1 - x_0 \cos \alpha)}{\cos \alpha}$$

and

$$a_2^2 = \frac{\cos \alpha}{\cos \alpha - x_0} \frac{(\cos \alpha - x_0)(1 - x_0 \cos \alpha)}{\cos \alpha}$$

$$= 1 - x_0 \cos \alpha$$

Using (1), we know that

$$(s - x_0)^2 = (\cos \alpha - x_0)(\sec \alpha - x_0)$$

$$s^2 - 2sx_0 = 1 - x_0(\cos \alpha + \sec \alpha)$$

$$x_0(\cos \alpha + \sec \alpha - 2s) = 1 - s^2$$

$$x_0 = \frac{s^2 - 1}{2s - \cos \alpha - \sec \alpha}$$

We can therefore write the volume of E in terms of just $\beta = \cos \alpha$:

$$\text{volume}(E) = K a_1 a_2^{d-1}$$

$$= K \sqrt{f(\beta)}$$

$$= K \sqrt{\frac{(\beta - x_0)(1 - x_0 \beta)^d}{\beta}}$$

$$= K\sqrt{\frac{\left(\beta - \frac{s^2-1}{2s-\beta-\beta^{-1}}\right)\left(1 - \frac{s^2-1}{2s-\beta-\beta^{-1}}\beta\right)^d}{\beta}}$$

$$= K\sqrt{\left(1 + \frac{s^2-1}{\beta^2 - 2s\beta + 1}\right)\left(1 + (s^2-1)\frac{\beta^2}{\beta^2 - 2s\beta + 1}\right)^d}$$

$$= K\sqrt{\frac{(\beta-s)^2}{\beta^2 - 2s\beta + 1}\left(\frac{(s\beta-1)^2}{\beta^2 - 2s\beta + 1}\right)^d}$$

$$= K\sqrt{\frac{(\beta-s)^2(s\beta-1)^{2d}}{(\beta^2 - 2s\beta + 1)^{d+1}}}$$

$$= K\frac{(\beta-s)(s\beta-1)^d}{(\beta^2 - 2s\beta + 1)^{(d+1)/2}}$$

To find the $\beta \in (-1,0)$ that minimizes $f(\beta)$,

$$f'(\beta) = \frac{(\beta^2 - 2s\beta + 1)^{-d/2}(s\beta-1)^d}{(\beta^2 - 2s\beta + 1)^{3/2}(s\beta-1)}\left[-d(\beta-s)^2(s\beta-1) + sd(\beta^2 - 2s\beta + 1)(\beta-s) - (s^2-1)(s\beta-1)\right]$$

$$= \frac{(\beta^2 - 2s\beta + 1)^{-d/2}(s\beta-1)^d}{(\beta^2 - 2s\beta + 1)^{3/2}(s\beta-1)}d(\beta-s)\left[(-s\beta+1)(\beta-s) + s(\beta^2 - 2s\beta + 1)\right] - (s^2-1)(s\beta-1)$$

$$= \frac{(\beta^2 - 2s\beta + 1)^{-d/2}(d\beta-1)^d}{(\beta^2 - 2s\beta + 1)^{3/2}(s\beta-1)}(1-s^2)(d\beta^2 - s(d-1)\beta - 1)$$

Setting $f'(\beta) = 0$,

$$\frac{(\beta^2 - 2s\beta + 1)^{-d/2}(s\beta-1)^d}{(\beta^2 - 2s\beta + 1)^{3/2}(s\beta-1)}(1-s^2)(d\beta^2 - s(d-1)\beta - 1) = 0$$

Now $s > 1$, since p' is outside the unit circle, so $1 - s^2 \neq 0$.

$$d\beta^2 - s(d-1)\beta - 1 = 0$$

$$\beta = \frac{s - \sqrt{s^2 + \frac{4d}{(d-1)^2}}}{\frac{2d}{d-1}}$$

E is described by $\dfrac{(x_1 - x_0)^2}{a_1^2} + \dfrac{x_2^2}{a_2^2} + \dfrac{x_3^2}{a_2^2} + \cdots + \dfrac{x_d^2}{a_2^2} = 1$, where

$$x_0 = \frac{s^2-1}{2s - \beta - \beta^{-1}} \tag{4}$$

$$a_1^2 = (\beta - x_0)(\beta^{-1} - x_0) \tag{5}$$

$$a_2^2 = 1 - x_0\beta \tag{6}$$

□

Lemma 6. *The minimum ellipsoid that encloses the unit ball B and is incident on a point lying outside this unit ball has its major axis along the line joining the point and the center of the unit ball.*

Proof. Let o be the center of the unit ball B and p a point lying outside B. Suppose the major axis of the minimum ellipsoid E is not along op. Then we can rotate this ellipsoid so that the major axis is aligned with op and the ellipsoid encloses the unit ball and is incident on p. Now we can reduce the major axis and all the minor axes so that the new ellipsoid E' touches the unit ball and passes through p. The volume of this new ellipsoid E' is smaller than the volume of E.   □

By Lemma (3), Lemma (4) and Lemma (6) we can assume that the new point is on x_1 axis. Then we will find the minimum ellipsoid that touches this ball and its major axis is along the line joining the origin and the new point.

The second problem that we have to solve is the following one.

Problem: Given an ellipsoid E_A of full volume in k-dimensional space $(k < d)$ and a point lying outside the k-flat containing E_A, find an ellipsoid of minimum and full volume in $k + 1$-dimensional space that spans E_A and p.

We first use a scaling transfomation, just as in Problem 1, to transform the E_A into a unit ball, E'_A; the same transformation is also applied to the point p to obtain p'. The minimum volume ellipsoid that spans E'_A and p' is of the form:

$$\sum_{i=1}^{k-1} x_i^2 + \alpha_1 x_k^2 + \left(\sum_{i=1}^{k-1} \alpha_i x_i\right)x_k = 1,$$

since we must get the unit ball on setting $x_k = 0$ in the above equation. The α_i's are determined from two further conditions that the above ellipsoid must satisfy: (1) the point p' is incident on the spanning ellipsoid and (2) the spanning ellipsoid must be of minimum volume.

Once all the α_i's are determined, we apply the inverse of the scaling transformation that gave us the unit ball to obtain the spanning ellipsoid in the original $k + 1$-dimensional space.

To obtain an initial ellipsoid of full volume, we will have to solve one of the above problems for each new input point. Once we obtain an ellipsoid of full volume, we only have to solve the first problem for each new input.

Algorithm: ApproximateEllipsoid

1. Obtain an initial ellipsoid of full volume, solving Problem 1 or Problem 2.
2. Once we have an approximate ellipsoid of full volume, for each new input point not inside the current approximate ellipsoid we do the following:
   2.1 By rotation, translation and scaling, transform the approximate ellipsoid to the unit ball
   2.2 Apply the appropriate rotation so that the new point P aligns on x-axis
   2.3 Calculate β, a_1 and a_2 using the formulas given in Lemma (5)

In the next section we construct an example input sequence to show that the approximation ratio can be unbounded.

4   Approximation Ratio of Volume

4.1   A Special Exact Ellipsoid in d-Dimensional Space

Exact Ball

Lemma 7. *Suppose P_d is a set of $d+1$ points in $\mathbb{R}^d$ and s is the distance between any two points. Then the radius R_d of the ball passing through all those points is*

$$R_d = \frac{s\sqrt{d}}{\sqrt{2}\sqrt{d+1}}$$

Lemma 8. *Suppose $p_0, p_1, p_2, p_3, \cdots, p_d$ are equidistant $d+1$ points on an d-dimensional space and B is the d-dimensional ball through $p_0, p_1, p_2, p_3, \cdots, p_d$. If we denote the plane through $p_1, p_2, p_3, \cdots, p_d$ by P' then P' splits the diameter through p_0 by the ratio $\dfrac{d-1}{d+1}$.*

Proof. (Algebraic proof)
Suppose $p_0 = (x_1, x_2, \cdots, x_d)$ is equidistant from d points $p_1 = (s/\sqrt{2}, 0, 0, \cdots, 0)$, $p_2 = (0, s/\sqrt{2}, 0, \cdots, 0)$, $\cdots$ $p_d = (0, 0, \cdots, s/\sqrt{2})$ in $\mathbb{R}^d$. Then from the proof of Lemma 7 we know that $x_1 = x_2 = x_3 = \cdots = x_d = \frac{s}{d\sqrt{2}}(1 \pm \sqrt{1+d})$. Let AB be a diameter of the ball where $A = p_0$. If $N = (c, c, ..., c)$ is the point where AB intersects with P' then $|Np_1| = |Np_2| = \cdots = |Np_d| = R_{d-1}$ is the radius of the disk in $d-1$ dimensional space. Now,

$$(c - s/\sqrt{2})^2 + c^2 + c^2 \cdots + c^2 = R_{d-1}^2$$

$$(c - s/\sqrt{2})^2 + (d-1)c^2 = \left(\frac{s\sqrt{d-1}}{\sqrt{2d}}\right)^2$$

$$c = \frac{s}{d\sqrt{2}}$$

Since $A = \left(\dfrac{s}{d\sqrt{2}}(1+\sqrt{1+d}), \dfrac{s}{d\sqrt{2}}(1+\sqrt{1+d}), \cdots\right)$ and $N = \left(\dfrac{s}{d\sqrt{2}}, \dfrac{s}{d\sqrt{2}}, \cdots\right)$ we can write,

$$AN = \sqrt{d\left(\frac{s}{d\sqrt{2}}(1 + \sqrt{1+d}) - \frac{s}{d\sqrt{2}}\right)^2}$$

$$= \frac{s}{\sqrt{2d}}\sqrt{1+d}$$

$$BN = AB - AN$$

$$= 2R_d - s\frac{\sqrt{d+1}}{\sqrt{2d}}$$

$$= \frac{2s\sqrt{d}}{\sqrt{2}\sqrt{d+1}} - s\frac{\sqrt{d+1}}{\sqrt{2d}}$$

$$= s\frac{d-1}{\sqrt{2}\sqrt{d}\sqrt{d+1}}$$

Therefore, $\dfrac{BN}{AN} = \dfrac{d-1}{d+1}$

□

A special ellipsoid

Lemma 9. *The exact ellipsoid incident on* $(0,-1,0,\cdots,0)$, $(0,1,0,\cdots,0)$, $(0,0,1,\cdots,0)$, $(0,0,0,1,\cdots,0)$, $(0,0,0,0,1,\cdots,0)$, $\cdots$ *and* $((d+1)l,0,0,\cdots,0)$ *must also pass through* $(-(d-1)l,0,0,\cdots,0)$. *Moreover, the length of all the semi-minor axes of this exact ellipsoid is bounded.*

Proof. Consider the exact ball that we constructed in Lemma 8. If we scale (with scaling factor greater than 1) the d-dimensional space in x_0 direction then we get an ellipsoid. All the minor axis of this ellipsoid is $\dfrac{d\sqrt{d}}{\sqrt{2}\sqrt{d+1}}$ and the ratio of BN to AN is $\frac{BN}{AN} = \alpha = \frac{d-1}{d+1}$. Therefore if the exact ellipsoid passed through $((d+1)l,0)$, then it must pass through $(-(d-1)l,0)$. □

Lemma 10. *There exists an input point sequence for which the ratio of the minor axis of approximate ellipsoid to that of the exact ellipsoid becomes unbounded.*

Proof. Consider d points $(-1,0,0,\cdots,0)$, $(1,0,0,\cdots,0)$, $(0,1,0,\cdots,0)$, $(0,0,1,\cdots,0)$, $\cdots$, $(0,0,0,\cdots,1)$. Let E_0 be the exact ellipsoid through these d points. Since the approximate ellipsoid through these d points is identical with the exact one, the ratio of their volumes is 1 in this case. Now we start adding points on the positive x_1-axis.

If we add a point $(p_m,0,0,\cdots,0)$ on the positive x_1-axis then the exact ellipsoid will pass through $(0,1,0,\cdots,0)$, $(0,0,1,\cdots,0)$, $(0,0,0,\cdots,1)$ and $(p_m,0,0,\cdots,0)$. For any p_m, the length of all the minor axes of the exact ellipsoid is bounded (Lemma 9).

We denote the original plane by P_0. The transformation $S_0 T_0$ will transform E_0 to the unit ball where T_0 is a translation and S_0 is a scaling. Let $P_1 = S_0 T_0 P_0$ be the transformed plane. We can choose a point $(p',0)$ on this transformed plane such that the minor axis of the spanning ellipsoid, which passes through $(p',0)$ and encloses the unit circle, is greater than 1. Suppose ST is the transformation which transforms this new ellipsoid to the unit ball where S is a scaling and T is a translation. Let $P_2 = STP_1$. Take $(p',0)$ on P_2 and find the ellipsoid that passes through $(p',0)$ and encloses the unit ball.

If we repeat this process n times, then each time we increase all the minor axes of the ellipsoid same amount. Denote the approximate ellipsoid on P_n by E_n. The approximate ellipsoid on the original plane can be found by applying the transforamtion $(T^{-1}S^{-1})^n T_0^{-1} S_0^{-1} E_d$. Since S is a shrinking along all the minor axes, S^{-1} is an expansion along minor axes. Therefore we can expand all the minor axes of approximate ellipsoid as much as we want by adding a sufficient number of points. Since all the minor axes of the exact ellipsoid are fixed, we conclude that the ratio of the minor axis of approximate ellipsoid to that of the exact ellipsoid is unbounded. □

Theorem 1. *There exists an input point sequence of points for which the approximation ratio of the ellipsoid volumes becomes unbounded.*

Proof: Let a_{1A}, a_{2A}, and Volume$_A$ denote the semi-major axis, one of the semi-minor axes and the volume of the approximate ellipsoid respectively in the transformed plane, while a_E, b_E, and Volume$_E$ denote the same quantities respectively for the exact ellipsoid. By the method described in Lemma 10 we can construct an input point sequence for which $\dfrac{b_A}{b_E}$ is unbounded. Since relative volumes are preserved by an elliptic transformation, the ratio $\dfrac{\text{Volume}_A}{\text{Volume}_E}$ is identical for the ellipsoids in the original plane. Now $\dfrac{\text{Volume}_A}{\text{Volume}_E} = \dfrac{Ka_Ab_A^{d-1}}{\pi a_Eb_E^{d-1}} = \dfrac{a_A}{a_E}\dfrac{b_A}{b_E} \geq C\dfrac{b_A}{b_E}.$

Since $\dfrac{b_A}{b_E}$ is unbounded, we conclude that $\frac{\text{Volume}_A}{\text{Volume}_E}$ is unbounded for this input sequence. $\qquad\square$

Though the ratio of the volumes can become unbounded, there exists a direction in which the width of the approximate ellipsoid is within a factor 2 of the width of the exact ellipsoid. We prove this in the next section.

5 Boundedness of the Width

The idea is fairly straight forward. Given some minimum spanning ellipsoid (exact or approximated) of a point set, we need to be able to find 2 points in that point set whose distance between themselves in some direction is at least a constant fraction of the width of the ellipsoid in that direction. The following proof shows that we can always find 2 such points and a direction.

The proof hinges on the fact that the convex hull of the point set will always contain the center of the minimum ellipsoid, approximated or not. So, to start, we prove an equivalent version of this; that any halfspace that contains the center of a minimum spanning ellipsoid of a point set, also contains a point from that set. We start with the case of the exact minimum spanning ellipsoid and then use induction to extend this to the approximated case.

Lemma 11. *If E is some exact minimum spanning ellipsoid that contains some point set Q, then any closed halfspace that contains the center, c, of E must also contain a point $p \in Q$.*

Proof. Assume that there exists some closed halfspace, H, such that $c \in H$, but $H \cap Q = \emptyset$. Without loss of generality, we can assume c lies on the boundary of H. Let $\epsilon > 0$ be equal to the distance from H to Q. We can translate E by ϵ in the direction of the vector normal to H, while still ensuring $Q \subset E$. After the translation, Q no longer has any points on the boundary of E, so E cannot be the minimum ellipsoid that contains Q, a contradiction. $\qquad\square$

Lemma 12. *Let $Q = \{q_1, q_2, ..., q_k\}$ be the set of points used to find E_0, the initial minimum spanning ellipsoid. Let E_n be the approximated ellipsoid after the $(n+k)$-th point, p_n has been seen. Any halfspace that contains the center c_n of the ellipsoid E_n also contains at least one point $p \in P = Q \cup \{p_1, p_2, ..., p_n\}$.*

Proof. From Lemma 11 we know that the base case $(n = 0)$ is true. Let us assume that the $(n-1)$-th case was true. If a new point $p_n \notin E_{n-1}$ is seen, then a new minimum ellipsoid E_n is found that contains both E_{n-1} and p_n. If c_{n-1} is the center of E_{n-1} and $\overline{p_n c_{n-1}}$ is the line segment whose endpoints are p_n and c_{n-1}, then, from the construction of E_n we see that $c_n \in \overline{p_n c_{n-1}}$, where c_n is the center of E_n. Any halfspace H that has a non-empty intersection with $\overline{p_n c_{n-1}}$ (ie. $H \cap \overline{p_n c_{n-1}} \neq \emptyset$), must contain either p_n, c_{n-1} or both. If H contains p_n then clearly it contains a point from P, since $p_n \in P$. Further more, if H contains the c_{n-1}, center of E_{n-1}, then it must contain some point from $P - \{p_n\} \subseteq P$. Thus, any halfspace that contains $c_n \in \overline{p_n c_{n-1}}$ must also contain a point $p \in P$. □

With these 2 Lemmas, we can now give the main proof.

Definition 1. *Let $\Psi_v(P)$ be a real-valued function that returns the width of a point set P in direction v.*

Theorem 2. *If E is a minimum volume ellipsoid (exact or approximated) that contains the point set P, then there must exist 2 points $\{p_i, p_j\} \subseteq P$ and a direction n, such that $\Psi_n(E) \leq 2\Psi_n(\{p_i, p_j\})$.*

Proof. Since E is the minimum ellipsoid (exact or approximated) containing P, we can find a point $p_i \in P$ that lies on the boundary of E. Let c be the center of E and n be the vector normal to E at the point p_i. Let H be the halfspace whose normal is n, does not contain p_i and whose boundary contains c. From Lemma 12, we can find another point $p_j \in P \cap H$. Clearly, $\Psi_n(\{p_i, p_j\}) \geq \Psi_n(\{p_i, c\})$. Since $2\Psi_n(\{p_i, c\}) = \Psi_n(E) \leq 2\Psi_n(\{q_i, q_j\})$, the points p_i and p_j fulfill the requirements; that is, in direction n, the ellipsoid E has a bounded width by a constant factor 2. □

Theorem 3. *There exists a direction at which the approximation ratio is bounded.*

Proof. Suppose E_E is the exact minimum ellipsoid and E_A is the approximate ellipsoid containing P. By theorem 2, we get $\Psi_n(E_A) \leq 2\Psi_n(\{p_i, p_j\})$. Clearly we can see that $\Psi_n(E_E) \geq \Psi_n(\{p_i, p_j\})$. Therefore $\dfrac{\Psi_n(E_A)}{\Psi_n(E_E)} \leq 2$.

□

6 Conclusions

It is surprising that the approximation ratio of volumes is unbounded. We showed that this ratio becomes unbounded by taking a sequence of points in one direction. We believe that the approximation ratio is bounded for evenly distributed points. Experiments with this implementation for randomly input point sets in two dimensions seem to support this vew.

References

1. Bagchi, Chaudhary, Eppstein, Goodrich: Deterministic sampling and range counting in geometric data streams. ACM Transactions on Algorithms (TALG) 3 (2007)
2. Ben-Tal, A., Nemirovski, A.: Lectures on Modern Convex Optimization: Analysis, Algorithms, and Engineering Applications. MPS/SIAM Series on Optimization, vol. 2. SIAM, Philadelphia (2001)
3. Feigenbaum, J., Kannan, S., Zhang, J.: Computing diameter in the streaming and sliding-window models. Algorithmica 41(1), 25–41 (2004)
4. Henzinger, M., Raghavan, P., Rajagopalan, S.: Computing on data streams. Technical Report SRC-TN-1998-011, Hewlett Packard Laboratories, May 7 (1998)
5. Hershberger, Suri: Adaptive sampling for geometric problems over data streams. In: PODS: 23rd ACM SIGACT-SIGMOD-SIGART Symposium on Principles of Database Systems (2004)
6. Isenburg, Liu, Shewchuk, Snoeyink: Illustrating the streaming construction of 2D delaunay triangulations (short). In: COMPGEOM: Annual ACM Symposium on Computational Geometry (2006)
7. Mukhopadhyay, A., Greene, E., Sarker, A.: Approximate spanning ellpse in the streaming model. Technical report, University of Windsor, School of Computer Science (June 2009)
8. Post, M.J.: A minimum spanning ellipse algorithm. In: FOCS, pp. 115–122. IEEE, Los Alamitos (1981)
9. Zarrabi-Zadeh, H., Chan, T.: A simple streaming algorithm for minimum enclosing balls. In: Proceedings of the 18th Canadian Conference on Computational Geometry (CCCG 2006), pp. 139–142 (2006)

Author Index

Printing: Mercedes-Druck, Berlin
Binding: Stein+Lehmann, Berlin